Rigidity Theory and Applications

FUNDAMENTAL MATERIALS RESEARCH

Series Editor: M. F. Thorpe, *Michigan State University*
East Lansing, Michigan

ACCESS IN NANOPOROUS MATERIALS
Edited by Thomas J. Pinnavaia and M. F. Thorpe

DYNAMICS OF CRYSTAL SURFACES AND INTERFACES
Edited by P. M. Duxbury and T. J. Pence

ELECTRONIC PROPERTIES OF SOLIDS USING CLUSTER METHODS
Edited by T. A. Kaplan and S. D. Mahanti

LOCAL STRUCTURE FROM DIFFRACTION
Edited by S. J. L. Billinge and M. F. Thorpe

RIGIDITY THEORY AND APPLICATIONS
Edited by M. F. Thorpe and P. M. Duxbury

A Continuation Order Plan is available for this series. A continuation order will bring delivery of each new volume immediately upon publication. Volumes are billed only upon actual shipment. For further information please contact the publisher.

Rigidity Theory and Applications

Edited by

M. F. Thorpe and P. M. Duxbury

Michigan State University
East Lansing, Michigan

Kluwer Academic / Plenum Publishers
New York, Boston, Dordrecht, London, Moscow

Proceedings of a Conference on Rigidity Theory and Applications,
held June 14–17, 1998, at the Park Place Hotel, Traverse City, Michigan

ISBN 0-306-46115-3

233 Spring Street, New York, N.Y. 10013

10 9 8 7 6 5 4 3 2 1

A C.I.P. record for this book is available from the Library of Congress

Printed in the United States of America

SERIES PREFACE

This series of books, which is published at the rate of about one per year, addresses fundamental problems in materials science. The contents cover a broad range of topics from small clusters of atoms to engineering materials and involve *chemistry*, *physics*, *materials science*, and *engineering*, with length scales ranging from Ångstroms up to millimeters. The emphasis is on basic science rather than on applications. Each book focuses on a single area of current interest and brings together leading experts to give an up-to-date discussion of their work and the work of others. Each article contains enough references that the interested reader can access the relevant literature. Thanks are given to the Center for Fundamental Materials Research at Michigan State University for supporting this series.

M.F. Thorpe, Series Editor
E-mail: thorpe@pa.msu.edu

PREFACE

Although rigidity has been studied since the time of Lagrange (1788) and Maxwell (1864), it is only in the last twenty-five years that it has begun to find applications in the basic sciences. The modern era starts with the important theorem of Laman (1970), which made the subject rigorous in two dimensions, followed by the development of computer algorithms that can test over a million sites in seconds and find the *rigid regions*, and the associated *pivots*, leading to many applications.

This workshop on *Rigidity Theory and Applications* was organized to bring together leading researchers studying the underlying theory, and to explore the various areas of science where applications of these ideas are being implemented. This is certainly the first workshop of this type, and it was interesting to see how researchers from different fields (mathematics, computer science, statistical physics, experimental physics, biochemistry, and ceramics) struggled with the language and concepts of the other disciplines. Technical language is always a barrier, but with a little effort and patience, this was largely overcome. This focused workshop was held at the picturesque and historic Park Place Hotel in Traverse City, Michigan, USA from 14-17th June 1998. All participants were by invitation only and 23 gave presentations, most of which resulted in contributions which form this book.

These proceedings provide a unique tutorial snapshot of the state of the study of *rigidity in the basic sciences* in 1998. One perhaps unexpected outcome of this activity is to show that the insights of Maxwell were quite accurate, and that the *constraint counting* approach associated with his name, which ignores the possibility of redundant elements, is a remarkably good starting point in many cases. A number of papers at this workshop used the Maxwell approach to gain insight into glasses and ceramic materials. The concept of a *floppy mode*, a distortion of the system with no associated cost in energy, proved to be useful in all the applications covered at the workshop.

The ideas of Laman have filtered through applied mathematics and computer science to produce powerful methods to enumerate the rigid clusters in large systems and the recent applications of this work in glasses and proteins are covered in this book. Many questions remain. The theory of rigidity in three dimensions is not as robust as we would like, and this is important as most systems of interest in science are three- rather than two-dimensional in nature. The number of people working on problems associated with rigidity remains quite small, and the language rather specialized, so that an effort is needed to integrate these approaches into the mainstream of science. This is beginning to happen as words like *rigid*, *flexible*, *constraint* and *floppy mode* are being used increasingly in science and engineering, and we hope that this book will help.

We would like to thank Michigan State University for financing this workshop and the *Center for Fundamental Materials Research* at MSU for contributing to the cost of producing these proceedings. The efforts of Lorie Neuman and Janet King, who organized the workshop and proceedings, are greatly appreciated as was the advice and help of the *Advisory Committee* members: Punit Boolchand, Leslie Kuhn, and Walter Whiteley.

Michael F. Thorpe
Phillip M. Duxbury
East Lansing, Michigan

CONTENTS

RIGIDITY THEORY

APPLICATIONS TO NETWORKS

APPLICATIONS TO GLASSES

APPLICATIONS TO PROTEINS

Rigidity Theory and Applications

GENERIC AND ABSTRACT RIGIDITY

Brigitte Servatius and Herman Servatius

Department of Mathematics
Syracuse University
Syracuse NY, 13244-1150

Rigidity

We are all familiar with frameworks of rods attached at joints. A rod and joint framework gives rise to a simple mathematical model consisting of line segments in Euclidean 3-space with common endpoints. A *deformation* is a continuous one-parameter family of such frameworks. If a framework has only trivial deformations, e.g. translations and rotations, then it is said to be *rigid*. Before giving a more precise mathematical formulation, we can use simple geometry to explore these ideas.

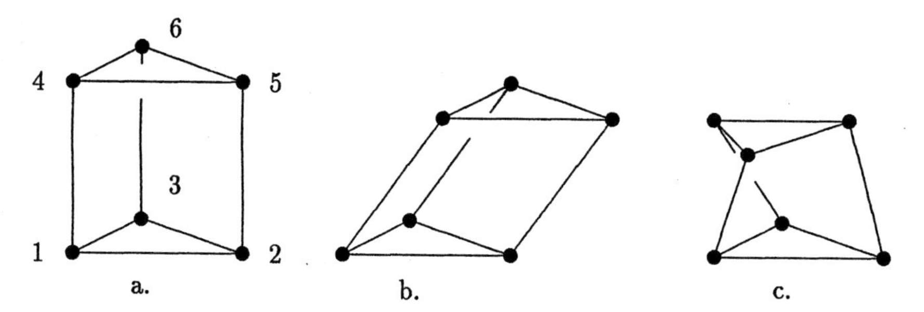

Figure 1: Deformations of the triangular prism in 3D.

Consider the triangular prism of Figure 1a. It has an obvious deformation in which the bottom triangle is held fixed and the three posts rotate simultaneously about their lower endpoints, remaining parallel throughout. As the posts move, any two of them form the sides of a parallelogram, and so the triangles formed by their upper points are congruent, see Figure 1b. Another deformation is to keep the planes of the two triangles parallel and "screw down" the top triangle, as in Figure 1c. We can try to

Rigidity Theory and Applications
Edited by Thorpe and Duxbury, Kluwer Academic/Plenum Publishers, 1999

roughly count the deformations by piecing together the framework. Starting with the bottom triangle, 123, without loss of generality we may assume that it is fixed, and then we need not concern ourselves with trivial deformations. Adding segment 14, we must have that point 4 is constrained to move on a sphere, which has two degrees of freedom. For any position of point 4, adding segments 25 and 45 constrains point 5 to lie on the intersection of two spheres, i.e. a circle, with one degree of freedom. Lastly we add segments 36, 46 and 56, so that point 6 is constrained to lie on the intersection of three spheres, and so is fixed by what has already been chosen. Altogether we have three degrees of freedom for deformations of the triangular prism. That this is a rough calculation can be seen from Figure 2, which shows a "flattened" triangular prism which

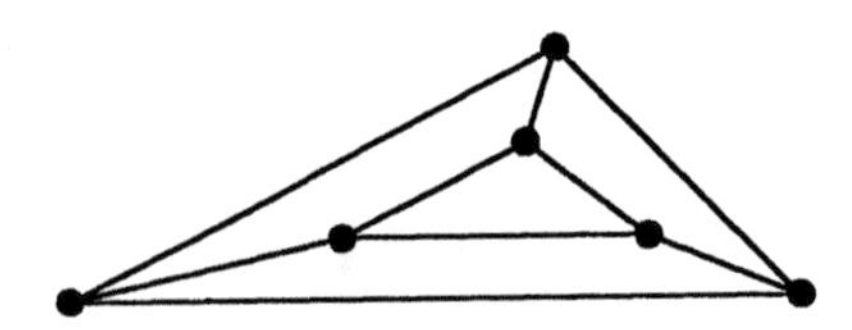

Figure 2: A rigid prism.

is rigid because the inner triangle is held tightly in place by the spokes. Notice that the rigidity of the flattened prism is unstable in the sense that, if the framework is not precisely in the plane, then a deformation is possible.

It is a different question entirely to consider a framework which is constrained to move only in the plane. Beyond being of theoretical interest, such frameworks arise in special engineering applications, as well as geometric questions arising from computer aided design, [12]. Regarding the Figure 1a as a drawing of a plane framework, it also has an obvious deformation. Again let us count the degrees of freedom. As before let the bottom triangle be fixed. Adding segment 14 constrains 4 to move on a circle, which has one degree of freedom. Adding segments 25 and 45 constrains 5 to move on the intersection of two circles, which is a single point. The same with adding edges 36 and 46. Thus the framework with one segment missing has only one degree of freedom. Adding the last edge constraint should remove that last degree of freedom and yield a framework which is rigid in the plane. So in the plane we expect a triangular prism framework to be rigid, but there nevertheless exist flexible ones, while in 3-space we expect a triangular prism framework to be flexible, yet there exist rigid ones.

We now make precise what we mean by a framework. A *graph* (V, E) consists of a vertex set $V = \{1, 2, \ldots, n\}$ and edge set E, where E is a collection of unordered pairs of vertices. A *framework* is a triple $(V, E, \mathbf{p})$ where (V, E) is a graph and $\mathbf{p} = \{\mathbf{p}_1, \ldots, \mathbf{p}_n\}$ is a list of distinct points of m dimensional Euclidean space corresponding to the vertices of V.

If $\{i, j\}$ is an edge of (V, E), then a *deformation* of the framework is a continuous one parameter family $\mathbf{p}(t) = (\mathbf{p}_1(t), \ldots, \mathbf{p}_n(t))$ with $\mathbf{p}(0) = \mathbf{p}$, such that the distance from $\mathbf{p}_i(t)$ to $\mathbf{p}_j(t)$ is kept fixed if $\{i, j\} \in E$,

$$(\mathbf{p}_i(t) - \mathbf{p}_j(t)) \cdot (\mathbf{p}_i(t) - \mathbf{p}_j(t)) = c_{ij}, \text{ for all } \{i, j\} \text{ edges of } G, \tag{1}$$

The framework $(V, E, \mathbf{p})$ is said to be *rigid* if all deformations are locally trivial, that is, $\mathbf{p}(t)$ is congruent to $\mathbf{p}$ for all t near 0. Equivalently, a deformation is trivial if it preserves the distance between any two points, whether they are adjacent or not.

Finding deformations, or even solving the system of equations 1 is very difficult in general, and we have already seen some delicate special cases. One successful approach is to not to look for deformations directly but to look for their first derivatives. If $\mathbf{p}$ is a framework and $\mathbf{p}(t)$ is a deformation, then, by Equation 1, its derivative $d\mathbf{p}/dt = \mathbf{p}'$ must satisfy

$$(\mathbf{p}_i - \mathbf{p}_j) \cdot (\mathbf{p}'_i - \mathbf{p}'_j) = 0, \text{ for all } \{i, j\} \in E \quad (2)$$

For $t = 0$ this is a system of $|E|$ linear equations with the nm unknowns being the coordinates of the $\mathbf{p}'_i$. If, as we shall hereafter assume, the framework contains at least $m + 1$ points in general position, then there are always $m(m + 1)/2$ solutions of Equations 2 corresponding to the derivatives of trivial deformations. If the system of equations (2) has no other solutions, then we say the framework $(V, E, \mathbf{p})$ is *first order rigid*, or *infinitesimally rigid.* A solution to Equations 2 is called an *infinitesimal flex* or just a *flex.* Infinitesimal rigidity is a natural approximation to rigidity, and the connection between them must be carefully examined.

If $(V, E, \mathbf{p})$ is infinitesimally rigid, then any deformation must have its initial velocities coinciding with that of a trivial deformation, or, in other words, any deformation which fixes enough vertices to prevent trivial deformations must have initial velocity 0 at every vertex. If every deformation could be parameterized so that the initial velocities were non-zero, then it would be obvious that infinitesimal rigidity implies rigidity, however, in 1992 the bar and joint framework of Figure 3 was discovered which is a cusp point in its configuration space [1]. The embedding $\mathbf{p}$ of a framework $(V, E, \mathbf{p})$

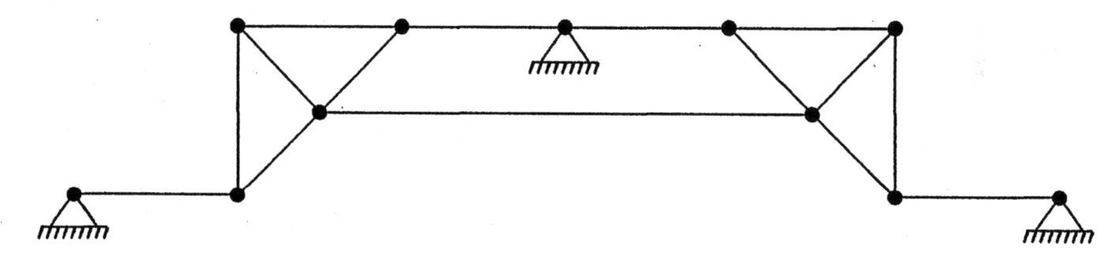

Figure 3: A third order rigid framework that is a mechanism

may be regarded as defining a single point of mn dimensional space, and the subspace of $\mathbb{R}^{mn}$ comprising all solutions of the equations (1) is called the *configuration space of* $(V, E, \mathbf{p})$. Pinning sufficiently many vertices to prevent trivial deformations, indicated in Figure 3 by the grounds, a framework with a one dimensional configuration space is called a *mechanism.* A framework which is a cusp mechanism must have all deformations with initial velocity 0. (The actual motion of the cusp framework of Figure 3 can currently be viewed at http://www.wpi.edu/~ hservat/index.html.) Nevertheless, any deformation $\mathbf{p}(t)$, which can be assumed to be analytic, must have some derivative which is non-zero at $t = 0$, and it can be shown that the lowest order derivative of $\mathbf{p}(t)$ which is non-zero at $t = 0$ satisfies Equation 2, which means the framework is not infinitesimally rigid and we have the following theorem.

THEOREM 1 *Infinitesimal rigidity implies rigidity.*

EXAMPLE: The 2D framework of Figure 4 is infinitesimally rigid. If we have a flex, then, using isometries, we may assume that triangle 456 is fixed and $\mathbf{p}'_4 = \mathbf{p}'_5 = \mathbf{p}'_6 = 0$. Equation 2 applied to bars $\{1, 4\}$ and $\{2, 5\}$ implies that $\mathbf{p}'_1$ and $\mathbf{p}'_2$ are both horizontal. Then, since bar $\{1, 2\}$ is horizontal, $(\mathbf{p}_1 - \mathbf{p}_2) \cdot (\mathbf{p}'_1 - \mathbf{p}'_2) = 0$ implies $\mathbf{p}_1 = \mathbf{p}_2$. Therefore, using the constraints of bars $\{1, 3\}$ and $\{2, 3\}$ we have that $\mathbf{p}'_3 - \mathbf{p}'_1 = \mathbf{p}'_3 - \mathbf{p}'_2$ is simultaneously perpendicular to bars $\{1, 3\}$ and $\{2, 3\}$, hence $\mathbf{p}'_1 = \mathbf{p}'_2 = \mathbf{p}'_3$, all horizontal. But the constraint that $\mathbf{p}'_3 - \mathbf{p}'_6 = \mathbf{p}'_3$ be perpendicular to bar $\{3, 6\}$ implies that $\mathbf{p}_1 = \mathbf{p}_2 = \mathbf{p}_3 = 0$ and the flex is trivial. □

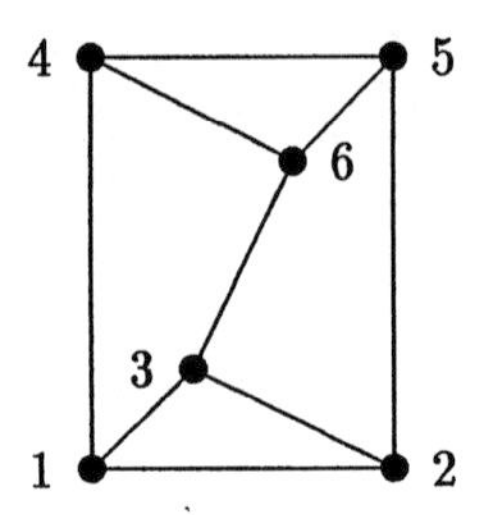

Figure 4: An infinitesimally rigid triangular prism in 2D.

The converse of Theorem 1 is not true – if a framework is rigid, it may nevertheless have an infinitesimal flex.

EXAMPLE: The flattened triangular prism of Figure 5 is rigid in 3-space, but has an infinitesimal flex which assigns zero velocity to the vertices of the outer triangle, and velocities to the inner triangle, all of which are perpendicular to the plane of the prism. □

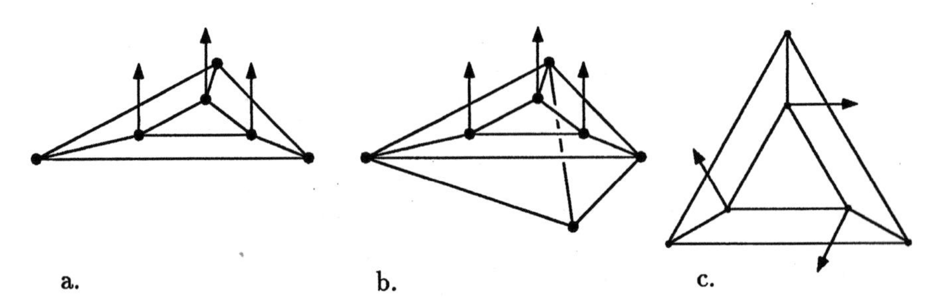

Figure 5: Flexes on a flat prism.

Actually, any framework in 3-space all of whose vertices lie in a plane has many flexes of the type of Figure 5a, since each equation of (2) is equivalent to

$$(\mathbf{p}_i - \mathbf{p}_j) \cdot \mathbf{p}'_i = (\mathbf{p}_i - \mathbf{p}_j) \cdot \mathbf{p}'_j$$

which says geometrically that the initial velocities of the endpoints of any bar have equal projections in the direction of the bar, as in Figure 6. This geometric interpretation is

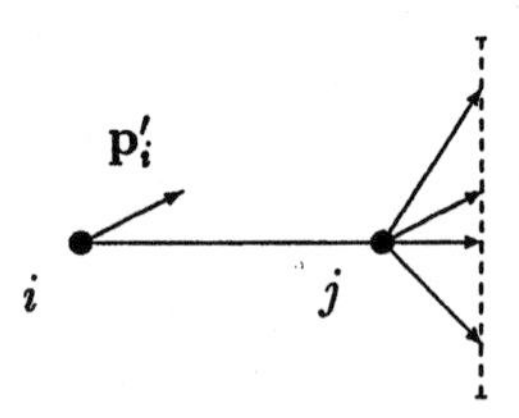

Figure 6: The infinitesimal edge condition.

very useful in the construction of small examples.

EXAMPLE: The framework of Figure 5a violates our assumption that a framework in m-space has at least $m+1$ points in general position, so we augment the flat prism in 3D by another vertex as in Figure 5b in which case the flex implies that the framework of Figure 5b is not infinitesimally rigid in 3D.

If the flattened prism is regarded as a 2D framework, then the flex of Figure 5a is not valid, however, there is still the infinitesimal flex of Figure 5c, which fixes the outer triangle and is an infinitesimal rotation of the inner triangle. □

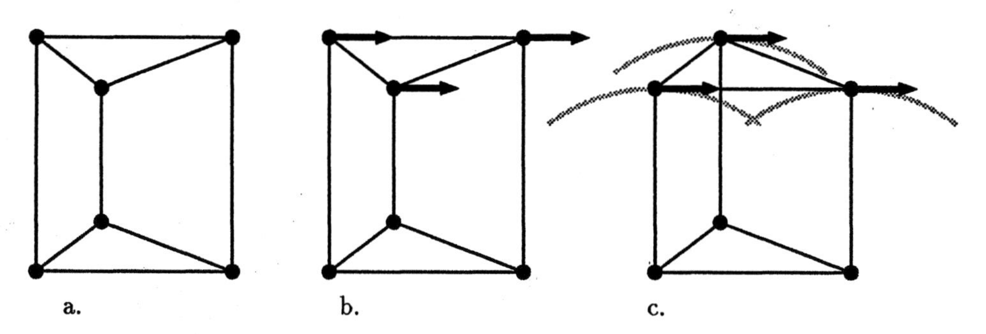

Figure 7: An infinitesimally flexible, rigid prism, and a prism mechanism.

EXAMPLE: Another example of a rigid framework with an infinitesimal flex is the 2D embedding of the triangular prism illustrated in Figure 7a, which, despite the flex indicated by the arrows in Figure 7b, is rigid. Holding the lower triangle fixed, the outer equal-length vertical bars force every point of the upper triangle to move in a circular path whose radius is equal to the length of those bars. The middle vertical bar however forces the lower point to move in a circle of strictly smaller radius and so the framework is rigid in the plane. The framework of Figure 7c has an infinitesimal flex which comes from a deformation. □

One can generalize the idea of infinitesimal rigidity to try to detect rigid but not infinitesimally rigid frameworks. The most successful is to look for an infinitesimal flex which is compatible with an assignment of formal initial accelerations, $\mathbf{p}_i''$. Taking the derivative of the equations in (2) we have

$$(\mathbf{p}_i - \mathbf{p}_j) \cdot (\mathbf{p}_i'' - \mathbf{p}_j'') + (\mathbf{p}_i' - \mathbf{p}_j') \cdot (\mathbf{p}_i' - \mathbf{p}_j') = 0, \text{ for all } \{i,j\} \in E \tag{3}$$

which are taken to be the constraints on $\mathbf{p}_i''$. If, given a framework $(V, E, \mathbf{p})$, there exists vector assignments to the vertices $\mathbf{p}'$ and $\mathbf{p}''$ such that $\mathbf{p}'$ is a flex and $\mathbf{p}'$ and $\mathbf{p}''$ satisfy the equations in (3), then we say the framework has a *second order flex*. A framework all of whose second order flexes have trivial first order flexes is said to be *second order rigid*, and it can be shown that second order rigidity implies the rigidity of the framework, [2].

EXAMPLE: Consider the flat prism in 2D pictured in Figure 5c. Assuming without loss of generality that $\mathbf{p}'$ and $\mathbf{p}''$ are zero on the outer triangle, it is easy to calculate that the only solution to the system of equations (2) and (3) has $\mathbf{p}'$ zero on all vertices and $\mathbf{p}''$ as pictured by the arrows of Figure 5c, so the framework is 2'nd order rigid but not infinitesimally rigid. □

It is natural to hope, therefore, that a hierarchy of higher order rigidities could be developed such that, given any rigid framework, it is k'th order rigid for some k. This can be done, but it is not effective for detecting rigidity since, for example, the cusp framework of Figure 3, which is not rigid, is 3'rd order rigid.

In general, we can regard frameworks which are rigid but not infinitesimally rigid as being in some sense singular, and we see from the examples that the flexes, while not being realizable in an ideal motion of the framework, will certainly give rise to sagging or swaying in any physical model of bars and joints. Thus, for civil engineering applications, infinitesimal rigidity is a superior concept both in terms of the applicability as well as the computational complexity.

The Rigidity Matrix

Given a framework $(V, E, \mathbf{p})$, determining if it is infinitesimally rigid involves solving the system of equations (2), which is a system of $|E|$ linear equations in nm unknowns. Naturally we can form the matrix of this system, $R(V, E, \mathbf{p})$, whose rows correspond to the edges and whose columns correspond to the coordinates of the vertices. Since $V = \{1, 2, \ldots, n\}$, we order the coordinates of the vertices as

$$\{(\mathbf{p}_1)_1, \ldots, (\mathbf{p}_1)_m, (\mathbf{p}_2)_1, \ldots, (\mathbf{p}_2)_m, \ldots (\mathbf{p}_n)_1, \ldots, (\mathbf{p}_n)_m\}$$

and order the edges lexicographically. Then the row of the matrix corresponding to edge $\{i, j\}$ has the coordinates of the vector $(\mathbf{p}_i - \mathbf{p}_j)$ in the columns from $im + 1$ to $im+m$, $(\mathbf{p}_j - \mathbf{p}_i)$ in the columns from $jm+1$ to $jm+m$, and zero elsewhere. $R(V, E, \mathbf{p})$ is an $|E| \times nm$ matrix.

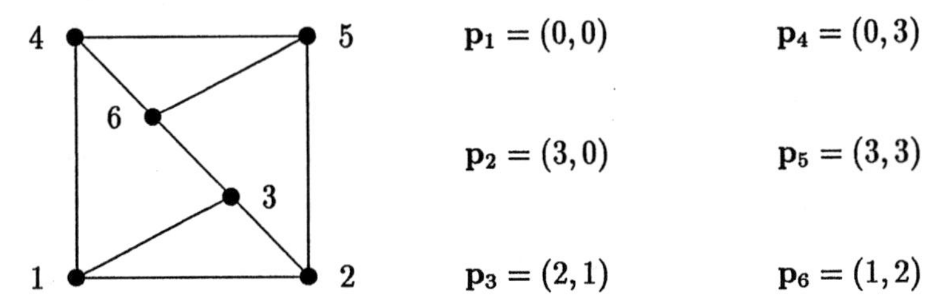

Figure 8: An embedding of a 2D prism.

EXAMPLE: Consider the 2D framework of Figure 8. The rigidity matrix is:

$$R(V, E, \mathbf{p}) = \begin{bmatrix} -3 & 0 & 3 & 0 & 0 & 0 & 0 & 0 & 0 & 0 & 0 & 0 \\ -2 & -1 & 0 & 0 & 2 & 1 & 0 & 0 & 0 & 0 & 0 & 0 \\ 0 & -3 & 0 & 0 & 0 & 0 & 0 & 3 & 0 & 0 & 0 & 0 \\ 0 & 0 & -1 & 1 & 1 & -1 & 0 & 0 & 0 & 0 & 0 & 0 \\ 0 & 0 & 0 & -3 & 0 & 0 & 0 & 0 & 0 & 3 & 0 & 0 \\ 0 & 0 & 0 & 0 & -3 & 0 & 0 & 0 & 0 & 0 & 0 & 3 \\ 0 & 0 & 0 & 0 & 0 & 0 & -3 & 0 & 3 & 0 & 0 & 0 \\ 0 & 0 & 0 & 0 & 0 & 0 & -1 & 1 & 0 & 0 & 1 & -1 \\ 0 & 0 & 0 & 0 & 0 & 0 & 0 & 0 & -2 & -1 & 2 & 1 \end{bmatrix}$$

and it is not difficult, owing to the the large number of zero entries, to show that this matrix is of rank $9 = 2|V| - 3$, hence its framework is infinitesimally rigid, hence rigid. □

Since the rigidity matrix is large even for small examples, it is common to use a more compact notation by assigning m-dimensional vector entries to an $|E| \times n$ matrix, so that, e.g., $R(V, E, \mathbf{p})$ for a triangular prism is:

$$\begin{bmatrix} (\mathbf{p}_1 - \mathbf{p}_2) & (\mathbf{p}_2 - \mathbf{p}_1) & \mathbf{0} & \mathbf{0} & \mathbf{0} & \mathbf{0} \\ (\mathbf{p}_1 - \mathbf{p}_3) & \mathbf{0} & (\mathbf{p}_3 - \mathbf{p}_3) & \mathbf{0} & \mathbf{0} & \mathbf{0} \\ (\mathbf{p}_1 - \mathbf{p}_4) & \mathbf{0} & \mathbf{0} & (\mathbf{p}_4 - \mathbf{p}_1) & \mathbf{0} & \mathbf{0} \\ \mathbf{0} & (\mathbf{p}_2 - \mathbf{p}_3) & (\mathbf{p}_3 - \mathbf{p}_2) & \mathbf{0} & \mathbf{0} & \mathbf{0} \\ \mathbf{0} & (\mathbf{p}_2 - \mathbf{p}_5) & \mathbf{0} & \mathbf{0} & (\mathbf{p}_5 - \mathbf{p}_2) & \mathbf{0} \\ \mathbf{0} & \mathbf{0} & (\mathbf{p}_3 - \mathbf{p}_6) & \mathbf{0} & \mathbf{0} & (\mathbf{p}_6 - \mathbf{p}_3) \\ \mathbf{0} & \mathbf{0} & \mathbf{0} & (\mathbf{p}_4 - \mathbf{p}_5) & (\mathbf{p}_5 - \mathbf{p}_4) & \mathbf{0} \\ \mathbf{0} & \mathbf{0} & \mathbf{0} & (\mathbf{p}_4 - \mathbf{p}_6) & \mathbf{0} & (\mathbf{p}_6 - \mathbf{p}_4) \\ \mathbf{0} & \mathbf{0} & \mathbf{0} & \mathbf{0} & (\mathbf{p}_5 - \mathbf{p}_6) & (\mathbf{p}_6 - \mathbf{p}_4) \end{bmatrix}$$

Setting $\mathbf{q}_{ij} = (\mathbf{p}_i - \mathbf{p}_j)$ we have the even more compact form:

$$\begin{bmatrix} \mathbf{q}_{12} & \mathbf{q}_{21} & \mathbf{0} & \mathbf{0} & \mathbf{0} & \mathbf{0} \\ \mathbf{q}_{13} & \mathbf{0} & \mathbf{q}_{31} & \mathbf{0} & \mathbf{0} & \mathbf{0} \\ \mathbf{q}_{14} & \mathbf{0} & \mathbf{0} & \mathbf{q}_{41} & \mathbf{0} & \mathbf{0} \\ \mathbf{0} & \mathbf{q}_{23} & \mathbf{q}_{32} & \mathbf{0} & \mathbf{0} & \mathbf{0} \\ \mathbf{0} & \mathbf{q}_{25} & \mathbf{0} & \mathbf{0} & \mathbf{q}_{52} & \mathbf{0} \\ \mathbf{0} & \mathbf{0} & \mathbf{q}_{36} & \mathbf{0} & \mathbf{0} & \mathbf{q}_{63} \\ \mathbf{0} & \mathbf{0} & \mathbf{0} & \mathbf{q}_{45} & \mathbf{q}_{54} & \mathbf{0} \\ \mathbf{0} & \mathbf{0} & \mathbf{0} & \mathbf{q}_{46} & \mathbf{0} & \mathbf{q}_{64} \\ \mathbf{0} & \mathbf{0} & \mathbf{0} & \mathbf{0} & \mathbf{q}_{56} & \mathbf{q}_{65} \end{bmatrix}$$

In this last form we see a similarity between $R(V, E, \mathbf{p})$ and the vertex edge adjacency matrix of the graph (V, E). In fact, if $m = 1$, then the rows of $R(V, E, \mathbf{p})$ are constant multiples of the corresponding rows of the adjacency matrix of (V, E). This similarity will be exploited in the sections on generic rigidity.

If $(V, E, \mathbf{p})$ has an infinitesimal flex, then that flex is a solution to Equation 2 and so, written as a column vector, is an element of the nullspace of $R(V, E, \mathbf{p})$. Since we are assuming that every framework has at least $m + 1$ points in general position, the trivial deformations constitute a subspace of dimension $m(m + 1)/2$, so the graph is infinitesimally rigid if that is the dimension of the kernel of $R(V, E, \mathbf{p})$.

THEOREM 2 *A framework on n vertices is infinitesimally rigid in m-space if and only if its rigidity matrix has rank $mn - \frac{m(m+1)}{2}$.*

We can use all the tools of linear algebra to determine infinitesimal rigidity. For example, since the row and column rank of a matrix are the same, one way to compute the rank of a matrix is to compute its cokernel - that is, the subspace of vectors $\mathbf{u}$ with $\mathbf{u}R(V, E, \mathbf{p}) = \mathbf{0}$. These row vectors correspond to an assignment of scalars u_{ij} to the edges of the framework so that, at each vertex i, we have

$$\sum u_{ij}(\mathbf{p}_i - \mathbf{p}_j) = \mathbf{0},$$

where the sum is taken over all j adjacent to i. Such a $\mathbf{u}$ is called a *resolvable stress* on the framework, and may be interpreted as an assignment of spring constants to the edges such that the resulting force at each vertex is $\mathbf{0}$, leaving the framework at equilibrium. Since a resolvable stress is a row dependency of the rigidity matrix, the existence of a resolvable stress is equivalent to the condition that one of the edges of the framework may be deleted without affecting its infinitesimal rigidity.

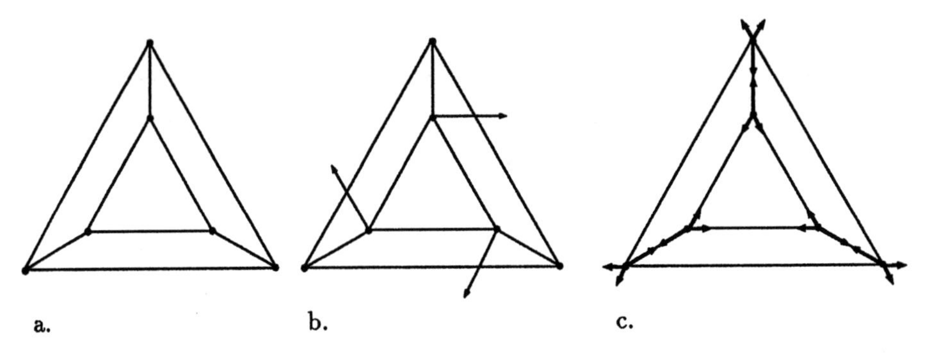

Figure 9: A deformation and a stress on a 2D Triangular Prism.

EXAMPLE: Although the 2D triangular prism of Figure 9a has enough edges to be infinitesimally rigid, we have seen that it has a flex. So the rank of $R(V, E, \mathbf{p})$ is strictly less than $6 \cdot 2 - 3 = 9$ and hence there must be a dependence among the edges. This stress is illustrated in Figure 9c, in which the stress on an edge is indicated by two arrows at its endpoints which point inward if the stress is positive, and outward if it is negative. □

Historically, the the study of stresses, resolvable and unresolvable, came before the study of infinitesimal rigidity and began with the work of Maxwell, see [9]. See also [2] for a modern treatment.

Limiting Frameworks

Practically, one may object that the embeddings we are considering allow edges to cross. This may present difficulties, especially in the plane, if we are building physical frameworks. If our applications involve virtual frameworks, however, then it may be essential to allow such crossing. For example, in computer aided design, uniquely specifying a 3D rendering of a framework involves rigidity, see [12]. Obviously, specifying the 2D rendering on a computer screen of a 3D framework involves 2D rigidity of frameworks with crossing edges. In fact, such rendered frameworks can have vertices which coincide, and theoretically, we should not exclude this case as well.

If we allow vertices to coincide, nothing in the previous pages is altered with respect to either rigidity or infinitesimal rigidity if the vertices which occupy the same location in $\mathbb{R}^m$ are non-adjacent. Allowing adjacent vertices to coincide is more problematic. In ordinary rigidity, if two adjacent vertices coincide, then the constraint between those vertices states that the distance between them is zero, so those two vertices together have only two degrees of freedom, whereas two vertices joined by an edge of positive length have three degrees of freedom. In infinitesimal rigidity, the situation is reversed. If $\mathbf{p}_1 = \mathbf{p}_2$, then $(\mathbf{p}_1 - \mathbf{p}_2) \cdot (\mathbf{p}_1' - \mathbf{p}_2') = 0$, the infinitesimal condition on $\mathbf{p}_1'$ and $\mathbf{p}_2'$, is automatically satisfied, so, infinitesimally, two vertices joined by an edge of length zero have four degrees of freedom, instead of three, and the row of the rigidity matrix corresponding to that edge is zero.

A more interesting and useful concept is the limit of frameworks. Specifically, we are interested in the case where one point $\mathbf{p}_i$ approaches another point $\mathbf{p}_j$ along a direction defined by unit vector $\mathbf{q}$, see Figure 10 where the directions of the unit vectors $\mathbf{q}$ are indicated by the directions of the arrowheads. Of course, the limit of the rigidity matrices will have a zero row corresponding to edge $\{i, j\}$. To avoid this we define the *normalized rigidity matrix*, $NR(V, E, \mathbf{p})$, which is obtained from the ordinary rigidity matrix by dividing each of the rows by the norm of the vectors in that row, i.e., if we write $NR(V, E, \mathbf{p})$ in the compact form then every non-zero entry is a unit vector.

It is easy to see that if $\mathbf{p}$ is a limit framework with $\mathbf{p}_a$ approaching $\mathbf{p}_b$ along direction $\mathbf{q}$, then $NR(V, E, \mathbf{p})$ is the matrix whose $\{i, j\}$'th row is $(\mathbf{p}_i - \mathbf{p}_j)/|\mathbf{p}_i - \mathbf{p}_j|$ in columns $m(i-1)+1$ to mi and $(\mathbf{p}_j - \mathbf{p}_i)/|\mathbf{p}_i - \mathbf{p}_j|$ in columns $m(j-1)+1$ to mj if i is adjacent to j and $\{i, j\} \neq \{a, b\}$ and $\mathbf{q}$ and $-\mathbf{q}$ in columns $m(a-1)+1$ to ma and $m(b-1)+1$ to mb respectively.

If the limit framework has a set of edges which is *independent*, that is, it corresponds to a set of independent rows in the rigidity matrix, then the continuity of the determinant function implies that there exists a nearby ordinary framework whose corresponding edges are also independent.

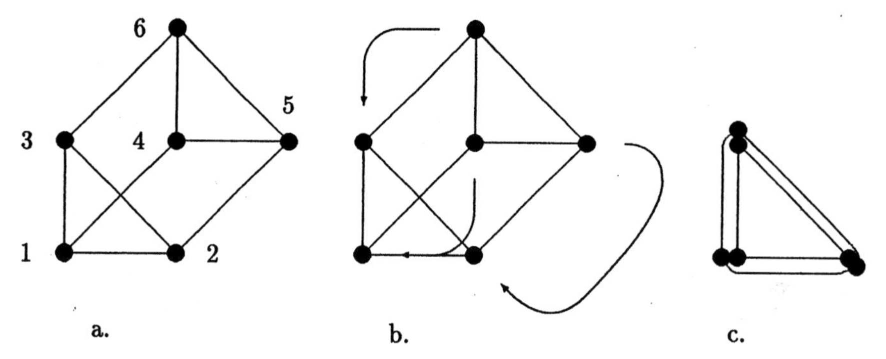

Figure 10: The limit of a framework.

Example: The framework of Figure 10a with $\mathbf{p}_1 = (0,0)$, $\mathbf{p}_2 = (1,0)$, $\mathbf{p}_3 = (0,1)$, $\mathbf{p}_4 = (1,1)$, $\mathbf{p}_5 = (2,1)$, and $\mathbf{p}_6 = (1,2)$, is another flexible 2D prism, so, since it has 9 edges, there must be a dependence among the edges, and every 9×9 minor of its rigidity matrix is zero.

$$R(V, E, \mathbf{p}) = \begin{bmatrix} -1 & 0 & 1 & 0 & 0 & 0 & 0 & 0 & 0 & 0 & 0 & 0 \\ 0 & -1 & 0 & 0 & 0 & 1 & 0 & 0 & 0 & 0 & 0 & 0 \\ -1 & -1 & 0 & 0 & 0 & 0 & 1 & 1 & 0 & 0 & 0 & 0 \\ 0 & 0 & 1 & -1 & -1 & 1 & 0 & 0 & 0 & 0 & 0 & 0 \\ 0 & 0 & -1 & -1 & 0 & 0 & 0 & 0 & 1 & 1 & 0 & 0 \\ 0 & 0 & 0 & 0 & -1 & -1 & 0 & 0 & 0 & 0 & 1 & 1 \\ 0 & 0 & 0 & 0 & 0 & 0 & -1 & 0 & 1 & 0 & 0 & 0 \\ 0 & 0 & 0 & 0 & 0 & 0 & 0 & -1 & 0 & 0 & 0 & 1 \\ 0 & 0 & 0 & 0 & 0 & 0 & 0 & 0 & 1 & -1 & -1 & 1 \end{bmatrix}$$

If we look at the limit as $\mathbf{p}_1 \to \mathbf{p}_4$, $\mathbf{p}_2 \to \mathbf{p}_5$, and $\mathbf{p}_3 \to \mathbf{p}_6$ along the paths indicated Figure 10b, then we get the limiting framework indicated in Figure 10c, which has normalized rigidity matrix

$$NR(V, E, \lim \mathbf{p}) = \begin{bmatrix} -1 & 0 & 1 & 0 & 0 & 0 & 0 & 0 & 0 & 0 & 0 & 0 \\ 0 & -1 & 0 & 0 & 0 & 1 & 0 & 0 & 0 & 0 & 0 & 0 \\ -1 & 0 & 0 & 0 & 0 & 0 & 1 & 0 & 0 & 0 & 0 & 0 \\ 0 & 0 & \alpha & -\alpha & -\alpha & \alpha & 0 & 0 & 0 & 0 & 0 & 0 \\ 0 & 0 & -\alpha & \alpha & 0 & 0 & 0 & 0 & \alpha & -\alpha & 0 & 0 \\ 0 & 0 & 0 & 0 & 0 & -1 & 0 & 0 & 0 & 0 & 0 & 1 \\ 0 & 0 & 0 & 0 & 0 & 0 & -1 & 0 & 1 & 0 & 0 & 0 \\ 0 & 0 & 0 & 0 & 0 & 0 & 0 & -1 & 0 & 0 & 0 & 1 \\ 0 & 0 & 0 & 0 & 0 & 0 & 0 & 0 & \alpha & -\alpha & -\alpha & \alpha \end{bmatrix}$$

with $\alpha = \sqrt{2}/2$. Since the matrix has rank 9, we conclude that there exists an ordinary embedding of the prism in 2-space whose edges are independent. We will see in the section on generic rigidity in the plane a nice way to prove that this matrix is of full rank. □

Generic Rigidity

Clearly, analyzing the rigidity and the infinitesimal rigidity of a framework requires one to study both the incidences of the segments as well as their placement in space. In combinatorial rigidity, [5], we try to separate these aspects and study to what extent can we judge the rigidity of a framework merely from knowing the number of points and segments and their incidences, in other words, just from knowing the underlying graph of the framework.

If we wish to avoid singular frameworks, then we want to avoid dependencies that arise from only from the embedding. We say that a given framework $(V, E, \mathbf{p})$ is *generic* if all frameworks sufficiently near $\mathbf{p}$ are have the same infinitesimal rigidity properties as $(V, E, \mathbf{p})$. (When we speak of the distance between frameworks, it is as points in $\mathbb{R}^{nm}$.) An embedding $\mathbf{p}$ is said to be *generic* if each framework $(V, E, \mathbf{p})$ is generic, for any graph (V, E) on V. It is not hard to show that almost all embeddings in m-space are generic [5]. In fact, it can be shown that the non-generic embeddings form an algebraic set of codimension greater than 2 in $\mathbb{R}^{nm}$, so the generic embeddings form an open connected dense subset of $\mathbb{R}^{nm}$, and hence two generic frameworks with the same graph are either both infinitesimally rigid or they are both not infinitesimally rigid. Thus we may define a *graph* (V, E) to be *generically rigid* in dimension m if it has a generic embedding in m-space which is rigid.

We also say that a framework is generically rigid if its graph is generically rigid, however it should be noted that this is only a statement about the graph of the framework. In other words, a generically rigid framework may be a non-generic framework, and may be non-rigid.

EXAMPLE: The graph of the triangular prism, which has been the leitmotiv of this article, is generically rigid in dimension 1. In one dimension, the normalized rigidity matrix is just the incidence matrix of the graph and, in fact, the concepts of rigidity, infinitesimal rigidity, and generic rigidity all coincide. □

From the argument of the previous example, we have that a graph is generically rigid in dimension 1 if and only if it is connected, and that every framework in dimension 1 is generic.

EXAMPLE: The graph of the triangular prism is generically rigid in dimension 2. The 2 dimensional rigidity matrix of a triangular prism is 9 by 12. We have previously seen that the triangular prism has an infinitesimally rigid embedding in the plane, see Figure 4. This embedding has only trivial flexes, and the kernel of the rigidity matrix has dimension 3, and so its rank is $12 - 3 = 9$, which implies that the rigidity matrix has a 9 by 9 submatrix of non-zero determinant. Since generic embeddings are an open dense subset of m-space, and since the determinant is a continuous function, there must be a nearby generic embedding of the triangular prism for which the corresponding submatrix of its rigidity matrix is also non-zero. Hence the rigidity matrix of the generic framework is also of rank 9, and so the generic framework is infinitesimally rigid Thus the graph of the triangular prism is generically rigid in the plane. □

The argument of the previous example proves the following.

THEOREM 3 *An m dimensional framework which is infinitesimally rigid is also generically rigid.*

A similar argument shows the following useful tool.

THEOREM 4 *An m dimensional framework which has an infinitesimally rigid limit framework is generically rigid.*

EXAMPLE: Since the graph of the triangular prism is generically rigid in the plane, every framework on that graph is generically rigid in the plane. However, it is possible to embed a prism in the plane so that it is not infinitesimally rigid, see Figure 7a, or even so that it is deformable, see Figure 7c. □

EXAMPLE: The graph of the triangular prism is not generically rigid in 3-space. We have seen that it has an embedding into 3-space which is rigid, however that embedding was not infinitesimally rigid. An infinitesimally rigid embedding into 3-space would allow only trivial flexes, so the rigidity matrix should have a kernel of dimension 6 and have rank $3 \cdot 6 - 6 = 12$, which is impossible since it has only 9 rows. □

THEOREM 5 *A graph with n vertices which is generically rigid in dimension m has at least $m \cdot n - \frac{m(m+1)}{2}$ edges.*

The relationship between rigidity, infinitesimal rigidity and generic rigidity is summarized in the Venn diagram of Figure 11, where the frameworks inside the generic

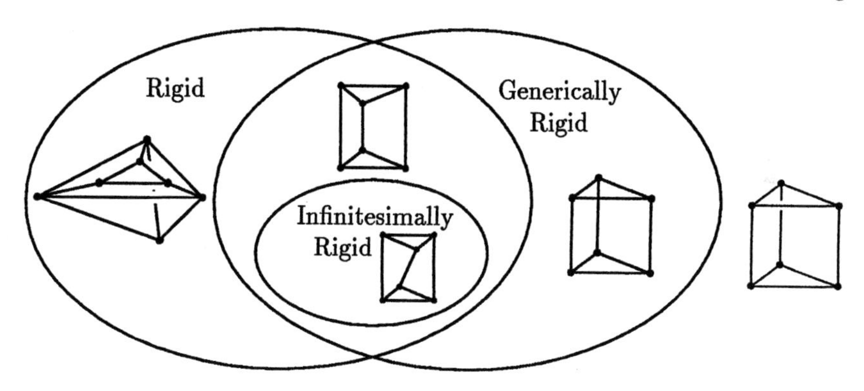

Figure 11: The rigidity map.

region are 2D, and those outside are 3D. Since generic rigidity depends on the graph alone, it is natural to look for purely graph theoretic, or at least combinatorial, characterizations. In dimension 1 we have seen that generic rigidity is equivalent to connectivity. We will see in the next section that there is a completely analogous development in dimension 2 which simultaneously makes two dimensional rigidity accessible via graph theory, and gives rise to new theoretical tools for graph theory itself. In dimension three and higher, such a characterization has so far eluded the mathematicians.

Generic Rigidity in the Plane

Theorems 3 and 5 imply that a graph $G = (V, E)$ which is generically rigid in the plane must have at least $2n - 3$ independent edges, that is, $2n - 3$ edges which span the vertices and are distributed "wisely". If there are more than $2k - 3$ edges connecting a subset of k vertices, the corresponding rows of the rigidity matrix must be dependent. Thus, in order for a graph to be generically rigid in the plane, there must be a subset $F \subseteq E$ satisfying the following two conditions:

[L1:] $|F| = 2n - 3$

[L2:] For all $F' \subseteq F$, $F' \neq \emptyset$, $|F'| \leq 2k - 3$, where k is the number of vertices which are endpoints of edges in F'.

That these two conditions are also sufficient was proved by Laman in 1970, [7].

THEOREM 6 (Laman's Theorem)$G = (V, E)$ *is generically rigid in the plane if and only if there is a subset* F *of* E *satisfying* **L1** *and* **L2**.

Laman's Theorem is a combinatorial characterization of generic rigidity in the plane. We can use it to determine if a graph is generically rigid without embedding the vertices into $\mathbb{R}^2$ and calculating the rank of the rigidity matrix. Unfortunately, Laman's conditions are not easy to check for large graphs because we are required to check all subsets $F \subseteq E$ of the correct cardinality, and then all subsets thereof. A direct algorithmic approach to generic rigidity using Laman's Theorem would clearly be exponential in the number of vertices. However, there is a striking similarity between Laman's Theorem and the following theorem of Nash-Williams [10].

THEOREM 7 (Nash-Williams) *A graph is the edge disjoint union of two spanning trees if and only if* $|E| = 2n - 2$ *and, for all* $E' \subseteq E$, $|E'| \leq 2k - 2$, *where* k *is the cardinality of the set of endpoints of* E'.

Edmonds [4] provided a polynomial time algorithm for decomposing a graph into two spanning trees and we can easily get a polynomial algorithm for testing a graph for generic rigidity: For each set $F \subseteq E$, $|F| = 2n-3$, Edmonds' algorithm yields the union of two trees after adding any one edge to F. Recski [11] showed that it is enough to double every edge of F, which decreases the number of times Edmonds' algorithm needs to be applied. Crapo [3] observed that the rigid subset we are looking for is actually the union of three trees such that every vertex is contained in exactly two of them, and such that no two non-trivial subtrees have the same underlying vertex set. Crapo calls such a decomposition a *proper* 3T2 *decomposition.* It is easy to see the connection between Nash-Williams' Theorem and Crapo's 3T2 decomposition: Remove one edge of a graph which is the union of two spanning trees, and you have a 3T2 decomposition of the remaining graph (not necessarily proper, since two subtrees may span the same set of vertices.)

Crapo's algorithm [3, 13] always produces a 3T2 decomposition such that one of the trees is spanning, however there are many 3T2 decompositions where all three trees have approximately the same size. It is an open question if Crapo's algorithm can be altered to yield such a balanced decomposition, which would improve the running time.

Tay in [13] showed directly, without using Laman's Theorem, that Crapo's condition characterizes generic rigidity in the plane and we sketch his proof here.

THEOREM 8 (Tay) *A graph* $G = (V, E)$ *is generically rigid and independent in the plane if and only if it has a proper 3T2 decomposition.*

PROOF: Assume G has $2n-3$ edges and there exists a generic embedding $\mathbf{p}$ of V into $\mathbb{R}^2$ such that $R(G)$ has rank $2n-3$. Then there exists a $(2n-3)\times(2n-3)$ submatrix R' of R with $\mathrm{Det}(R') \neq 0$, in fact R' can be taken to be R with the last three columns deleted (the last two columns are clearly always dependent on the previous ones). Expanding the determinant along the odd columns, we can find submatrices R_1 and R_2 of R', R_1 an $(n-1)\times(n-1)$ matrix consisting of the odd columns of R' and rows corresponding to the edge set E_1 and R_2 an $(n-2)\times(n-2)$ matrix consisting of the even columns of R' and rows corresponding to edge set $E_2 = E - E_1$, such that $\mathrm{Det}(R_1)\mathrm{Det}(R_2) \neq 0$. R_1 is the incidence matrix of the subgraph of G induced by E_1 with the last column deleted. Since R_1 has full rank, this induced subgraph is a (spanning) tree. Likewise E_2 induces two (edge disjoint) trees. So we have a 3T2 decomposition which must be proper because we assumed G to be generically independent.

Given a proper 3T2 decomposition, we would like to produce an infinitesimally rigid embedding. We will start with a limit framework. If T_1, T_2 and T_3 are the trees of the 3T2 decomposition, then the sets of vertices of their pairwise intersections form a partition of V. Denote the vertices of $T_1 \cap T_2$ by $\bar{V}_3$, those of $T_2 \cap T_3$ by $\bar{V}_1$, and those of $T_3 \cap T_1$ by $\bar{V}_2$. At least two of these vertex sets are non-empty. Define a map $\mathbf{p}$ of the vertices by sending $\bar{V}_1$ to $(0,0)$, $\bar{V}_2$ to $(0,1)$, and $\bar{V}_3$ to $(1,0)$. It follows that all edges of T_1 which with distinct endpoints connect $(1,0)$ and $(0,1)$, and likewise for T_2 and T_3, see Figure 12.

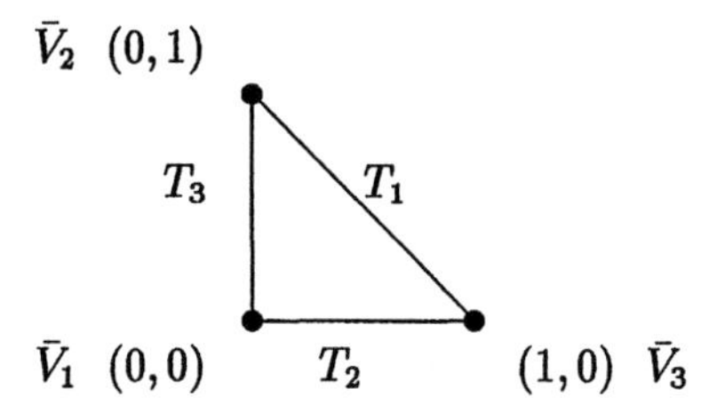

Figure 12: A limit framework for a 3T2 graph.

We want to show that $(V, E, \mathbf{p})$ is a limit framework such that all the edges of T_i, infinitesimal or not, are parallel. To do this we inductively peel apart the vertex sets $\bar{V}_1$, $\bar{V}_2$ and $\bar{V}_3$, one direction at a time, involving edges of one tree at a time. We can do this since, at any stage, the induced subgraph on those vertices mapped to some single point, say $\bar{V}_3$, consists of edges of just two of the trees, T_1 and T_2, and the intersection of that induced subgraph with one of the trees, say T_1, which join those vertices is disconnected, (otherwise there would be two subtrees with the same span.) We can then separate the connected components, in this case, in the horizontal direction.

Lastly we need to show that the rigidity matrix of the limit framework is of full rank. Reorder the rows of the rigidity matrix so the edges belonging to T_1 come first, then those of T_2, and then those of T_3. Reorder the columns in three sections by taking the columns corresponding to the first coordinates of $\bar{V}_2$ and the second coordinates of $\bar{V}_3$ in the first section, those corresponding to the first coordinates of $\bar{V}_1$ and $\bar{V}_3$ in the second section, and the second coordinates of $\bar{V}_1$ and $\bar{V}_2$ in the third.

$$\begin{matrix} T_1 \\ T_2 \\ T_3 \end{matrix} \begin{bmatrix} I_1 & ? & ? \\ 0 & I_2 & ? \\ 0 & 0 & I_3 \end{bmatrix},$$

The matrix then has the block structure shown, with I_i being the incidence matrix of T_i, and so it has full rank. □

In the proof, each 3T2 decomposition gives a representation of the triangle as a limit framework. For the 3T2 decomposition of the triangular prism indicated in Figure 13a the limit framework is the one described in Figure 10. The 3T2 decomposition of the

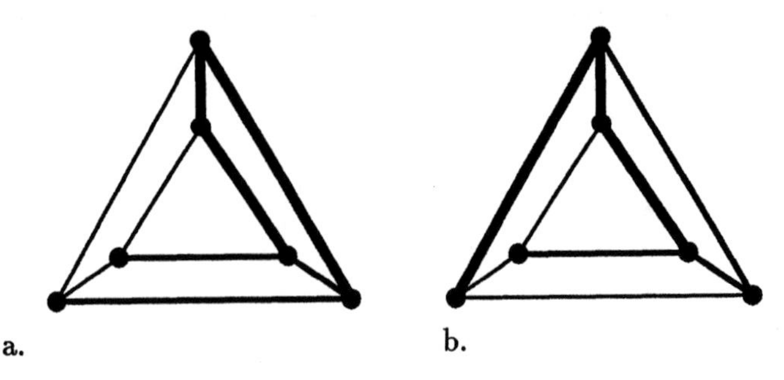

Figure 13: 3T2 decompositions of the triangular prism.

triangular prism of Figure 13b results in a limit framework that corresponds to the result of the vertical projection of the "screwing down" deformation of Figure 1c.

Tay's proof, [14, 15] can be generalized to hinged panel structures in m-space.

Generic Rigidity in 3-space

The most famous general result for generic rigidity in dimension three is that the obvious analogue of Laman's Theorem is not true. In dimension three a graph with n vertices needs at least $3n-6$ edges. We say that a graph G on n vertices satisfies *Laman's condition* if G has a subset F of $3n-6$ edges which spans all n vertices, and such that for any subset F' of F, $|F'| \leq 3|V(F')|-6$, where $V(E)$ denotes the set of all vertices which are endpoints of some edge in an edge set E. In other words, G has enough edges to be rigid but does not have any obviously overbraced subgraph. It is straightforward to show that every generically rigid graph in dimension 3 satisfies Laman's condition, however, the condition is not sufficient to insure generic rigidity.

EXAMPLE: The "double banana" graph of Figure 14a is not generically rigid in dimension 3, but does satisfy Laman's condition. This graph is obviously flexible in any

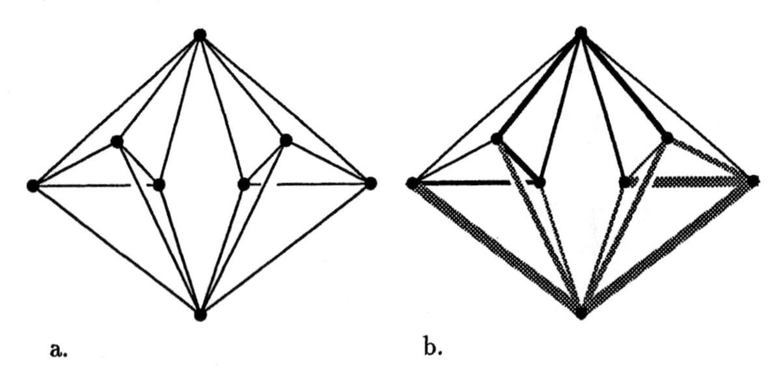

Figure 14: The double banana.

embedding since the two "bananas" can twist about the two vertex cutset. On the other hand it is easy to check that no subgraph is overbraced. □

While Laman's condition does not characterize generic rigidity in 3 space, the other characterizations of planar rigidity are not known to fail. It may also be possible to augment Laman's condition. The obvious question, which remains unanswered, is whether every 3-connected graph satisfying Laman's condition is rigid.

The most promising approach is via tree decompositions. The three dimensional analogue to the 3T2 condition is to decompose the graph into 6 trees, each vertex belonging to exactly 3 of the trees. The remaining task here is to define "proper" correctly, since requiring merely that no two subtrees have the same span is not enough, since again the double banana is a counterexample, see Figure 14b.

For a complete discussion of the current state of affairs in 3D generic rigidity as well as possible approaches see [5].

Abstract Rigidity

It is possible to study infinitesimal and generic rigidity without reference to a specific rigidity matrix by looking only at how sets of edges depend on one another. Lovász and Yemini [8] were the first to explicitly state that matroid theory is the appropriate tool for studying generic rigidity. More recent treatments can be found in [5] and [17].

Recall that both infinitesimal and generic rigidity in dimension 1 are equivalent to the connectivity of the graph. Given a set E of edges of a complete graph, define the *closure* of that set, $\langle E \rangle$, as the set of all edges both of whose endpoints lie in the same connected component of the subgraph generated by E. It is easy to see that this closure operator is defined on the set of all subsets of the edges of a complete graph and satisfies

C1 $T \subseteq \langle T \rangle$;

C2 If $R \subseteq T$, then $\langle R \rangle \subseteq \langle T \rangle$;

C3 $\langle \langle T \rangle \rangle = \langle T \rangle$.

C4 If $s, t \in (E - \langle T \rangle)$, then $s \in \langle T \cup \{t\} \rangle$ if and only if $t \in \langle T \cup \{s\} \rangle$.

Any set operator, defined on the set of all subsets of a fixed set S and which satisfies these conditions is said to form a *matroid* M on the set S. Given any matrix, there is a matroid defined on the set of its rows by setting the closure of any subset X of rows to be the set of all rows which are linear combinations of X. In this way, using the rigidity matrix, we can define a matroid on the set of edges of a framework in any dimension. In particular, for any embedding $\mathbf{p}$ of n vertices into m space, we can form a matroid $\mathcal{F}(\mathbf{p})$ on the edges of a complete graph called an *infinitesimal rigidity matroid.* If $\mathbf{p}$ is a generic embedding, then $\mathcal{G}_m(n) = \mathcal{F}(\mathbf{p})$ is called the *m-dimensional generic rigidity matroid on n vertices.*

$\mathcal{G}_1(n)$ is the *connectivity matroid* previously described. In general, for $(V, K, \mathbf{p})$ a framework on the complete graph, the closure of a set E of edges in $\mathcal{F}(\mathbf{p})$ consists of those edges which depend on E, that is, those pairs of vertices which are not infinitesimally expanded or contracted in any flex of $(V, E, \mathbf{p})$. In particular, the closure of any rigid set of edges is complete.

EXAMPLE: The closure of the set of edges of an infinitesimally rigid triangular prism in 2D is the set of edges of the complete graph on six vertices. If the prism has an infinitesimal flex, then its closure is itself. The closure of the double banana, Figure 15a, is the double banana together with the edge joining the two "hinge" vertices, Figure 15b. □

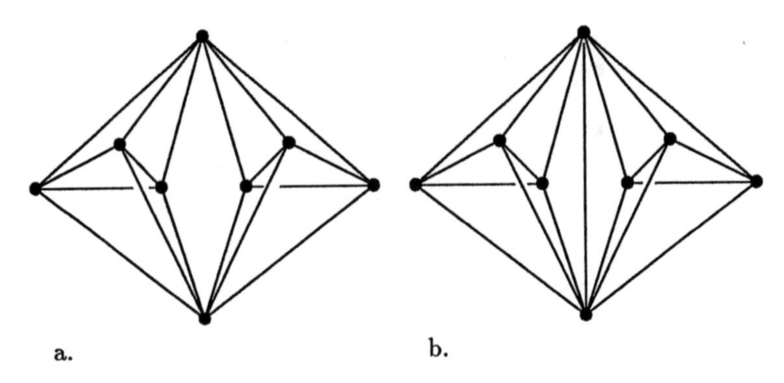

Figure 15: The closure of the double banana.

Recall that $V(E)$ denotes the set of all vertices which are endpoints of edges in E. For a set X of vertices, let $K(X)$ denote the edges of the complete graph on X. Observe that all infinitesimal and generic rigidity matroids satisfy the following extra conditions.

C5 If $E, F \subseteq K$ and $|V(E) \cap V(F)| < m$, then $\langle E \cup F \rangle \subseteq (K(V(E)) \cup K(V(F)))$.

C6 If $\langle E \rangle = K(V(E))$, $\langle F \rangle = K(V(F))$ and $|V(E) \cap V(F)| \geq m$, then $\langle E \cup F \rangle = K(V(E \cup F))$.

Briefly, the 5'th property comes from the fact that, if two subgraphs meet too few vertices, then there is an (infinitesimal) motion twisting the two graphs along their intersection, like in the case of the double banana. The 6'th property expresses that if two rigid subgraphs meet in sufficiently many vertices, then their union is rigid.

We call any matroid $\mathcal{A}_m$ on the edges of a complete graph which satisfies the two extra conditions an *m-dimensional abstract rigidity matroid.* A set of edges is said to be *rigid* in $\mathcal{A}_m$ if its closure is complete, $\langle E \rangle = K(E)$. Generally, a set of edges E is said to be *independent* if the closure of any proper subset of E is a proper subset of the closure of E, otherwise E is said to be *dependent.* A maximally independent set is called a *basis.* The *rank* of a set is the cardinality of its largest independent subset. A minimally dependent set is called a *cycle,* and a *cocycle* is a minimal set which intersects every basis.

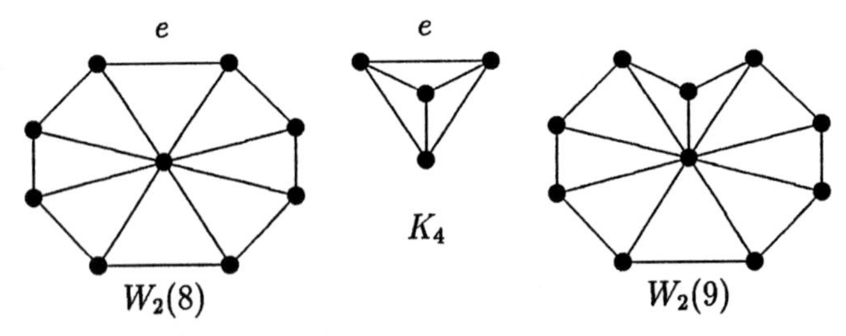

Figure 16: 2 dimensional wheels - cycles $\mathcal{A}_2$.

The connectivity matroid is the only abstract rigidity matroid in dimension 1, however there are many abstract rigidity matroids in higher dimensions. Nevertheless, they do contain many features in common. For example, it is not hard to show that every abstract rigidity matroid in dimension m has rank $m|V| - \binom{m+1}{2}$. In dimension 1, the cycles of the abstract rigidity matroid are the usual cycles in a graph. In dimensions greater than two, there are quite a variety of different cycles. In two dimensions, the edge sets of *wheels* are always cycles in $\mathcal{A}_2$, see Figure 16. The 3-dimensional wheels of Figure 17 must always be dependent in $\mathcal{A}_3$, although they need not be cycles.

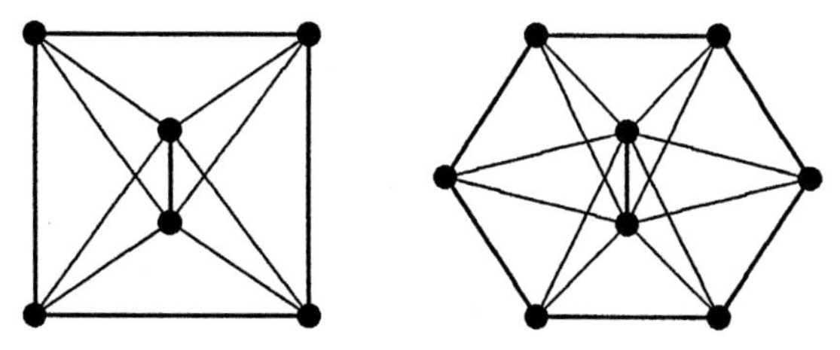

Figure 17: 3 dimensional wheels - dependent in $\mathcal{A}_3$.

Infinitesimal rigidity matroids have some obvious small cycles and cocycles. In the 1-dimensional generic rigidity matroid on the complete graph, i.e. the connectivity matroid, the triangle is the smallest cycle, and the star of any vertex is a cocycle. In dimension m, it is clear from elementary row operations that the set of edges of any complete graph on $m+2$ vertices, K_{m+2}, is a cycle, and also that the star of any vertex v minus any set A of $m-1$ edges incident to v, denoted by $S_A(v)$, is a cocycle, These facts are not only true for abstract rigidity matroids, they characterize them.

THEOREM 9 *A matroid $\mathcal{M}$ on the edge set of K_n is an m-dimensional abstract rigidity matroid if and only if all of the K_{m+2}'s are cycles and all of the $S_A(v)$'s are cocycles.*

From this theorem we have immediately that every infinitesimal rigidity matroid is an abstract rigidity matroid, however the converse is not true. We will close this section by constructing an abstract rigidity matroid which is not an infinitesimal rigidity matroid.

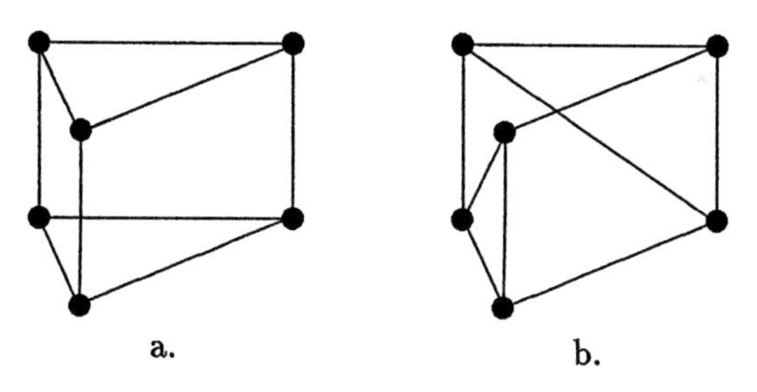

Figure 18: Prisms

The generic rigidity matroid on 6 vertices in the plane has rank $2 \cdot 6 - 3 = 9$, so every basis, that is, every a minimal rigid subgraph which spans the vertices, has 9 edges. Thus the edge set of any triangular prism is a basis in $\mathcal{G}_2(n)$ for $n \geq 6$. The set of edges of the triangular prism $(V, E, \mathbf{p})$ of Figure 18a is not a basis in $\mathcal{F}(\mathbf{p})$ since it is infinitesimally flexible, although the edge set of another prism on the same embedding may be a basis, Figure 18b. We want to construct an abstract rigidity matroid all of whose prisms are dependent.

We define a matroid on the edges of $K_6 = (V, E)$ as follows. Let $\mathcal{B}$ denote the collection of all subsets of E which are bases of $\mathcal{G}_2(6)$ with the exception of those which correspond to subgraphs isomorphic to the triangular prism. It is straightforward to verify that $\mathcal{B}$ is the collection of bases of a matroid, $\mathcal{M}$. To see that the matroid with bases $\mathcal{B}$ is an abstract rigidity matroid we note that the only difference between its closure operator and that of $\mathcal{G}_2(6)$ is that the closure of a prism, or a prism minus an edge is that prism, and no such edge set has either a separating vertex or is the union of two rigid subgraphs, so the axioms are satisfied.

To see that $\mathcal{M}$ is not an infinitesimal rigidity matroid, we have to show that there is no embedding $\mathbf{p}$ of K_6 into $\mathbb{R}^2$, so that the corresponding infinitesimal rigidity matroid is isomorphic to $\mathcal{M}$.

Any non-trivial infinitesimal motion of a prism, which we may assume to be zero on one of the triangles, must extend to an infinitesimal isometry of $\mathbb{R}^2$, which is either an infinitesimal translation or rotation, see Figure 19. If it is a translation the lines con-

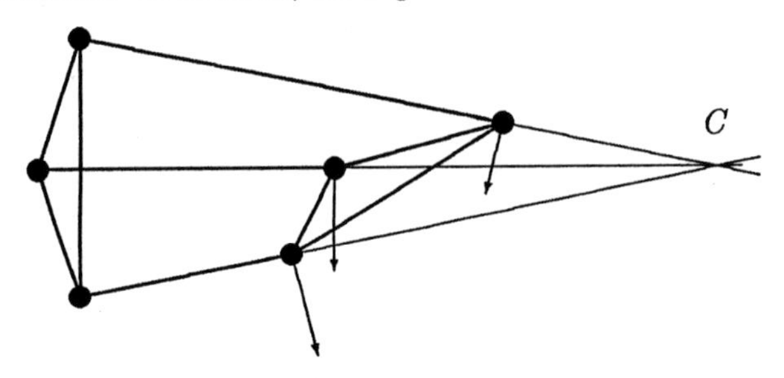

Figure 19: An infinitesimal motion on a prism

necting the triangles are parallel, while if it is a rotation, the lines must all pass through the center of rotation. Thus, in order for the edge set of a prism to be infinitesimally dependent, the three lines connecting the two triangles must be projectively concurrent. Since the 3 diagonal points of a complete quadrangle are never collinear in the projective plane, there is no general embedding of 6 points in $\mathbb{R}^2$ such that all prisms are dependent, and so $\mathcal{M} \neq \mathcal{F}(\mathbf{p})$ for any embedding $\mathbf{p}$.

If the edges E of a graph G are independent in some infinitesimal rigidity matroid $\mathcal{F}(\mathbf{p})$ in dimension m, then E must be independent in the generic rigidity matroid, as well as the edges of any subgraph isomorphic to G, and we write $\mathcal{F}(m) \leq \mathcal{G}_m(n)$. If the same were true for all abstract rigidity matroids, then this would yield a combinatorial characterization of $\mathcal{G}_m(n)$. This is known to be true in dimension less than three, and known to be false in dimension greater than 3.

CONJECTURE 1 (The Maximal Conjecture) *If $\mathcal{A}_3$ is any 3-dimensional abstract rigidity matroid on n vertices, then $\mathcal{G}_3(n) \geq \mathcal{A}_3$.*

References

[1] R. Connelly and H. Servatius, *Higher-order rigidity—what is the proper definition?,* Discrete Comput. Geom. 11, no. 2, 193–200, (1994).

[2] R. Connelly, W. Whiteley, *Second-order rigidity and prestress stability for tensegrity frameworks,* SIAM J. Discrete Math. 9,no. 3, 453–491, (1996).

[3] H. Crapo, *On the generic rigidity of structures in the plane,* Advances in Applied Math., (1993), to appear.

[4] J. Edmonds, *Minimum partition of a matroid into independent subsets,* J. Res. Nat. Bur. Standards B **69**, 67–72, (1965).

[5] J. Graver, B. Servatius and H. Servatius "Combinatorial Rigidity", Graduate Studies in Math., AMS, (1993).

[6] J. E. Graver, B. Servatius and H. Servatius, *Abstract rigidity in m-space,* in "Jerusalem Combinatorics '93", 145–151, Contemp. Math., 178, AMS, (1994).

[7] G. Laman, *On Graphs and rigidity of plane skeletal structures,* J. Engrg. Math. **4**, 331–340, (1970).

[8] L. Lovász and Y. Yemini, *On Generic rigidity in the plane,* SIAM J. Alg. Disc. Methods **3**, 91–98, (1982).

[9] J. C. Maxwell, *On reciprocal figures and diagrams of forces,* Philos. Mag. **4** 27, 250–261, (1864).

[10] C. St. J. A. Nash-Williams, *Edge disjoint spanning trees of finite graphs,* J. London Math. Soc., **36**, 445-450, (1961).

[11] A. Recski, *A network theory approach to the rigidity of skeletal structures. Part 1. Modeling and interconnection,* Discrete Applied Math. **7**, 313–324, (1984).

[12] B. Servatius and W. Whiteley, *Constraining plane configurations in CAD: The combinatorics of direction length designs,* SIAM J. Discrete Math., 1998 to appear.

[13] T. S. Tay, *A new proof of Laman's theorem,* Graphs Combin. 9, no. 4, 365–370, (1993).

[14] T. S. Tay, *Linking $(n-2)$-dimensional panels in n-space. I. $(k-1,k)$-graphs and $(k-1,k)$-frames,* Graphs Combin. 7, no. 3, 289–304, (1991).

[15] T. S. Tay, *Linking $(n-2)$-dimensional panels in n-space. II. $(n-2,2)$-frameworks and body and hinge structures,* Graphs Combin. 5, no. 3, 245–273, (1989.)

[16] W. Whiteley, *Matroids and rigid structures,* in "Matroid Applications", Neil White, ed. Encyclopedia of Mathematics and its Applications **40**, Cambridge University Press, 1–53, (1992).

[17] W. Whiteley, *Some matroids from discrete applied geometry,* in "Matroid Theory", pp 171–311, Contemp. Math., 197, AMS, (1996).

RIGIDITY OF MOLECULAR STRUCTURES:

GENERIC AND GEOMETRIC ANALYSIS

Walter Whiteley*

Department of Mathematics and Statistics
York University
Toronto Ontario, M3J-1P3

INTRODUCTION

In chemistry and physics, one encounters 'molecular models' in which pairs of 'atoms' are connected by a 'bond' of fixed length and fixed angles between the various bonds of each atom (Figure 4). In practice, both the physical models and the mathematical models for these molecules build the atoms and their 'half' of the bonds as rigid objects (i.e. balls with spikes), while the connection of the two atoms along a bond functions as a hinge: a constraint between the two balls which allows only a rotation of one ball relative to the other, about the line down the middle of the bond.

Our purpose in this paper is to draw together some mathematical techniques for analyzing these structures. From some previous results for the rigidity of various types of structures, we extract both specific methods to predict the behaviour of molecular models with generic positions on given graphs, and methods for analyzing particular geometric configurations. The results on which we draw range over:

(a) classical results on the rigidity of frameworks and panel structures built on convex polyhedra (Cauchy's Theorem[2,24], Alexandov's Theorem[1,24]);
(b) a basic isomorphism between infinitesimal motions of hinge-panel structures built on faces and edges of a geometric spherical complex with static self-stresses of the framework on the vertices and edges of the same geometric complex[7];
(c) the projective polarity between: (i) hinge-panel structures (with the hinges of each rigid body lying in a single plane): e.g. the faces and edges of a polyhedron; and (ii) molecular structures (with the hinges of each rigid body concurrent in a central point): e.g. the vertices and edges of a dual polyhedron[19,28];
(d) projective polarities of statics and kinematics between a bar-and-joint framework in 3-space and the polar 'sheet structure' with thin plane sheets rigid in their plane[27]

* Work supported, in part, by grants from NSERC (Canada).

Rigidity Theory and Applications
Edited by Thorpe and Duxbury, Kluwer Academic/Plenum Publishers, 1999

The underlying theory explicitly belongs to the areas of analytic and synthetic projective geometry. While some partial version could be presented using the more common Euclidean 'free vectors', we have chosen to present a version using the more appropriate 'line-bound vectors' and their analytic counterparts: the 2-extensors of Grassmann or Cayley algebra[21]. We hope that the examples, both algebraic and visual, will be sufficient to convince the reader that the resulting visual and algebraic simplicity is worth the effort of learning this less common vocabulary.

Where possible, we will present results as they apply directly to molecular structures. Throughout the paper, we will focus on simple examples such as the benzene ring and the fullerine[5] molecules. Our analysis describes a simple mathematical model abstracted away from the complexity of chemistry with multiple bonds, distorted angles and stored energy. We make no claim that the rigidity or flexibility of this abstract model directly explains the chemical or physical behaviour of an actual molecule. However, if our analysis indicates that the mathematical model is shaky (infinitesimally flexible), then there will be a 'virtual velocity' of a path of positions with the same (minimal) energy under simple models of the energy. These shakes are the 'zero-modes of the dynamic matrix at temperature 0' in the language of people working on models of glasses and crystals.

This paper makes explicit results which were implicit in the previous literature. The prior form and the underlying translations are left implicit unless this underlying source gives an additional insight into the methods. A major difficulty in writing this (and perhaps in reading it) is the need to provide consistent 'polar' versions of know results. Certain definitions are difficult to give or to remember correctly, without the vision of another 'familiar' object and some facility with the combinatorial duality which comes from projective polarity. It is a continuing and pleasant surprise that such projective play makes visible connections between otherwise very different physical objects. I offer these results in the spirit of the task of mathematics 'to make the invisible visible'.

This is one slice of a much larger theory of the static and first-order kinematics of structures built in 3-space. Other papers in this volume present other important aspects of the underlying theory. A recent handbook chapter[30] surveys the literature with basic definitions and results in the core mathematical theory as well as references to the broader literature. A more complete presentation of the combinatorial portions of the general mathematical theory appears in a recent book[11] and paper[31].

GENERAL HINGE STRUCTURES

We start with a general structure formed of rigid bodies and hinges in 3-space. To analyze the class of structures with the 'same' overall combinatorics, the definition is split into: a combinatorial level of a graph indexing the bodies and which pairs are linked along a hinge; and a geometric level, in which we assign lines (or line segments) in 3-space to the hinges. We begin by introducing an appropriate algebra for these line segments.

We do not use standard vectors for the hinges, since these vectors do not easily distinguish one line from a parallel line. Instead we use the Grassman-Cayley algebra[7,21,22] developed for weighted objects in affine or projective space. Specifically, given two Euclidean points $\mathbf{a} = (a_1, a_2, a_3), \mathbf{b} = (b_1, b_2, b_3)$, or equivalently, two projective points $\overline{a} = (a_1, a_2, a_3, 1)$, $\overline{b} = (b_1, b_2, b_3, 1)$ we take the six 2×2 minors of the matrix:

$$\begin{bmatrix} a_1 & a_2 & a_3 & 1 \\ b_1 & b_2 & b_3 & 1 \end{bmatrix}$$

written as the 6-tuple 2-*extensor* $\overline{a} \vee \overline{b}$, in the order

$$\overline{a} \vee \overline{b} = (a_1 - b_1, a_2 - b_2, a_3 - b_3, a_2 b_3 - a_3 b_2, a_1 b_3 - a_3 b_1, a_1 b_2 - a_2 b_1).$$

In Euclidean terms, this 6-tuple is $\overline{ab} = \overline{a} \vee \overline{b} = (\mathbf{a}-\mathbf{b}, \mathbf{a}\times\mathbf{b})$. From either presentation we observe that $\overline{b} \vee \overline{a} = -\overline{a} \vee \overline{b}$ and $\overline{a} \vee \overline{a} = \mathbf{0}$

We will not verify it here, but this 6-tuple depends only on the line in question and the Euclidean distance between $\mathbf{a}$ and $\mathbf{b}$. With the same line but a different distance, we end up with a scalar multiple of this 6-tuple. Conversely, any non-zero multiple of this 2-extensor represents a segment of this same line. We have an algebra for what were classically called *line-bound vectors*[14,15]. These 6-tuples are also called *Plücker coordinates for the line*[14,15].

We can also find 6-tuples using coordinates for projective points at infinity, written in the form $(c_1, c_2, c_3, 0)$, for one or both of the points. This disturbs the immediate interpretation of the 2-extensors in terms of line segments but continues to represent the line. If we take two points at infinity, the resulting 2-extensor will represent a line at infinity. The collection of all 6-tuples created from pairs of points is called the set of *2-extensors* L^2.

While all pairs of points (lines) create 6-tuples, not all 6-tuples come from pairs of points. A 6-tuple $\mathbf{d} = (d_{1,4}, d_{2,4}, d_{3,4}, d_{2,3}, d_{1,3}, d_{1,2})$ represents a line (and belongs to L^2) if and only if it satisfies the Plücker condition[15]:

$$d_{1,4}d_{2,3} - d_{2,4}d_{1,3} + d_{1,2}d_{3,4} = 0.$$

General 6-tuples represent *screws* in 3-space. Any screw, or 6-tuple, can be written as the sum of two 2-extensors for two lines, which are classically selected to represent a rotation about an axis and a translation along the line of this axis[15]. The set of all 6-tuples is called the space of *2-tensors* Λ^2. We are really working in projective space, but the reader can assume, if desired, that all the 2-extensors in the structures presented here represent finite lines.

Finally, we note that two 2-extensors $\mathbf{d}, \mathbf{e}$ are coplanar if and only if:

$$\mathbf{d} * \mathbf{e} \;=\; d_{1,4}e_{2,3} - d_{1,3}e_{2,4} + d_{1,2}e_{3,4} + d_{2,3}e_{1,4} - d_{2,4}e_{1,3} + d_{3,4}e_{1,2} = 0.$$

This equation, of which the Plücker condition is the special case $\mathbf{d} * \mathbf{d}$, comes from taking points $\overline{d'}\overline{d''} = \mathbf{d}$ and $\overline{e'}\overline{e''} = \mathbf{e}$ and placing them into a 4×4 determinant which computes the oriented volume of the tetrahedron $\overline{d'}\overline{d''}\overline{e'}\overline{e''}$.

$$\det \begin{bmatrix} d_1' & d_2' & d_3' & 1 \\ d_1'' & d_2'' & d_3'' & 1 \\ -- & -- & -- & -- \\ e_1' & e_2' & e_3' & 1 \\ e_1'' & e_2'' & e_3'' & 1 \end{bmatrix}.$$

This determinant (volume) is zero if and only if the four points are coplanar. The Laplace expansion of this determinant by the first two rows gives the equation.

Definition 1. A *hinge structure* is a graph $G = (B, E)$ and an assignment of 2-extensors (line-bound vectors) to the directed edges $\mathbf{H} : E \to L^2$, written $\mathbf{H}(i,j) = \mathbf{H}_{i,j}$, with the property that $\mathbf{H}_{j,i} = -\mathbf{H}_{i,j}$. Together these can be written as $G; \mathbf{H}$.

Given a hinge structure, we study the infinitesimal motions or *shakes*, essentially velocities assigned to the points of the bodies, so that these would not change any distances within the body at first-order. Such infinitesimal motions can be recorded by giving a screw center $\mathbf{S}_i \in \Lambda^2$ to each body (see [7,22] for details). We note that a line at infinity as a center corresponds to a translation of the body along the direction perpendicular to all the planes through this line at infinity (Figure 3C).

The condition that the motions $\mathbf{S}_i, \mathbf{S}_j$ of the two bodies B_i, B_j preserves a common hinge $\mathbf{H}_{i,j}$ is written as a single vector equation[7]:

$$\mathbf{S}_i - \mathbf{S}_j = \lambda_{i,j}\mathbf{H}_{i,j} \qquad \text{for some scalar } \lambda_{i,j} \text{ for the undirected edge.}$$

Definition 2. An *infinitesimal motion* of a hinge structure $(B, E); \mathbf{H}$ (or *motion* for the rest of the paper) is an assignment of screw centers $\mathbf{S} : B \to \Lambda^2$ such that for every edge $(i, j) \in E$ the *hinge condition* is satisfied: $\mathbf{S}_i - \mathbf{S}_j = \lambda_{i,j}\mathbf{H}_{i,j}$.

A motion of a hinge structure is *trivial* if all screw centers are identical: for some fixed screw $\mathbf{C}$ and for all $i \in B$, the center $\mathbf{S}_i = \mathbf{C}$. A hinge structure is *infinitesimally rigid* (or *rigid* in this paper) if all motions are trivial. Otherwise it is *shaky*.

Example 1. Consider a graph which is a single cycle of length k: $B_1, B_2, \ldots B_k, B_1$. If we have an infinitesimal motion of hinge-body structure on this graph $G; \mathbf{H}$, $\mathbf{S}_1, \mathbf{S}_2, \ldots, \mathbf{S}_k$, then the entire set of hinges satisfies a system of linear equations.

Given $\mathbf{S}_1 - \mathbf{S}_2 = \lambda_{1,2}\mathbf{H}_{1,2}$, $\mathbf{S}_2 - \mathbf{S}_3 = \lambda_{2,3}\mathbf{H}_{2,3}$, $\ldots$, $\mathbf{S}_k - \mathbf{S}_1 = \lambda_{k,1}\mathbf{H}_{k,1}$, the k equations telescope to give the *cycle condition*:

$$\sum_{i=1}^{i=k} \lambda_{i,i+1}\mathbf{H}_{i,i+1} = \mathbf{0} \qquad \text{with the usual cycle convention } k+1 = 1.$$

This vector equation represents six linear equations in the k scalars $\lambda_{i,i+1}$ of the cycle. The motion is trivial if and only if all of these scalars $\lambda_{i,i+1}$ are zero.

Conversely, given any set of scalars $\lambda_{i,j}$ satisfying the cycle condition, we can define a motion by picking S_1 arbitrarily then constructing $\mathbf{S}_m$:

$$\mathbf{S}_m = \mathbf{S}_1 + \sum_{i<m} \lambda_{i,i+1}\mathbf{H}_{i,i+1}.$$

The cycle condition guarantees the consistency of $\mathbf{S}_k$ and $\mathbf{S}_1$ on the final hinge $\mathbf{H}_{k,1}$.

If the 6-tuples $\mathbf{H}_{i,i+1}$ are linearly independent vectors, then the structure will be rigid. If these extensors are linearly dependent, then the structure will be shaky! Clearly, any cycle of length $k > 6$ will have linearly dependent 6-tuples and will therefore be shaky for all choices of the hinge line.

We have not yet confirmed that there are six independent 6-tuples for lines in 3-space. The six edges of the tetrahedron with Euclidean vertices $(0,0,0)$, $(1,0,0)$, $(0,1,0)$, $(0,0,1)$ give the six 2-extensors which are the rows of the matrix

$$\begin{bmatrix} 1 & 0 & 0 & 0 & 0 & 0 \\ 0 & 1 & 0 & 0 & 0 & 0 \\ 0 & 0 & 1 & 0 & 0 & 0 \\ 0 & 1 & -1 & 1 & 0 & 0 \\ 1 & 0 & -1 & 0 & 1 & 0 \\ 1 & -1 & 0 & 0 & 0 & 1 \end{bmatrix}.$$

Clearly these rows are independent, and a cycle of length $k \leq 6$ using some or all of these six lines would be rigid. ∎

Example 2. Some sets of six lines in 3-space are linearly dependent (Figure 1). The set of lines satisfying such a linear equation is called a *linear line complex*[6,15]. Such objects are the subject of entire books[14] in projective geometry (and implicitly, in statics and kinematics). We do not draw the general case of six dependent 2-extensors, but a special case is six lines meeting a single line (Figure 1, $k = 6$). Dependent sets of five lines include lines meeting two skew lines or two coplanar lines in 3-space ($k = 5$ in Figure 1). For $k = 4$, dependent sets include four lines meeting three skew lines (i.e. four lines in the ruling of an hyperboloid); four lines through a common point (e.g., projectively, four parallel lines); and the dual case of four lines in a plane are illustrated (Figure 1).

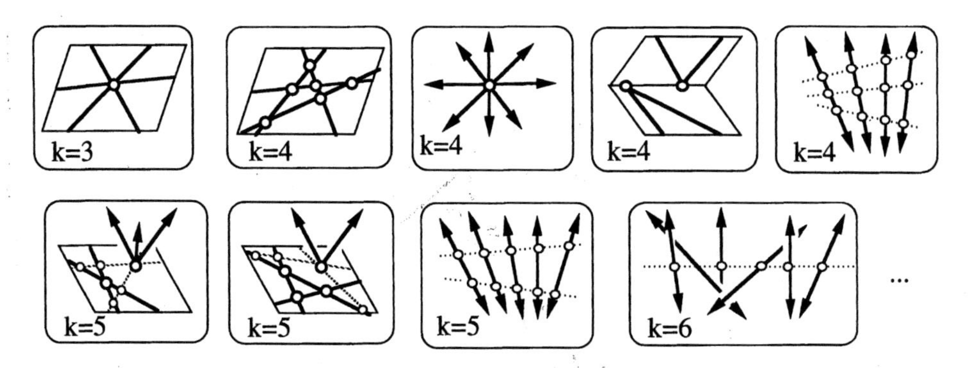

Figure 1. Examples of dependent sets of lines in 3-space[6]. Used as hinges in a cycle, each of these dependent sets will produce a shaky hinge structure.

It is even possible to find three dependent lines (Figure 1): take three lines which are coplanar and concurrent, such as the lines joining the Euclidean points $(0,0,0)$ to $(1,0,0)$, $(1,1,0)$ and $(1,1,0)$. These three 2-extensors generate the linearly dependent rows of the matrix:

$$\begin{bmatrix} 1 & 0 & 0 & 0 & 0 & 0 \\ 0 & 1 & 0 & 0 & 0 & 0 \\ 1 & 1 & 0 & 0 & 0 & 0 \end{bmatrix}$$

A hinge-body cycle of length three on these three hinges will be shaky. ∎

As this example illustrates, for a given graph, there may be some rigid hinge structures and some shaky hinge structures. In general, if we select hinges generated by algebraically independent pairs of points, we will get the maximum possible rank for the set of hinge equations, and the minimum space of motions. This is the *generic* behaviour. (Technically, we have an algebraic variety in $(\mathbb{P}^3)^{|V|}$ which describes the realizations for which the space of motions increases, and all other structures are generic.) For example, for a cycle of length 6, the generic hinge structures are rigid, and the singular hinge structures, with the six lines in a linear line complex, are shaky.

The generic behaviour depends only on the graph. If some realization of a graph as a hinge structure is rigid, the underlying graph is called *generically rigid*, and almost all realizations as a hinge structure are rigid). Otherwise, all realizations of the graph are shaky. In this terminology, the previous example shows that:

Proposition 3. *A hinge structure $G;\mathbf{H}$ built on a cycle of length $k > 6$ is shaky for all choices of* $\mathbf{H}$. *A hinge structure built on a cycle of length $k \leq 6$ is rigid for almost all choices of* $\mathbf{H}$.

There is a graph-theoretic algorithm for detecting whether the hinge structure on a graph is generically rigid[19,22,28]. Given the graph, the algorithm can be efficiently implemented with complexity $O(|B|^2)$, using techniques such as bipartite matching.

Tay's Theorem[17] 4. *A generic hinge structure on a graph $G = (B, E)$ is rigid if and only if, replacing each edge of G by 5 edges, the resulting multigraph contains six disjoint spanning trees on the vertices B.*

Notice that in Figure 2 B,C, some of the original edges of the tetrahedron are not covered five times – confirming that the corresponding hinge structure on the underlying graph is over constrained. Specifically, we can remove the two opposite edges (hinges) which do not appear in the trees of (C).

We can analyze the rigidity of all connected hinge structures entirely in terms of the scalars $\lambda_{h,i}$ rather than the centers $\mathbf{S}_i$. This switch will reduce the dimensions of the spaces being studied, replacing the 6-space of trivial motions by the zero vector.

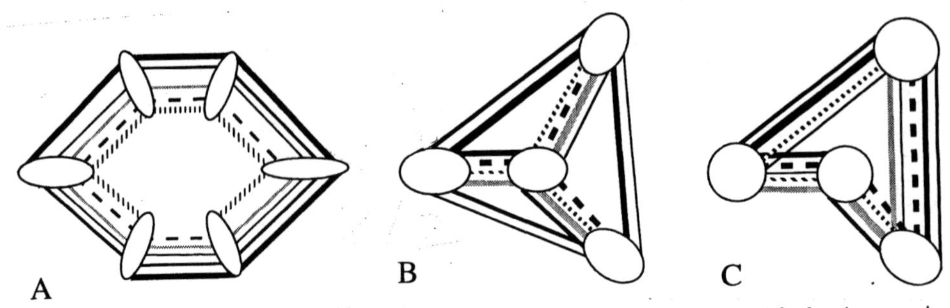

Figure 2. Examples of graphs which are generically rigid as hinge structures, with the six spanning trees in each multigraph.

It will also permit an effective visual computation of the non-trivial shakes. In some works on robotics or flexible proteins, this would be described as 'switching to internal coordinates'.

Definition 3. A *bending* of a hinge structure $(B, E); \mathbf{H}$ is an assignment of bending scalars $\lambda_{i,j}$ to the undirected edges such that, for every oriented body-hinge cycle C: $\sum_{(i,j)\in C} \lambda_{i,j}\mathbf{H}_{i,j} = \mathbf{0}$.

Because $(i,j),(j,i)$ is a (short) cycle, we have $\lambda_{i,j}\mathbf{H}_{i,j} + \lambda_{j,i}\mathbf{H}_{j,i} = \mathbf{0}$. Since $\mathbf{H}_{(i,j)} = -\mathbf{H}_{j,i}$, we have $\lambda_{i,j} = \lambda_{j,i}$. The scalar belongs to the undirected edge.

Proposition[7] 5. *For any connected graph $G = (B, E)$, there is an isomorphism between bendings of a hinge structure and motions of the hinge structure which set $\mathbf{S}_0 = \mathbf{0}$ for a selected body B_0.*

Proof. Consider any motion $\mathbf{S}$. For each edge $\{i,j\}$ we define $\lambda_{i,j}$ as the scalar which satisfies $\mathbf{S}_i - \mathbf{S}_j = \lambda_{i,j}\mathbf{H}_{i,j}$. By the same argument used in the cycle example, for every cycle C: $\sum_{(i,j)\in C} \lambda_{i,j}\mathbf{H}_{i,j} = \mathbf{0}$. These scalars do define a bending.

Conversely, consider any bending $\lambda_{i,j}$. We define a motion by setting $\mathbf{S}_0 = \mathbf{0}$ and $\mathbf{S}_m = \sum_{(i,j)\in P} \lambda_{i,j}\mathbf{H}_{i,j}$ for any path P of edges from B_0 to B_m.

This is well-defined, because any two paths P, P' will together form a cycle taken as P, P'^{-1}, and the cycle condition guarantees that

$$\sum_{(h,i)\in P} \lambda_{h,i}\mathbf{H}_{h,i} = \sum_{(j,k)\in P'} \lambda_{j,k}\mathbf{H}_{j,k}.$$

All paths to B_m define the same motion $\mathbf{S}_m$.

Notice that this isomorphism takes any trivial motion to the zero bending and a motion is non-trivial if the bending is non-zero. ∎

Finally, because we have written the hinges using a projective notation, a projective transformation (a linear transformation on the projective coordinates of the points), which take points to points and lines to lines, will act as a linear transformation on the 6-tuples (up to a scalar multiplication for the final 2-extensor, reflecting the 'weights' given to the corresponding points). This will also take a bending to a bending, since the cycle conditions are linear equations.

Theorem 6. *The rigidity of a hinge structure is invariant under projective transformations of the lines of the hinges.*

For example, we checked that the six edges of a particular tetrahedron were independent. Since all (non-coplanar) tetrahedra are projectively equivalent, this independence applies to all tetrahedra in 3-space, including tetrahedra with some edges at infinity. 'Hinges at infinity' do correspond to real built structures: slide joints in which one body can translate relative to the second along a fixed direction correspond to a hinge between to bodies along the line at infinity where all the normal planes to this slide meet the plane at infinity (Figure 3).

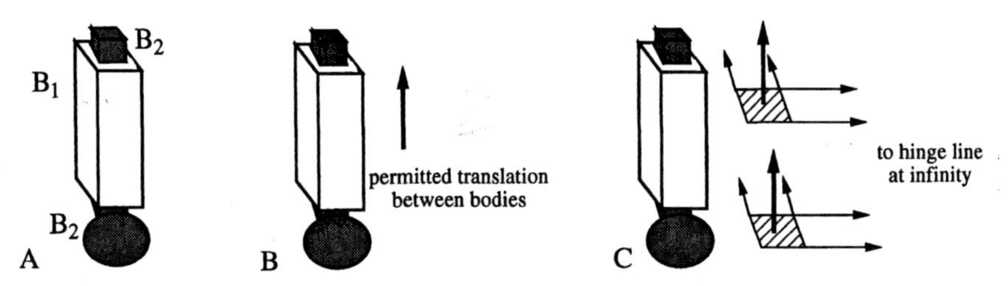

Figure 3. A slide joint permits a translation between the bodies (A,B) corresponds to a 'hinge' at infinity where the planes perpendicular to the translation all intersect (C).

MOLECULAR STRUCTURES

A standard chemist's model of a molecule has rigid bodies located as the atoms, and the single bonds as hinges between the atoms. Double bonds link two connected atoms into a single rigid unit. These are specialized hinge structures in which all the hinges (single bonds) at a given atom are concurrent in a point. With the exception of how atoms with only one bond or no bonds are handled, these are equivalent to the 'bond-bending' networks used in modeling glasses[13] and to the 'frames' of related mathematical work[9]. These will be our focus for the remainder of the paper.

Definition 4. A *molecular structure* is a hinge structure $G; \mathbf{H}$ in which, for each body or *atom* B_i, all the hinges or *bonds* are concurrent in a single point $\overline{p}_i$.

A *non-degenerate molecular structure* has distinct points for the distinct bodies. We can present such a non-degenerate molecular structure in simplified form as $G; \overline{\mathbf{p}}$, $\overline{\mathbf{p}} : V \rightarrow I\!P^3$ and write the hinge for a directed edge (j,k) as the 2-extensor $\mathbf{H}_{j,k} = \overline{p}_j\overline{p}_k$, with $\mathbf{H}_{k,j} = \overline{p}_k\overline{p}_j = -\overline{p}_j\overline{p}_k = -\mathbf{H}_{j,k}$. [$I\!P^3$ stands for projective 3-space, or the set of all non-zero 4-tuples which we are using to represent the points.]

For a connected graph G, a *twist* of the molecular structure $(V,E); \overline{\mathbf{p}}$ is an assignment of scalars $\lambda_{j,k}$ to the edges such that, for each oriented cycle C in the graph,

$$\sum_{(j,k)\in C} \lambda_{j,h}\mathbf{H}_{j,k} = \mathbf{0}.$$

A twist is *non-trivial* if and only if it is not the zero assignment.

For a given graph, different realizations $\overline{\mathbf{p}}$ will give different rigidity properties, as the following example illustrates.

Example 7. Consider the benzine ring (cyclohexane) formed by a skew hexagon of six carbon atoms and six hinges (single bonds) (Figure 4). With the given bond lengths (all equal) and the six equal angles between the bonds of an atom, there are two distinct isomers of the molecule – the 'chair' (A) and the 'boat' (C). We will see that chair is rigid while the boat is shaky. [In fact, the boat is flexible, but a demonstration of this is beyond the scope of these methods.]

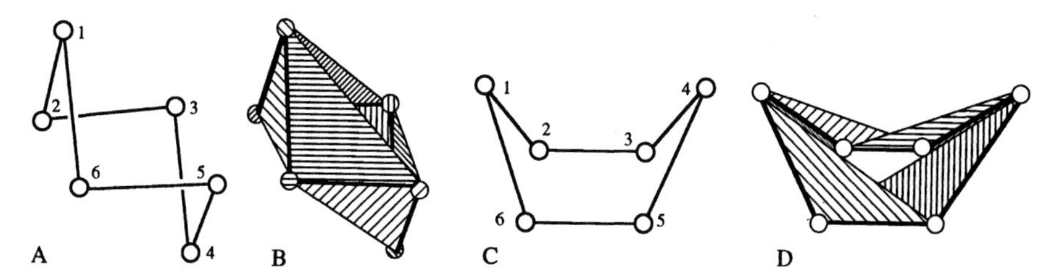

Figure 4. Isomers of the benzine ring: the chair (A) (B) is rigid; the boat (C,D) is flexible.

For such bodies, with two concurrent hinges, we can replace the 'atom' with a rigid triangle, hinged to its neighbors along two of its edge (B,D). Structurally, the chair configuration is equivalent to a convex octahedron - and is rigid (see Cauchy's Theorem below).

We show that the boat configuration is shaky. From Example 2 this ring of length six is shaky if and only if, as a molecular structure, $\sum_{i=1}^{i=6} \lambda_{i,i+1}\overline{p}_i\overline{p}_{i+1} = \mathbf{0}$.

In Figure 5, we give a direct visual presentation of the twist of the boat configuration, viewed from above. Notice that the atoms lie in two planes (A). For a selected orientation of the cycle , we draw the combined *twist-extensor* $\mathbf{L}_{i,i+1} = \lambda_{i,i+1}\overline{p}_i\overline{p}_{i+1}$ as a vector beside the edge (B). Visually, the sum of two twist-extensors on concurrent lines appears as the vector sum of the arrows – through the common point (C).

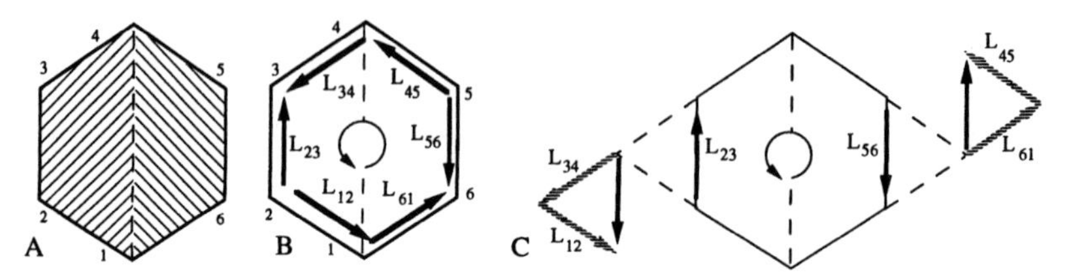

Figure 5. A check (C) that the twist-extensors (B) for the boat add to zero around the cycle.

First, the arrows in Figure 5(B) do satisfy the cycle equation. If we add the extensors for 12 and 34 at the point of intersection of these lines (Figure C), we get an extensor parallel to and equal in direction and magnitude to the extensor for 56. Similarly, the sum of the extensors for 45 and 61, at the point of intersection of these lines (Figure C), is parallel to and equal in direction and magnitude to the extensor for 23. These four extensors on parallel lines all add together, as vectors, at their common point of intersection (at infinity) - and give the zero 2-extensor as required.

Now, up to a single scalar multiplier, these twist-extensors are unique for this non-coplanar ring. The extensors lie on six lines all meeting a single line (the dotted line 14 in (A)) and form a linear line complex[6,14] with five independent 2-extensors (Figure 1). The cycle equation corresponds to a linear system with a 6×6 matrix of rank 5 – so the space of solutions (the twists) are a 1-parameter space generated by this particular solution. ∎

Example 8. We can describe the general geometric condition for such a molecular ring of length six to be shaky. [Recall that the boat is actually flexible turning through a path of shaky positions – for which we have only analyzed one point.]

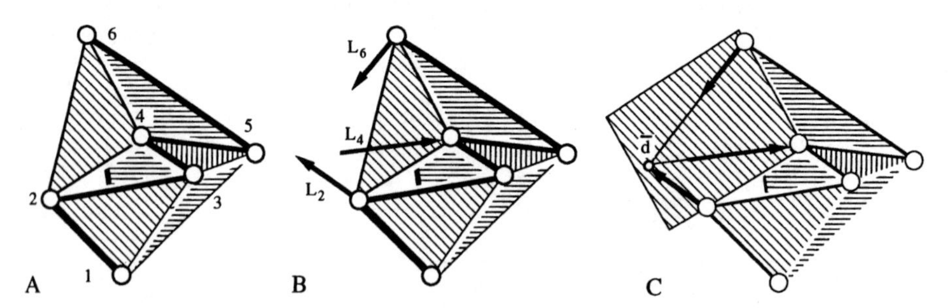

Figure 6. The geometric condition for a molecular cycle of length six (A) to be shaky is that the four planes 123, 345, 561 and 246 are concurrent in a point $\overline{d}$ (C).

The cycle is shaky if and only if the cycle condition $\sum_{i=1}^{i=6} \lambda_{i,i+1}\overline{p}_i\overline{p}_{i+1} = \mathbf{0}$ has a non-zero solution (Figure 6 A). Grouping the terms in this equation, we have:

$$[\lambda_{1,2}\overline{p}_1\overline{p}_2 + \lambda_{2,3}\overline{p}_2\overline{p}_3] \;+\; [\lambda_{3,4}\overline{p}_3\overline{p}_4 + \lambda_{4,5}\overline{p}_4\overline{p}_5] \;+\; [\lambda_{3,6}\overline{p}_5\overline{p}_6 + \lambda_{6,1}\overline{p}_6\overline{p}_1] = \mathbf{0}$$

The sum $[\lambda_{1,2}\overline{p}_1\overline{p}_2 + \lambda_{2,3}\overline{p}_2\overline{p}_3]$ is a new 2-extensor $\mathbf{L}_2$ through the point $\overline{p}_2$ in the plane $\overline{p}_1\overline{p}_2\overline{p}_3$ (Figure 6B). Careful choice of the scalars $\lambda_{1,2}, \lambda_{2,3}$ will make the sum any chosen line through $\overline{p}_2$ in this plane. Similarly, the other two terms are $\mathbf{L}_4$ through the point $\overline{p}_4$ in the plane $\overline{p}_3\overline{p}_4\overline{p}_5$ and L_6 through the point $\overline{p}_6$ in the plane $\overline{p}_5\overline{p}_6\overline{p}_1$. The equation becomes $\mathbf{L}_2 + \mathbf{L}_4 + \mathbf{L}_6 = \mathbf{0}$. It is well known that three 2-extensors in 3-space can only add to $\mathbf{0}$ if they lie on concurrent, coplanar lines in projective space (Figure C). This final plane is contains the points $\overline{p}_2$, $\overline{p}_4$, $\overline{p}_6$ and must be the plane $\overline{p}_2\overline{p}_4\overline{p}_6$. We call this point of concurrence $\overline{d}$.

In summary, if the ring is shaky, then the four planes $\overline{p}_1\overline{p}_2\overline{p}_3$, $\overline{p}_3\overline{p}_4\overline{p}_5$, $\overline{p}_5\overline{p}_6\overline{p}_1$ and $\overline{p}_2\overline{p}_4\overline{p}_6$ must all meet in the common point $\overline{d}$. This is a projective geometric condition which is easily checked and was first worked out in Bricard's analysis of flexible and shaky octahedra[3]. [In Figure 4 D, the point 4 is the point $\overline{d}$ on the four planes.]

We will not give the details but the converse is true. If the four planes meet in a single point $\overline{d}$, then the analysis can be reversed and non-zero scalars $\lambda_{i,i+1}$ can be chosen to satisfy the cycle condition. The molecular ring is shaky.

In developing this geometric analysis, we chose one set of four planes from the ring. An alternative grouping of terms leads to an equivalent set of four planes: $\overline{p}_2\overline{p}_3\overline{p}_4$, $\overline{p}_4\overline{p}_5\overline{p}_6$, $\overline{p}_6\overline{p}_1\overline{p}_2$ and $\overline{p}_1\overline{p}_3\overline{p}_5$. It is a theorem of projective geometry that, if one of the sets is concurrent in a point $\overline{d}$, then the other set is also concurrent in a (probably different) point $\overline{d}'$. For the boat configuration, the point $\overline{d}'$ is vertex 1. ∎

Clearly, the molecular structures are geometrically special forms of hinge structures. We conjecture that this specialization from general hinge structures to molecular structures with all hinges of a body concurrent does not alter the generic behaviour. The following is a translation of a conjecture of Tay and Whiteley[19].

Conjecture 9. *A generic molecular structure on a graph $G = (B, E)$ is rigid if and only if, replacing each edge of G by 5 edges, the resulting multi-graph contains six disjoint spanning trees on the vertices B.*

Remark 10. A recent analysis by Jacobs[13] of a related form of these structures gives additional evidence for this conjecture, but does not yet provide a full proof. Certainly, the conjecture is true for small simple cycles of the type discussed in Examples 2 and 7, and in the work of Franzblau[9]. This conjecture leads to an efficient algorithm of complexity $O(|V|^2)$ (see Jacobs[13]).

This conjecture should be contrasted with the long standing, fundamental problem of characterizing the combinatorial patterns of generically rigid bar-and-joint frameworks in 3-space[11,31]. Even the conjectures for that problem would not lead to efficient algorithms for the general problem. ∎

POLAR HINGE-PANEL STRUCTURES AND SPHERES

Molecular structures have not been widely studied in rigidity theory. However, their polar form has been studied for several centuries. The classic example is a convex polyhedron, made of rigid faces with hinges along the edges, studied by Cauchy[2] and Dehn[8]. This is a hinge structure, with the added feature that for each body (face) all the hinges lie in a single plane. Such objects are common as an abstraction from various built structures with pre-fabricated blocks. They also appear in children's building toys for constructing polyhedra, such as Polydron.

Definition 5. A *hinge-panel structure* is a hinge structure $G, \mathbf{H}$ in which, for each body or *panel* B_i, all the hinges lie in a single plane $\overline{P}_i$ – the *face-plane*.

In 3-space, a projective polarity is a map from the points to the planes, and the planes to points (in projective space) such that if $\overline{p}$ is a point on plane $\overline{Q}$ then the polar plane $\overline{P}^*$ is a plane through the polar point $\overline{q}^*$ (Figure 7). If we apply such a projective polarity in 3-space to a molecular structure, the point for an atom becomes a plane for the panel, and the concurrent bonds of an atom become coplanar hinges for each panel. That is, the polar of a molecular structure is a hinge-panel structure. We offer a specific geometric construction of such a projective polarity.

Example 11. Given a unit sphere centered at **c**, the *polarity in the sphere* takes each plane **Q**, at a distance r from the center, to the point $\mathbf{q}^*$ at a distance $\frac{1}{r}$ from the center, along the normal ray from the center to the plane (Figure 7 A). Conversely, it takes the point **q** at a distance r from the center to the plane $\mathbf{Q}^*$ at a distance $\frac{1}{r}$ from **c** along the ray **cq**. This takes a tangent plane to the unit sphere to its point of contact with the sphere (C)! Such a polarity is a common map used to take a platonic solid such as the cube, with vertices on the unit sphere, to its dual the octahedron, with faces tangent to the unit sphere.

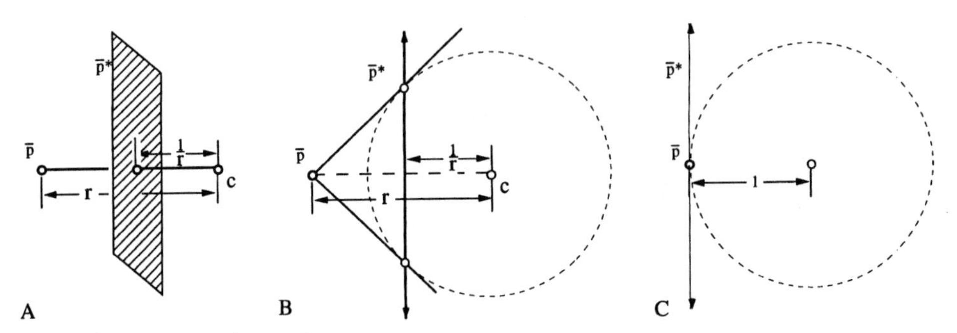

Figure 7. A polarity in 3-space interchanges points and planes (A,B). Shown in plane cross-section (C) there is a sythetic construction from the unit sphere, and a point on the sphere goes to the tangent plane at that point (D).

This polarity also takes the center **c** to the plane at infinity and conversely. Without the full projective space certain points and planes would not have a polar. Algebraically, if the center of the sphere is at the origin, this polarity exchanges the finite points $(a, b, c, 1)$ with the planes $(a, b, c, -1)$ not through the origin. ∎

At the level of 2-extensors, a polarity of the points and planes in $\mathbb{P}^3$ induces a linear transformation on the space of screws, taking 2-extensors to 2-extensors. Provided the hinges have been transformed by the induced linear transformation for the polarity, this algebraic map takes a twist of the molecular structure to a bending of the polar hinge-panel structure with identical scalars. We state the following corollary without proof.

Corollary 12. *A molecular structure* $(V, E); \overline{\mathbf{p}}$ *is shaky (resp. rigid) if and only if the polar hinge-panel structure* $(B, E); \overline{\mathbf{P}}^*$ *is shaky (resp. rigid)* .

If the hinge-panel structure is based on a 'spherical polyhedron', that is the bodies are the plane faces of the polyhedron and the hinges are (pieces of) the edges,

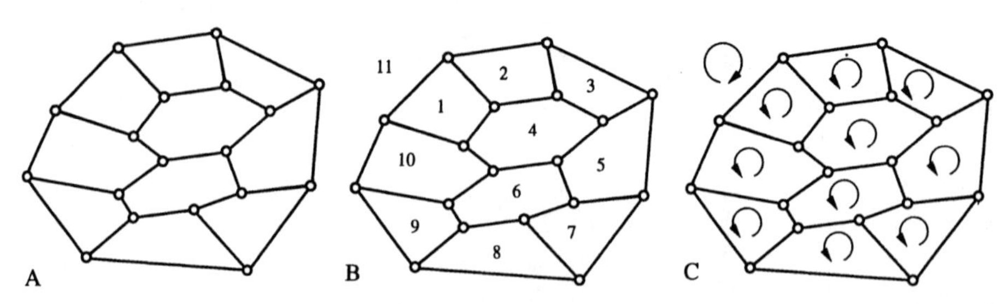

Figure 8. Any planar drawing of a 2-connected graph (A) induces plane regions as faces of an abstract sphere (B) which can be oriented in a consistent way (C).

then there are several classical mathematical results which apply. We begin with some definitions.

Definition 6. An *abstract spherical polyhedron* is formed by a 2-connected planar graph (V, E) drawn in the plane, without self-intersection (Figure 8 A). The *vertices* of the polyhedron are the vertices of the graph. The *edges* of the polyhedron are the edges of the graph. The *oriented faces* F of the polyhedron are the polygonal regions created by the drawing (B), oriented counterclockwise, plus the external polygonal region, oriented counterclockwise (C).

Note that every edge is now in two oriented faces, with opposite orientations. Also around each vertex, there is a single cycle created by the dual path of face and edges incident with the vertex. Topologically, the abstract spherical polyhedron is a *manifold*. We need a name for the class of realizations we will study in this paper.

Definition 7. A *proper spherical polyhedron* is a realization of an abstract spherical polyhedron with points in 3-space for the vertices, such that: the vertices for any face are coplanar but not collinear, the vertices of any edge are distinct, the faces at any edge are distinct, and the faces at any vertex are not collinear. A *strictly convex* polyhedron is a spatial polyhedron such that, for each face, the face plane places all other vertices into a single open half-space.

As an aside, we note that strictly convex polyhedra correspond to 3-connected planar graphs (Steinitz' Theorem). It is not know which 2-connected planar graphs have realizations as proper spatial polyhedra[29].

The following is a variant of the classical theorem[2], in the language of rigidity as presented by Dehn[8,24].

Cauchy's Theorem 13. *For any strictly convex polyhedron, the hinge-panel structure with the faces as panels and the edges as hinges is infinitesimally rigid.*

There is an extension this theorem, in which each convex face is further triangulated with diagonals, using the natural vertices.

Alexandrov's Theorem[1,24] 14. *For any strictly convex polyhedron, with each face triangulated with diagonals, the hinge-panel structure with the triangular faces built as panels with all the edges (including the triangulating diagonals) as hinges is infinitesimally rigid.*

Notice that these are not just generic results about the underlying graphs, but they are geometric results about a particular class of realizations. Results of Bricard[3] show that even an octahedron can be build as a shaky non-convex panel structure and an example of Connelly[4] shows that there are finitely flexible, non-convex embeddings of a triangulated sphere.

There is one broader generic result which we state without giving the precise definitions. The intuitive sense of '2-surface' is something like a sphere, torus, multi-holed torus, etc.

Fogelsanger's Theorem[10] 15. *The hinge-panel structure built on the faces and edges of any triangulated 2-surface is generically rigid as a hinge-panel structure.*

These results for hinge-panel structures have a polar form for molecular structures. If we take the polar of a convex polyhedron, about a sphere with its center inside the polyhedron, the polar object is also a convex polyhedron, with a face for each of the original vertices, and a vertex for each of the original faces. At the abstract level, this new polyhedron, with faces and vertices interchanged, is called the *dual abstract polyhedron*. The dual of a triangulated polyhedron is an *abstract simple polyhedron*, with each vertex of valence 3.

Polar Cauchy's Theorem 16. *For any strictly convex polyhedron, the molecular structure with the vertices as atoms and the edges as bonds is infinitesimally rigid.*

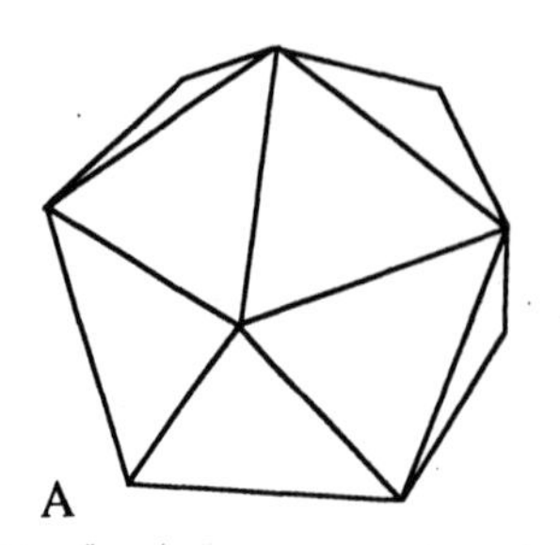

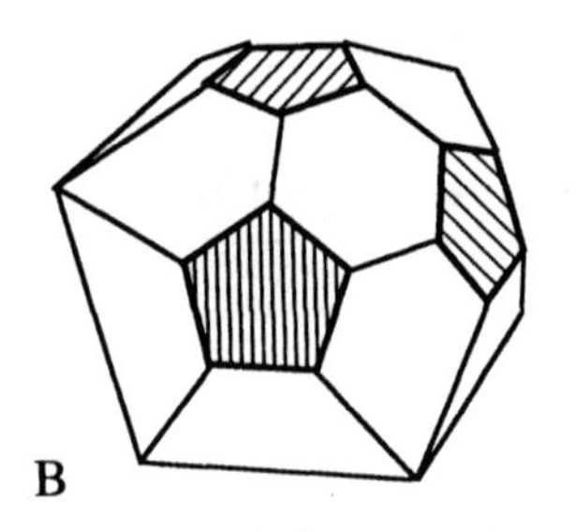

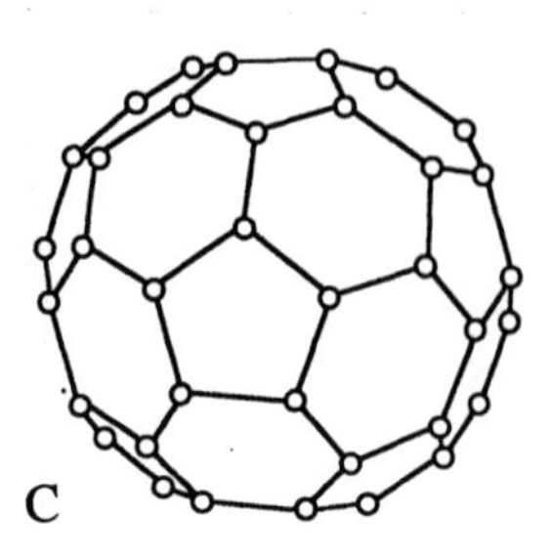

Figure 9. The fullerine molecule C_{60} (C) is formed by placing carbon atoms at the vertices of a truncated icosahedron (A,B).

Example 17. Consider the fullerine molecule C_{60} (or Buckyball) which has carbon atoms at the vertices of the truncated icosahedron[5] (Figure 9). This is a convex simple polyhedron. Even with single bonds along each of the edges, this is infinitesimally rigid. Since carbon atoms form four bonds, many of the edges of fullerine are actually double bonds, increasing the rigidity enormously. However, in fully hydrogenated fullerine $C_{60}H_{60}$, with an attached hydrogen at each carbon (or similar structures with fluorine at each carbon), the bonds are all single. The molecule is infinitesimally rigid even in this extreme situation. ∎

Remark 18. Consider a perturbed hinge-panel structure on a polyhedron, so that the 'vertices' are broken (the hinges in this 'vertex cycle' are no longer concurrent) but the faces retain the 'convexity' property that all of the polyhedron lies on one side of each of these faces. Then the rigidity is not lost. If one inspects the proof of Cauchy's Theorem[24], the infinitesimal rigidity still follows. Such perturbations typically increase the rigidity of hinge-panel structures.

In the polar form, this means that if we have a a molecule built on the graph of a convex polyhedron, with the vertices and edges on the convex hull of the points, but the 'face polygons' possibly not coplanar, the proof of infinitesimal rigidity still holds. Again, such perturbations typically increase the rigidity of molecules. ∎

THE CYCLE CONDITIONS ON SPHERES AND DISCS.

Many of these convex polyhedra are over-constrained as hinge-panel structures and as the polar molecular structures. For example, on a tetrahedron we can remove two opposite hinges and still have a generically rigid hinge-panel structure (Figure 2C). In a polar fashion, we can also remove some bonds, and still have an infinitesimally rigid molecular structure. What happens if we remove larger pieces?

Example 19. Consider a triangular panel in a spherical hinge structure. If we remove this triangle and leave the surrounding panels and hinges in place, the other panels 'hold' the shape of the missing triangle (Figure 10A). We can repeat this operation, as long as the deleted triangles do not share a vertex.

For example, on an icosahedron, we can remove up to four vertex-disjoint triangles (B). Since there are 12 vertices and each triangle occupies three vertices, $\frac{12}{3} = 4$ is the maximum number of vertex disjoint triangles. The hexagonal ring of Example 2 is obtained from the octahedron in by removing two opposite faces in this way (Figures 4,6).

If we remove two triangles sharing an edge (C), the hinge-panel structure on a sphere becomes shaky. This is equivalent to a theorem about triangulated bar-and-joint frameworks[26]. If the triangulated sphere is 4-connected in a vertex sense (any two vertices are connected by four vertex-disjoint paths) then, for generic realizations, the resulting shake has non-zero scalars on all of the remaining hinges. ∎

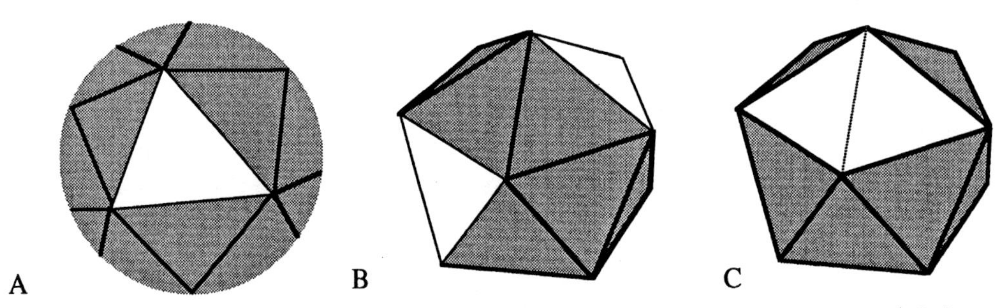

Figure 10. Removing triangles from a hinge-panel sphere (white regions in A) will leave the rigidity unchanged (A,B) unless the removed triangles share an edge or vertex (C).

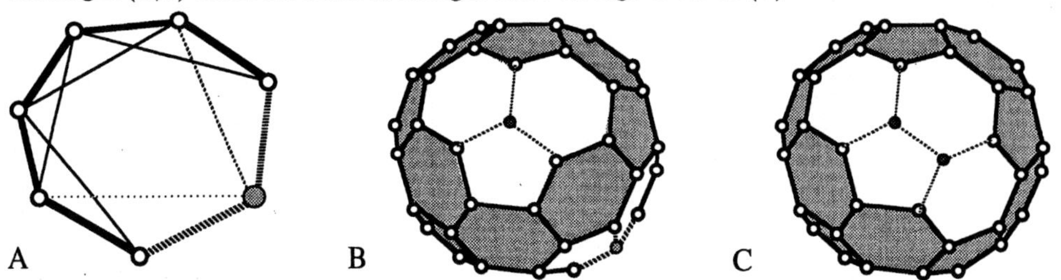

Figure 11. Removing 3-valent atoms from a molecular sphere will leave the rigidity unchanged in each face plane (A) and therefore in the molecular structure (B) unless the removed vertices share an edge or a face (C).

Example 20. We have polar results for 3-valent vertices in molecular structures. If we remove an atom from a face cycle, the remaining constraints in that face hold the substructure rigid in its face (Figure 11A). The net effect is that removing a 3-valent atom lying in 3-faces does not disturb the rigidity. We can repeat this provided that the removed atoms do not share a face (B). The hexagonal ring of Example 2 and Figure 4 is also created by deleting opposite vertices from a molecular cube.

If we remove two 3-valent atoms sharing an bond, a simple molecular structure on a sphere will become shaky. This is the polar of the result cited above for hinge-panel structures (and bar-and-joint frameworks[26]). ∎

Results such as these are derived from a more careful analysis of the cycle conditions for molecular structures built on spherical polyhedra – polyhedral molecules and modified structures formed by deleting some vertices from such polyhedral molecules. In the remainder of the paper, we will present the underlying techniques, both combinatorial and geometric, along with further applications.

We recall that an assignment of scalars $\lambda_{j,k}$ to the bonds of a molecular structure is a twist if and only if, for every vertex-edge cycle C: $\sum_{(j,k)\in C} \lambda_{j,k}\mathbf{H}_{j,k} = \mathbf{0}$. If we have a molecular structure built on the vertices and edges of any spherical polyhedron, then the homology of the sphere gives us a special subset of cycles which generates all cycles: the cycles C_m of vertices and edges around each face F_m (Figure 12).

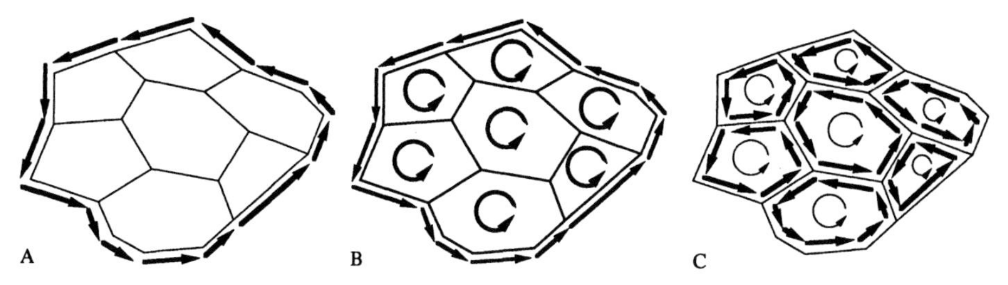

Figure 12. On a sphere, or a disc, a general vertex-edge cycle (A) is a linear combination of face cycles on the interior (B), with cancellation on the interior edges which occur with opposite directions on its two faces (C).

Remark 21. In algebraic topology[16], the *first homology* of a 2-surface is precisely the study of the all oriented vertex-edge cycles (as a vector space) in relation to the subspace of oriented cycles C_i generated by the boundaries of the faces. Our observation that the face cycles generate all cycles says that the first homology of the sphere is 0. If we work with another surface, such as a torus, then there will be some cycles which are not linear combinations of the face-cycles: the first homology of the torus is not 0.

For a topological disc formed from polygons, the general vertex-edge cycles are also linear combinations of the face cycles. In fact, Figure 12 is the picture of a disc – what is cut out of a sphere by a simple vertex-edge cycle. ∎

The topological result for cycles on spheres or discs extends to an algebraic result about the cycle conditions for a twist:

If the scalars $\lambda_{j,k}$ satisfy the cycle condition for each face of a sphere or disc, then they satisfy all the cycle conditions.

The same cancellation of arrows we see in Figure 12 C extends to cancellation for the twist-extensors on the edges because $\mathbf{H}_{j,k} = -\mathbf{H}_{k,j}$ and $\lambda_{j,k}\mathbf{H}_{j,k} = -\lambda_{j,k}\mathbf{H}_{j,k}$.

This reduction of the space of all cycles to face cycles as generators can be used to simplify the analysis for a sphere, or structures derived from spheres. This will permit simple counting arguments to estimate the space of twists of the corresponding molecule. There counts are analogous to those traditionally applied to bar-and-joint frameworks.

We can write the cycle constraints of a spherical (or disc) molecule in an $|E|\times 6|F|$ *cycle matrix* which expresses the homogeneous equations for each of the face cycles:

$$[\dots\ \lambda_{j,k}\ \dots] \begin{array}{c} \vdots \\ \{j,k\} \\ \vdots \end{array} \overset{\begin{array}{ccccccc} C_1 & \dots & C_h & \dots & C_i & \dots & C_{|F|} \end{array}}{\begin{bmatrix} \vdots & \ddots & \vdots & \ddots & \vdots & \ddots & \vdots \\ \mathbf{0} & \dots & \mathbf{H}_{j,k} & \dots & \mathbf{H}_{k,j} & \dots & \mathbf{0} \\ \vdots & \ddots & \vdots & \ddots & \vdots & \ddots & \vdots \end{bmatrix}} = [\mathbf{0}\ \ \dots\ \ \mathbf{0}\ ].$$

Notice that the columns of this matrix may not be independent. For a spherical structure in which each edge appears in exactly two faces, oriented oppositely, there is a standard relationship. If the columns are grouped as blocks A_i of width 6 for each cycle C_i, they satisfy the six equations: $\sum_{i\in F} A_i = 0_{|E|\times 1}$. This global relationship simply reminds us that if the scalars satisfy all but one of the face cycle conditions on a sphere, then they will satisfy the last one.

Example 22. Consider a standard section of the fullerine type cylinders (also called carbon nanotubes): an arrangement of 'boat configurations' (carbon hexagonal rings) in three space tiling a disc or a cylindrical tube.

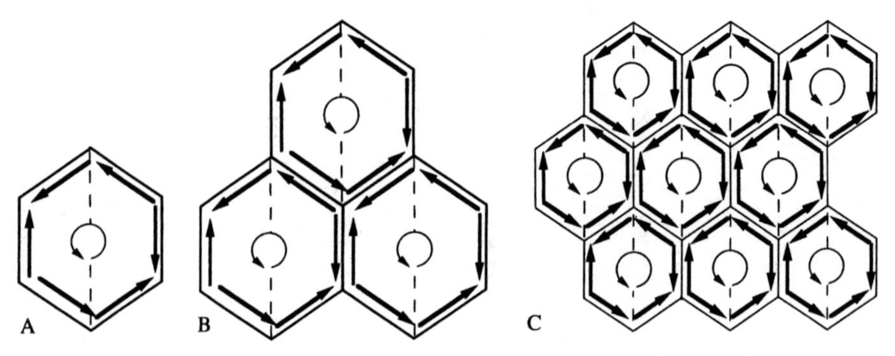

Figure 13. A series of boat configurations (A), packed into a disc (B) or a cylindrical tube (C, after identification), forms a shaky molecular structure.

A single boat configuration has (up to scalar multiplication) the single solution illustrated in Figure 13A, with the twist-extensor $\mathbf{L}_e = \lambda_e \mathbf{H}_e$ drawn as a vector beside its edge e. If we use the two rules:

(i) each twist-extensor in one cycle is reversed for the adjacent cycle; and
(ii) a given twist-extensor has a unique extension to its boat configuration;

then we transmit this initial solution into a unique solution on the adjacent cycles around a vertex (B). This process clearly extends further to a unique shake (up to a single scaling) for any disc formed from these rings (C). It also extends to a cylinder formed by identifying the opposite left and right-hand sides of patterns such as Figure 13C.

If any of these nanotubes is built with all bonds single then it is shaky. If at least one of these bonds is double – and that twist-extensor is forced to $\mathbf{0}$ – then the entire disc or tube is infinitesimally rigid, as a molecular structure. ∎

Remark 23. As this example illustrates, 'local' twist vectors around one cycle in repeating patterns (cylinders, crystals etc.) can often be repeated using the underlying pattern to give a coherent twist for the larger pattern. Such repeating patterns, which often have special geometry (as the nano-tubes do) are open to direct geometric analysis using these tools.

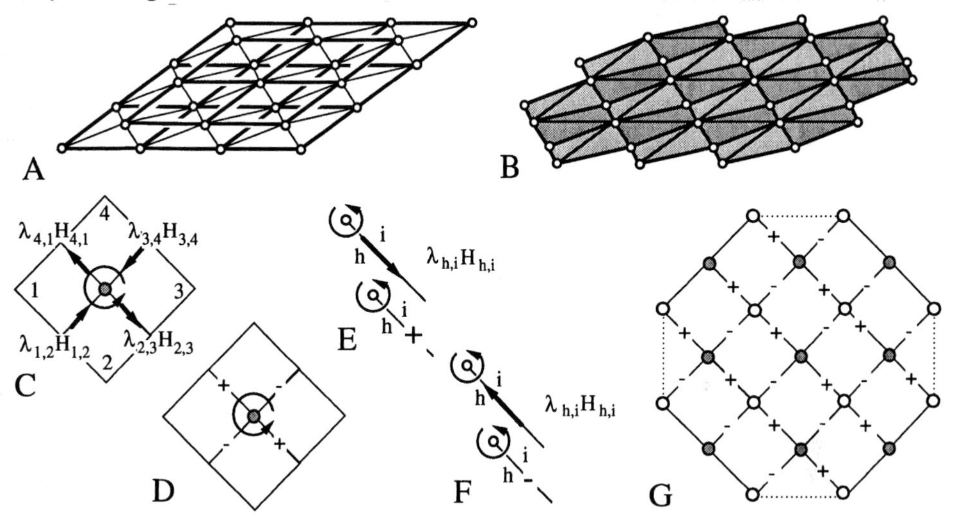

Figure 14. For a standard truss (A) we can extract a structure of hinged tetrahedra (B). The bending can be recorded with either extensors (C) or signs (D) following a standard convention (E,F). It is easy to see that the local solution around one vertex (D) extends across the sheet (G).

Figure 14 illustrates an analogous situation for a hinge structure on a disc, extracted from the standard half-octahedral truss[25]. A bending is developed from a local solution around a single vertex cycle and extended across the disc. Precisely because these structures are not generic, the visual, geometric methods can be effectively applied. Some crystals and related structures are modeled as hinge structures, because they contain both atoms with strong bond angles (the bodies) and atoms with weak bond angles which, in pairs, form the hinges between the bodies. ∎

POLYHEDRAL MOLECULES.

So far, the form of the matrix only depended on the topology (sphere or disc) of the graph underlying the molecular structure. If we also assume that a face F_i of the polyhedron lies in a plane $\overline{P}_i$, the extensor equation $\sum_{(j,k)\in C_i} \lambda_{j,k}\mathbf{H}_{j,k} = \mathbf{0}$ reduces to a set of three equations. Equivalently, the six columns for the cycle of this face reduce to three independent columns, under column reduction.

Example 24. Consider extensors in the plane $z = 0$. Since the points all have the form $\overline{a} = (a_1, a_2, 0, 1), \overline{b} = (b_1, b_2, 0, 1)$, the minors for the 2-extensor $\overline{ab}$ of two points in the plane have the form: $(a_1 - b_1, a_2 - b_2, 0, a_1 b_2 - a_2 b_1, 0, 0)$. They form a subspace of dimension 3. In particular, the vector equation $\sum_{(j,k) \in C_i} \lambda_{j,k} \mathbf{H}_{j,k} = \mathbf{0}$ reduces to three independent linear equations. Provided the extensors are not all collinear, these three equations are independent, representing a *static equilibrium* in the plane $z = 0$. The first two equations say that there is no net translation component, and the final equation says that there is no net torque about the z-axis.

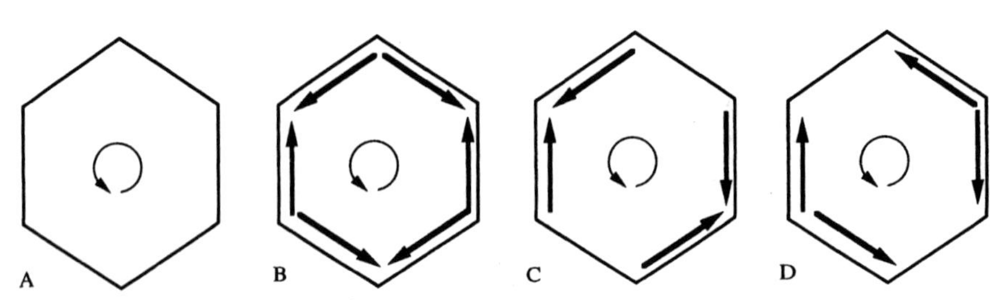

Figure 15. For a plane hexagonal ring (A), the space of molecular twists is generated by three independent twists (B,C,D).

Consider a plane hexagonal ring (Figure 15 A). With six variables and three independent homogeneous linear equations, the system $\sum_{(j,k) \in C_i} \lambda_{j,k} \mathbf{H}_{j,k} = \mathbf{0}$ has a 3-parameter solution space. Figure 15 B,C,D illustrates one set of three independent generators. We note that the sum of the twists in Figure 15 C,D is the plane projection of the twist of the boat configuration. ∎

While the algebra for the plane $z = 0$ is easier, the underlying principle for any plane is the same: the six equations for the 2-extensors reduce to three equations. For the three extensors for non-collinear points $\overline{a}, \overline{b}, \overline{c}$ in the plane, we have three equations which are satisfied by all 2-extensors $\mathbf{H}$ in this plane:

$$(\overline{ab}) * \mathbf{H} = 0 \qquad (\overline{bc}) * \mathbf{H} = 0 \qquad (\overline{ca}) * \mathbf{H} = 0.$$

We can bound the dimension of the space of possible twists of the molecular structure, using estimates for the rank of this cycle matrix. For a disc (or sphere), our first estimate is now rank $\leq 3|F|$.

Corollary 25. *For a molecular structure on a sphere or disc (V, E, F) with each face in a plane, the dimension of the space of twists is bounded below by $|E| - 3|F|$.*

Example 26. Consider the molecular structures in Figure 16, with the labeled polygons each coplanar. Since the exterior is unlabeled, these are viewed as topological discs and the exterior polygon is *not* assumed to lie in a plane.

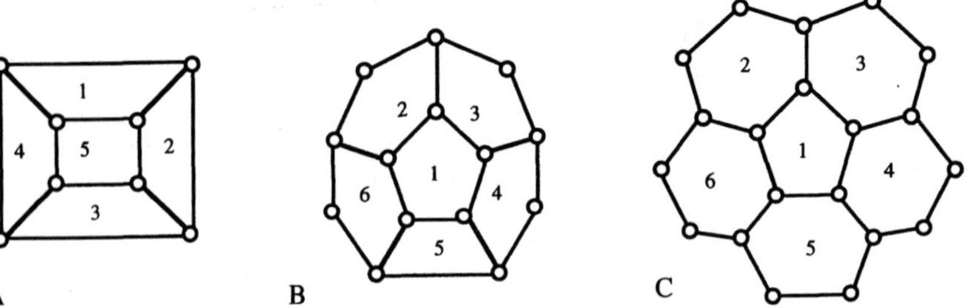

Figure 16. For a molecular structure on a sphere or disc, with plane faces, we have a lower bound on the dimension of the space of stresses.

For the structure in Figure 16A, we have $|E| = 12$ and $|F| = 5$, so our estimate is $\dim(T) \geq |E| - 3|F| = 12 - 15 = -3$. This tells us nothing useful. We shall see below, that, in general, a polyhedral molecular structure on this disc is rigid, but we need a better estimate for this situation.

For Figure 16B, we have $|E| = 19$ and $|F| = 6$, so our estimate is $\dim(T) \geq |E| - 3|F| = 19 - 18 = 1$. This structure is always shaky.

For Figure 16C, we have $|E| = 25$ and $|F| = 6$, so $\dim(T) \geq |E| - 3|F| = 25 - 18 = 7$. This structure has at least a 7-dimensional space of twists. ∎

SIMPLY CONNECTED MANIFOLDS

We will explore two improvements to this estimate of the space of twists of a molecular structure:

(i) extend this estimate to apply to manifolds formed by removing more pieces of the sphere: simply-connected manifolds (Figure 17);

(ii) give a refined estimate for simple manifolds, in which all vertices are 3-valent, which is exact for generic realizations with plane faces and quick to compute.

To derive these improved estimates we rely on an underlying correspondence with statics which provides additional insights for the analysis of specific configurations and provided the option of translating any results, old or new, from the statics of bar-and-joint frameworks to the rigidity results for molecular structures on simply-connected manifolds.

We first give a careful definition of the class of objects for (i). We will then move to the statics of a related structure (next section), which provides the mathematical basis for (i) and (ii).

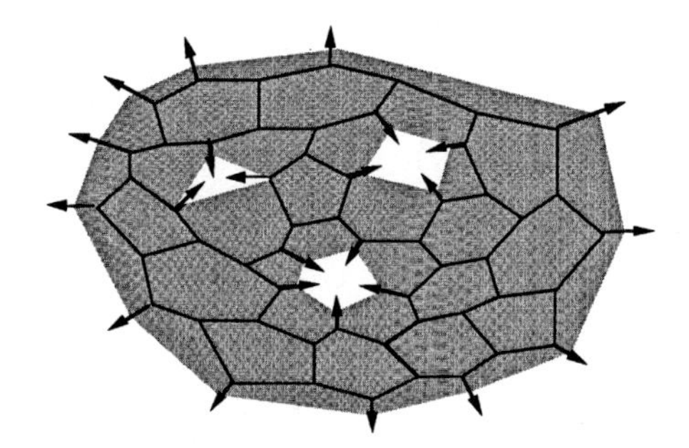

Figure 17. A simply-connected dual manifold is formed from an abstract spherical manifold by deleting selected vertices and their adjacent edges.

Definition 8. An *abstract simply-connected dual manifold* $(V; E; F)$ is formed from an abstract spherical polyhedron $(V, A; E, E'; F, F')$ by removing a set A of vertices no two on the same face, to create *holes*. We remove the edges E' attached to the vertices in A but not the attached faces F'. If we remove only one vertex, we have a *topological disc*.

As Figure 17 illustrates, we think of the dual manifold as still containing a cycle of faces surrounding the removed vertex. We draw a set of 'boundary edges' to separate them around the hole, but do not have designated vertices at which these edges must meet. Geometrically, we could have removed several vertices – and mentally decided it was 'one' hole (e.g. Figure 10C). Provided that the edges entering this hole are then abstractly identified to one 'missing vertex', the abstract structure fits our definition. The faces and the boundary edges form the *face cycle* of the hole and the *size* of the hole is the length of this cycle. For simplicity, we will assume that each removed vertex has valence at least 3, so all holes have size ≥ 3.

We call these 'dual manifolds' because this unusual pattern of retaining designated faces for the hole is the dual of the usual process of defining 'manifolds with boundary' by retaining the vertex-edge cycle around the removed faces, but allowing those 'boundary edges' to belong to only one face of the manifold.

These objects, including the original sphere, are called 'simply-connected' because, if we take any closed vertex-edge-vertex cycle, the removal of the edges in this cycle will disconnect the faces $F \cup F'$ of the manifold. This property will be critical to the our final correspondence below.

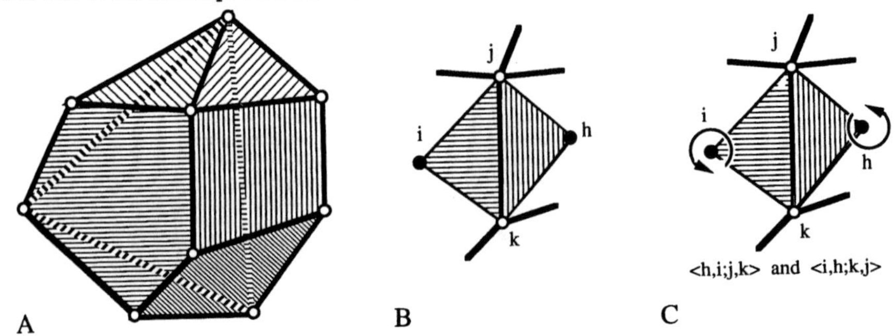

Figure 18. Patches for edges of a simply-connected dual manifold joint two vertices and separate two face cycles.

Finally, we note that every 'edge' in E of a dual manifold connects two vertices and separates two faces (Figure 18 A,B). We assume that the faces of the sphere are all consistently oriented (C). We denote the edge as a *oriented patch* $< h, i; j, k >=< i, h; k, j >$ where the edge from vertex j to k appears in the cycle of face i and the edge from k to j appears in the cycle of face i (Figure 18C). This patch will be used to switch the indexing between edges as pairs of vertices and edges as pairs of faces, in our dinal correspondence.

STATICS OF SHEETWORKS

Our analysis of the cycles in molecular twists has been building towards an 'inverse' structure indexed by a set F of faces built in planes, with 2-extensors for the edges at which selected pairs of face planes meet. Those familiar with a projective presentation of the statics of bar-and-joint frameworks may recognize that we are constructing the polar of this more standard theory[7,18]. What follows is a translation of some standard results for bar-and-joint frameworks into this polar pattern, which we have called the *statics of sheet structures*[27]. Intuitively, the 'sheets' will behave like thin sheets of cardboard: rigid within their plane, free to warp out of their plane, and creating a rigid line segment when two sheets in different planes are attached along this segment[27].

Definition 9. A *sheet framework*, or *sheetwork* for this paper, is a graph $G = (F, E)$ with a configuration $\overline{\mathbf{P}}$ realizing the *faces* F as planes $\overline{P}(i) = \overline{P}^i$ in affine 3-space such that the two faces of any edge are distinct non-parallel planes. In addition, as a convention, for each edge $(h, i) \in E$, we select $\mathbf{H}^{h,i}$ as a nonzero 2-extensor along the line of intersection of $\overline{P}^h, \overline{P}^i$, with $\mathbf{H}^{i,h} = -\mathbf{H}^{h,i}$. [Within the full Grassmann-Cayley algebra, we would take the wedge of the planes $\overline{P}^h \wedge \overline{P}^i$ for the line of intersection and convert this 'dual extensor' back point coordinates with a further isomorphism. This additional layer of algebra is not essential, so we have replaced it by the arbitrary convention given above.]

A *self-stress* on a sheetwork is an assignment of a scalar $\omega_{h,i}$ to each edge $\{h, i\} \in E$, with $\omega_{h,i} = \omega_{h,i}$ such that, for each face $i \in F$: $\sum_{h|\{h,i\}\in E} \omega_{h,i}\mathbf{H}^{h,i} = \mathbf{0}$.

The homogeneous linear equations defining a self-stress can be expressed using the $|E| \times 6|F|$ *projective rigidity matrix* of the sheetwork:

$$[\dots \ \omega_{j,k} \ \dots] \begin{array}{c} \vdots \\ \{h,i\} \\ \vdots \end{array} \overset{\begin{array}{ccccccc} 1 & \dots & h & \dots & i & \dots & |F| \end{array}}{\begin{bmatrix} \vdots & \ddots & \vdots & \ddots & \vdots & \ddots & \vdots \\ \mathbf{0} & \dots & \mathbf{H}^{h,i} & \dots & \mathbf{H}^{i,h} & \dots & \mathbf{0} \\ \vdots & \ddots & \vdots & \ddots & \vdots & \ddots & \vdots \end{bmatrix}} = [\mathbf{0} \ \dots \ \mathbf{0}].$$

Definition 10. Given a sheetwork $(F, E); \overline{\mathbf{P}}$, an *equilibrium load* is an assignment of 2-extensors to the vertices $\mathbf{L} : V \to L^2$ such that (i) for each face i the extensor $\mathbf{L}^i$ lies in the plane $\overline{P}^i$; and (ii) $\sum_{i \in F} \mathbf{L}^i = \mathbf{0}$.

A *resolution* of the equilibrium load $\mathbf{L}$ on a sheetwork $(F, E); \overline{\mathbf{P}}$ is an assignment ω to the undirected edges E, such that for each face i:

$$\sum_{h|\{h,i\} \in E} \omega_{h,i} \mathbf{H}^{h,i} + \mathbf{L}^i = \mathbf{0}.$$

A sheetwork is *statically rigid* if every equilibrium load has a resolution. A sheetwork is *isostatic* if every equilibrium load has a unique resolution.

A self-stress is the same as a resolution of the zero equilibrium load. Simple linear algebra shows that a framework is isostatic if and only if it is statically rigid and has only the zero self-stress. Provided we have at least 3 faces in the sheetwork, not all collinear, the space of equilibrium loads has dimension $3|F| - 6$.

Equilibrium Lemma 27. *For any sheetwork $(F, E); \overline{\mathbf{P}}$ without all faces collinear, the space of equilibrium loads has dimension $3|F| - 6$.*

Proof. Clearly the condition that the load $\mathbf{L}^i$ lies in the plane $\overline{P}^i$ leaves us with a possible space of $\mathbf{L}^i$ of dimension 3 for each face. This space of dimension $3|F|$ is reduced by the six equilibrium equations $\sum_{h|\{h,i\} \in E} \omega_{h,i} \mathbf{H}^{h,i} + \mathbf{L}^i = \mathbf{0}$. It remains to prove that these are six independent equations on the possible loads.

Since the sheets are not all collinear, there are three sheets $\overline{P}^1, \overline{P}^2, \overline{P}^3$, which meet only in a point $\overline{q}$. The extensors for the corresponding 3 lines of intersection are $\mathbf{H}^{1,2}, \mathbf{H}^{2,3}, \mathbf{H}^{3,1}$, and these extensors are linearly independent. (The sum of extensors in plane 2 is an extensor in plane 2 – and therefore not a multiple of $\mathbf{H}^{3,1}$.)

Consider any equilibrium load: $\mathbf{L}^1 + \mathbf{L}^2 + \mathbf{L}^3 = \mathbf{0}$. As we have stated earlier, these must be concurrent and this point of concurrence must be $\overline{q}$. $\mathbf{L}^1$ is now a unique linear combination of the two extensors $\mathbf{H}^{1,2}\mathbf{H}^{3,1}$ through $\overline{q}$ in the plane $\overline{P}^1$: $\mathbf{L}^1 = a\mathbf{H}^{1,2} + b\mathbf{H}^{3,1}$. Similarly, $\mathbf{L}^2 = c\mathbf{H}^{1,2} + d\mathbf{H}^{2,3}$ and $\mathbf{L}^3 = e\mathbf{H}^{2,3} + f\mathbf{H}^{3,1}$. The equilibrium equation now reads:

$$[a\mathbf{H}^{1,2} + b\mathbf{H}^{3,1}] + [c\mathbf{H}^{1,2} + d\mathbf{H}^{2,3}] + [e\mathbf{H}^{2,3} + f\mathbf{H}^{3,1}] = \mathbf{0}.$$

The independence of the three extensors yields: $a = -c$, $b = -f$, $d = -e$. This gives a three parameter space of equilibrium loads and the equilibrium equations had rank six on these three faces. The same rank of six extends to any extension of these three faces with additional faces.

We conclude that the space of solutions to the equilibrium equations has dimension $3|F| - 6$. ∎

Isostatic Counts Theorem 28. *A sheetwork $(F, E); \overline{\mathbf{P}}$, with at least three faces, is isostatic if and only if two of the following hold:*

(a) $(F,E);\overline{\mathbf{P}}$ *has* $|E| = 3|F| - 6$ *edges;*
(b) $(F,E);\overline{\mathbf{P}}$ *has only the zero self-stress;*
(c) $(F,E);\overline{\mathbf{P}}$ *is statically rigid.*

Proof. (b) and (c) together are effectively the definition of an isostatic sheetwork.

(a) and (b). For any sheetwork with three non-collinear faces, we know that the space of equilibrium loads has dimension $3|F|-6$. We also observe that the rows of our projective rigidity matrix are also equilibrium loads of the form $\mathbf{L}^h = \mathbf{H}^{h,i}, \mathbf{L}^i = \mathbf{H}^{i,h}$, $\mathbf{L}^k = \mathbf{O}, k \neq h, i$. If there is only the zero self-stress, these rows are independent, and they must span the full $|E| = 3|F| - 6$ dimensional space of equilibrium loads. The sheetwork is statically rigid and therefore isostatic.

(a) and (c). These $|E|$ rows span the $3|F|-6$ dimensional space of equilibrium loads for the statically rigid sheetwork. With $|E| = 3|F|-6$, the rows must be independent and there is only the zero self-stress. The sheetwork is isostatic. ∎

Corollary 29. *A sheetwork on a simple spherical polyhedron is statically rigid if and only if it has only the zero self-stress.*

Proof. How many edges does a simple spherical polyhedron with $|F|$ faces have? This can be computed from Euler's formula for spherical polyhedra $|V|-|E|+|F| = 2$. We have the added condition $3|V| = 2|E|$ since each vertex has valence 3 and each edge is attached to two vertices. Together, these give:

$$3|V| - 3|E| + 3|F| = 6 \quad \rightarrow \quad 2|E| - 3|E| + 3|F| = 6 \quad \rightarrow \quad |E| = 3|F| - 6.$$

The rest follows from the Isostatic Counts Theorem. ∎

A careful analysis of Alexandrov's Theorem actually shows static rigidity of any strictly convex simple polyhedron realized as a sheetwork[27]. This gives one class of isostatic sheetworks. Below, we will need to know that there exist isostatic sheetworks formed by adding edges to any reasonable collection of sheets (planes).

Proposition 30. *A complete graph* $K_{|F|}$ *realized either as*
(a) *three non-collinear faces, if* $|F| = 3$; *or*
(b) *with at least four non-concurrent faces, if* $|F| \geq 4$;

is statically rigid as a sheetwork.

Proof. For three faces, we have effectively done this in the previous lemma, since we showed that the equilibrium loads form a space of dimension $3 = 3 \times 3 - 6$, and the three rows of the corresponding matrix clearly have dimension 3. For larger sets, the result can be proven by an induction which we will not give here. The interested reader can look at standard expositions for bar-and-joint frameworks and translate points to planes, lines to lines etc. ∎

Definition 11. In a graph $G = (F,E)$ an *oriented cut set* C is a minimal subset of oriented edges C such that its removal separates the 'faces' F (and edges) of the graph into two components, with all edges of C oriented into one component.

Cut Lemma 31. *For any sheetwork* $(F,E);\mathbf{P}$, *if* $C \subset E$ *is an oriented cut set then* $\sum_{(h,i)\in C} \omega_{h,i}\mathbf{H}^{h,i} = \mathbf{0}$.

Proof. Consider one connected component D of the disconnected graph created by the cut set C. We can add up the stresses on all of the edges of all of the faces of this component:

$$\sum_{m\in D} \Big[\sum_{(h,m)\in F} \omega_{h,m}\mathbf{H}^{h,m} \Big] = \sum_{m\in D} \mathbf{0} = \mathbf{0}.$$

However, on all edges inside the component, we have added over both faces of the edge (with opposite extensors) – giving a cancellation of these extensors. We are left with the simple addition $\sum_{(h,i)\in C} \lambda_{h,i}\mathbf{H}^{h,i} = \mathbf{0}$. ∎

There are many further results for sheetworks which can be translated from classical statics of frameworks. However, there is just one further construction which we will need for our correspondence.

Definition 12. A *proper simply-connected manifold* is an abstract dual simply-connected manifold $(V;E;F)$ formed from the polyhedron $(V,A;E,E';F,F')$, realized in $\mathbb{P}^3$ with a plane $\overline{P}^i$ for each face $i \in F \cup F'$ and a point $\overline{p}_j$ for each vertex $j \in V$, such that

(i) for each vertex $j \in A$ on face $i \in F \cup F'$, $\overline{p}_j$ lies on $\overline{P}^i$;
(ii) for each edge $< h,i;j,k >$, $\overline{p}_j$ and $\overline{p}_k$ are distinct points and $\overline{P}^h$, $\overline{P}^i$ are distinct planes
(iii) for each hole $a \in A$, the faces of the hole are not collinear.

Consider a proper simply-connected manifold built as a sheetwork. Around each hole, we have a cycle of faces (and planes). We will plug each hole of the manifold with an isostatic block – a sheetwork which is just large enough to be statically rigid, but not large enough to create any stresses entirely within the block.

Definition 13. For a sheetwork on the faces and edges of a proper simply-connected manifold, the *block sheetwork* is formed by converting each hole to an isostatic sheetwork or *block* on the existing boundary faces of the hole (plus one added face if the cycle has length $k > 3$ and all faces are concurrent).

Remark 32. If the faces of a cycle of length $k \geq 4$ are concurrent (but not collinear), then we add an extra face on a plane not containing this point. We either have exactly three non-collinear faces, or we have some four faces which are not concurrent. By Proposition 30, the complete graph on these $|F^*|$ faces is statically rigid, and some subgraph of size $3|F^*| - 6$ creates an isostatic sheetwork on these faces. The required block sheetworks can be constructed.

For any cycle of length k, without an added sheet, we must add $3k - 6$ edges to form the isostatic block. If we added an extra face making $k + 1$ faces, then we have $3(k + 1) - 6$ edges in the graph of the isostatic block. ∎

A CORRESPONDENCE BETWEEN TWISTS AND SELF-STRESSES

The self-stresses of a sheetwork on a simply-connected manifold (V,E,F) have a striking resemblance to the twists of a molecular structure on the same manifold. The self-stresses satisfy a condition for every face and the twists satisfy a similar equation on every cycle. With the help of the block sheetworks for any holes, we can make this into a precise correspondence. With this correspondence, and the counts developed for sheetworks, we can provide the improved estimates for the twists of our molecular structures. This is the polar of a correspondence between statics of bar-and-joint frameworks and bendings of hinge-panel structures[7].

Theorem 33. *Given a proper simply-connected manifold $M = (V,E,F); \overline{\mathbf{P}}, \overline{\mathbf{p}}$ on the abstract sphere $(V,A;E,E';F,F')$, the space of twists of the molecular structure on the vertices and edges of M is isomorphic to the space of self-stresses of the block sheetwork on the faces, edges and holes of M.*

Proof. The isomorphism is based on taking exactly the same scalars for the edges, along with a switching of indexes for the edges, from pairs of vertices to pairs of faces of a given patch.

Assume we have a twist assigning $\lambda_{j,k}$ to each patch $< h,i;j,k > = < i,h;k,j >$ between vertices of the molecular structure. Around each face $m \in F$ of the molecular structure, we have $\sum_{(j,k)\in C_m} \lambda_{j,k}\mathbf{H}_{j,k} = \mathbf{0}$. This scalar $\lambda_{j,k}$ becomes the scalar $\omega_{h,i}$ on the dual edge (h,i) between the faces h,i. It is easy to see that, for the same face: $\sum_{(h,m)\in F_m} \omega_{h,m}\mathbf{H}^{h,m} = \mathbf{0}$.

Moreover, each hole $a \in A$ of the abstract simply-connected manifold has a surrounding vertex-edge cycle C_a (the *link* of a in the terminology of topology). For this cycle, we have $\sum_{(j,k)\in C_a} \lambda_{j,k}\mathbf{H}_{j,k} = \mathbf{0}$. This assignment becomes an equilibrium load on the faces of an isostatic block on this hole and there is a unique resolution ω_a^o on the added edges within the block. Together with the incoming load from ω, the extensors corresponding to ω_a reach an equilibrium on every face of this block.

Carried out over all blocks, we now have a set of scalars ω, ω^o on all edges of the block sheetwork, which reach an equilibrium on every face of the original manifold and of the blocks. We have a self-stress on the block sheetwork for the proper manifold.

Conversely, assume we have a self-stress ω, ω^o on the block sheetwork of the proper manifold. If we simply convert the scalar $\omega_{h,i}$ to the scalar $\lambda_{j,k}$ whenever $< h,i;j,k >$ is the patch of the manifold, we have a possible twist on the molecular structure.

Consider any simple vertex-edge cycle C in the abstract simply-connected manifold. By the simple-connectivity, this cycle of edges forms a cut-set of the faces of the manifold – and of the block sheetwork. By the Cut Lemma, the scalars of the self-stress ω on this vertex-edge cycle satisfy $\sum_{(h,i)\in C} \omega_{h,i}\mathbf{H}^{h,i} = \mathbf{0}$. With these scalars switched to the indexing by vertices, we have $\sum_{(j,k)\in C} \lambda_{j,k}\mathbf{H}_{j,k} = \mathbf{0}$, as required.

We conclude that the scalars $\lambda_{j,k}$ satisfy all the cycle conditions and we have a twist. Because of the unique extension from scalars $\omega_{h,i}$ on the edges E of the manifold to the larger block manifold sheetwork, we have a full isomorphism. ∎

We can combine all of this into a lower bound for the space of twists of a molecular structure for a proper simply-connected manifold.

Manifold Count Theorem 34. *Given a proper simply-connected manifold* $M = (V,E,F); \overline{\mathbf{P}}, \overline{\mathbf{p}}$ *with* n *holes, the space of twists on the molecular structure on the manifold satisfies* $\dim(T) \geq |E| - \big[3|F| + 6(n-1)\big]$.

If the corresponding block sheetwork is statically rigid, then the space of twists has dimension $\dim(T) = |E| - \big[3|F| + 6(n-1)\big]$.

Proof. For convenience, we assume that each hole t of size $m_t > 3$ has faces in at least four planes, not all concurrent. [If this is false, we would add an additional to create the block. The reader can verify that this change will wash through the counts which follow.]

We replace each hole of size m_t by an appropriate statically rigid sheet block with the m_t faces and $|E_t| = 3|m_t| - 6$ new edges not in E. Summing over the original faces and edges and over these added blocks, we have $|E^*| = |E| + \sum_{t=1}^{t=n}(3m_t - 6)$ edges and $|F^*| = |F| + \sum_{t=1}^{t=n}(m_t)$ faces. By our general counts for sheetworks we know that we have a space of self-stresses of dimension at least

$$
\begin{aligned}
|E^*| - \big[3|F^*| - 6\big] &= |E| + \sum_{t=1}^{t=n}(3m_t - 6) - \left[3\left(|F| + \sum_{t=1}^{t=n}(m_t)\right) - 6\right] \\
&= |E| - \big[3|F| + 6(n-1)\big].
\end{aligned}
$$

By Theorem 33, the space of self-stresses is isomorphic to the space of twists of the molecular structure on the same manifold. We conclude that the dimension of the space of twists is at least $|E| - \big[3|F| + 6(n-1)\big]$.

If the block sheetwork is statically rigid, then the estimate of the space of self-stresses is sharp. This translates immediately to a sharp estimate of the space of twists. ∎

In general, this will still be an underestimate of dim(T). For discs ($n = 1$) this is the same count we started with: $|E| - 3|F|$. For complete spheres (no holes), this is $|E| - [3|F| - 6]$, a count that is never positive on 3-connected spheres.

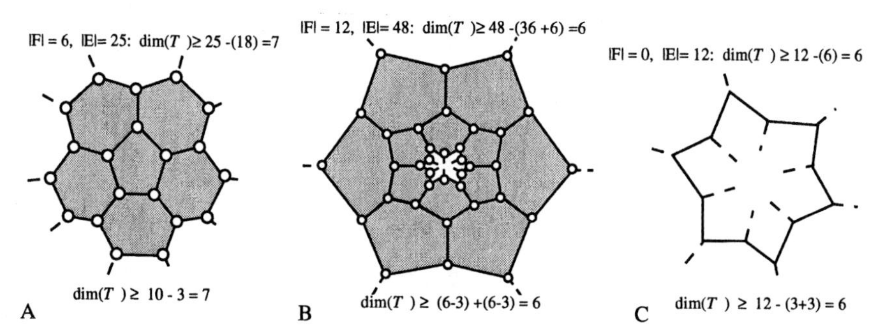

Figure 19. Some simply-connected manifolds, with the lower bounds on their spaces of twists computed from the Manifold Count Theorem (above) and the Simple Count Corollary (below).

Example 35. In Figure 19, we show the three manifolds for analysis.
(A) For the manifold in Figure 19A, we have $|F| = 6$, $|E| = 25$, and $n = 1$ so dim$(T) \geq 25 - (18 + 0) = 7$. This is the structure for the fullerine bowl C_{20}.
(B) For the manifold in Figure 19B, we have $|F| = 12$, $|E| = 48$, and $n = 2$ so dim$(T) \geq 48 - (36 + 6) = 6$.
(C) For the manifold in Figure 19C, we have $|F| = 0$, $|E| = 12$, and $n = 2$ so dim$(T) \geq 12 - (6) = 6$. Note that this is the same answer we would get treating this directly as a ring of 12 atoms following Example 2. ∎

Simple Count Corollary 36. *Given a proper simple simply-connected manifold $M = (V, E, F); \overline{\mathbf{P}}, \overline{\mathbf{p}}$, with n holes of size $\{m_t\}_{t=1}^{t=n}$, the space T of twists on the molecular structure on the manifold has dimension* dim$(T) \geq \sum_{t=1}^{t=n}(m_t - 3)$. *For realizations with the faces on generic planes,* dim$(T) = \sum_{t=1}^{t=n}(m_t - 3)$.

Proof. To capture the counts, we can consider part of each block to be a simple 'skin' of edges (and vertices) placing a disc over each hole. With the cycle of edges entering the hole and this skin extension, we have counted $m_t + m_t - 3$ edges of the block. When complete, this portion forms an entire simple sphere. In addition, to complete an isostatic block on a hole of size m_t we must add a second (bottom) skin of $m_t - 3$ more edges.

When we have done this for each hole, we have $3|F + F'| - 6$ edges for the simple sphere, and $\sum_{t=1}^{t=n}(m_t - 3)$ extra edges from completing the block for each hole. Therefore, the estimate is

$$|E^*| - [3|F+F'|-6] = [3|F+F'|-6] + \sum_{t=1}^{t=n}(m_t - 3) - [3|F+F'|-6] = \sum_{t=1}^{t=n}(m_t - 3).$$

If we take generic planes for the faces, then the corresponding block sheetwork can be chose to include a simple spherical polyhedron - and be statically rigid. ∎

Example 37. If we reexamine the manifolds of Figure 19, we get simpler calculations, but the same answers (see the bottoms of the Figure 19).

Another illustration would be the manifold in Figure 17, which is also simple. The holes have size $m_1 = 3$, $m_2 = 4$, $m_3 = 5$ and $m_4 = 12$. With these assumptions, we have dim$(T) \geq (0) + (1) + (2) + (9) = 12$. Note that this description assumes that one set of four vertices on the 'outer' hole are coplanar (they are counted as belonging to a single face in the figure). ∎

CONCLUDING REMARKS

It would certainly be possible to translate all of the necessary results for the statics of sheetworks into equivalent results on twists, and their extensions to some corresponding version of the block sheetworks. One advantage of an explicit isomorphism to statics lies in the easy access to results for bar-and-joint frameworks which have accumulated over one hundred and fifty years.

We can use the sharp version of the Count Theorem, or the Simple Count Theorem, if the wide array of techniques developed for the geometric analysis of statics will let us verify that the block sheetwork is statically rigid. At the other extreme, if the block sheetwork is independent, then the correspondence of Theorem 33 guarantees that the molecular structure is rigid.

It would also be possible to replace the 2-extensor $\overline{p} \vee \overline{q}$ with the pairs of vectors $\mathbf{p}-\mathbf{q}, \mathbf{p}\times\mathbf{q}$ and write out all of the results in a 'Euclidean' form. To do that would be to hide the underlying projective geometry, and the very basis for imagining that polarity would reveal important connections!

Real glass or crystals do not have the topology of the sphere or a 2-manifold. In fact, they may have the homology of a 3-disc or part of the 3-sphere with cells, faces, edges and vertices. In a fashion analogous to our study of spherical molecules, it is possible to use that underlying homology to organize the cycle equations, find a basis for the space of all cycles, and extract appropriate information about the twists of the large molecular structure (Figure 20). The polar form of this analysis can also be applied to certain crystals, modeled as hinge structures.

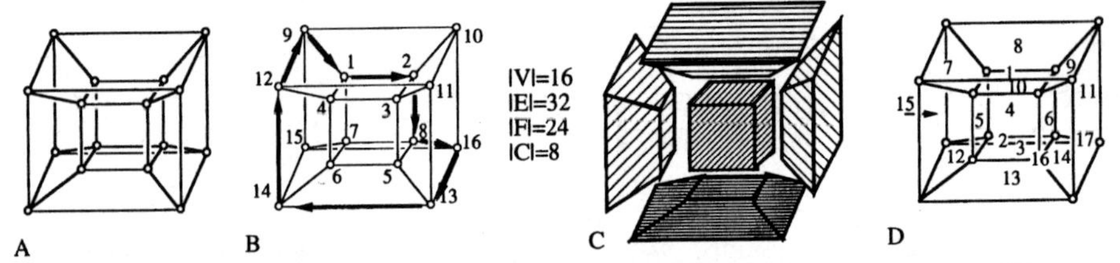

Figure 20. The homology of a hypercube (A) lets us organize the vertex-edge cycles (B) onto the faces of the cells (C) and further reduce this to a generating set of 17 face-cycles (D).

Example 38. Consider a moleculė built on a projected hypercube (Figure 20A). There are 16 vertices (B), 36 edges, 24 faces, and 8 cells (C) (3-dimensional polyhedra). [The 4-dimensional version of Euler's formula for polytopes gives the check on our counts: $|V|-|E|+|F|-|C| = 16-36+24-8 = 0$.] General vertex-edge cycles are generated by the cycles on the 24 faces, by general principles. Moreover, these face cycles are not independent: each cell gives a dependence among the face cycles of this cell. Finally, in the algebraic topology version of the 'inclusion/exclusion counting principle', the dependencies of these 8 cycles are themselves dependent (captured by the third homology of the complex being 1). The net conclusion is that there are 17 independent face cycles to be checked (Figure D). ∎

We would welcome specific examples where these (projective) geometric techniques would give additional insights.

Acknowledgments. This paper explicitly draws results developed over several decades of collaboration with Henry Crapo (Paris), Janos Baracs (Montréal), Tiong-Seng Tay (Singapore) and Neil White (Gainesville). Important portions of the basic theory of hinge structures first appeared in joint publications with Henry Crapo[7] and

Tiong-Seng Tay[19]. The critical role played by the (unpublished) observations and conjectures of Janos Baracs, a structural engineer and geometer, is less obvious - but crucial to all of our collaborations.

BIBLIOGRAPHY

[1] A.D. Alexandrov; *Konvex polyeder*; German Translation, Akademie-Verlag, 1958.

[2] A. Cauchy; Deuxiéme memoire sur les polygons; *J. École Polytechnique*, XVIe Cahier (1831), 87-98.

[3] R. Bricard; Memoire sur la theorie de l'octaedre articulé; J. Math. Liouville(5) 3 (1897), 113-148.

[4] R. Connelly; A flexible sphere; *Math. Intelligencer* 1 (1982), 130-1311.

[5] F. Chung and S. Sternberg ; Mathematics and the Buckyball; *American Scientist* 81 (1993), 56-71.

[6] H. Crapo and A. Cheung; A combinatorial perspective on algebraic geometry; *Advances in Mathematics* 20 (1976), 388-414.

[7] H. Crapo and W. Whiteley; Statics of frameworks and motions of panel structures, a projective geometric introduction; *Structural Topology* 6 (1982), 43-82.

[8] M. Dehn; *Über die Starrheit konvexer Polyeder*; *Math. Annal.* 77 (1914), 466-473.

[9] D. Franzblau; Combinatorial algorithm for a lower bound on frame rigidity; *SIAM J. DIscrete Math* 8 (1995), 338-400.

[10] A. Fogelsanger; Triangulated minimal 2-cycles are generically rigid; PhD. Thesis, Cornell University, Ithaca New York, 1988.

[11] J. Graver, B. Servatius and H. Servatius; Combinatorial Rigidity; Graduate Studies in Mathematics, Amer. Math Soc., 1993.

[12] H. Gluck; Almost all simply connected surfaces are rigid; in *Springer-Verlag Lecture Notes #438: Geometric Topology*, Berlin, (1975), 225-239.

[13] D. Jacobs; Generic rigidity in three-dimensional bond-bending networks; *J. Physics A* to appear.

[14] C.M. Jessop; **A Treatise on the Line Complex**; Cambridge, 1904. Chelsea reprint, 1969.

[15] F. Klein; *Elementary Mathematics from an Advanced Standpoint: Geometry*; English Translation, Dover, (1948).

[16] J. Munkres; **Elements of Algebraic Topology**; Addison-Wesley, Reading Mass; 1984.

[17] T-S Tay; Rigidity of multigraphs I: linking rigid bodies in n-space; *J. Combinatorial Thoery Series B* 26 (1984), 95-112.

[18] T-S Tay, N. White and W. Whiteley; Skeletal rigidity of simplicial complexes; *European J. Combinatorics I, II* 16 (1995), 381-403, 502-523.

[19] T-S. Tay and W. Whiteley; Recent progress in the generic rigidity of structures; *Structural Topology* 9 (1984), 31-38.

[20] T-S. Tay and W. Whiteley; Generating isostatic frameworks; *Structural Topology* 11 (1985), 21-69.

[21] N. White; A tutorial on Grassmann-Cayley Algebra; in *Invariant Methods in Discrete and Computational Geometry* N/ White (ed), Kluwer, 1995.

[22] N. White and W. Whiteley; The algebraic geometry of motions in bar and body frameworks; *SIAM J. Algebraic and Discrete Methods* 8 (1987), 1-32.

[23] W. Whiteley; Motions, stresses and projected polyhedra; *Structural Topology* 7 (1982), 13-38.

[24] ______; Infinitesimally rigid polyhedra I: statics of frameworks; *Transactions Amer. Math. Soc.* 285 (1984), 431-465.

[25] ______; Infinitesimal motions of a bipartite frameworks ; *Pacific J. of Mathematics* 110 (1984), 233-255.

[26] ______; Infinitesimally rigid polyhedra II: modified spherical frameworks; *Transactions Amer. Math. Soc.* 306 (1988), 115-139.

[27] ______; Rigidity and polarity I: statics of sheet frameworks; *Geometriae Dedicata* 22 (1987), 329-362.

[28] ______; The union of matroids and the rigidity of frameworks; *SIAM J. Discrete Methods* 1 (1988), 237-255.

[29] ______; Problems on the realizability and rigidity of polyhedra; in *Shaping Space: A Polyhedral Approach* (M. Senechal and G. Fleck eds.), Birkhauser, Boston, 1988.

[30] ______; Rigidity and scene analysis; in *Handbook of Discrete and Computational Geometry* J. O'Rourke and J. Goodman (eds), CRC Press 1997, 893-916.

[31] ______; Matroids from Discrete Geometry; in *Matroid Theory*, J. Bonin, J. Oxley and B. Servatius (eds.), AMS Contemporary Mathematics, 1996, 171-313.

TENSEGRITY STRUCTURES: WHY ARE THEY STABLE?

R. CONNELLY

June 14, 1998

ABSTRACT. A particular definition of stability for tensegrity structures is presented, super stability. This is a stronger case of prestress stability that applies to many examples of tensegrities found in nature.

1. Introduction: The Basic Object.

A basic question in dealing with tensegrity structures is: What are they? Here several definitions are briefly described with some of their basic characteristics, but it is proposed that one particularly strong condition for stability, what has been called, prestress stability, is of central importance. Furthermore, it is claimed that a particular type of prestress stability, what we call super stability, is especially relevant.

There are also several different categories or types of models for tensegrities. The following, though, is a good starting point for the basic object. The question of the stability or rigidity of this object will be dealt with next, and is a basic part of the whole picture.

Start with a finite configuration of labeled points $p_i, i = 1, \dots, n$ in R^d. This is denoted as

$$p = (p_1, p_2, \dots, p_n).$$

Define the graph $G = (V, E)$ of the tensegrity as some graph on the set of vertices $V = \{1, 2, \dots, n\}$, where E is a set of unordered edges (without loops or multiple edges) of G such that each edge is designated as either a *cable*, *strut*, or *bar*. An edge of G is often referred to as a *member*. The whole tensegrity is written as $G(p)$.

The basic general idea of these definitions is:

Cables can shrink in length, but not increase.

Struts can increase in length, but not decrease.

Bars stay the same length.

But the specific form that these conditions take can be quite varied. Figure 1 shows how the tensegrity is denoted graphically.

2. Stability.

The whole point of a tensegrity is what sort of stability is to be considered. There are many inequivalent, but related definitions of rigidity and/or stability. They usually

Rigidity Theory and Applications
Edited by Thorpe and Duxbury, Kluwer Academic/Plenum Publishers, 1999

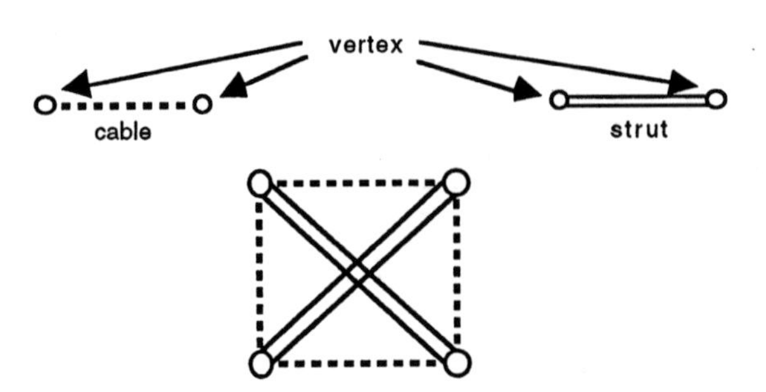

Snelson's X tensegrity. This is one of the simplist tensegrities.

Figure 1

involve the following rigidity constraints. The configuration p is fixed and we consider the following constraints on other configurations q:

$$|q_i - q_j| \leq |p_i - p_j|$$

for $\{i,j\}$ a cable

$$|q_i - q_j| \geq |p_i - p_j|$$

for $\{i,j\}$ a strut.

-Rigidity: Any configuration q, sufficiently close to p and satisfying the constraints is congruent to p.

-Infinitesimal Rigidity (= Static Rigidity): This uses a linearized form of the constraints given by a matrix inequality $R(p)p' \leq 0$. If this has only the minimal trivial solutions, given by derivatives of one parameter families of congruences, $G(p)$ is said to be *infinitesimally rigid.*

-Generic Rigidity: This is just a property of the underlying graph, G. If there is some configuration p (with algebraically independent coordinates) such that $G(p)$ is infinitesimally rigid, then G is said to be *generically rigid.*

-Second Order Rigidity: This uses the first-order motion p' and asks for an extension p'' satisfying the appropriate equations.

-Prestress Stability: For $G(p)$ this means that there are potential functions for each strut and cable, such that the sum over all nearby configurations q has a quadratic local minimum only when q is congruent to p. (This will be discussed in more detail in the next section.)

-Global Rigidity: $G(p)$ is *globally rigid* if any configuration q in R^d satisfying the constraints is congruent to p.

-Super Stability: $G(p)$ is *super stable* if there is a particular positive semi-definite quadratic form defined on all configurations q, in all dimensions, that is minimized when q is an affine image of p, together with other conditions mentioned below.

3. Prestress Stability.

I claim that the most important notions of stability are prestress stability and super stability, which will be described more carefully.

Start with the "ideal" configuration $p = (p_1, p_2, \ldots, p_n)$ and define real valued potential functions, whose domain is the positive reals

$$f_{i,j} : R^1{}_+ \rightarrow R^1$$

for each $\{i,j\}$ an edge of G.

The total potential for any configuration q is

$$E(q) = \sum_{\{i,j\}} f_{i,j}(|q_i - q_j|^2).$$

If E has a non-degenerate local minimum when $q = p$, which is unique up to rigid congruences, then we say $G(p)$ is *prestress stable.*

The $f_{i,j}$ determine the stress-strain characteristics of the cable or strut $\{i,j\}$, and we assume that they all are differentiable and have a strictly positive second derivative where they are defined.

This potential function defines an (equilibrium) stress as follows:

$$\omega_{i,j} = \frac{1}{2} f'_{i,j}(|p_i - p_j|^2).$$

These can be interpreted as tensions (when positive) for the cables and compressions (when negative) for the struts.

Lemma. *When p is a critical point for the potential function E, then the stresses above provide a vector equilibrium at each vertex. In other words the following vector equation holds at each vertex: For each i,*

$$\sum_j \omega_{i,j}(p_i - p_j) = 0.$$

Figure 2 shows some examples of prestress stable tensegrities.

See [1] for a more complete description of prestress stability.

4. The Hessian.

Suppose that $\omega = (\dots, \omega_{i,j}, \dots)$ is an equilibrium stress as above. Define the (symmetric) stress matrix Ω so that the $\{i,j\}$-th entry, when $i \neq j$ is $-\omega_{i,j}$, and so that the row and column sums are 0.

Define the rigidity matrix $R(p)$ as that e-by-dn matrix, where the row corresponding to the edge $\{i,j\}$ has all its entries 0, except for two blocks of entries, the row vector $p_i - p_j$ corresponding to vertex i, and $p_j - p_i$ corresponding to vertex j. The number of struts and cables of G is e, and the number of vertices is n. The ambient space is R^d.

For each cable or strut $\{i,j\}$, define the stiffness coefficient as

$$c_{i,j} = \frac{1}{4} f''_{i,j}(|p_i - p_j|^2),$$

and let D be the e-by-e diagonal matrix, where the $\{i,j\},\{i,j\}$ entry is $c_{i,j} > 0$

Then a calculation shows that the Hessian of the potential function is the symmetric matrix:

$$H(E) = \Omega \otimes I^d + R(p)^t D R(p),$$

where $()^t$ is the transpose operation and I^d is the d-by-d identity matrix. The symbol $\otimes$ represents the tensor product of two matrices. In this case it repeats the action of Ω on each set of coordinates.

The second term is called the stiffness matrix and it is always positive semi-definite.

Figure 2 shows some examples of prestress matrices. Notice that the Figures 2a is not rigid in 3-space. Its stress matrix is positive semi-definite, but there are realizations of the framework in higher dimensional spaces, where affine transformations serve as flexes.

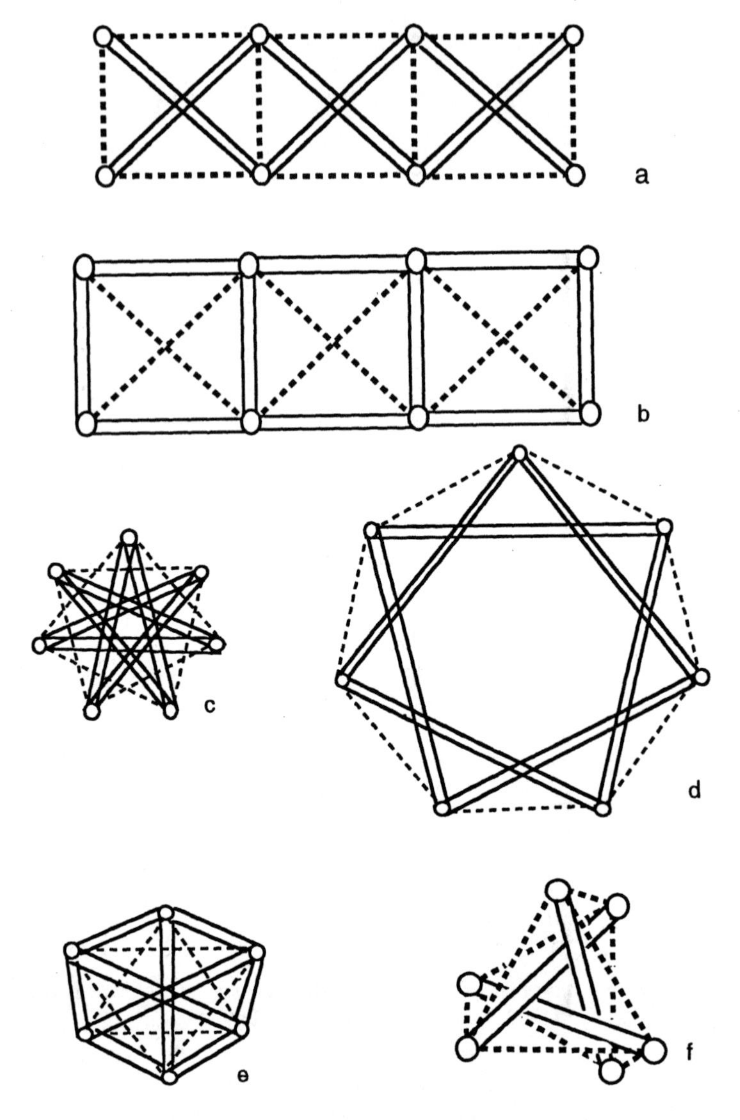

Figure 2

Figures 2a, 2b, 2c, 2d and 2e are all infinitesimally rigid in the plane, and so they are automatically prestress stable, since the stiffness matrix will dominate the quadratic form if the stresses are chosen small enough.

The Snelson octet truss, Figure 2f, has a positive semi-definite stress matrix, but is not infinitesimally rigid in three-space.

5. Super Stability.

Prestress stability has some disadvantages, especially when the stress matrix Ω has some negative eigenvalues. Here it is proposed that a particular sort of prestress stability be considered.

We say the tensegrity framework $G(p)$ with n vertices in R^d is *super stable* if there is an equilibrium stress $\omega = (\dots, \omega_{i,j}, \dots)$ with a stress matrix Ω such that

(1) The stresses $\omega_{i,j}$ are positive for cables and negative for struts.
(2) For $i \neq j$ the (i,j) entry of Ω is $-\omega_{i,j}$, and $(1,1,\dots,1)\Omega = 0$. (This determines the symmetric stress matrix Ω from the stress ω.)

(3) As a quadratic form Ω is positive semi-definite.
(4) The rank of Ω is $n - d - 1$ (the largest possible).
(5) There are no affine (infinitesimal) flexes of $G(p)$.

For example for the Snelson X tensegrity, the stress matrix becomes:

$$\Omega = \begin{pmatrix} 1 & -1 & 1 & -1 \\ -1 & 1 & -1 & 1 \\ 1 & -1 & 1 & -1 \\ -1 & 1 & -1 & 1 \end{pmatrix}$$

which is positive semi-definite of rank 1. Note that if the cables and struts are reversed, the stress matrix is NOT positive definite.

6. Properties and Advantages of Super Stability.

If $G(p)$ is super stable with respect to the equilibrium stress ω in R^d, then the following properties hold:

(1) $G(p)$ is globally rigid in any R^k, for $k \geq d$. That is any other configuration q in R^k that satisfies the distance constraints is congruent to p.
(2) $G(p)$ is prestress stable, stabilized by any positive multiple of the stress w.
(3) If q is any non-singular affine image of p, $G(q)$ is super stable with respect to same equilibrium stress w.
(4) If q is any non-singular projective transformation of p, then there is a suitably altered equilibrium stress ω' (possibly with changes in sign) such that $G'(q)$ is super stable with respect to ω', where G' changes the assignment of struts and cables appropriately.
(5) The subgraph of G determined by the cables alone must be connected.
(6) The graph G determined by struts and cables is vertex $(d+1)$-connected.

Property (2) is the key for many physical situations.. Suppose a tensegrity has some given equilibrium stress ω determined by the physics of the materials involved. If the tensegrity remains in a particular configuration for an extended period of time, we can reasonably expect that there is an equilibrium stress ω that forms the prestress in prestress stability. But what happens if ω is increased with no corresponding increase in the stiffness coefficients $\{c_{i,j}\}$? The stress matrix Ω dominates. If Ω has any negative eigenvalue, the structure will not remain at rest, even though it is an equilibrium configuration. The slightest perturbation will cause the structure to change its shape drastically. This can be what influences many physical tensegrities to be super stable. No matter how strong the equilibrium stress is relative to the stiffness coefficients, it will remain stable as long as the members themselves do not fail.

When the structure is super stable, increasing the prestress "stiffens" it. When there is a negative eigenvalue in the stress matrix increasing the prestress will cause a catastrophic failure.

For example, in Figure 2b, the array of squares with struts on the outside can only have an equilibrium stress with a negative eigenvalue. This is despite the fact that the tensegrity itself is infinitesimally rigid in the plane. If the prestress is increased sufficiently with no corresponding increase in the stiffness coefficients $c_{i,j}$, the structure will fail, even when it is constrained to stay in the plane.

On the other hand, the tensegrity in Figure 2a can only have a prestress that has positive eigenvalues. The rank of the stress matrix is low, however, and this prohibits it from being super stable strictly from the definition. It fails the condition (6) above. But it does retain many of the features of a super stable tensengrity. Each one of the squares is super stable and so they must retain their shape, even when they are allowed to move in three-space. Figures 2d and 2f are super stable.

But notice that Snelson octet truss in Figure 2f has only three struts and nine cables. In a sense, it is underbraced. For example, in order to be infinitesimally rigid, it is necessary that the total number of members in the graph G be at least one more than $3n-6 = 12$ for three-space. See for example, Roth and Whiteley [5] for a good discussion of this. There are exactly 12 members, so this count alone implies that the stiffness matrix is of too small a rank to allow prestress stability by itself. But it turns out that the stress matrix has no negative eigenvalues, and so it takes over the stability, even for large stresses. Notice that the underlying graph is the same as the regular octahedron, and it is not infinitesimally rigid since it has an equilibrium stress and a triangulated sphere (such as the octahedron) is not infinitesimally rigid precisely when it does have an equilibrium stress.

One might be tempted to think that the underbraced nature of many tensegrities detracts from their stability, but that is not the case.

Property (5) is helpful in detecting many examples where the stress matrix must have a negative eigenvalue. For example, the tensegrity, Figure 2c, has only the two cable triangles. Imagine moving the two triangles separately, but individually rigidly quite far apart. The quadratic form associated to the stress matrix, must be negative in that configuration. So there must be a negative eigenvalue. When any equilibrium stress (with signs properly chosen for each member) is increased, the structure will become unstable even though it also is infinitesimally rigid.

Property (1) can also be used directly to detect negative eigenvalues. The tensegrity in Figure 2c has the same graph as the tensegrity in Figure 2d, and, indeed, the cables

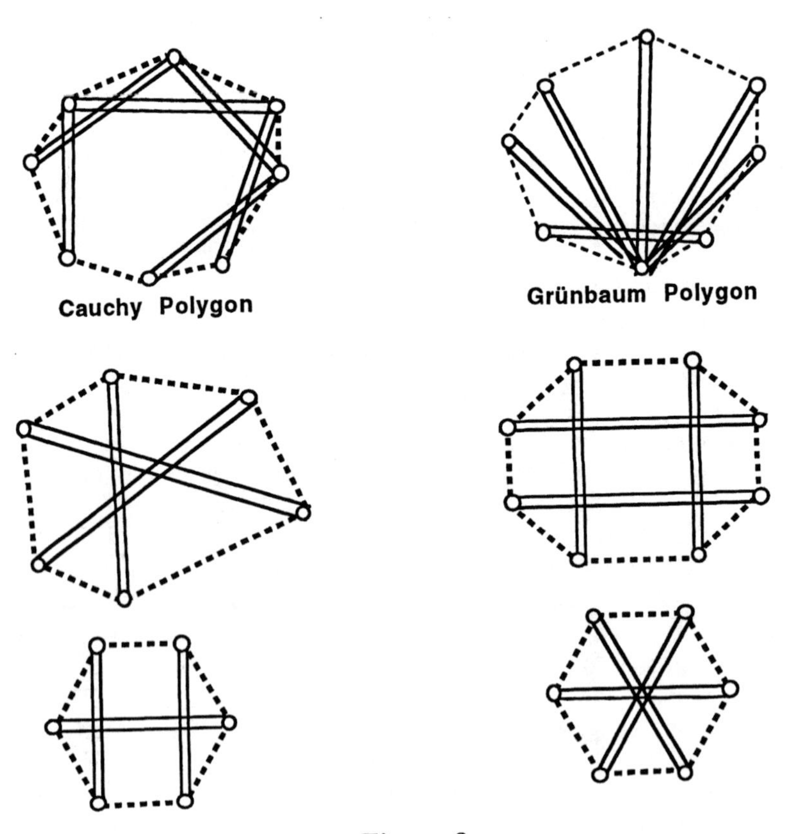

Figure 3

in both graphs are the same lengths. But the struts in Figure 2d are longer than those in Figure 2c, so the stress matrix for Figure 2c must have a negative eigenvalue, while tensegrity in Figure 2d is super stable.

It can be expected that there are many instances in nature where super stable tensegrities appear. There are many examples and a strong case is made for the importance of tensegrities in a biological setting in the paper by Ingber [4].

7. Starting Examples for Super Stability.

The next question is: How does one construct super stable tensegrities? One category of super stable tensegrities to start further constructions is the following.

The Polygon Theorem. *Any convex planar polygon with an equilibrium stress (and corresponding stress matrix Ω), that is positive on the external edges and negative (or zero) otherwise, is super stable with respect to Ω.*

(For a proof see R. Connelly, [2])

Figure 3 shows some examples where the Polygon Theorem applies to guarantee super stability. The first two, the Cauchy polygon and Grünbaum polygon are part of a general class of polygons, and they happen to be infinitesimally rigid as well. As long as the vertices of the configuration p are part of a convex polygon, then the combinatorial structure, given by the graph G, will always give a super stable tensegrity $G(p)$. On the other hand, the other polygons in the Figure depend on there being a particular conditions on the configuration, just to insure that there is a non-zero equilibrium stress. For example, the polygon on the middle left must have the vertices lie on an a conic (in addition to being convex). The other three polygons are drawn as regular polygons.

Another source of examples is the catalogue of highly symmetric super stable tensegrities developed with Allen Back. See Connelly, Back [3]. See also our website at
http://math.lab.cit.cornell.edu/visualization/tenseg/tenseg.html

8. Combining by Superposition.

Start with two different tensegrities, each with its own equilibrium stress, and superimpose some of their vertices. The combined stress will be the sum of the stress matrices, regarded as quadratic forms. It can even turn out that some of the stresses may cancel, eliminating it as a cable or strut.

Figure 4 shows an example:

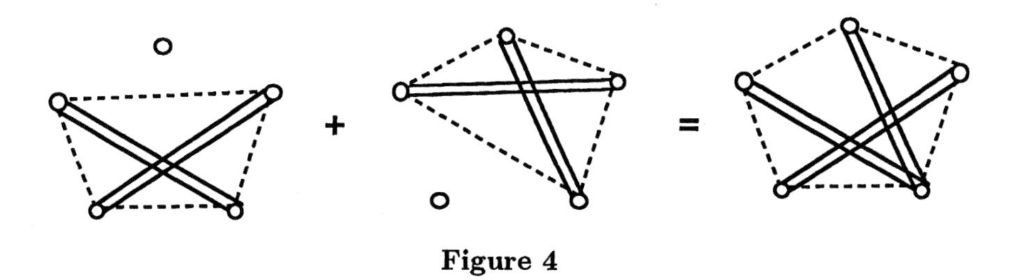

Figure 4

Note that when one uses this technique and adds the stress matrices, each subgraph, where each stress is non-zero, must overlap with the others in such a way that the whole configuration is forced to be in the appropriate dimensional Euclidean space. In the example of Figure 2a, each of the smaller squares is forced to lie in a two-dimensional space, but two successive squares have only two vertices in common. So there is a "hinge" where the whole configuration can rotate into higher dimensions.

Figure 5 shows some examples where the same unit is repeated with the effect that there is an extended truss. The top and bottom frameworks are super stable with the

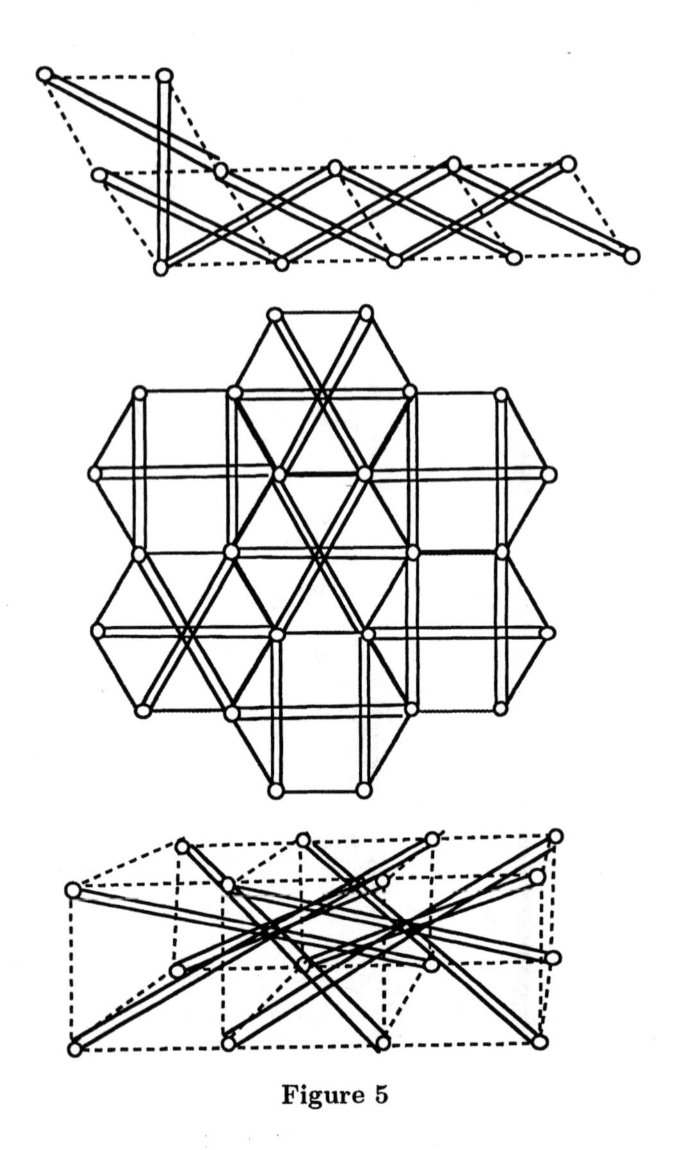

Figure 5

sum of stresses on the appropriate units as the stabilizing stress. For the middle example, if we take as the stabilizing stress the sum of the stresses on each hexagon, the rank of the stress matrix will be one less than what is required in condition (4) of Section 5. Nevertheless, in all the cases of Figure 5, the configuration is such that it is forced to stay rigid in the plane even considering it as being in three-space.

REFERENCES

1. R. Connelly and W. Whiteley, Second-Order Rigidity and Prestress Stability for Tensegrity Frameworks **9** (1996), no. 3, 453–491.
2. R. Connelly, *Rigidity and Energy*, Invent. Math. **66** (1982), 11–33.
3. R. Connelly and A. Back, *Mathematics and Tensegrity*, American Scientist, March-April (1998), 142–151.
4. D. Ingber, *The Architecture of Life*, Scientific American, January (1998), 48–57.
5. B. Roth and W. Whiteley, *Tensegrity Frameworks*, Trans. Amer. Math. Soc. **265** (1981), 419–446.

THE ROLE OF TENSEGRITY IN DISTANCE GEOMETRY

Timothy F. Havel

Biological Chemistry and Molecular Pharmacology
Harvard Medical School, Boston, MA 02115, USA

THEY'RE NOT JUST TOYS!

The theory of tensegrity frameworks, as developed by Connelly and colleagues [3, 5, 6, 34], was originally motivated by their sheer aesthetic appeal (for an introduction and account, see [4]). While this appeal can hardly be denied, it seems to me that this feature may to some extent have distracted from the fact that they also play a very fundamental role in the geometry of Euclidean space. In this paper I shall attempt to justify this statement by describing some relations between tensegrity theory and Euclidean distance geometry. These relations were encountered in the course of applying distance geometry to problems in molecular conformation, as detailed in my monograph with Gordon Crippen [7]. The bottom line is that the *globally* rigid tensegrities [3], in a certain sense made precise below, represent the "boundaries" of the conformation space (i.e. internal configuration space) of a system of N points in a Euclidean space of arbitrarily high dimension. For this reason, and because the forces among a system of particles can be viewed as a stress in a tensegrity framework, tensegrity theory would seem to have a great deal to say about the classical N-body problem in mechanics. I shall further suggest that these ideas might profitably be generalized to quantum N-body problems as well, where one must work in a much larger product space.

EUCLIDEAN DISTANCE GEOMETRY

Distance geometry may be defined as the invariant theory of metric vector spaces. The classical text on the subject is Blumenthal's *Distance Geometry* [1]. Applied to Euclidean space, the first fundamental theorem of states that the squared distances $d^2(\mathrm{a}, \mathrm{b})$ among a set of N points $\mathcal{A} = \{\mathrm{a}, \mathrm{b}, \ldots, \mathrm{z}\}$ constitute a complete system of invariants for the (direct sum of copies of the) Euclidean group [7, 8]. Thus one can describe any Euclidean point configuration, up to reflection, translation and rotation, by these distances. An intrinsic characterization of Euclidean distances, in turn, was first given by Karl Menger [25], using what Blumenthal subsequently named *Cayley-Menger determinants.* These will be denoted by $D(\mathrm{a}_1, \ldots, \mathrm{z}_1; \mathrm{a}_2, \ldots, \mathrm{z}_2)$

$$\equiv \frac{(-1)^N}{2^{N-1}} \det \begin{pmatrix} 0 & 1 & 1 & \cdots & 1 \\ 1 & d^2(\mathrm{a}_1, \mathrm{a}_2) & d^2(\mathrm{a}_1, \mathrm{b}_2) & \cdots & d^2(\mathrm{a}_1, \mathrm{z}_2) \\ 1 & d^2(\mathrm{b}_1, \mathrm{a}_2) & d^2(\mathrm{b}_1, \mathrm{b}_2) & \cdots & d^2(\mathrm{b}_1, \mathrm{z}_2) \\ \vdots & \vdots & \vdots & \ddots & \vdots \\ 1 & d^2(\mathrm{z}_1, \mathrm{a}_2) & d^2(\mathrm{z}_1, \mathrm{b}_2) & \cdots & d^2(\mathrm{z}_1, \mathrm{z}_2) \end{pmatrix}, \tag{1}$$

where $\mathcal{A}_1, \mathcal{A}_2 \subseteq \mathcal{A}$ with $\mathcal{A}_1 = \{\mathrm{a}_1, \ldots, \mathrm{z}_1\}$, $\mathcal{A}_2 = \{\mathrm{a}_2, \ldots, \mathrm{z}_2\}$, of cardinalities N_1, N_2, respectively. In the special case that $\mathrm{a} \equiv \mathrm{a}_1 = \mathrm{a}_2, \ldots, \mathrm{z} \equiv \mathrm{z}_1 = \mathrm{z}_2$, we shall abbreviate the corresponding symmetric determinant by $D(\mathrm{a}, \ldots, \mathrm{z})$. Note in particular that $D(\mathrm{a}, \mathrm{b}) = d^2(\mathrm{a}, \mathrm{b})$.

Menger's characterization of Euclidean distances may now be stated simply as follows. Let $d : \mathcal{A} \times \mathcal{A} \to \Re$ with $d(\alpha, \beta) = d(\beta, \alpha) \geq 0$ and $d(\alpha, \alpha) = 0 \ \forall \alpha, \beta \in \mathcal{A}$; then there exists a set of points $\mathbf{p}(\mathrm{a}), \ldots, \mathbf{p}(\mathrm{z}) \in \Re^n$ such that $d(\alpha, \beta) = \|\mathbf{p}(\alpha) - \mathbf{p}(\beta)\|$ if and only if for all $\mathcal{A}_1 \subseteq \mathcal{A}$ as above:

$$D(\mathrm{a}_1, \ldots, \mathrm{z}_1) \begin{cases} \geq 0 & \text{if } N_1 \leq n+1 \\ = 0 & \text{otherwise} \end{cases} \tag{2}$$

In fact, if $d(\alpha, \beta) = 0$ implies $\alpha = \beta$, it is sufficient for the "otherwise" case to hold for all $N_1 = n + 2$, with the exception of a single type of counterexample on $n + 3$ points [1, 7].

Another characterization of Euclidean distances, which we shall have occasion to refer to in the following, was subsequently given by Schoenberg [27, 28], using what Blumenthal called *Schoenberg's quadratic form*:

$$Q(t_\mathrm{a}, \ldots, t_\mathrm{z}) \equiv -\tfrac{1}{2} \sum\nolimits_{\binom{\mathcal{A}}{2}} t_\alpha t_\beta D(\alpha, \beta) \qquad \left(\sum\nolimits_{\mathcal{A}} t_\alpha = 0\right) \tag{3}$$

This form is positive semi-definite whenever $D(\alpha, \beta)$ are the squared distances in a Euclidean space, in which case the rank of the form n is equal to the minimum dimension in which these distances can be embedded. Interestingly, Schoenberg seems to have also been the first to study the kinematics of frameworks by distance-theoretic methods [29].

Since invariant theory does not usually deal with inequalities (or algebraically non-closed fields, for that matter!), we convert the inequalities into equalities by means of slack variables $v(\mathrm{a}_1, \ldots, \mathrm{z}_1)$. These may be interpreted as the *oriented volumes* spanned by subsets of the points, and lead to the following (conjectured!) complete system of algebraic relations among the invariants of the *proper* Euclidean group:

$$\begin{aligned} D(\mathrm{a}_1, \ldots, \mathrm{z}_1) &= 0 & (N_1 > n+1) \\ D(\mathrm{a}_1, \ldots, \mathrm{z}_1; \mathrm{a}_2, \ldots, \mathrm{z}_2) &= v(\mathrm{a}_1, \ldots, \mathrm{z}_1) v(\mathrm{a}_2, \ldots, \mathrm{z}_2) & (N_1 = N_2 = n+1) \\ D(\mathrm{a}_1, \ldots, \mathrm{z}_1) &= v^2(\mathrm{a}_1, \ldots, \mathrm{z}_1) & (N_1 < n+1) \end{aligned} \tag{4}$$

The oriented volumes, of course, satisfy some relations of their own, but I have shown that these are consequences of the $N_1 = N_2 = n + 1$ case above [7]. To obtain the relations for the improper Euclidean group, the $N_1 = N_2 = n + 1$ case is restricted to symmetric determinants. For work on proving this conjecture, see Ref. 8.

THE DISTANCE CONE

This invariant-theoretic ansatz to Euclidean geometry leads naturally to the consideration of an associated $\binom{N}{2}$-dimensional Euclidean vector space, which may be identified with the space of symmetric matrices with zero diagonal under the Frobenius inner product [15]. The matrices of squared distances

$$\mathbf{D} = [D_{\alpha\beta} | \alpha, \beta = 1, \ldots, N] = [\,\|\mathbf{p}_\alpha - \mathbf{p}_\beta\|^2 |\, \mathbf{p}_\alpha, \mathbf{p}_\beta \in \Re^{N-1};\, \alpha, \beta = 1, \ldots, N] \quad (5)$$

in an $(N-1)$-dimensional Euclidean space constitute a closed convex cone $\mathcal{D}^N_{N-1}$ in this larger space, which is extra-ordinarily rich in mathematical structure. For example, the extreme rays of this cone, which correspond to those (squared) distance matrices that are embeddable in one dimension, form a closed subcone $\mathcal{D}^N_1$ which is canonically homeomorphic to an $(N-2)$-dimensional projective space [7, p. 124].

More generally, Hayden, Tarazaga and Wells [23, 32] have shown that each face of $\mathcal{D}^N_{N-1}$ consists of those distance matrices obtained by affine transformation of the coordinates corresponding to any one distance matrix in that face. This in turn implies that the faces of this (nonpolyhedral) cone may be classified combinatorially by affinely realizable N-point oriented matroids, the ranks of which are equal to the minimum dimension of the Euclidean space in which the distance matrices of the face can be embedded. The Pythagorean theorem implies that the sum of any two distance matrices $\mathbf{D}_1, \mathbf{D}_2$ in the cone can be embedded in $\dim(\mathbf{D}_1+\mathbf{D}_2) \le \dim(\mathbf{D}_1)+\dim(\mathbf{D}_2)$ dimensions. It would be interesting to study the relation between the oriented matroids associated with $\mathbf{D}_1$, $\mathbf{D}_2$ and $\mathbf{D}_1 + \mathbf{D}_2$.

MOLECULAR CONFORMATION

The applications of distance geometry to molecular conformation are based on the fact that a "molecule" is largely determined by a list of its (essentially) invariant interatomic distances. For example, the length of a single covalent bond between e.g. a pair of carbon atoms varies less than a percent from its equilibrium value of ca. 1.53 Å under ambient conditions. Similarly, the geminal distances (between atoms both bonded to a third atom) are usually essentially fixed, and in many cases (e.g. across double bonds) longer-range distances may be fixed as well. Of course, such distance information does not determine the *chirality* (orientation) of the molecule's rigid asymmetric tetrahedra of atoms, and hence this information must be supplemented by the signs of the invariant oriented volumes of such tetrahedra; these are called *chirality constraints*. The *conformation* of the molecule is the set of spatial structures it can assume, wherein these invariant distances and chiralities necessarily have their prescribed values [7]. For further discussion of these issues and consideration of the group-theoretic aspects of labeling molecules by such constraints, the reader is referred to Dreiding [14].

Although the above distance and chirality constraints are insufficient to determine the conformation in general, it turns out that the conformation can similarly be specified with reasonable precision by means of a suitable list of *bounds* on the values of the noninvariant distances, possibly supplemented by the chiralities of certain noninvariant volumes. For example, the van der Waals repulsion between nonbonded pairs of atoms ensures that they always remain at least some predetermined distance apart, and in many cases these distance constraints, together with the invariant distances, actually determine the conformation reasonably well. The state a molecule with respect to rotation about a single bond, which is parametrized by the dihedral angle between the two planes defined by the bond together with two additional atoms at each end, can also be specified by means of distance and chirality constraints [7]. Finally, bounds on the values of the longer-range distances can in principle be used to select subpopulations of conformations from the statistical ensemble that is present in solution, such that the exchange rate of the structures within a subpopulation is much less than the exchange rate between them.

In many cases, particularly in biological macromolecules such as proteins, the conformation can approach a single unique spatial structure even though those distances that are known to be invariant *a priori* allow a vast range of possibilities. In such cases one must resort to sophisticated experimental techniques, such as x-ray crystallography and nuclear magnetic resonance (NMR) spectroscopy, in order to determine the conformation of the molecule in question. In particular, the NMR technique enables one to infer that the distances between certain pairs of atoms are small compared to the overall molecular dimensions, i.e. that the atoms are "in contact". For this reason distance geometry is widely used as a means of determining protein conformation from NMR measurements [20, 33]. Other applications of biological significance, including "homology modeling" of protein conformation, and the docking of ligands to receptor proteins, may be found in my forthcoming article in the Encyclopedia of Computational Chemistry [22].

GEOMETRIC REASONING

The analysis of a "distance geometry description" (list of distance and chirality constraints) of the conformational state of a molecule involves deriving new geometric facts about the molecule from those that are given explicitly by the description, a process known more generally as *geometric reasoning*. There are two basic approaches to geometric reasoning in distance geometry, namely inductive and deductive.

The inductive approach involves computing a random sample of spatial structures from the conformation space defined by the input distance geometry description. Such a random sample is called a *conformational ensemble*. Providing this ensemble is sufficiently large and unbiased, any geometric features that are common to all these structures can safely be inferred to be necessary consequences of the input. This approach to geometric reasoning has the disadvantage of being probabilistic, but the significant advantages of being applicable to very large problems ($>$ 1000 atoms), and of taking all the constraints (including chirality) into account at once. For details concerning the algorithms used to compute such conformational ensembles, the reader is referred to Refs. 20 and 22.

The deductive approach, in its most general form, expresses the hypotheses and conclusion in terms of (bounds on) polynomials in the squared distances, e.g. $D(\mathrm{a},\mathrm{b}) + D(\mathrm{b},\mathrm{c}) - D(\mathrm{a},\mathrm{c}) = 0$ if the lines $\overline{\mathrm{ab}}$ and $\overline{\mathrm{bc}}$ are perpendicular. It then uses the algebraic relations among the distances, expressed in terms of Cayley-Menger determinants as above, to rewrite the conclusion in terms of the hypothesis polynomials. Examples of the use of this approach, together with computer algebra programs, to prove elementary theorems in Euclidean geometry may be found in Ref. 18. Both because of the size of many problems of interest, as well as the importance of *in*equality constraints, this approach is not widely applicable to molecular problems (an exception being my recent work on cyclohexane [21]). A simpler numerical approach that can handle inequalities and can be extended to large problems involves deriving new *limits* on the values of the distances from the given distance bounds. This process, known as *bound smoothing*, is described in the next section.

TRIANGLE INEQUALITY BOUND SMOOTHING

The three-point Cayley-Menger determinant factorizes as follows:

$$\begin{aligned} D(\mathrm{a},\mathrm{b},\mathrm{c}) \;=\; & \tfrac{1}{4}\,(d(\mathrm{a},\mathrm{b})+d(\mathrm{a},\mathrm{c})+d(\mathrm{b},\mathrm{c}))\,(d(\mathrm{a},\mathrm{b})+d(\mathrm{a},\mathrm{c})-d(\mathrm{b},\mathrm{c})) \\ & (d(\mathrm{a},\mathrm{b})-d(\mathrm{a},\mathrm{c})+d(\mathrm{b},\mathrm{c}))\,(-d(\mathrm{a},\mathrm{b})+d(\mathrm{a},\mathrm{c})+d(\mathrm{b},\mathrm{c})) \end{aligned} \tag{6}$$

This shows that the nonnegativity of this determinant is equivalent to satisfying all three triangle inequalities among the three points. The distance limits implied by the triangle inequality are the infimum and supremum over all metric spaces $\mathcal{M}_3 = \mathcal{M}_3(\mathcal{A}, d)$ whose distances d lie within the given lower and upper bounds $\bar{\ell}$, $\bar{u}$, namely $\forall\ \{\alpha,\beta\} \in \binom{\mathcal{A}}{2}$:

$$\begin{aligned} \ell(\alpha,\beta) &= \ell(\beta,\alpha) \equiv \inf_{\mathcal{M}_3}\left(d(\alpha,\beta)\mid \bar{\ell}\le d\le \bar{u}\right) \\ \text{and}\quad u(\alpha,\beta) &= u(\beta,\alpha) \equiv \sup_{\mathcal{M}_3}\left(d(\alpha,\beta)\mid \bar{\ell}\le d\le \bar{u}\right) \end{aligned} \tag{7}$$

These *triangle inequality limits*, in turn, can be characterized as being the "loosest" set of limits that are tighter than the bounds and satisfy a system of linear inequalities [10], namely $\forall\ \{\alpha,\beta,\gamma\} \in \binom{\mathcal{A}}{3}$:

$$u(\alpha,\beta) \le u(\alpha,\gamma) + u(\beta,\gamma) \quad \text{and} \quad \ell(\alpha,\beta) \ge \ell(\alpha,\gamma) - u(\beta,\gamma) \tag{8}$$

These inequalities have been used to show that the triangle inequality limits can also be characterized combinatorially as the lengths of the shortest paths in a certain digraph [10]. This digraph consists of two strongly connected components, each with N nodes, and the two-way arcs within each component have length equal to the corresponding upper bounds $\bar{u}$. The arcs between the two components, on the other hand, all go from the "left" to the "right" component, and have length equal to the *negative* of the lower bounds $\bar{\ell}$. This reduction to a shortest paths problem shows that the triangle inequality limits can be computed in polynomial-time. By utilizing the sparsity of the generally available distance constraints, it is in fact possible to compute these limits for proteins containing a thousand or more atoms [17]. Contoured plots of the resulting matrices of limits have been found to be a very useful way to survey the conformation of proteins in a wholly coordinate-free fashion (see Fig. 1).

TETRANGLE INEQUALITY BOUND SMOOTHING

Although they can be computed very rapidly, the triangle inequality limits are a very "loose" approximation to true *Euclidean limits*, which are the ranges of the distances over all sets of points in Euclidean space whose distances satisfy the bounds. More generally, if we let $\mathcal{M}_M$ denote the set of metric spaces wherein each K-point

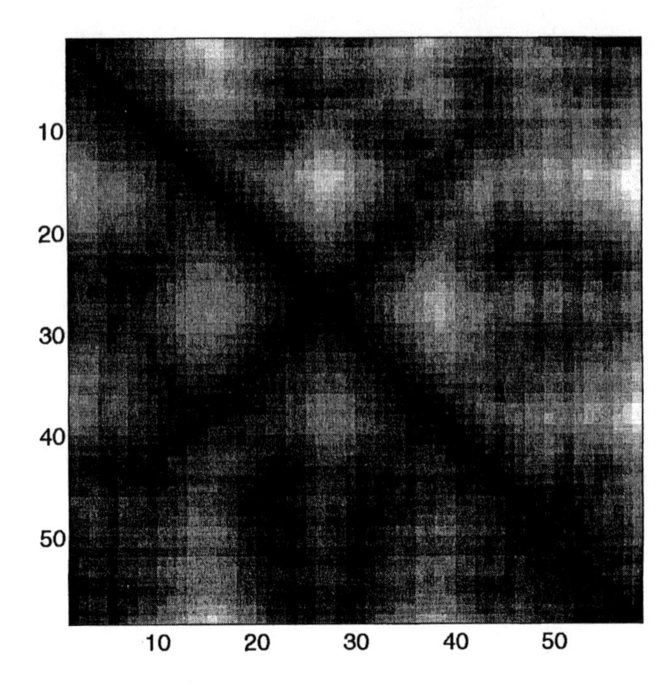

Figure 1: Contoured plot of the lower (below diagonal) and upper (above diagonal) triangle inequality limits among the 58 α-carbons of the protein pancreatic trypsin inhibitor. These limits were computed from an input set of distance bounds that included approximately 500 interproton contacts along with the distance constraints implied by the known chemical structure. The overall structural features of the protein chain can be read directly from this plot. For example, the dark anti-diagonal band starting at about residue 26 corresponds to a β-turn, at which the chain folds back anti-parallel to itself.

subset can be embedded in a Euclidean space for all $K \leq M$, then the M-limits are defined $\forall \ \{\alpha, \beta\} \in \binom{\mathcal{A}}{2}$ as:

$$\begin{aligned} \ell(\alpha,\beta) \;&=\; \ell(\beta,\alpha) \;\equiv\; \inf_{\mathcal{M}_M} \left(d(\alpha,\beta) \mid \bar{\ell} \leq d \leq \bar{u} \right) \\ \text{and} \qquad u(\alpha,\beta) \;&=\; u(\beta,\alpha) \;\equiv\; \sup_{\mathcal{M}_M} \left(d(\alpha,\beta) \mid \bar{\ell} \leq d \leq \bar{u} \right) \end{aligned} \tag{9}$$

Clearly, the N-limits are the Euclidean limits, while the triangle limits are the 3-limits.

The next logical step in the sequence is the limits implied by the distances in the space $\mathcal{M}_4$, i.e. whose 4-point Cayley-Menger determinants are also nonnegative. Unlike the 3-point Cayley-Menger determinants, the 4-point determinants do not factorize. Nevertheless, such a determinant can be expanded as a quadratic function in any one of its six squared distances, e.g.

$$D(\mathrm{a,b,c,d}) \;=\; A(\mathrm{a,b})\, d^4(\mathrm{c,d}) + B(\mathrm{a,b;\,c,d})\, d^2(\mathrm{c,d}) + C(\mathrm{a,b;\,c,d})\;, \tag{10}$$

where:

$$A(\mathrm{c,d}) \;=\; -\tfrac{1}{8}\det \begin{pmatrix} 0 & 1 & 1 \\ 1 & 0 & d^2(\mathrm{a,b}) \\ 1 & d^2(\mathrm{a,b}) & 0 \end{pmatrix} \tag{11}$$

$$B(\mathrm{a,b;\,c,d}) \;=\; -\tfrac{1}{4}\det\begin{pmatrix}0 & 1 & 1 & 1\\ 1 & 0 & d^2(\mathrm{a,b}) & d^2(\mathrm{a,d})\\ 1 & d^2(\mathrm{a,b}) & 0 & d^2(\mathrm{b,d})\\ 1 & d^2(\mathrm{a,c}) & d^2(\mathrm{b,c}) & 0\end{pmatrix} \tag{12}$$

$$C(\mathrm{a,b;\,c,d}) \;=\; \tfrac{1}{8}\det\begin{pmatrix}0 & 1 & 1 & 1 & 1\\ 1 & 0 & d^2(\mathrm{a,b}) & d^2(\mathrm{a,c}) & d^2(\mathrm{a,d})\\ 1 & d^2(\mathrm{a,b}) & 0 & d^2(\mathrm{b,c}) & d^2(\mathrm{b,d})\\ 1 & d^2(\mathrm{a,c}) & d^2(\mathrm{b,c}) & 0 & 0\\ 1 & d^2(\mathrm{a,d}) & d^2(\mathrm{b,d}) & 0 & 0\end{pmatrix} \tag{13}$$

The discriminant of this quadratic is

$$B^2(\mathrm{a,b;\,c,d}) - 4A(\mathrm{a,b})C(\mathrm{a,b;\,c,d}) \;=\; D(\mathrm{a,b,c})\,D(\mathrm{a,b,d})\;, \tag{14}$$

and is therefore nonnegative whenever the triangle inequalities among the four points are satisfied. Under this assumption, the two roots of the quadratic are always nonnegative, and correspond (up to an isometry) to a pair of quadrilaterals related by reflection of the point d (or c) in the line $\overline{\mathrm{ab}}$. We shall denote these lower and upper roots (regarded as functions of the distances) by

$$\begin{aligned} & L(d(\mathrm{a,b}), d(\mathrm{a,c}), d(\mathrm{a,d}), d(\mathrm{b,c}), d(\mathrm{b,d}))\\ \text{and}\quad & U(d(\mathrm{a,b}), d(\mathrm{a,c}), d(\mathrm{a,d}), d(\mathrm{b,c}), d(\mathrm{b,d}))\;, \end{aligned} \tag{15}$$

respectively.

The tetrangle limits among each subset of four points (only!) can be shown to be characterized by the linear inequalities (8) together with a system of nonlinear inequalities [7, 9, 13], namely $\forall\ \{\alpha,\beta,\gamma,\delta\} = \{\mathrm{a,b,c,d}\}$:

$$\begin{aligned} u^2(\gamma,\delta) \;&\le\; \max\begin{cases} U(\ell(\alpha,\beta), u(\alpha,\gamma), u(\alpha,\delta), u(\beta,\gamma), u(\beta,\delta))\\ U(u(\alpha,\beta), \ell(\alpha,\gamma), \ell(\alpha,\delta), u(\beta,\gamma), u(\beta,\delta))\\ U(u(\alpha,\beta), u(\alpha,\gamma), u(\alpha,\delta), \ell(\beta,\gamma), \ell(\beta,\delta))\end{cases}\\ \ell^2(\gamma,\delta) \;&\ge\; \min\begin{cases} L(u(\alpha,\beta), u(\alpha,\gamma), \ell(\alpha,\delta), \ell(\beta,\gamma), u(\beta,\delta))\\ L(u(\alpha,\beta), \ell(\alpha,\gamma), u(\alpha,\delta), u(\beta,\gamma), \ell(\beta,\delta))\\ L(\ell(\alpha,\beta), \ell(\alpha,\gamma), u(\alpha,\delta), \ell(\beta,\gamma), u(\beta,\delta))\\ L(\ell(\alpha,\beta), u(\alpha,\gamma), \ell(\alpha,\delta), u(\beta,\gamma), \ell(\beta,\delta))\end{cases} \end{aligned} \tag{16}$$

An alternative (and more readable!) diagramatic display of these limits is given in Fig. 2. Unfortunately, as we shall soon see, this "local" characterization does not imply that we have attained the global tetrangle limits defined by Eq. (9) with $M = 4$, as was the case for the triangle inequality limits. In addition, a neat combinatorial characterization is not known which would permit a closed algorithm to be given. It is nevertheless possible to iterate over all quadruples, substituting the bounds by the tetrangle limits whenever the latter are tighter, until no further changes occur. Extensive numerical experience [13] indicates that although this procedure may converge only asymptotically, the final limits obtained are independent of the order in which the quadruples are considered.

While pentangle and higher-order bound smoothing is certainly possible, very little work has been done along these lines.

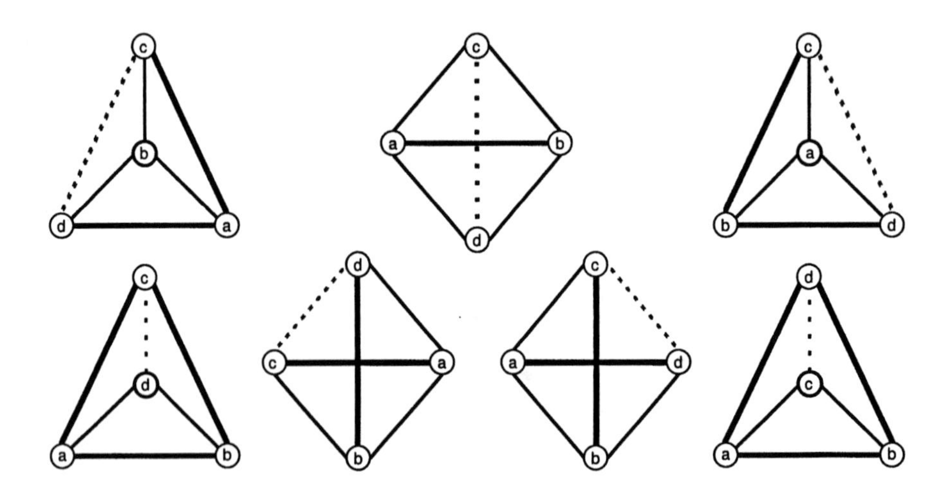

Figure 2: Diagrams depicting the combinations of bounds at which the various possible tetrangle inequality limits are attained among four points. The light solid lines represent upper bounds, while the heavy solid lines represent lower bounds, and the dashed lines indicate upper (top row) or lower (bottom row) tetrangle inequality limits.

THE TENSEGRITY CONNECTION

The connection between bound smoothing and tensegrity frameworks lies in the fact that the diagrams shown in Fig. 2 constitute *globally rigid tensegrity frameworks*, as defined by Connelly [3]. The upper bounds correspond to the cables, and the lower bounds to the struts, of the tensegrity, while the upper and lower limits correspond to struts and cables, respectively (reversed from the correspondence with the bounds). Because globally rigid tensegrities are by definition *uniquely realizable* (up to translation and rotation), not only is it impossible to make the distance between any pair of points connected by a cable or strut larger or smaller, respectively, but it is impossible to change the distance between any pair at all, at least without "breaking" some cable or strut, no matter how drastically the points are moved around! This means, in particular, that the distance between a given pair connected by a cable is an implicit *lower* limit on the possible values of that distance in any point configuration wherein the other cables and struts all remain intact, even if the cable between the given pair is "cut". Similarly, the distance between a given pair connected by a strut is an implicit *upper* limit on its possible values, even if the strut between the given pair is taken out so that the framework as a whole becomes flexible.

The process of iterating over all quadruples in order to obtain the limits determined by the bounds among all quadruples can also be interpreted in terms of tensegrity frameworks, namely as an operation known as *tensegrity exchange.* This consists of forming one rigid tensegrity out of another by merging a pair of points from each into a single pair (or equivalently, by connecting the pairs by two cables of zero length), where the pair is connected by a strut in one tensegrity and by a cable in the other (see Fig. 3), and then eliminating both the strut and the cable from the combined framework. We observe that the combined framework can be flexed by rotating one framework with respect to the other about the line connecting the merged pair, and that even if the combined framework is confined to the plane, it admits another realization given by reflection of one of the two frameworks in that line. Nevertheless, since these motions do not change any of the distances within the two frameworks, the cables and struts of the combined tensegrity still correspond to implicit lower and upper limits on their distances, respectively.

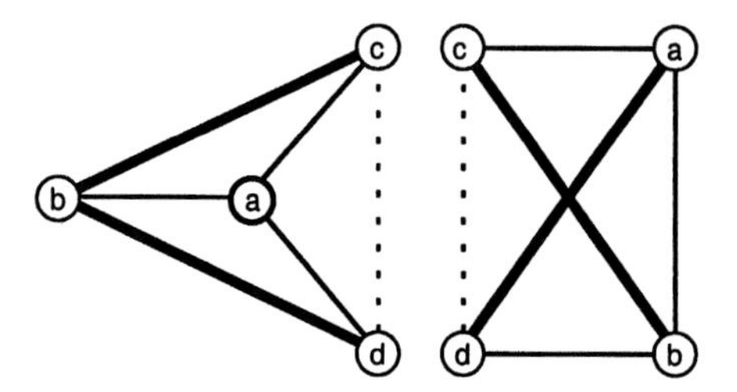

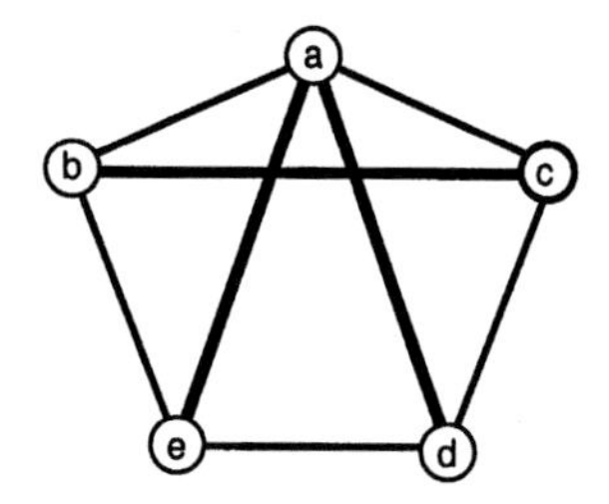

Figure 3: Diagrams illustrating the operation of tensegrity exchange (left), in which the two tensegrities shown are merged along the line $\overline{cd}$, and the Cauchy pentagon (right), which is the simplest example of a basic tetrangle limit on more than four points; see text for details.

This correspondence between the globally rigid tensegrities and the limiting combinations of bounds that must be considered in bound smoothing is important because it enables us to show that the tetrangle limits among each subset of four points separately does *not* determine the tetrangle inequality limits as a whole. This is in distinct contrast to the case with the triangle inequality limits, where subpaths of shortest paths are also shortest paths. The counterexample is a globally rigid tensegrity known as the Cauchy pentagon (Fig. 3), which is planar so that all the four-point Cayley-Menger determinants vanish together, and yet the $(1,5)$-distance is not contained in any of the four-point tensegrities shown in Fig. 2 (even after adding the implicit (b,d) and (c,e) struts). The Cauchy pentagon, moreover, is but one of an infinite sequence of convex planar polygonal globally rigid tensegrities, none of which properly contain a globally rigid subtensegrity. It follows that finding the true tetrangle limits among a set of points will require us to examine all subsets of points, so that the time required is (at least) exponential in the number of points.

Thus further progress in bound smoothing will require either that we find a set of globally rigid tensegrities whose combinatorics is in some sense nice, so that the limits determined by all possible tensegrity exchanges among them can be efficiently enumerated. Alternatively, one could use the fact that in three dimensions, the five-point and higher Cayley-Menger determinants must vanish, thereby providing two-sided inequalities on their values and hence completely new relations among the distances (for applications of these two-sided relations to the problem of embedding points on the line subject to distance and order constraints, see [11]). In addition, the lower bounds derived from the hard-sphere radii of the atoms limit the volume in which the atoms can be packed, so if this volume can be bounded in terms of the upper bounds on the distances among the atoms then one could in principle obtain some large lower limits among the distances. The inability of bound smoothing procedures based only on the *in*equalities among Cayley-Menger determinants (as above) to determine any lower limits larger than the largest lower bound seriously restricts the utility of these procedures in chemical applications.

ENERGY FORMS AND GLOBAL RIGIDITY

Connelly's proofs of the global rigidity of tensegrity frameworks are based upon the properties of certain quadratic forms, which he calls *energy forms* [3]. Let $\big(\mathbf{p} : \mathcal{A} \to \Re^n, \mathrm{L} \subseteq \binom{\mathcal{A}}{2}, \mathrm{U} \subseteq \binom{\mathcal{A}}{2}\big)$ be a tensegrity framework, where $\mathbf{p}$ gives the positions of the joints and L, U are the struts and cables, respectively. A *self-stress* $s \mapsto s(\alpha,\beta) : \binom{\mathcal{A}}{2} \to \Re$ is defined to satisfy

$$\forall\, \beta \in \mathcal{A} : \quad \sum\nolimits_{\alpha \in \mathcal{A} \setminus \beta} s(\alpha,\beta)(\mathbf{p}(\alpha) - \mathbf{p}(\beta)) \;=\; \mathbf{0}\,, \tag{17}$$

where:

$$s(\alpha,\beta) \begin{cases} \le 0 & \text{if } \{\alpha,\beta\} \in \mathrm{L} \setminus \mathrm{U} \\ \ge 0 & \text{if } \{\alpha,\beta\} \in \mathrm{U} \setminus \mathrm{L} \\ = 0 & \text{if } \{\alpha,\beta\} \notin \mathrm{L} \cup \mathrm{U} \end{cases} \tag{18}$$

The corresponding energy form is given by

$$E_{\mathbf{S}} \;\equiv\; \sum\nolimits_{\binom{\mathcal{A}}{2}} s(\alpha,\beta)\, D(\alpha,\beta) \;=\; \mathrm{Tr}(\mathbf{P}^\top \mathbf{S} \mathbf{P})\,, \tag{19}$$

where $\mathbf{P}^\top = [\mathbf{p}^\top(\mathrm{a}), \ldots, \mathbf{p}^\top(\mathrm{z})]$ and the *stress matrix* is defined by:

$$\mathbf{S} \;=\; [S_{\alpha\beta}]\,, \quad S_{\alpha\beta} \;\equiv\; \begin{cases} -s(\alpha,\beta) & \text{if } \alpha \ne \beta \\ \sum_{\mathcal{A}\setminus\alpha} s(\alpha,\beta) & \text{otherwise} \end{cases} \tag{20}$$

This form has the following straightforward but important properties [3]:

1. The matrix $\mathbf{P}$ of coordinates is a critical point of $E_{\mathbf{S}} = E_{\mathbf{S}}(\mathbf{P})$.
2. At any such critical point $\mathbf{P}$, the form $E_{\mathbf{S}}(\mathbf{P}) = 0$ vanishes.
3. The set of critical points of $E_{\mathbf{S}}$ is invariant under affine transformations.
4. The dimension of the affine span of any tensegrity framework $(\mathbf{P}, \mathrm{L}, \mathrm{U})$ with s as a self-stress is less than $\mathrm{null}(\mathbf{S})$.
5. Any tensegrity framework $(\mathbf{P}, \mathrm{L}, \mathrm{U})$ with s as a self-stress and $\dim(\mathbf{P}) < \mathrm{null}(\mathbf{S}) - 1$ is related to a full-dimensional one by an orthogonal projection.

Moreover, is easily seen from the signs of the stresses that $E_{\mathbf{S}}(\mathbf{P}') > E_{\mathbf{S}}(\mathbf{P})$ implies that some cable has grown longer, or some strut shorter, in $\mathbf{P}'$ than it was in $\mathbf{P}$. Since if the quadratic form $E_{\mathbf{S}} = E_{\mathbf{S}}(\mathbf{P})$ is positive semi-definite, every critical point is necessarily a global minimum, in this case any other realization $\mathbf{P}'$ of a full-dimensional tensegrity framework $(\mathbf{P}, \mathrm{L}, \mathrm{U})$ must be related to $\mathbf{P}$ by a (possibly singular) affine transformation. Thus if the affine flexings can be eliminated, the tensegrity framework is uniquely realizable, and hence globally rigid as above.

TENSEGRITY AND THE DISTANCE CONE

Now let us reconsider these same results from the perspective of the distance geometry. I will first note that we may view the energy form $E_{\mathbf{S}} = E_{\mathbf{S}}(\mathbf{D})$ as a *linear* form on distance space, rather than a quadratic form on the coordinates. Second, I note that Schoenberg's quadratic form *is* Connelly's energy form relative to the stress matrix $\mathbf{S} \equiv \mathbf{t}^\top\mathbf{t}$, regarded now as a function of $\mathbf{t} \in \Re^N$ with $\sum_\alpha t_\alpha = 0$ (see Eq. (3)). Third, minimization of a positive semi-definite energy form yields a hyperplane tangent to the distance cone at the minimum $\mathbf{D}$, whose normal is exactly the stress matrix $\mathbf{S}$ (with its diagonal set to zero). The intersection of a tangent hyperplane with the distance cone is a face of the cone, which is maximal exactly when the rank of the corresponding stress matrix is $N - 1$. More generally, any positive semi-definite stress matrix can be decomposed into a sum of rank one positive semi-definite matrices (canonically, using an eigenvalue decomposition). Since minimizing the energy form with respect to $\mathbf{D} \in \mathcal{D}_{N-1}^N$ yields a value of zero, it follows that $\mathbf{D}$ is also a minimum of the forms determined by each of its rank one components. Each of these rank one components, in turn, (canonically) defines a hyperplane at $\mathbf{D}$, and the joint intersection of these hyperplanes with $\mathcal{D}_{N-1}^N$ is the maximal face of $\mathcal{D}_{N-1}^N$ containing $\mathbf{D}$.

Finally, the signs of any given stress determine the struts L and cables U of a tensegrity (struts for negative signs, cables for positive). They also restrict the orientations of the hyperplanes determined by the stress, and hence what faces of the distance cone the hyperplanes can be tangent to. Conversely, the set of all hyperplanes tangent to the full distance cone at a given face will be consistent only with certain sets of struts and cables, and in particular only with certain minimal sets thereof. It follows that there exist relations between the sets of struts and cables of tensegrities which admit positive semi-definite stress matrices, and the facial structure of the distance cone. This imparts a combinatorial structure to the cone, in addition to the partial ordering induced by inclusion among the faces and the above-mentioned oriented matroid structure. In the hope of stimulating research on these matters, I will pose a couple of specific questions:

1. Does the dual cone of $\mathcal{D}_{N-1}^N$ consist exactly of the positive semi-definite stress matrices?

2. Is the manifold of faces whose normals have a fixed sign pattern (simply) connected?

It should be noted that the relation between the signs of the self-stresses in a given realization of a tensegrity and the signs of the oriented volumes spanned by its joints (which determine an oriented matroid) have received some attention already [34], and that it is known that the realization spaces of oriented matroids are not connected in general [2].

SOME POSSIBLE EXTENSIONS OF RIGIDITY THEORY

Cayley-Menger determinants can be interpreted geometrically as the Gramians of sets of null vectors in a Minkowski space $\mathcal{V}(1, n+1)$ two dimensions higher than the Euclidean space in which the corresponding distance matrix can be embedded [12, 19, 30]. This same construction has been utilized to construct an isomorphism of the conformal group $\mathcal{C}(n)$ with the group of isometries $O(1, n+1)$ [24]. In this construction, the null vector corresponding to the row/column of one's in the matrix of Cayley-Menger determinants constitutes the "point at infinity" from inversive geometry, and the remaining (Euclidean) points of the determinant necessarily lie on an affine hyperplane parallel to it. More generally, one may regard a vector $\mathbf{v}(\mathrm{a}) \in \mathcal{V}(1, n+1)$

with positive square as a *sphere* of radius $r^2(\mathrm{a}) = \mathbf{v}(\mathrm{a}) \cdot \mathbf{v}(\mathrm{a})$ in Euclidean space, and an inner product of a pair of such vectors as the "power" of the spheres, namely

$$\mathbf{v}(\mathrm{a}) \cdot \mathbf{v}(\mathrm{b}) \;=\; \tfrac{1}{2}\left(r^2(\mathrm{a}) + r^2(\mathrm{b}) - d^2(\mathrm{a},\mathrm{b})\right) \;, \tag{21}$$

(for an elegant exposition, see Pedoe [26]). This extension of Euclidean point geometry to *Möbius sphere geometry* invites a collateral the extension of rigidity theory to the same geometry, in which one would attempt to discover conditions under which the specification of the powers of certain pairs of spheres locally determined their arrangement (up to conformal transformation).

A second direction in which I believe rigidity theory might profitably be extended is to the kinematics of multiple two-state quantum systems. This is of central importance in "quantum computing", where the systems are usually spin $= \frac{1}{2}$ particles (see [31, 35]). Although a two-state system is generally described by a "spinor" $[a, b]^\top \in \mathcal{C}^2$ with $a = \cos(\vartheta/2)e^{\imath\varphi/2}$ and $b = \sin(\vartheta/2)e^{-\imath\varphi/2}$, this can also be thought of as a unit vector in a three-dimensional Euclidean space, via the identification

$$\begin{bmatrix} a \\ b \end{bmatrix} \begin{bmatrix} a^\dagger & b^\dagger \end{bmatrix} \;=\; \tfrac{1}{2} \begin{bmatrix} 1+\cos(\vartheta) & \sin(\vartheta)e^{\imath\varphi} \\ \sin(\vartheta)e^{-\imath\varphi} & 1-\cos(\vartheta) \end{bmatrix} \;\equiv\; \tfrac{1}{2} \begin{bmatrix} 1+z & x+\imath y \\ x-\imath y & 1-z \end{bmatrix} \;. \tag{22}$$

The state of the joint (combined) system, however, is given by a direct product of these vectors ... which is not generally decomposable into a product of well-defined states for the individual particles. Such states are referred to as "entangled", and their physical interpretation is among the deepest mysteries of quantum mechanics. The natural group action on the joint system is the direct product of the rotation groups of the individual vectors, which preserves the (non)decomposability of the joint state. The question of whether two joint states are in the same orbit or not has been addressed using invariant theory [16], but even for a single pair of spin $\frac{1}{2}$ particles, a total of 21 (!) algebraically independent invariants have been found so far. Extending these results to larger spin systems is challenging, but perhaps a "local" theory, analogous to the theory of infinitesimal / static rigidity in frameworks, will be more tractable.

Acknowledgements

This work was supported by NSF grant BIR 9511892. I would like to thank the mathematicians working in rigidity theory, in particular Bob Connelly, Henry Crapo, Neil White and Walter Whiteley, for enriching my career as an applied scientist with their results.

References

[1] L. M. Blumenthal. *Theory and Applications of Distance Geometry.* Cambridge Univ. Press, Cambridge, U.K. (1953). Reprinted by the Chelsea Publishing Co. (1970).

[2] J. Bokowski and B. Sturmfels. *Computational Synthetic Geometry.* Lect. Notes Math. 1355. Springer-Verlag, Berlin, F.R.G. (1989).

[3] R. Connelly. *Invent. Math.*, 66:11–33 (1982).

[4] R. Connelly and A. Back. *Am. Sci.*, 86:142–151 (1998).

[5] R. Connelly and M. Terrell. *Structural Topology*, 21:59–78 (1995).

[6] R. Connelly and W. Whiteley. *SIAM J. Discrete Math.*, 9:453–491 (1996).

[7] G. M. Crippen and T. F. Havel. *Distance Geometry and Molecular Conformation.* Research Studies Press, Taunton, U.K., ISBN 0-86380-073-4 (1988).

[8] J. P. Dalbec. *Ann. Math. Artif. Intel.*, 13:97–108 (1995).

[9] A. W. M. Dress and T. F. Havel. *Adv. Math.*, 62:285–312 (1986).

[10] A. W. M. Dress and T. F. Havel. *Discrete Applied Math.*, 19:129–144 (1988).

[11] A. W. M. Dress and T. F. Havel. *SIAM J. Disc. Math.*, 4:535–549 (1990).

[12] A. W. M. Dress and T. F. Havel. *Found. Phys.*, 23:1357–1374 (1993).

[13] P. L. Easthope and T. F. Havel. *Bull. Math. Biol.*, 51:173–194 (1989).

[14] P. Floersheim, K. Wirth, M. K. Huber, D. Pazis, F. Siegerist, H. R. Haegi, and A. S. Dreiding. *Symmetries and Properties of Non-Rigid Molecules*, volume 23 of *Studies in Physical and Theoretical Chemistry, ed. A. Maruani and J. Serre.* Elsevier Scientific Publ. Co., Amsterdam, Holland (1983).

[15] J. C. Gower. *Linear Algebra Appl.*, 67:81–97 (1985).

[16] M. Grassl, M. Rötteler, and T. Beth. *Phys. Rev. A* (1998), in press (available from `http://xxx.lanl.gov/abs/quant-ph/9712040`).

[17] T. F. Havel. In D. Nobel and T. L. Blundell, editors, *Progress in Biophysics and Molecular Biology*, volume 56, pages 43–78. Permagon Press, Oxford, England (1991).

[18] T. F. Havel. *J. Symbolic Comput.*, 11:579–593 (1991).

[19] T. F. Havel. In *Invariant Methods in Discrete and Computational Geometry*, pages 245–256. Kluwer Academic, Amsterdam (1995).

[20] T. F. Havel. In *Encyclopedia of Nuclear Magnetic Resonance*, pages 1701–1710. J. Wiley & Sons (1996).

[21] T. F. Havel. *Lect. Notes in Artif. Intellig.*, 1360:102–114 (1997).

[22] T. F. Havel. In *Encylopedia of Computational Chemistry.* J. Wiley & Sons (1998). In press.

[23] T. L. Hayden, J. Wells, W.-M. Liu, and P. Tarazaga. *Lin. Alg. Appl.*, 144:153–169 (1991).

[24] D. Hestenes. *Acta Appl. Math.*, 23:65–93 (1991).

[25] K. Menger. *Math. Annal.*, 100:75–163 (1928).

[26] D. Pedoe. *A Course of Geometry for Colleges and Universities.* Cambridge Univ. Press, Cambridge, England (1970).

[27] I. J. Schoenberg. *Ann. Math.*, 36:724–732 (1935).

[28] I. J. Schoenberg. *Annals Math.*, 38:787–793 (1937).

[29] I. J. Schoenberg. *Nederl. Akad. Wetensch. Proc. Ser. A (= Indag. Math.)*, 72 (=31):43–63 (1969).

[30] J. J. Seidel. *Simon Stevin*, 29:32–50, 65–76 (1952).

[31] S. S. Somaroo, D. G. Cory, and T. F. Havel. *Phys. Lett. A*, 240:1–7 (1998).

[32] P. Tarazaga, T. L. Hayden, and J. Wells. *Lin. Alg. Appl.*, 232:77–96 (1996).

[33] G. Wagner, S. Hyberts, and T. F. Havel. *Ann. Rev. Biophys. Biomol. Struct.*, 21:167–198 (1992).

[34] N. L. White and W. Whiteley. *SIAM J. Alg. Discrete Meth.*, 4:481–511 (1983).

[35] C. P. Williams and S. H. Clearwater. *Explorations in Quantum Computing.* Springer-Verlag, New York, NY (1998).

COMPARISON OF CONNECTIVITY AND RIGIDITY PERCOLATION

Cristian F. Moukarzel and Phillip M. Duxbury

Department of Physics/Astronomy and
Center for Fundamental Materials Research,
Michigan State University,
East Lansing, MI 48824-1116

1. INTRODUCTION

Connectivity percolation has devotees in mathematics, physics and in myriad applications [1-4]. Rigidity percolation is a more general problem, with connectivity percolation as an important limiting case. There is a growing realization that the general rigidity percolation problem exhibits a broader range of fundamental phenomena and has the potential for many new applications of percolation ideas and models. In connectivity percolation, the propagation of a scalar quantity is monitored, while in rigidity the propagation of a vector is, in general, considered. In both cases, one or more conservation laws hold. Moreover, connectivity percolation is rather special and appears to be, in many cases, quite different than other problems in the general rigidity class. We illustrate these differences by comparing connectivity and rigidity percolation in two cases which are very well understood, namely diluted triangular lattices [5,6] and trees [7].

A large part of the intense fundamental interest in percolation is due to the fact that percolation is like a phase transition, in the sense that there is a critical point (critical concentration) and non-trivial scaling behavior near the critical point [1-4]. This analogy carries over to the rigidity percolation problem. However, although the connectivity percolation problem is usually second order (including on trees), the rigidity percolation problem is first order in mean field theory and on trees [7]. However on triangular lattices, both connectivity and rigidity percolation are second order, though they are in *different universality classes* [5,6]. The emphasis of most of the analysis in the literature and in this presentation is the behavior at and near the critical point.

In this paper, we discuss (Section 2) ideas which apply to connectivity and rigidity percolation on diluted lattices. In Section 3, we discuss and compare the specific case of connectivity and rigidity percolation on trees. In Section 4 we summarize the matching algorithms which may be used to find the percolating cluster in both connectivity and rigidity cases. The behavior in the connectivity and rigidity cases on site diluted triangular lattices is then compared. Section 5 contains a summary and discussion of

Rigidity Theory and Applications
Edited by Thorpe and Duxbury, Kluwer Academic/Plenum Publishers, 1999

the similarities and differences between connectivity and rigidity percolation in more general terms.

2. RIGIDITY AND CONNECTIVITY OF RANDOM GRAPHS

In mathematical terms, percolation is the study of the connectivity properties of random graphs as a function of the number of edges in the graph. We are usually interested in the asymptotic limit of graphs with an infinite number of vertices. In the physics community, we usually study this process on a regular lattice (e.g. square, simple cubic, triangular) and consider the effect of adding or removing edges which lie between the vertices of these graphs. However the mean field limit is equivalent to considering connections between all sites on the lattice, no matter their Euclidean separation. The behavior of this "infinite range" model is equivalent to the limit of infinite dimensional lattice models (e.g. infinite dimensional hypercubes) and tree models. The reason for the equivalence of the critical behavior of these models is that they are all dominated by long range "rings", whereas on regular lattices in lower dimensions short range rings can be very important. Field theory models seek to add short range loops to these models in a systematic and non-perturbative (in some sense) way.

We consider lattices consisting of sites which are defined to have z neighbors (i.e. they have coordination number z). We then add edges between neighbors randomly with probability p. Once this process is finished we are left with a random graph with each site having average coordination $C = zp$. In studying percolation, we are always asking the question "Is it possible to transmit some quantity (i.e. a scalar, or a vector) across a graph". If it is possible to transmit the quantity of interest, we say that the network is above the "percolation threshold" p_c (or equivalently $C_c = zp_c$) for that quantity. If it is impossible to transmit the quantity of interest we are below p_c. At p_c the part of the random graph which transmits the quantity of interest is *fractal* **provided** the percolation transition is second order. The probability that a bond is on this "percolating backbone" is P_B, and is one of the key quantities in the analysis of percolation problems on regular lattices. If the percolation transition is "second order",

$$P_B \sim (p - p_c)^{\beta'} \quad \text{as} \quad p \to p_c^+ \tag{1}$$

with $\beta' > 0$. On trees it is more difficult to define P_B. Nevertheless we are able to analyze the problem effectively using "constraint-counting" ideas [7].

Let us for a moment put aside thinking in terms of percolation and instead develop constraint counting ideas originally discussed by Maxwell, and which have been developed extensively in the engineering, math and glass communities [7,8]. These ideas have not been applied to connectivity percolation till recently and are enriching to both the connectivity and rigidity cases. At a conceptual level constraint counting is deceptively simple. To illustrate this, we assign to each site of our lattices a certain number of *degrees of freedom.* In connectivity percolation, the transmission of a scalar quantity is of interest, therefore each site is assigned one degree of freedom. However if we consider a point object in two dimensions from the point of view of the transmission of forces, it has two degrees of freedom (two translations). In d dimensions point masses have d degrees of freedom. However extended objects have rotational degrees of freedom, so that when we have clusters which are mutually rigid, we must also consider these rotational degrees of freedom. We call such objects bodies and they have $d(d+1)/2$ degrees of freedom. From a model viewpoint, we allow the number of degrees of freedom of a free site to be a control variable which we label g, with $g = 1$ the connectivity case. With equal generality we may say that G is the number of degrees of freedom of a rigid cluster or body. It is easy to see that there is a vast array of models with $g \neq 1$, and with $G \neq 1$ and much of the physics and applications of these models have yet to be examined [7].

Now we have a model in which a free site has g degrees of freedom, so that a lattice with N sites(and no edges) has a total of $F = Ng$ degrees of freedom, or "floppy modes" (i.e. modes which have zero frequency due to the fact that there is no restoring force). Constraint counting consists in simply saying that each time an *independent* edge is added to the lattice, the number of degrees of freedom (zero frequency modes) is reduced by one if this edge is not "wasted" (see later). For example, the minimum number $E_{\mathbf{min}}$ of constraints needed to make a rigid cluster out of a set of N sites is

$$E_{\mathbf{min}} = Ng - G, \tag{2}$$

which holds for the case in which edges are put on the graph in such a way that none is wasted. In general, if E edges are added to the graph, the number of degrees of freedom (or floppy modes) which remains is

$$F = Ng - E + R. \tag{3}$$

Note the additional term R on the right hand side. This term is key in understanding the relation between constraint counting and percolation, and in finding algorithms for rigidity percolation. R is the number of "redundant bonds" or "wasted" edges which are added to the lattice. An edge does not reduce the number of floppy modes if it is placed between two sites which are already mutually rigid, in which case this edge is "redundant". The simplest examples of subgraphs containing a redundant bond on a triangular lattice are illustrated in Fig. 1 for the connectivity ($g = 1$) and $g = 2$ rigidity cases.

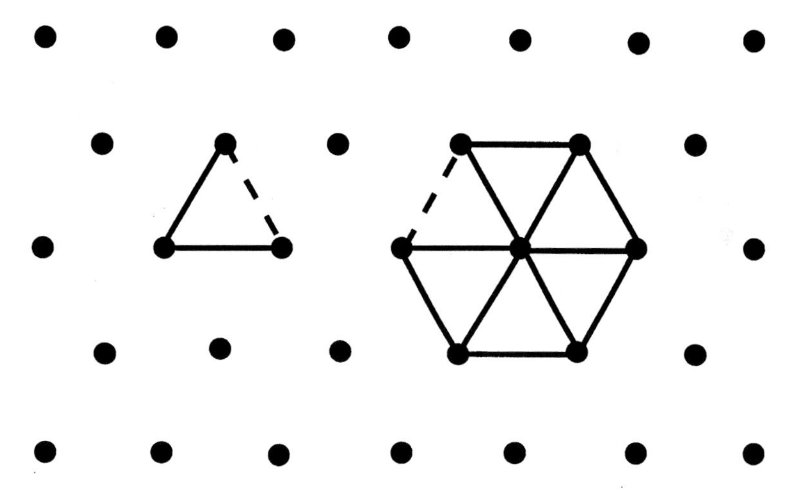

Figure 1: The simplest subgraphs on a triangular lattice which contain a redundant bond (dashed). Connectivity case (left), $g = 2$ rigidity case (right).

Note that any one of the bonds in these structures could be labeled as the redundant one. However once any one of them is removed, all of the others are necessary to ensure the mutual rigidity of the structure. In fact the set of all bonds which are mutually redundant form an "overconstrained" or "stressed cluster". This is because, in the rigidity case we can think of each edge as being a central force spring, which means that there is a restoring force only in tension and compression. Then an overconstrained cluster of such springs (with random natural lengths) is internally stressed due to the redundant bond. In the connectivity case each bond is like a wire which can carry current or fluid flow. The simplest overconstrained cluster is then a loop which can support an internal "eddy" current. Rigid structures which contain no redundant bonds are minimally rigid or "isostatic". In connectivity percolation isostatic structures are trees, whereas in ($g > 1$) rigidity percolation isostatic structures must always contain loops (see Fig. 1).

In percolation problems, we are interested in the asymptotic limit of very large graphs ($N \to \infty$), and it is more convenient to work with intensive quantities, so we define $f(p) = F/gN$ and $r(p) = 2Rg/zN$ which leads to

$$f(p) = 1 - \frac{z}{2g}(p - r(p)), \tag{4}$$

where the number of edges $E/N = zp/2$. The normalization on $r(p)$ is chosen this way because $r(p)$ is now the probability that a bond is redundant (times g). Note that we normalize the number of floppy modes by g to be consistent with previous work [5-7]. We have shown that $f(p)$ acts as a free energy for both connectivity and rigidity problems, so that if $\partial f(p)/\partial p$ undergoes a jump then the transition is first order. The behavior of this quantity is directly related to the probability that a bond is overconstrained P_{ov} via the important relation [6]

$$\frac{\partial f}{\partial p} = -\frac{z}{2g}(1 - P_{ov}) \tag{5}$$

If the transition is second order, the second derivative $\partial^2 f/\partial p^2 \sim (p - p_c)^{-\alpha}$, where α is the specific heat exponent [6].

On both triangular lattices and on trees, we also calculate the infinite cluster probability. This is composed of the backbone plus the dangling ends. Dangling ends are rigidly connected to the backbone but *do not* participate in the transmission of the quantity of interest. Examples of dangling ends in the connectivity and rigidity cases are illustrated in Fig. 2.

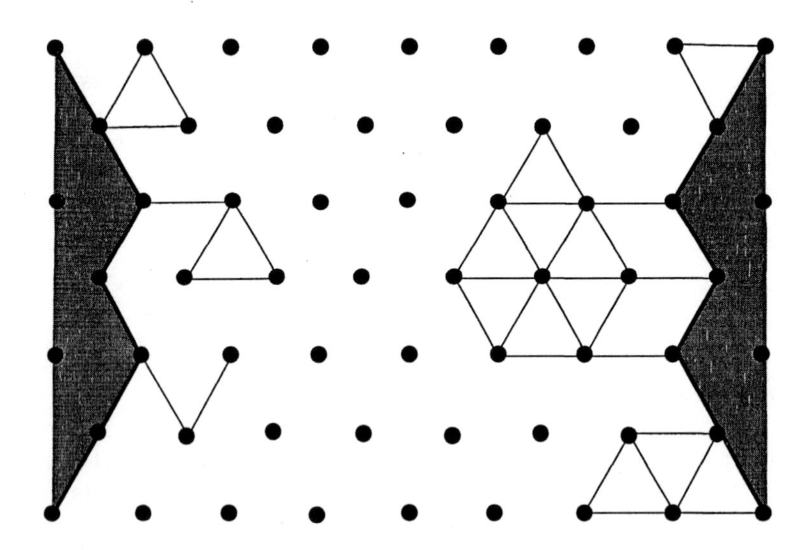

Figure 2: Examples of dangling ends (thin lines) connected to a backbone (shaded). Connectivity case (left) and $g = 2$ rigidity case (right).

3. TREES

In our previous analysis of tree models, we have considered propagation of rigidity outward from a rigid boundary (this is the same as a strong surface field). Here we also add a bulk field which induces rigidity. The results are essentially the same whether the boundary or the bulk field is used. In percolation problems, the bulk field corresponds to nuclei embedded in the lattice and tethered to a rigid substrate. This is the sort of construction envisioned by Obukov [9], and used extensively in the construction of models for connectivity percolation. The mean-field equation for the order parameter on trees is then,

$$T_0 = h + (1-h)\sum_{l=g}^{z-1}\binom{z-1}{l}(pT_0)^l(1-pT_0)^{z-1-l}, \tag{6}$$

where T_0 is the probability that a site is connected to the infinite rigid cluster through one branch of a tree (see the paper by Leath for more details on the derivation), and h is the probability that a site is rigidly connected to the rigid substrate (h is sometimes called a "ghost field"). On trees, the probability that a bond is overconstrained is,

$$\frac{N_0}{N_B} = T_0^2 \tag{7}$$

so that we have the key equation,

$$\frac{\partial f}{\partial p} = -\frac{z}{2g}(1-T_0^2). \tag{8}$$

Other useful formulae are the probability that a bond is overconstrained,

$$P_{ov} = T_0^2, \tag{9}$$

and the probability that a bond is on the infinite cluster,

$$P_{inf} = T_0^2 + 2T_0T_1. \tag{10}$$

In the connectivity cases, an infinitesimal field h or any finite order parameter at the boundary is sufficient to allow a percolation transition to occur on trees. In contrast in $g \neq 1$ cases, there must be a finite h, or a finite boundary field before a transition occurs. These differences are illustrated in Fig. 3.

It is clearly seen from these figures that the rigidity transition is first order on trees, while the connectivity one is second order. The results presented in Fig. 3 are simple to obtain. We iterate the mean field equation (6) until a stable fixed point is reached (there are similar equations for T_1 etc.) and we then evaluate Eqs. (7) - (9). We identify the point at which the stable solution becomes nonzero as p_s, the spinodal point, for reasons discussed below. In the connectivity case $p_s = p_c$ because the transition is second order.

We also want to find the total number of redundant bonds $r(p)$ and the total number of floppy modes $f(p)$. In order to find these quantities, we integrate Eq. (7) and then use Eq. (3). However, the integration of Eq. (7) leads to one free constant. This constant depends on the situation we wish to model. If we wish to model a regular lattice, then we impose the constraint,

$$r(1) = 1 - \frac{2g}{bz}, \tag{11}$$

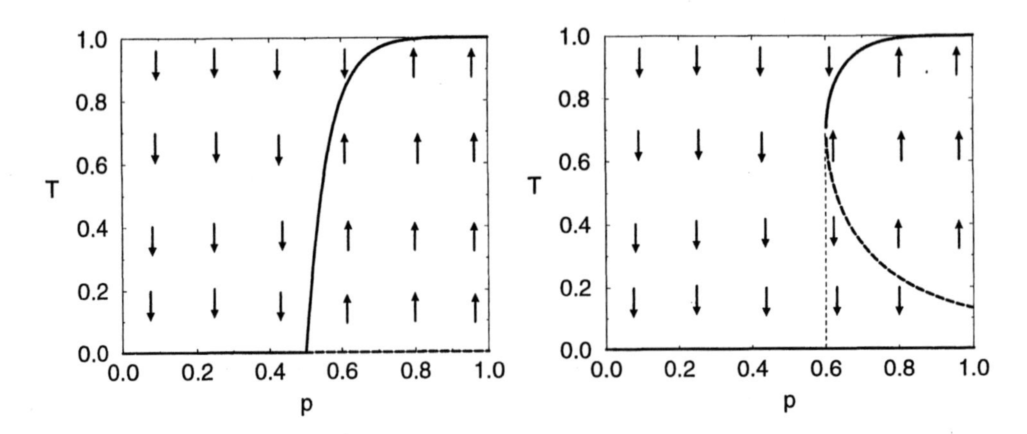

Figure 3: Domains of attraction of the mean field equations (6) with $h = 0$: (left) A typical connectivity behavior (this example is $z = 3$); and (right) a typical rigidity behavior (this example is $z = 6$, $g = 2$). Dark lines are stable (attractive) solutions and shaded lines repulsive.

for example in the case of central force springs on a triangular lattice 1/3 of the springs are redundant when $p = 1$. However on trees,

$$r(1) = 1 - \frac{g}{b(z-1)}. \tag{12}$$

In using tree models to provide approximations to rigidity percolation on regular lattices, we impose the constraint (11) [7]. Then we find that r(p) approaches zero at a critical point p_c, which is NOT the same a p_s if $g > 1$. Moreover, if $g > 1$, this p_c is very close to the Maxwell estimate $p_c^{\text{Maxwell}} = 2g/bz$, and is *the same* as that found numerically for infinite range (random bond) models [7]. Typical results found for tree models are presented in Fig. 4.

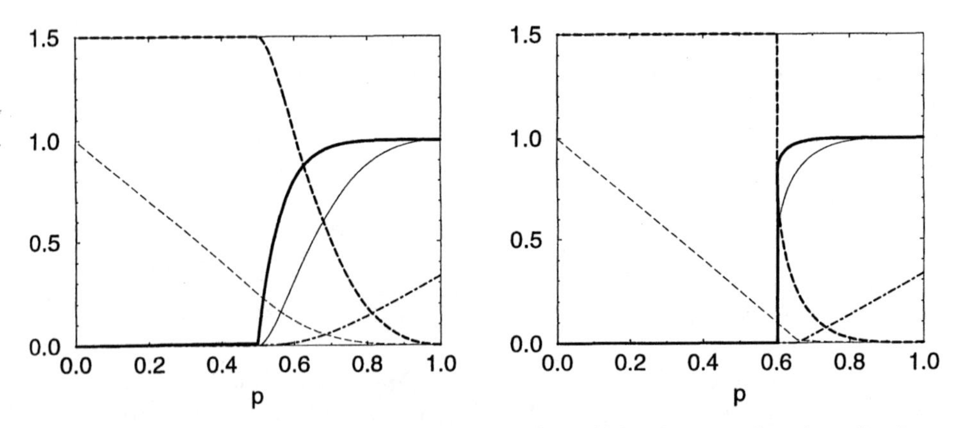

Figure 4: The order parameter, floppy modes and its derivative as a function of p for a typical connectivity $z = 3, g = 1$(left), and rigidity $z = 6, g = 2$(right) cases, both with $b = 1$. $f(p)$(thin dashed), $r(p)$(dot dashed), $-df/dp$(heavy dashed), P_{ov}(thin solid), P_{inf}(heavy solid)

In the connectivity case we find $p_c = p_s$, while in rigidity cases $p_s < p_c$. For the rigidity case of Fig. 4, $p_s = 0.605$, while $p_c = 0.655$.

4. TRIANGULAR LATTICES

We consider connectivity percolation($g = 1, b = 1, z = 6$) and central force percolation($g = 2, b = 1, z = 6$) on site diluted triangular lattices. In connectivity percolation the scaling behavior near the percolation threshold has been proven on trees and has been extensively tested on regular lattices using large scale numerical simulations. We have carried out a similar program for rigidity percolation as summarized below for triangular lattices. However doing numerical simulations in the rigidity case has required a breakthrough in algorithm development, and this has only occurred recently through contact with the mathematical computer science community. The methods developed for rigidity have even improved some aspects of algorithmic methods [12] for the connectivity case, as discussed below.

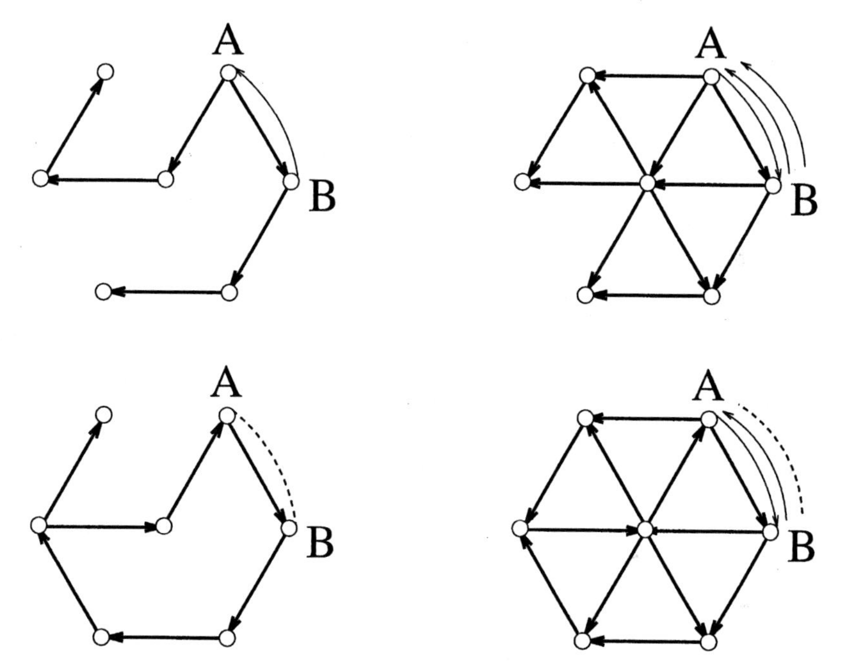

Figure 5: Examples of successful (top figures) and failed (bottom figures) matches in the connectivity ($g = G = 1$, left figures) and joint-bar rigidity ($g = 2, G = 3$, right figures) cases on triangular lattices. AB is the new bond that is being tested. Each site has g degrees of freedom and therefore accepts at most g incoming arrows. Each new bond carries with it G auxiliary arrows, which must also be matched. If any of the arrows cannot be matched (dashed), the new bond is redundant.

Before discussing the rigidity algorithms, it is important to point out the limitations of these methods. Firstly these algorithms are able to identify structures which can support stress (rigidity case), or which can carry a current (connectivity case). They do not find the actual current or stress, but rather those bonds which are able to transmit the load from an input node or set of nodes to an output node or set of nodes. In the connectivity percolation case it is trivial that the actual geometric realization of a given graph connectivity does not change the set of bonds which carry current from a fixed set of input nodes to a fixed set of output nodes. However in the rigidity case, there can be "special" realizations of a given graph which are responsible for singular rigid properties of "generically" rigid clusters. This may occur when there are "degenerate" constraints (e.g. parallel bonds on a lattice). The probability of such degenerate realizations is zero on geometrically disordered lattices and for this reason geometry, and the existence of these degenerate configurations, may be simply ignored in this case. In the mathematical community this problem is called "generic rigidity" and is the only one for which powerful algorithms exist. Thus the problem of rigidity on a *regular triangular lattice remains unsolved.* The results described below apply to triangular lattices whose sites have been randomly displaced (e.g. by 0.1 of a lattice constant). In addition, the powerful matching algorithms that we use apply to "joint-bar graphs in the plane" and to a subset of graphs in general dimensions (so-called body - bar problems) However, it turns out that glasses in $3-d$ correspond to a case which is solvable (they map to a body-bar problem), so this is one of those unusual cases where the practical case is actually theoretically convenient.

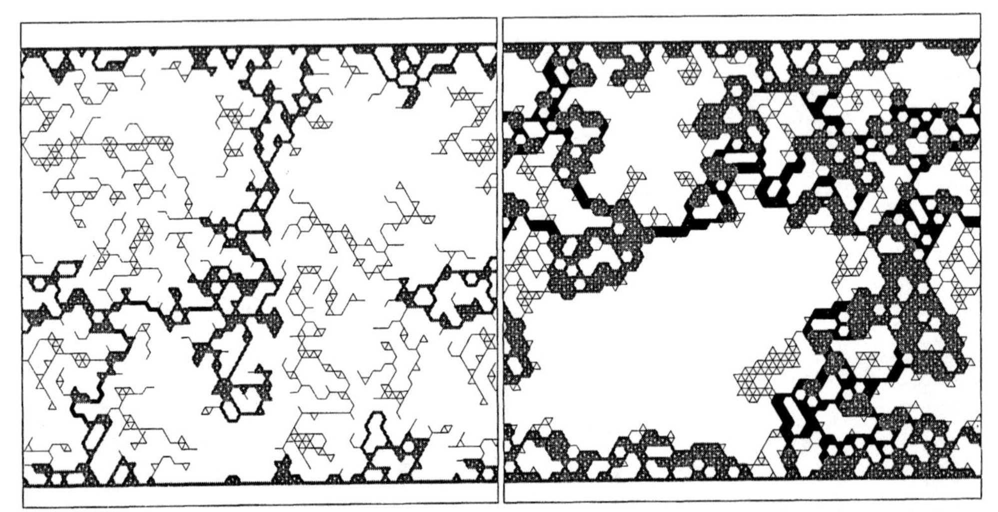

Figure 6: The infinite-cluster geometry for connectivity (left) and rigidity(right) percolation on site-diluted triangular lattices with $L = 64$. In each case dark wide lines are cutting bonds, wide lines are non-critical backbone bonds (blobs) and thin lines are dangling ends.

The matching algorithm is implemented as follows [10]:

> Start with an empty triangular lattice(no bonds) and assign to each node g degrees of freedom.
> *Then:*
>
> 1. Randomly add a bond to the lattice.
> 2. Test whether this bond is redundant with respect to the bonds which are currently in the lattice.
> 3. If the bond is redundant *do not add it to the lattice*, but instead store its location in a different array.
> 4. Return to 1.
>
> *End*

The key step is step 2. The algorithm to do this test is rigorous and based on Laman's theorem. It was developed by Bruce Hendrickson[10], who also provided a key service in explaining his algorithm to the physics community. Step 2. is performed by exact constraint counting. This is implemented by "matching" constraints (bonds) to degrees of freedom, with the restriction that the constraints can only match degrees of freedom at each end of the corresponding bond. Thus it is natural to represent this constraint counting by using arrows to indicate the degree of freedom to which each bond is matched. The idea is that, when a new bond is to be tested that bond and G copies are added to the lattice. If it is *impossible* to match these $G+1$ arrows to degrees of freedom of the graph then this new bond is redundant. If this task can be accomplished [1] the new bond is *independent*, which means that it is not wasted, and is left on the graph. In this case only G auxiliary arrows are removed. If the last edge is *redundant* all $G+1$ tested arrows are erased.

A successful and a failed match are illustrated in Fig. 5 for a connectivity (g=1) case and a rigidity (g=2) case. Note that the bond that is being tested carries with it G additional "copies" which account for the global degrees of freedom of a rigid cluster. In the connectivity case $G = 1$, while on central-force bar-joint networks in two dimensions $G = 3$.

A failed match occurs when a bond is unable to find a degree of freedom to "cover". This bond is then redundant. In trying to find a degree of freedom to which a redundant bond can be assigned or "matched", the algorithm identifies all bonds which are "overconstrained" of stressed with respect to that bond. This set of bonds is called a *Laman subgraph.* Note that if a redundant bond is already in a graph, it is not possible to add a new bond and test its redundancy with this method. This is the reason that the algorithm proceeds by adding bonds one at a time starting with an empty lattice. Any error in testing the redundancy of a bond invalidates the rest of the addition sequence. However since this algorithm is an integer method there is no problem with roundoff. It is easy to see that the matching algorithm is quite efficient, however it requires quite a bit of effort to fully optimize these methods.

Pictures of the infinite cluster for connectivity and rigidity percolation on a triangular lattice are presented in Fig. 6.

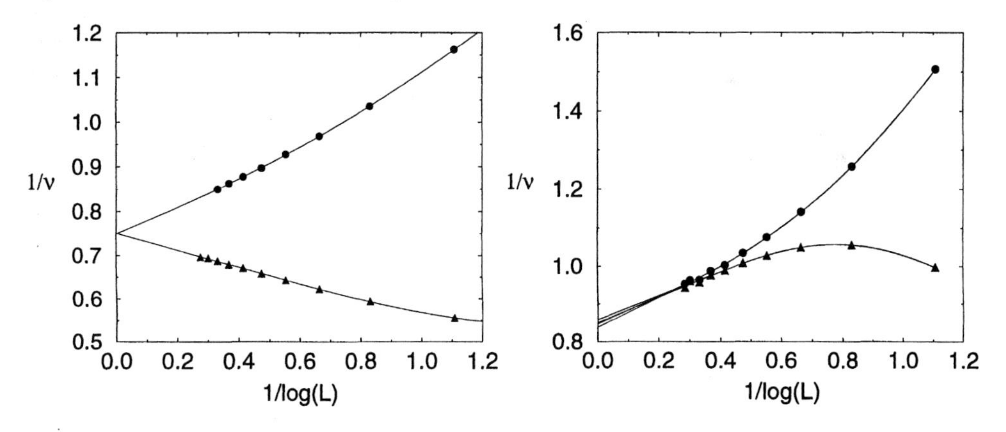

Figure 7: Analysis to find the correlation length exponent ν for connectivity (left figure) and rigidity percolation (right figure) on triangular lattices. In both figures the upper curve is a fit (including corrections to scaling) of $\delta p_c \sim L^{-1/\nu}$ (full circles) while the lower curve is for the number of cutting bonds $n_c \sim L^{1/\nu}$ (full triangles).

Because of the fact that we add bonds one at a time until the percolation point is reached, we are able to identify p_c *exactly* for each sample, and therefore measure the

[1] It is valid to rearrange previously existing arrows, provided this is done in such a way that all remain matched to some degree of freedom

components of the spanning cluster exactly at p_c. This eliminates the error associated with measurements at fixed values of p, since estimated exponents are known to depend very sensitively on p. This method was proposed and applied for the first time in ref. [5].

At p_c we identify three different types of bonds: *backbone bonds*, *dangling ends* and *cutting bonds*. These together form the *infinite cluster*. The cutting bonds are stressed (belong to the backbone), but they are "critical" because if one of them is removed, load is no longer transmitted across the infinite cluster. The results in Fig. 6 are for bus bars at the top and bottom of the figures and periodic boundary conditions in the other direction. In order to find the correlation length exponent, we used two relations: Firstly the size dependence of the threshold behaves as $\delta p_c \sim L^{-1/\nu}$ and secondly, the number of cutting bonds varies as $n_c \sim L^{1/\nu}$ [4]. An analysis of this data in the connectivity and rigidity cases is presented in Fig. 7.

It is seen from this figure that the rigidity case has a different exponent ($\nu = 1.16 \pm 0.03$ [5,6]) than the connectivity case ($\nu = 4/3$).. A finite-size-scaling plot of the infinite-cluster and backbone densities at p_c is presented in Fig. 8, along with the density of dangling ends.

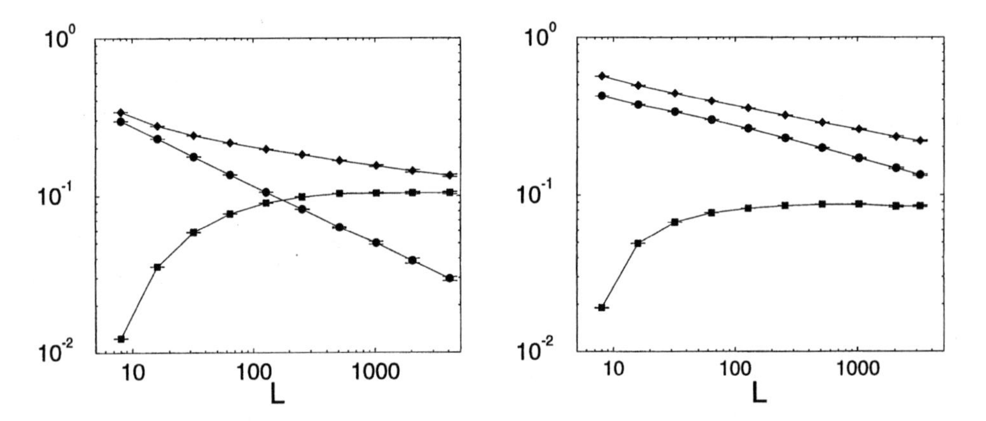

Figure 8: The density of backbone bonds(circles), infinite cluster bonds(diamonds) and dangling bonds(squares) for the connectivity ($g = 1$, left figure) and rigidity ($g = 2$, right figure) cases on site diluted lattices. Our method of single-bond addition allowed these measurements to be made *exactly* at p_c for each sample.

What is unambiguous from these figures is that the backbone density is decreasing algebraically $P_B \sim L^{-\beta'/\nu}$, and from this data we find that in the rigidity case $\beta' = 0.25 \pm 0.02$ [5,6]. It also appears that the infinite-cluster probability is decreasing algebraically, however that is difficult to reconcile with the behavior of the dangling ends. In the connectivity case much larger scale simulations have been done, and it is known that the algebraic decrease in the infinite cluster probability continues, so that the infinite cluster in the connectivity case is indeed fractal [4]. In the rigidity case, such large scale simulations are still lacking, so the possibility of an infinite cluster with finite density (about 0.1) remains.

5. SUMMARY

The development of tree models and the use of new algorithms in numerical studies have revolutionized our understanding of the geometry of rigidity percolation. In particular we now know that on triangular lattices, the rigidity transition is second order, but in a different universality class to the connectivity case. In contrast on trees the rigidity transition is first order while the connectivity transition is second order. The geometry of rigidity percolation is different from that of connectivity percolation due to the requirement of multiple connectivity in the rigidity case. However this is clearly not enough to ensure that the transition becomes first order. Perhaps a deeper question is whether the infinite cluster breaks up into two or an infinite number of subclusters when a critical bond is removed. In the connectivity case, the answer is clearly two. In the rigidity case it is infinity on trees and difficult to analyze precisely on triangular lattices. This and a host of other questions remain unanswered in this interesting class of problems.

ACKNOWLEDGEMENTS

PMD and CM thank the DOE for support under contract DE-FG02-90ER45418 and CM also acknowledges support from the Conselho Nacional de Pesquisa CNPq, Brazil.

REFERENCES

1. G. Deutscher, R. Zallen and J. Adler (Eds.), *Percolation Structures and Processes*, Bristol: Adam Hilger (1983)
2. G. Grimmett, *Percolation*, New York: Springer-Verlag (1989)
3. H. Kesten, *Percolation Theory for Mathematicians*, Boston: Birkhuser (1982)
4. D. Stauffer and A. Aharony, *Introduction to Percolation Theory* 2nd ed., London: Taylor & Francis (1992)
5. C. Moukarzel and P.M. Duxbury, Phys. Rev. Lett. **75**, 4055 (1995);
   C. Moukarzel, P.M. Duxbury and P.L. Leath, Phys. Rev. Lett. **78**, 1480 (1997);
   C. Moukarzel and P.M. Duxbury, preprint
6. D. Jacobs and M.F. Thorpe, Phys. Rev. Lett. **75**, 4051 (1995);
   D. Jacobs and M.F. Thorpe, Phys. Rev. **E53**, 3682 (1996)
7. C. Moukarzel, P.M. Duxbury and P.L. Leath, Phys. Rev. **E55**, 5800 (1997);
   P.M. Duxbury, D. Jacobs, M.F. Thorpe and C. Moukarzel, preprint
8. J.C. Phillips, J. Non-Cryst. Solids **43**, 37 (1981);
   M.F. Thorpe, J. Non-Cryst. Sol. **57**, 355 (1983)
9. S. Obukov, Phys. Rev. Lett. **74**, 4472 (1994)
10. B. Hendrickson, Siam J. Comput. **21**, 65 (1992);
    C. Moukarzel, J. Phys. **A29**, 8079 (1996);
    D. Jacobs, J. Phys. **A31**, 6653 (1998)
11. D. Jacobs and M.F. Thorpe, Phys. Rev. Lett. **80**, 5451 (1998);
    P.M. Duxbury, C. Moukarzel and P.L. Leath., Phys. Rev. Lett. **80**, 5452 (1998)
12. C. Moukarzel 1997, Int. J. Mod. Phys. C, to appear.

RIGIDITY PERCOLATION ON TREES

P. L. Leath and Chen Zeng

Department of Physics & Astronomy
Rutgers University
Piscataway, NJ 08854-8019

INTRODUCTION

Since the work of Feng and Sen[1] in 1984, it has been known that the geometry and properties of rigidity percolation are very different on lattices with and without bond-bending forces. So a number of groups[2,3,4] have carried out numerical simulations of generic rigidity percolation on site-diluted triangular lattices (at T=0), with only central forces, for samples up to 1000×1000 sites in size. These calculations agree on the existence (at a site concentration $p_* = 0.66$) of a second-order transition in the percolation of an elastic backbone P_{ov} (with exponent $\beta' = 0.25 \pm 0.03$) having a fractal dimension $D_B = 1.78 \pm 0.02$. But there is disagreement[5,6] over whether the spanning rigid cluster P_∞ (including the isostatic ends, which do not participate in the elasticity) has a weak first-order or a very sharp second-order transition at p_*. A numerical resolution of this controversy seems to require data on much larger networks than those presently feasible with present computers. In particular, Moukarzel et al[7] found that the fraction of sites in the isostatic ends (the equivalent of dangling ends of the percolation problem) of the spanning rigid cluster seemed to be building up to a finite discontinuity (see Fig. 1a), which corresponded to about 8.6% of the sites in the site-percolation case and 11% of the bonds in the bond-percolation case (see Fig. 1b). In addition, Obukov from a field theory had predicted a first-order phase transition[8]. And we were looking for hard evidence of the possible existence of a first-order rigidity transition.

Thus, we turned our attention to Cayley trees[9] (or Bethe lattices) where exact analytical calculations are possible and which have had much success (as a kind of mean field theory) in physics for describing some important features of phase and percolation transitions[10,11].

A simple Cayley tree is shown in Fig. 2, for the case of the number of nearest neighbors $z = 3$ (or the branching ratio $\alpha = z - 1 = 2$) for the number of bonds (constraints) connecting nearest neighbors $b = 1$, and for the fraction of occupied bonds or sites $p = 1$. [We shall use p to represent either the fraction of randomly occupied

Rigidity Theory and Applications
Edited by Thorpe and Duxbury, Kluwer Academic/Plenum Publishers, 1999

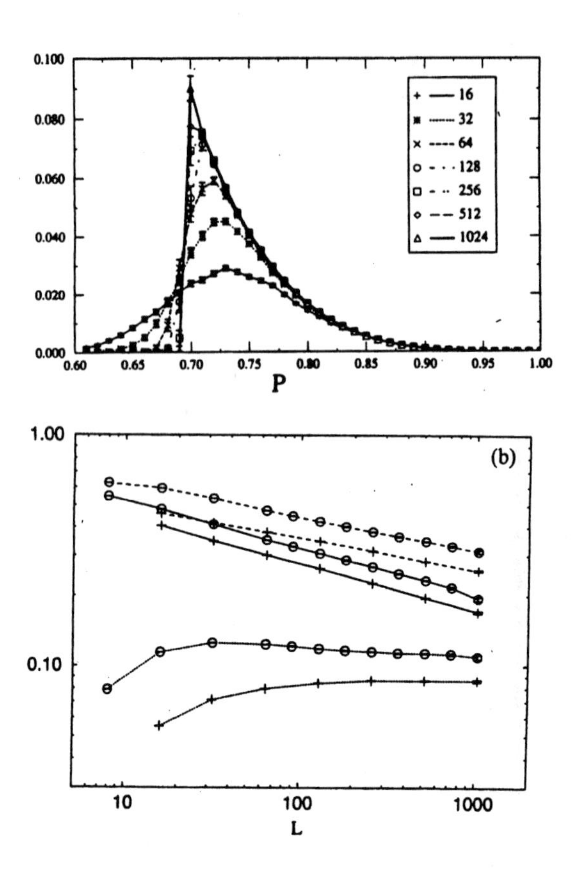

Figure 1. (a) The fraction of sites in the isostatic ends $P_I(p)$ versus p, the fraction of bonds present, for $L \times L$ triangular lattices, of size L = 16, 32, 64, 128, 256, 512, and 1024. (b) the rigid cluster (long dashes) $P_\infty(p_c)$, the backbone $P_{ov}(p_c)$ (solid lines), and the isostatic ends $P_I(p_c)$ (short dashes), for site -(plusses), and bond-(circles) diluted $L \times L$ triangular lattices, versus L.

bonds or sites, and p_c or p_* to be the rigidity threshold for either site- or bond-rigidity percolation]. These simple Cayley trees ($b = 1$) have no closed loops and are thus entirely floppy unless constrained by a boundary. So we consider the Cayley tree to be rigidly connected at its edge (or girth) to a single rigid boundary. This is equivalent to rigidly attaching all the edge sites to a single "ghost" site. Let us consider the piece of the Cayley tree near the rigid boundary (at infinity) shown in Fig. 3 for $\alpha = 2$, $b = 1$, and $p = 1$, where the heavy bar is the rigid boundary which extends around the entire tree. We shall only consider generic rigidity so that the counting of degrees of freedom and constraints shall be the rigorous calculation of the state of rigidity. (i.e. All bonds present shall be assumed to be linearly independent constraints.) And we shall let g be the number of degrees of freedom of a single free vertex, joint, or node of the tree. When the local embedding dimension is d, we have $g = d$ if the joints are simply points, or $g = d(d+1)/2$ if the joints are d-dimensional bodies with rotational as well

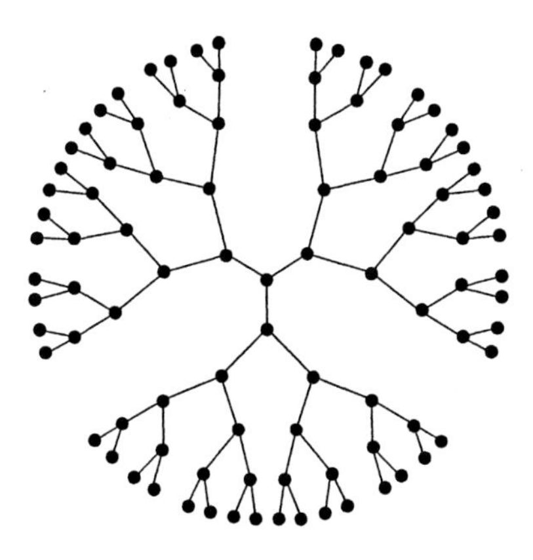

Figure 2. A section of a Cayley tree with $b = 1$, and $\alpha = (z - 1) = 2$.

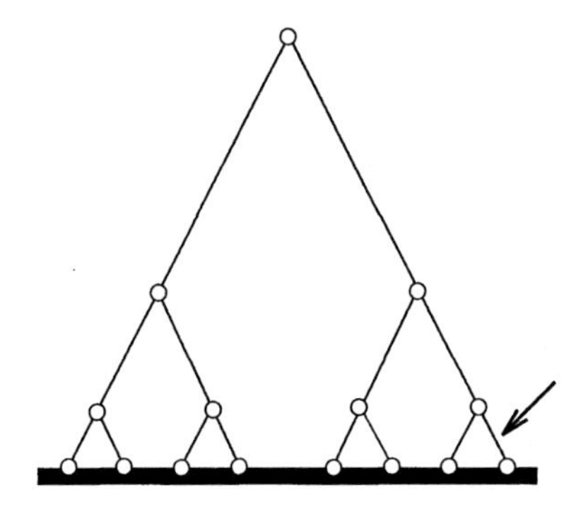

Figure 3. A rigid portion of a Cayley tree, with $g = 2$, $b = 1$, and $\alpha = 2$, near the rigid boundary (heavy line). The removal of the indicated bond breaks the rigid portion into six rigid pieces, as a small example of the house-of-cards effect.

as translational degrees of freedom. For the case of Fig. 3, if we assume $g = 2$, then the entire piece is rigidly attached to the boundary. But if the indicated bond is removed the structure becomes floppy and breaks into 6 flexibly connected rigid pieces. Indeed, when a bond is removed in this Cayley tree, the entire structure above it collapses into many connected, rigid pieces. This phenomenon is very unlike ordinary connectivity percolation, where removal of a single bond merely separates a cluster into at most two pieces. We shall call this phenomenon the "house-of-cards" effect. The question of whether the rigidity transition is first-order then can be posed in terms of the size of the house-of-cards effect. If the falling house-of-cards that results from the removal of a single bond or site (or a finite number of bonds or sites) causes the breaking of this infinite rigid cluster into an extensive number of pieces across the sample, and corresponding to a finite fraction of the lattice, then the transition will be first-order.

SITE-DILUTED CAYLEY TREES

First, we shall consider site-percolation for the case $b = 1$, and where p is the probability that a site is occupied (and hence connected by a single bond to all of its

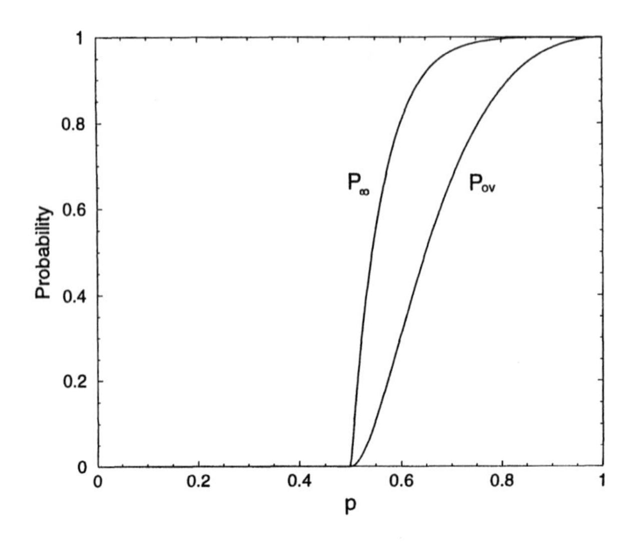

Figure 4. The infinite cluster bond probability $P_\infty(p)$ and the backbone, or overconstrained bond probability $P_{ov}(p)$ versus p, for bond dilution, with $g = 1$, $\alpha = 2$, and $b = 1$.

occupied nearest neighbors). Then $T_i^{(l)}$ is defined to be the probability that a site, which is l bonds away from the rigid boundary, has i degrees of freedom relative to that fixed boundary.

The simplest non-trivial case is ordinary connectivity percolation, which is obtained if we set $g = 1$, so that only one bond (constraint) is needed to connect a site to the spanning cluster. In this case, we find that $T_0^{(l+1)}$ (the probability that the site is connected by the sites below it, to the boundary is given by

$$T_0^{(l+1)} = p[1 - (1 - T_0^{(l)})^{z-1}] \ . \tag{1}$$

That is, in words, the probability that the site is connected to the boundary is p times the probability that not all of the sites below it are disconnected from the boundary. The value of T_0, as one heads away from the boundary toward the bulk, or as $l \to \infty$, converges to the solution of Eq. (1), without the l-labels, or

$$T_0 = p[(z-1)T_0 - (\frac{z-1}{2})T_0^2 + \ldots \pm T_0^{z-1}] \ . \tag{2}$$

This equation has a solution at $T_0(p) = 0$, plus a real non-negative solution that vanishes continuously (corresponding to a second-order transition) at $p = 1/(z-1) = \alpha^{-1}$, which is the well-known percolation threshold on Cayley trees[10].

For example, the solution for z=3 is

$$\begin{aligned} T_0 &= (2p-1)/p^2 \text{ for } p \geq 0.5 \\ &= 0 \qquad\qquad\quad \text{for } p \leq 0.5 \ . \end{aligned} \tag{3}$$

But for any Cayley tree with $b = 1$, and for any value of g, the probability $P_\infty(p)$ that a bond in the bulk is rigidly connected to the boundary (and thus is in the infinite cluster), and the probability $P_{ov}(p)$ that the bond is overconstrained (stressed) is given by

$$P_{ov} = T_0^2 \ , \tag{4}$$

and

$$P_\infty = T_0^2 + 2T_0T_1. \tag{5}$$

Thus, for ordinary percolation ($g = 1$), and for $z = 3$, one finds using Eq. (3) the values of P_∞ and P_{ov} shown in Fig. 4.

Moving on to the more general site-dilution case of rigidity percolation ($g > 1$) we find for $b = 1$ that the equivalent of Eq. (2) becomes

$$T_0 = p\sum_{l=g}^{\alpha}\binom{\alpha}{l}T_0^l(1 - T_0)^{\alpha-l} \ , \tag{6}$$

and

$$T_i = p\binom{\alpha}{g-i}T_0^{g-i}(1 - T_0)^{\alpha-g+i} \ , \quad \forall i \geq 1 \ . \tag{7}$$

In this case, since $g > 1$ where there is no term linear in T_0 on the right-hand side of Eq. (6), there is no possibility for a real, positive solution for T_0 which vanishes continuously at a critical point $p_c > 0$. For example, for the case $g = 2$ and $\alpha = 3$, Equation (6) becomes

$$T_0 = p[3T_0^2(1 - T_0) + T_0^3] \ , \tag{8}$$

which has the solution

$$T_0 = 0 \ , \left(3 \pm \sqrt{9 - 8/p}\right)/4 \ . \tag{9}$$

The solution $\left(3 + \sqrt{9 - 8/p}\right)/4$ is a stable solution which tends to one at $p = 1$. The solution $\left(3 - \sqrt{9 - 8/p}\right)/4$ is an unstable solution. Thus, as shown in Fig. 5, as p decreases from 1, $T_0(p)$ follows the stable solution $\left(3 + \sqrt{9 - 8/p}\right)/4$ down to the critical point at $p_* = 8/9$ at which T_0 has a square-root singularity, and discontinuously drops to the $T_0 = 0$ solution with a jump of $\Delta T = 3/4$. In addition T_1 and T_2 are seen from Eq. (6) to be the simple functions of T_0

$$T_1 = 3pT_0(1 - T_0)^2 \ , \tag{10}$$

and

$$T_2 = p(1 - T_0)^3 \ , \tag{11}$$

which also must have singularities and discontinuous jumps at $p_* = 8/9$. In addition we explored[3] all values of g, b, and α to see if we could find a crossover to a second-order rigidity transition, other than those cases which are essentially the ordinary connectivity percolation case. For example, Fig. 5 shows the trend of increasing z, or α, which shows that the first-order jump decreases but only approaches zero in the limit $z \to \infty$. Indeed we found no cases of rigidity percolation which were second-order (except for those that were equivalent to the ordinary connectivity cases).

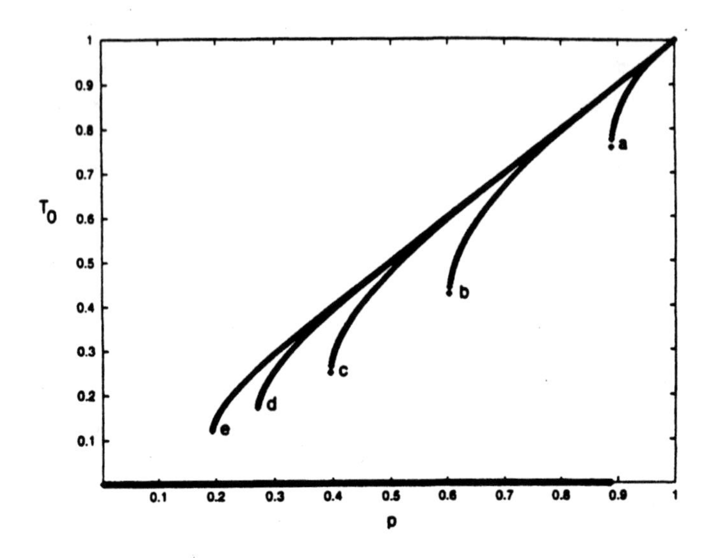

Figure 5. $P_\infty(p)$ for site-diluted trees with $g = 2$, $b = 1$, and $\alpha = 3, 5, 8, 12$, and 17 (which have decreasing values of p_c, as α increases).

BOND-DILUTED CAYLEY TREES

Bond-diluted Cayley trees, for the case $b \geq g$, would seem to offer new features since the extra local bonds provide local closed loops and a mechanism for local internal rigidity that would be present even with open boundaries. An example of a Cayley tree for (p = 1) with body joints $g = d(d+1)2$, $\alpha = 2$ and $b = 3$ is shown in Fig. 6, where all three bonds connecting nearest neighbors are assumed to be linearly independent constraints. For the cases with $b > 1$, the combinatorial counting of the bonds constraints as well as the branch counting is a bit more difficult and so we developed a transfer-matrix technique to keep track of the various factors. We shall not review the transfer-matrix technique here, but the details can be found in Ref. 3. Instead we will just review the results. First, Figure 7a shows T_0 versus bond concentration p for the case $g = 2$, $z = 6$ (or $\alpha = 5$), and $b = 1$. These results are qualitatively similar to the site-dilution case, but a vertical transition has been added to the graph from the free-energy calculations of Duxbury et al[12], which will be discussed below. Figure 7b shows the probabilities P_∞ and P_{ov} of rigid and overconstrained bonds respectively near this transition.

The first exploration[3] of the solutions for $b \geq g$, where local internal rigidity is possible, seemed to still always give a first-order jump in T_0 at p_*. But one new feature did appear (see Fig. 8) in that $T_1(p)$ exhibited a small peak just above p_*, which surely represents the effects of local internal rigidity.

More recently, in preparation for this conference, we have realized that for $b \geq g$ in Cayley trees with *open* boundaries, there is a second-order rigidity transition,

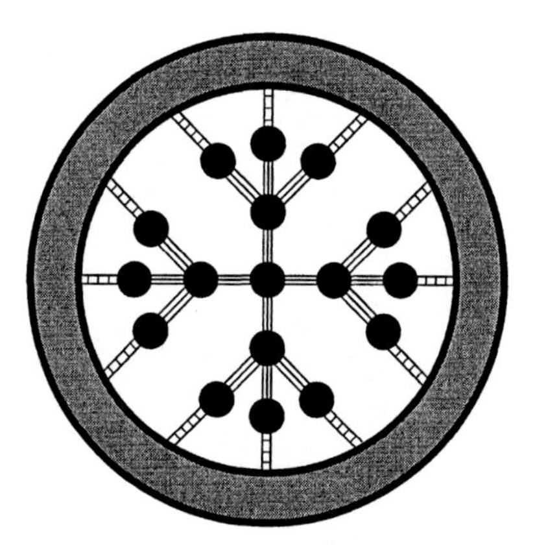

Figure 6. A Cayley tree with $\alpha = b = 3$, where the joints are bodies with $g = 3$ (for $d = 2$). The rigid connections of the edge joints to the rigid boundary are considered to be present with a probability e (but are shown here for the case $e = 1$)

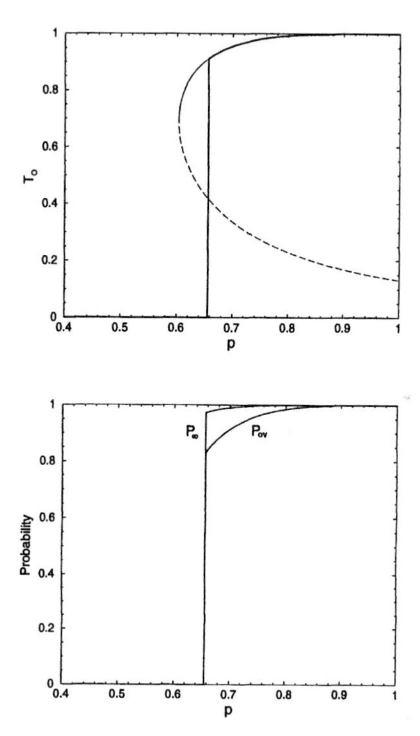

Figure 7. (a) Solutions for $T_0(p)$ vs. p, to the analog of Eq. (6) for bond dilution, are shown for $g = 2$, $b = 1$, and $\alpha = 5$. The upper solid curve and $T_0 = 0$ are the stable solutions, while the dashed curve is the unstable solution. The vertical transition at $p_c = 0.656$ is the transition predicted by Eq. (18) (b) $P_\infty(p)$ and $P_{ov}(p)$ vs p, for the same case.

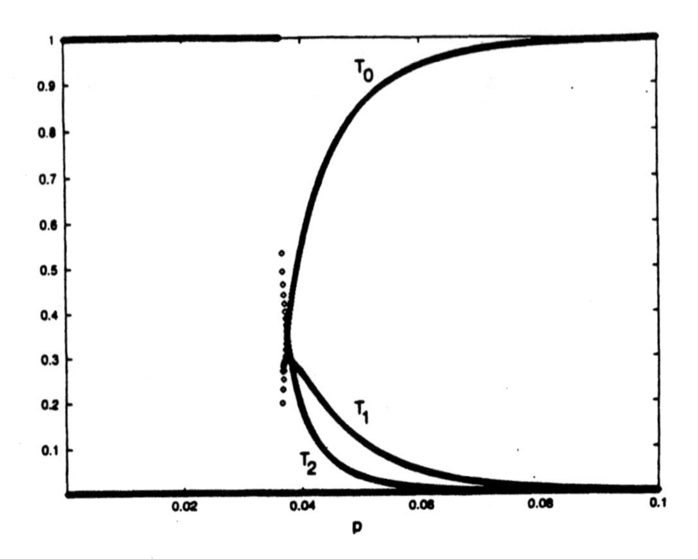

Figure 8. $T_0(p)$, $T_1(p)$ and $T_2(p)$ vs. p for $\alpha = 2$, $g = 2$, and $b = 40$. The transition is only weakly first-order, with an interesting nonmonotonic behavior of $T_1(p)$ which is due to the local, internal rigidity.

corresponding to the percolation of local rigid regions, consisting entirely of neighbors connected by at least g bonds. For example, for bond percolation with $b = g$, the second-order transition with open boundaries appears at $p_o^g = 1/(z-1)$. However, the equation with the rigid boundary[3] gives a first-order transition at $p < p_o$. So, we thought that perhaps a partially rigid boundary would show the crossover of the transition from first- to second-order.

Therefore, we constructed the model exemplified in Fig. 6, which shows an $\alpha = b = 3$ Cayley tree. The probability of bond of the tree being present is p, and the probability of a joint on the edge (girth) of the tree being rigidly connected to the rigid boundary is e. Thus when $g = 3$, we assume $T_0^{(1)} = e$, $T_1^{(1)} = T_2^{(1)} = 0$ and $T_3^{(1)} = 1 - e$ for the edge sites, with $0 \leq e \leq 1$. The transfer-matrix generalizations of Eq. (6) for this case were then iterated to find the stable solution for $T_0, T_1, ..., T_g$ for each value of e and p. The results, illustrated in Fig. 9, were that, for a given value of $0 < e < 1$, as p is increased from zero, the value of $T_0(p)$ is zero up to a critical point $p_*(e)$ where the stable value of $T_0(p)$ jumps to the upper curve, which is the stable solution for the rigid boundary ($e = 1$). At *precisely* the transition point $p = p_*(e)$, the iteration converged to a value of T_0 given by the *unstable* real solution, which is shown in Fig. 9. This transition point $p_*(e)$ is shown in Fig. 9, for $e = 0.1$. The unstable solution for $T_0(p)$ in Fig.9 with a rigid boundary, vanishes at $p_o = (1/3)^{\frac{1}{3}} = 0.693$, the second-order critical point for the tree with *open* boundaries. And as e is varied from 0 to 1, the rigidity transition (the vertical line) moves continuously from $p_o = 0.693$ to $p_*(e = 1)$ at the spinodal point, where there is a square-root singularity. For *open* boundaries, a second-order curve for T_0 with $p_o = (1/3)^{\frac{1}{3}}$ is observed. But for any $e > 0$, i.e., with any finite fraction e of edge sites connected to the rigid boundary, and for $p \geq p_o$, the only stable solution is the upper, first-order curve. Therefore simply by changing the boundary conditions on a Cayley tree (where boundaries are a finite fraction the tree)

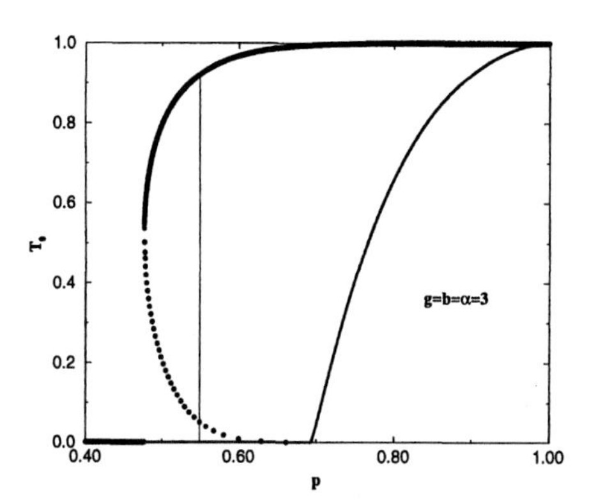

Figure 9. $T_0(p)$ vs. p for a partially rigid tree with $\alpha = b = g = 3$, and $e = 0.1$. The upper solid curve, and $T_0 = 0$ are the stable solutions when there is a rigid boundary, and the dotted curve is the unstable solution. The lower solid curve represents the second-order transition that occurs at $p_o = (1/3)^{\frac{1}{3}}$ when there are open boundaries ($e = 0$). For $e = 0.1$ the stable solution is $T_0 = 0$ for p less than the $p_c \approx 0.5489$ represented by the vertical line, and T_0 follows the upper solid curve for p greater than the p_c. The lower solid curve is a stable solution only for $e = 0$.

one can shift between the various stable solutions for $T_0(p)$ that exist for $e = 0$ or 1, but there is no new interpolating solution. Thus, which solution is stable for a Cayley tree and thus where the transition occurs, depends upon the boundary conditions which determine the number of overall constraints in the system.

FLOPPY-MODE FREE ENERGY

Recently, Duxbury et al.[12] have proposed for $b = 1$ bond percolation, that an effective free energy of the systems can be defined in terms of the number of floppy modes. Specifically, they define the number of floppy modes per degree of freedom $f(p)$ to be the intensive quantity proportional to the bulk free energy, namely

$$f(p) \equiv \frac{F}{gN} = 1 - \frac{z}{2g}[p - r(p)] \quad , \tag{12}$$

where F is the number of floppy modes, N is the number of sites, $z/2$ is the number bonds per site of the perfect lattice ($p = 1$), p is the fraction of occupied bonds, and $r(p)$ is the probability that a bond is redundant (overconstrained). Then, since there are the boundary conditions $f(0) = 1$, and $f(1) = 0$, or alternatively $r(0) = 0$, and $r(1) = 1 - 2g/z$, they find that

$$\int_0^1 \frac{\partial f(p)}{\partial p} dp = -1 \quad , \tag{13}$$

or, alternatively, that

$$r(1) = \int_0^1 \frac{\partial r(p)}{\partial p} dp = 1 - 2g/z \quad . \tag{14}$$

Furthermore, they argue that $\frac{\partial r(p)}{\partial p}$ is given by the fraction of overconstrained bonds

$$\frac{\partial r(p)}{\partial p} = P_{ov} \quad , \tag{15}$$

which for Cayley trees (from Eq. 4) with $b = 1$, becomes

$$\frac{\partial r(p)}{\partial p} = T_0^2(p) \quad . \tag{16}$$

In addition, they argue that, for $b = 1$,

$$r(p) = 0, \text{for } \; p \leq p_c \quad , \tag{17}$$

Therefore, for Cayley trees, with $b = 1$, they locate p_c by the integral equation

$$\int_{p_c}^{1} T_0^2 dp = 1 - 2g/z \quad , \tag{18}$$

and more generally write the free energy per site as

$$f(p) = 1 - \frac{z}{2g}[p - \int_{p_c}^{p} T_o^2(p)dp] \quad . \tag{19}$$

This is analogous to the Maxwell construction used for thermodynamic phase transitions. For example, this calculation for ordinary (connectivity) percolation gives the free-energy and its derivatives shown in Fig. 10. The second derivative $f^{(2)}(p)$, shown in Fig. 10c, is always positive and corresponds to the specific heat, which has a peak just above the second-order transition.

For rigidity percolation ($g > 1$) the first-order transition in $T_0(p)$, given by Eq. (18) is illustrated in Fig. 6 above. This corresponds to the free-energy and its derivatives show in Fig. 11, where $f^{(1)}(p)$ exhibits the first-order jump.

Both Figures 10 and 11 also show data points (open circles). These were obtained from numerical simulations on the random-bond model, a small example of which is illustrated in Fig. 12, for the case of $b = 1$ and $z = 3$. The agreement of these numerical simulations with the Cayley tree result is striking. But, in retrospect, the agreement is understandable since the random bond model (which was discussed in a previous talk by Michael Thorpe) is, in the limit of large N, very much like the Cayley tree. In the random-bond model, points are dispersed randomly, in a d-dimensional space, and bonds are drawn to connect each site to z other sites chosen at random (without regard to proximity). Thus the number of closed loops is of $O(1/N)$. Furthermore if one starts from a random site and follows the connections, they expand tree-like, except for these few closed loops which connect the branches. Nevertheless, the number of bonds in this lattice (and the bulk of all real lattices in d-dimensions) is strictly $Nzp/2$, so the transition is correctly predicted by Eq. (18).

On the other hand, real Cayley trees do not have the number of bonds $B = Nzp/2$ due to the enormous number of edge sites. However, the effective number of constraints in the system can be varied by changing the boundary conditions as we have shown above and thus the transition can be moved. It would seem that the floppy-mode free energy of Duxbury et al[12] would also correctly predict the position of the transition on Cayley trees, if by varying the boundary conditions, the number of constraints were to be set at $Nzp/2$.

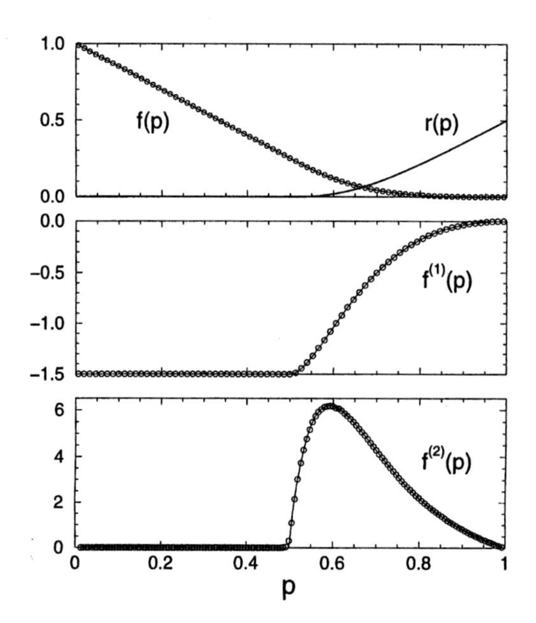

Figure 10. The floppy-mode free energy $f(p)$ (given by Eq. (19), its derivatives $f^{(1)}(p)$ and $f^{(2)}(p)$, and $r(p)$ vs. p, for $g = 1$, $b = 1$, and $\alpha = 3$. The second derivative $f^{(2)}(p)$ acts as the specific heat and is non-negative.

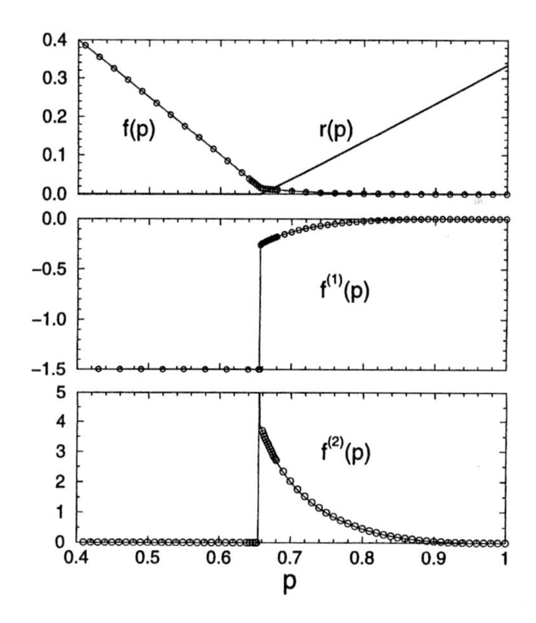

Figure 11. The floppy-mode free energy $f(p)$, its derivatives $f^{(1)}(p)$ and $f^{(2)}(p)$, and $r(p)$ vs. p for $g = 2$, $b = 1$, and $\alpha = 5$. The percolation threshold is given by $p_c = 0.656$. The open circle are from numerical simulations of Ref. 12 on the random bond model averaged over 2, 000 realizations with $N = 262,\ 144$ sites.

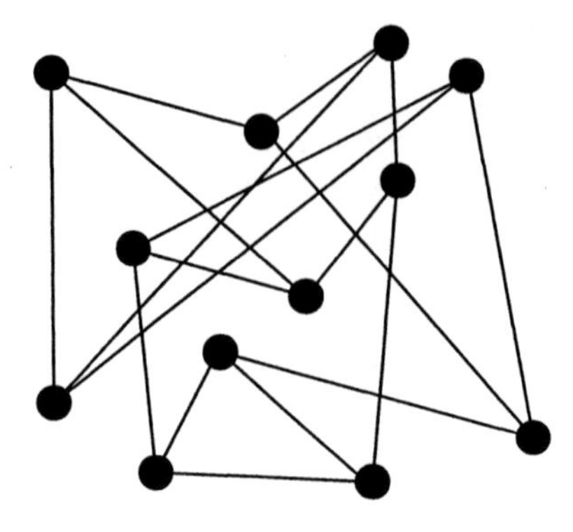

Figure 12. A sketch of an example of the random-bond model with $b = 1$, $z = 3$, and $N = 12$ sites.

CONCLUSIONS

Diluted Cayley trees of great generality having only central forces and a rigid boundary seem to always have a first-order rigidity percolation transition due to the enormous "house-of-cards" effect. In addition there is always a square-root singularity at the spinodal point except for those cases which corresponds to ordinary connectivity percolation. Diluted Cayley trees do not seem to have a simultaneous transition in $P_\infty(p)$ while having a second-order transition in $P_{ov}(p)$, which has been considered for two-dimensional lattices. Cayley trees (with sufficient numbers of small local closed loops, i.e. b being sufficiently large) do exhibit second-order rigidity percolation for the case of open boundaries. And for partially rigid boundaries the location of the first-order rigidity percolation transition is determined by the number of constraints at the rigid boundary.

Finally Duxbury et al[12] have developed a floppy-mode free energy and the related Maxwell construction to locate the first-order transition in the bulk of real lattices which works very accurately for the random-bond model, where it has been compared to numerical simulations.

ACKNOWLEDGMENTS

We would like to thank Philip Duxbury, Don Jacobs, Cristian Moukarzel, and Michael Thorpe for useful discussions.

REFERENCES

1. S. Feng and P. Sen, Phys. Rev. Lett. **52**, 216 (1985).
2. C. Moukarzel and P.M. Duxbury, Phys. Rev. Lett. **75**, 4055 (1995).
3. C. Moukarzel, P.M. Duxbury and P.L. Leath, Phys. Rev. Lett. **78**, 1480 (1997).
4. D.J. Jacobs and M.F. Thorpe, Phys. Rev. Lett. **75**, 4051 (1995); Phys. Rev. E **53**, 3682 (1996).
5. D.J. Jacobs and M.F. Thorpe, Phys. Rev. Lett. **80**, 5451 (1998).
6. P.M. Duxbury, C. Moukarzel, and P.L. Leath, Phys. Rev. Lett. **80**, 5452 (1998).

7. C. Moukarzel and P.M. Duxbury, unpublished.
8. S.P. Obukov, Phys. Rev. Lett. **74**, 4472 (1995).
9. C. Moukarzel, P.M. Duxbury, and P.L. Leath, Phys. Rev. E **55**, 5800 (1997).
10. M.E. Fisher and J.W. Essam, J. Math Phys. **2**, 609 (1961).
11. R.B. Stinchcombe, J. Phys. C **6**, L1 (1973); **7**, 179 (1974).
12. P.M. Duxbury, D.J. Jacobs, M.F. Thorpe, and C. Moukarzel, preprint (1998).

RIGIDITY AS AN EMERGENT PROPERTY OF RANDOM NETWORKS: A STATISTICAL MECHANICAL VIEW

Paul M. Goldbart

Department of Physics
University of Illinois at Urbana-Champaign
1110 West Green Street
Urbana, Illinois 61801-3080, U.S.A.

INTRODUCTION

The aim of this article is to give an informal and, I hope, "user-friendly" survey of our recent (and not so recent!) efforts to develop a theory that accounts for the structure and properties of certain random network forming media. These efforts are described in a number of publications, amongst which the primary ones are Refs. [1-7], along with a lengthy technical review[8] and a less formal discussion[9]. Two seminal papers underpin our effort: the formulation of the statistical mechanics of a randomly crosslinked macromolecule, due to Deam and Edwards;[10] and the construction and calculation of an order parameter capable of detecting spin glass order, due to Edwards and Anderson.[11]

The present article is organised as follows. I begin by outlining the basic issues associated with random networks: what are they made of and what distinguishes the solid state that they exhibit. Next, I catalogue the goals and results of our approach to random networks. I then give a tour of of the elements of our approach, focusing on the order parameter for the amorphous solid state and its physical content, and how to calculate the value of the order parameter (both semi-microscopically and via a Landau-theory approach). Finally, I turn to the issue of rigidity to shear deformations, a property that emerges as a consequence of the formation of the amorphous solid state, and then to some concluding remarks.

What are the basic issues?

The basic issues concern the equilibrium phase transition, exhibited by random networks, from a fluid state to an amorphous solid state. The general scenario for this transition is as follows. Take a macroscopic sample of matter formed by the random permanent "gluing together" of some of the constituent atoms. Enough "gluing" and the sample is solid: some (or all) of the constituents are spatially localised, having preferred locations about which they perform Brownian fluctuations; owing to the random nature of the "gluing," there is no long-range regularity (i.e. no crystallinity in the

Rigidity Theory and Applications
Edited by Thorpe and Duxbury, Kluwer Academic/Plenum Publishers, 1999

mean positions); but the sample is indeed solid, in the sense that it has a nonzero static shear modulus, albeit an amorphous solid. What would have happened if less "gluing" had been done? Fewer of the constituent atoms would be localised; the extent of the Brownian fluctuations of those that are would be larger; and the shear modulus would be smaller. Even less "gluing" and the trends continue until, as a critical density of "gluing' is approached: the fraction of localised constituents tends to zero; the Brownian fluctuations of the small fraction that are localised grow without bound; and the shear modulus vanishes continuously. Said another way, as a function of the density of random permanent "gluing' there is an equilibrium phase transition—the vulcanisation transition—between the fluid state of matter and the amorphous solid state of matter, and the aim of the work that I shall be describing is to provide a theory of this transition and, in doing so, a theory for the structure and properties of the solid state of certain random network forming media.

It should be emphasised that the transition from a liquid state to a solid state occurs at constraint densities such that the networks are enormously underconstrained. In the case of vulcanised macromolecular networks, e.g., there is on the order of one crosslink per macromolecule, although each macromolecule can have many thousands of degrees of freedom. Thus, even when solid the networks are by no means mechanically rigid, and are instead thermodynamically rigid: they owe their rigidity almost exclusively to the loss of entropy that arises when the sample shape is changed (at fixed volume).

The basic concepts and tools that I shall be using come from the subject of statistical mechanics, more specifically the statistical mechanics of systems having what is called *quenched randomness*.[12] Systems with quenched randomness are associated with information of two types: annealed variables, which are those that undergo random thermal (i.e. Brownian) motion; and quenched variables, which are needed for the precise specification of the microscopic constitution of the system, and vary from sample to sample but do not vary as the annealed variables of a given sample undergo thermal motion. In the context of random network forming media, the annealed variables are the coordinates of the atoms or molecules that constitute the system, and the quenched variables specify which atoms have been connected by permanent random chemical bonds during the network forming process.

Why statistical mechanics?

As physicists, our aim is to find some common and predictable properties of random networks. These properties should hold widely, regardless of the vast majority of the coordinates specifying the locations of the atoms in the system, and regardless of the vast majority of the details of the architecture of the networks. The statistical mechanics of J. W. Gibbs provides a prescription for how to eliminate the atomic coordinates from the problem (by suitably ensemble-averaging over configurations of these coordinates). And the strategy of averaging physically observable quantities (free energies, order parameters, correlators, etc.) over ensembles of disorder-realisations provides a prescription for eliminating the architectural variables. We invoke these two strategies in order to seek a description of the emergent physical properties of typical random networks that are specified by certain gross parameters: the constituent objects, the type and density of constraints imposed, etc. And, as we shall see, what results is a description in which these properties are characterised in terms of a modest amount of natural and readily digestible information. Such a scenario is commensurate with our aim of finding properties common (up to calculable fluctuations) to essentially all realisations of broad classes of networks and, furthermore, allows us to sidestep the awful prospect of dealing with individual configurations of individual realisations of random networks.

What are random network forming media?

By random network forming media I shall mean a variety of condensed matter systems, including:

vulcanised macromolecular systems i.e., systems of macromolecules into which a certain density of permanent random chemical bonds has been introduced (see Fig. 1) in addition to those bonds that give the macromolecules their integrity; as well as

endlinked macromolecular systems in which the constraints correspond to the permanent chemical bonding of some specified number (e.g. 3) of randomly selected ends of macromolecules (see Fig. 2); and

atomic or low-molecular-weight systems that have been *gelled*, i.e., constrained by the imposition of permanent covalent bonds between randomly selected constituent atoms or molecules (see Fig. 3)—of particular interest to the participants in this workshop.

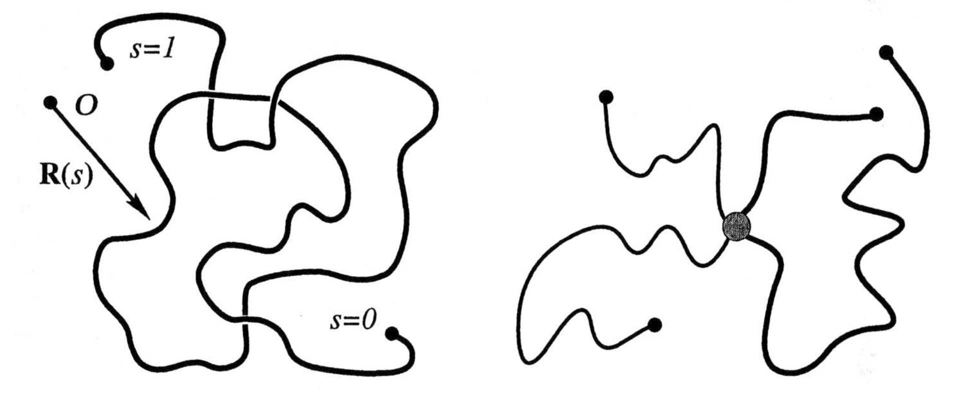

Figure 1. *Left*: The semi-microscopic configuration of a single macromolecule is specified by giving the location in space $\mathbf{R}(s)$ of the segment of the macromolecule a fraction s from one of its ends. O indicates the coordinate origin. *Right*: A pair of crosslinked macromolecules. One macromolecule is depicted by a thinner line than the other. The crosslink is indicated by the shaded circle. The details of the chemical bonding associated with a crosslink are idealised by a permanent constraint that serves to identify the locations of the pair of participating segments.

The relevant common features of these random network forming systems are that they begin life as fluid systems, by which we mean systems in which none of the atoms—or molecules or other basic units—has a preferred spatial location, and no chemical bond has a preferred spatial orientation. More specifically, the equilibrium state is one in which any given atom is equally likely to be found at anywhere in the sample and, similarly, any chemical bond is equally likely to be found pointing in any direction. A certain density of permanent random constraints is then imposed (e.g. by vulcanisation, endlinking or gelation). These constraints have the effect of eliminating degrees of freedom. In the case of vulcanisation, e.g., the imposition of a crosslink renders identical what had been two kinematically independent atomic positions. Indeed, the imposition of more and more constraints renders operative a smaller and smaller (classical statistical mechanical) configuration space. (We shall exclusively consider classical statistical

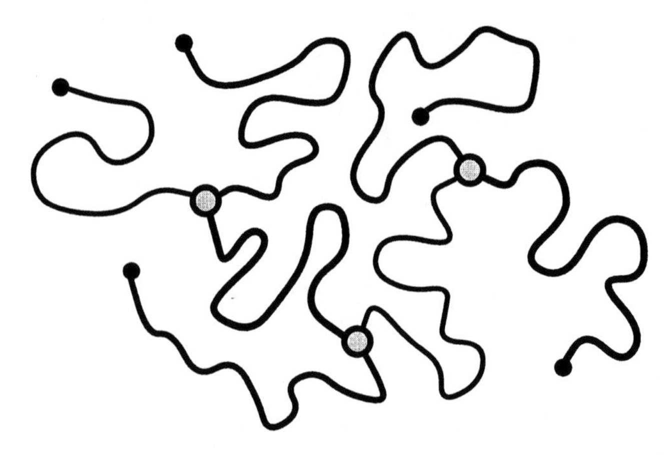

Figure 2. Semi-microscopic configuration of 7 macromolecules that are connected by 3 endlinks (shaded circles), corresponding to the case of 3–type endlinking. Some macromolecules are depicted by thinner lines than others. Unlinked ends are indicated by filled circles.

mechanics, and kinetic energy will play essentially no role.*) This forces the system to inhabit lower and lower dimensional hypersurfaces in the configuration space of the unconstrained system. However—and this point could not be more important—*the constraints do not explicitly break the translational invariance of the uncrosslinked system* (i.e. the feature that the statistical weights accorded to any pair of configurations that are simply translations of one another are identical).

In fact, in contexts such as the crosslinking of macromolecular systems the imposition of constraints can also have the effect of deconstructing configuration space according to the *topology* of the configuration, owing to the fact that the dynamics is unable to pass macromolecular chains through one another. Strictly speaking, the topololgy of the network should be regarded a a quenched random variable, i.e., something to be maintained in time as the annealed degrees of freedom equilibrate. However, no mathematical tools are available to accomplish this type of *anholonomic* constraining of the configurations, and we shall have to hope that admitting, *a priori*, topology-altering thermal fluctuations (even though they ought to be excluded) will not have a profound impact on the theory. This hope is not unreasonable, given the very light degree of crosslinking necessary to put the system in the regime of the amorphous solidification transition.

The glass transition?

Although the present approach to random network forming media was originally designed to address randomly crosslinked macromolecular systems, it has been extended with some success to address chemically gelled systems, such as vitreous silica gels, in which permanent covalent bonds are introduced at random between low-molecular-weight ingredients (see Fig. 3), thus building up a network *ab initio*, and not from any pre-existing chain-like structures.[13] It therefore seems worthwhile to ex-

*In conventional macromolecular settings the temperature is sufficiently high that quantum effects are of no importance. However, the observation that rigidity is primarily entropic in origin, it being a manifestation of the reduction in number of available configurations by shear deformations of the sample, raises the following interesting question. Consider the quantum-mechanical ground state of a random network, and bear in mind that there will be quantum fluctuations (i.e. zero-point motion and zero-point energy). Could the ground state be rigid, in the sense that a shear deformation would alter the ground state quantum fluctuations and thereby raise the ground state energy?

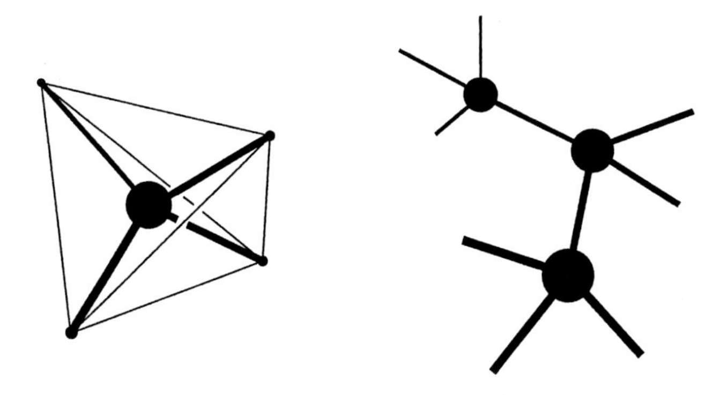

Figure 3. *Left*: An atom (filled circle) capable of forming up to 4 covalent bonds (thick lines) is envisaged as a flexible tetrahedral object (thin lines are a guide to the eye) consisting of a central point from which radiate 4 legs. *Right*: To form a network, atoms are connected at random by permanent covalent bonds. This is accomplished by constraining the relevant pairs of legs to be spatially superposed and oppositely oriented.

amine the extent to which the approach can be used to address glassy media formed by temperature-quenching, as is the mode of formation for many glassy materials. Although the temperature-quenching associated with the glass transition does not introduce constraints as recognisably (i.e. with quite the separation of time scales) as crosslinking or gelation does, it seems sensible to see whether progress can be made by regarding the temperature quench as selecting certain bonds to become much longer lived than they would be at high temperatures. Of course, the randomness is not extrinsic, in contrast with the case of vulcanised polymers, say, but if it is true that there is a certain degree of *self-induced* quenched randomness;[14, 15] caused by the temperature quench, then it may prove useful to idealise the situation, and regard these long-lived bonds as essentially permanent, thereby identifying certain quenched random variables.

From this perspective, crosslinked or gelled solids are the simplest form of glasses, inasmuch as they form random solids that, owing to a separation between the time scales for configuration relaxation and constraint breaking, are amenable to the techniques of equilibrium statistical mechanics.

Transition to the amorphous solid state and the emergence of rigidity

Whilst it is true that the constraints leave intact the *explicit* symmetry properties of the system, and at lower densities (of constraints) do little more (at least from the point of view of fundamental questions) than quantitatively modify the values of physical quantities (such as the viscosity), there comes a density at which the constraints have a radical effect. They cause the system to undergo a phase transition from the fluid state to a new phase of matter—the equilibrium amorphous solid state—in which a nonzero fraction of the atoms or molecules are spontaneously localised about random mean positions, about which they exhibit thermal fluctuations. This localisation of the particles indicates that the translational symmetry of the system is *spontaneously broken*, as it is when, say, a liquid crystallises. But—unlike the crystal—the amorphous solid state is random, in the sense that there is no periodicity associated with the spatial arrangement of the mean positions of the localised particles, and (although less significant fundamentally) the spatial extent of the localisation of the particles (as measured, e.g., by their r.m.s. displacements) varies randomly from particle to particle. In other words, the entire sample is one gigantic random unit cell. Moreover, although

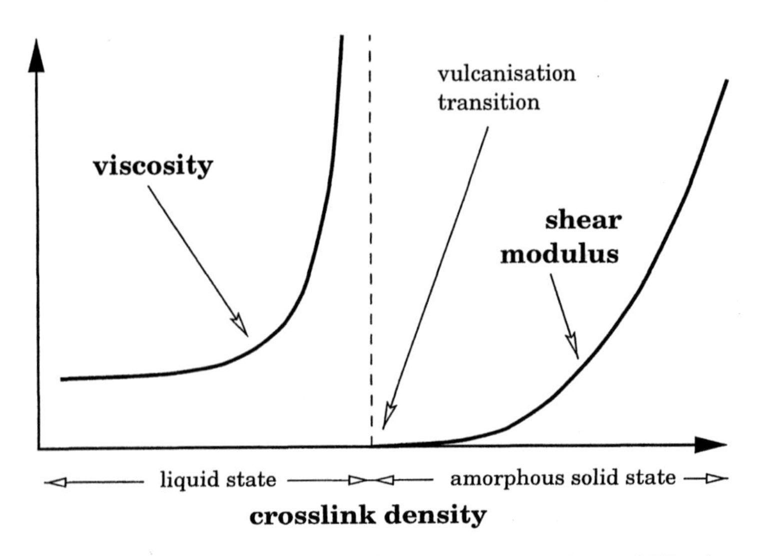

Figure 4. Sketch of the chief macroscopic consequences of the amorphous solidification transition. As the transition is approached the viscosity of the fluid state diverges and the shear modulus of the solid state vanishes.

translational and rotational symmetries are spontaneously broken, if one examines any progressively larger sequence of subvolumes (and, say, averages any quantity over the particles localised in that subvolume) one finds progressively smaller signals capable of indicating a preferred location or orientation of the subvolumes: from a macroscopic viewpoint the system retains its homogeneity and isotropy at the transition and, as we shall see shortly, it takes a rather carefully crafted "detector" to serve as an order parameter for the recognition and quantification of amorphous solidification.

Not only is the transition marked by a qualitative change in the *structure* of the system—particles are now localised—but also it is marked by a qualitative change in the *response* of the system to external perturbations (see Fig. 4). Specifically, the system is now solid, in the sense that its shear modulus is nonzero (and became nonzero continuously at the amorphous solidification transition): an infinitesimal (volume preserving) change in the shape of the sample, brought about after the constraints were imposed, causes a restoring force that persists indefinitely. There are other qualitative signatures of the amorphous solidification transition: the viscosity diverges as the transition is approached from the fluid side (see Fig. 4); and certain correlation become long-ranged (see Sec. 4.5 of Ref. 9). (It is worth noting that the determination of the viscosity requires a dynamical approach, which is outside the scope of Gibbs equilibrium statistical mechanics. For one approach to the dynamics of random networks see Ref. 16.) However, it is the rigidity, emerging as it does as a consequence of the spontaneous (but random) translational symmetry breaking that occurs at the transition, that is perhaps the most striking feature of the transition.

I should emphasise that this amorphous solidification transition is controlled not by a traditional thermodynamic parameter, such as temperature or pressure, but instead by the density of the random constraints that have been introduced.

Interlude: Randomness at several levels

There are three levels of randomness inherent to the statistical mechanics of random networks. First, there is the randomness in the *architecture* of the network: what objects are permanently bonded to what other objects? This is the quenched randomness, and one has to be mindful to average only physical observables over it, so as to represent the consequences of repeating the measurement of a physical quantity on a sequence of independently fabricated samples. (It is, e.g., not generally useful to average the statistical mechanical partition function over the quenched randomness, but it can be useful to average the free energy.) The distribution that governs this randomness is determined by the technique used to fabricate the network and, generally speaking, is a difficult object to quantify. In the case of one particular technique—that of instantaneous constraint imposition on what had been an unconstrained fluid at thermal equilibrium—the distribution that governs the randomness *can* readily be modelled, as realised by Deam and Edwards[10]: it is essentially the equilibrium statistical mechanical partition function of the constrained fluid, a quantity on whose logarithm one is already inclined to focus, owing to its close connection with the free energy. In fact, the connection between the distribution of the quenched randomness and the partition function for the constrained system is a surprising and useful feature, a feature that is very specific to systems involving quenched random constraints (as well as certain problems in the context of learning by neural networks with teachers.[17])

The architectural level of randomness is the level at which the subject makes contact with percolation theory. [For an introduction to the subject of percolation theory, see Ref. (18).] Indeed, as we shall see, the statistical mechanical approach to random networks connects precisely on to the mathematics of percolation theory at the stage at which one enquires about the fraction of particles that are localised: too few constraints, no particles are localised, and—in the sense of percolation—there is no infinite cluster, with all particles wandering throughout the container, given sufficient time; beyond the critical density of constraints, a nonzero fraction of the particles become localised, and an infinite cluster emerges. This is not at all surprising, but it is encouraging that the statistical mechanical approach does capture the connection between being part of an infinite cluster, being spontaneously localised, and being solid, as we shall see in the following section (see Fig. 5) and in the section on rigidity. The issue of spontaneous localisation is not accessible via percolation theory.

But only to see that increasing the constraint density yields an infinite cluster, and hence an amorphous solid, is only to see a corner of the picture. After all, what is becoming solid is a real physical system, with real degrees of freedom that are constantly undergoing thermal motion, and it is these degrees of freedom or, more precisely, their thermal motion, that provide the second level of statistics: statistical mechanics. A theory of random network forming media at nonzero temperature must account for the random thermal motion of the annealed variables, and this takes us beyond percolation theory, with its single ensemble that is capable only of scanning the collection of of various architectures.

The amorphous solidification transition has a profound impact on these annealed variables, inasmuch as some of them become localised rather than extended. Moreover, the transition is continuous: the localised fraction increases from zero as the amorphous solid state is entered. Furthermore, the typical length scale for r.m.s. displacements of the localised particles (so-called localisation length) diminishes continuously from infinity. These thermal fluctuations are an integral part of the theory of random network forming media, and how to characterise them leads to the third level of statistics.

Although it is macroscopically homogeneous, the amorphous solid state is microscopically inhomogeneous. This means that every localised atom "experiences" a distinct environment. As a consequence, rather than there being a unique (or at most a small, discrete set of) localisation lengths, there is instead a statistical continuum of them. Of course, it is likely than particles within a given region will have similar localisation lengths, especially if the region is smaller than the typical localisation length itself. However, as we scan the sample, we will find a statistical distribution of localisation lengths, indicating that it is useful to characterise the random r.m.s. displacements in the amorphous solid state by a distribution. More generally, we are recognising that, rather than characterise amorphous solids by parameters, it can be profitable to characterise them by distributions: the third level of statistics. Not only does this idea make sense on general grounds, but also, as we shall see below, once one finds the right order parameter for the amorphous solid state one sees that it is virtually imploring one to explore the distribution of localisation lengths. Furthermore, by comparing the results for such distributions from various model calculations and with independent molecular dynamics simulations (see Fig. 5 and Ref. [19]), we shall see that such distributions are remarkably universal, and we shall identify the origins of this universality.

GOALS AND RESULTS OF OUR APPROACH

What goals have we set ourselves?

Our central aim is to construct a theory of random network forming media having the following features: (i) The amorphous solid state, with its spontaneous random localisation of particles, should emerge from the theory as a natural consequence of the random constraints, and should not be put in "by hand." (ii) The theory should yield (and correlate) various physical properties, including the structure and response of the amorphous solid state. (iii) The theory should indicate what physical properties can be expected to be common to a range of network forming media, and should explain the origins of such universalities. (iv) The theory should serve as a natural launch-pad for refinements (such as the exploration of correlations and critical fluctuations in amorphous solids), as well as further developments (such as the exploration of dynamical properties).

By some measures our approach is, in fact, rather conservative, consisting as it does of the following sequence of steps that are basic to many problems in the physics of condensed matter:

1: Identify the relevant equilibrium phases.
2: Determine the essential distinctions between them, and construct an order parameter that quantifies these distinctions.
3: Select a semi-microscopic model.
4: Develop a mean field theory of this model; determine the value of the order parameter over the phase diagram; quantify the structure of the phases.
5: Compute responses to external static stimuli.
6: Progress to more advanced issues, including fluctuations, universality and dynamics.

To date, we have focused primarily (although not exclusively) on vulcanised macromolecular systems. Our reason for this is that such systems have the merit of possessing genuine quenched randomness (in the sense of a genuine and controllable separation between the time scales for the equilibration of the annealed variables and the breaking of the constraints). They therefore present the possibility of realising the equilibrium amorphous solid state, i.e., an utterly noncrystalline random solid state amenable to the techniques of equilibrium statistical mechanics.

I do not, however, view the macromolecular character of such networks as essential to the issue of random networks and rigidity; rather, I view such systems as providing perhaps the simplest setting in which to address the issues of rigidity and its emergence as a property of sufficiently constrained random systems.

What results have emerged?

I now catalogue the results that we have obtained on the properties of random networks, postponing until later (see the sections on order parameters as diagnostic tools and calculating the properties of the amorphous solid state) explanation of how these results were obtained.

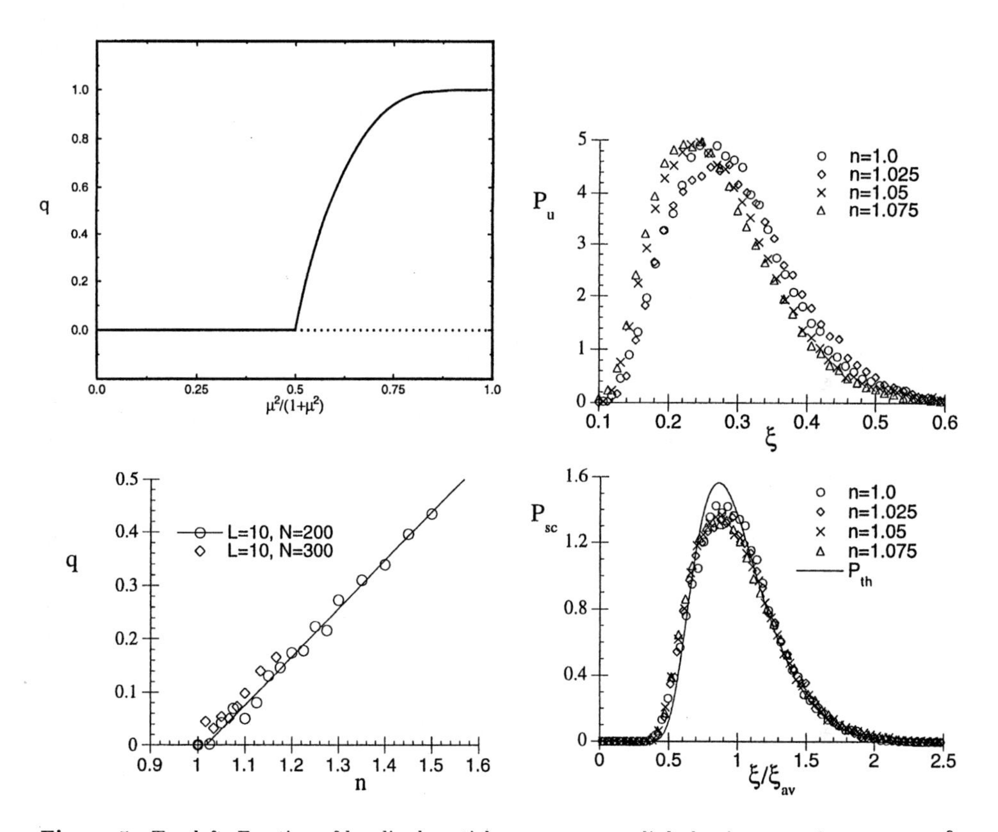

Figure 5. *Top left*: Fraction of localised particles q versus crosslink-density control parameter μ^2, within mean field theory. Nonzero q indicates the amorphous solid state. q varies linearly with μ^2 near the transition (which occurs at $\mu^2 = 1$), as does the (mean) crosslink density. *Bottom left*: Localised fraction q versus number of crosslinks per macromolecule n [molecular dynamics simulations by Barsky and Plischke (1997, unpublished)]. L is the number of monomers in each macromolecule; N is the number of macromolecules in the system. The straight line is a linear fit to the $N = 200$ data. Note the continuous phase transition at $n = 1$, as well as the linear variation of q with n, both features being consistent with the mean field theory. *Top right*: Unscaled distribution P_u of localisation lengths ξ (in units of the linear system size) [molecular dynamics simulations by Barsky and Plischke (1997, unpublished)]. The number of segments per macromolecule was 10; the number of macromolecules was 200. *Bottom right*: Distribution P_s of localisation lengths ξ (symbols), scaled with the sample-average localisation lengths ξ_{av} [molecular dynamics simulations by Barsky and Plischke (1997, unpublished)]. Note the collapse of the data on to a universal scaling distribution, and the quantitative agreement with the mean field prediction (solid line). The number of segments per macromolecule was 10; the number of macromolecules was 200.

We have considered macromolecular networks formed from flexible, semi-flexible and stiff macromolecules, and we have considered constraints imposed by crosslinking as well as by endlinking. On the basis of semi-microscopic mean field theories of these

systems we have ascertained that such systems exhibit an amorphous solidification transition having the following characteristics:
1: For densities of crosslinks smaller than a certain critical value (on the order of one crosslink per macromolecule) these systems exhibit a liquid state in which all particles (in the context of macromolecules, all monomers) are delocalised.
2: At the critical crosslink density there is a continuous thermodynamic phase transition to an amorphous solid state, this state being characterised by the emergence of random static density fluctuations.
3: In this state, a nonzero fraction q of the particles have become localised around random positions and with random (and finite) localisation lengths (i.e. r.m.s. displacements). In the case of crosslinked macromolecules q satisfies the simple equation

$$1 - q = \exp\left(-\mu^2 q\right), \tag{1}$$

where μ^2 is a parameter that controls the mean density of crosslinks. ($\mu^2 = 0$ means no crosslinks; $\mu^2 = 1$ means the critical density for solidification, and corresponds to roughly one crosslink per macromolecule.) The solution of Eq. (1), as well as a comparison with the simulation data of Barsky and Plischke (unpublished, 1997; see Ref. [19]), is shown in Fig. 5. Equation (1) coincides with that found by Erdős and Rényi in their study of random graph theory;[20] as it should, because this aspect of random graph theory is equivalent to the mean field theory for percolation. From this equation it is straightforward to establish that near the amorphous solidification transition the fraction of localised particles grows linearly with the excess crosslink density. A similar equation, with identical critical behaviour, holds for the case of endlinking (rather than crosslinking).
5: Near to the amorphous solidification transition the dependence of the statistical distribution of localisation lengths on the localisation length and the crosslink density is expressible in terms of a single universal scaling function π of a single variable. This scaling function satisfies the simple-looking integro-differential equation

$$\frac{\theta^2}{2}\frac{d\pi}{d\theta} = (1-\theta)\,\pi(\theta) - \int_0^\theta d\theta'\,\pi(\theta')\,\pi(\theta-\theta'), \tag{2}$$

together with the normalisation constraint $\int_0^1 d\theta\,\pi(\theta) = 1$. In fact, it is slightly more natural to work with the inverse square localisation length ξ^{-2} rather than the localisation length ξ itself, as well as the small parameter ϵ [$\equiv 3(\mu^2-1)$], which measures the amount by which the crosslink density control parameter μ^2 exceeds its critical value of unity. In terms of π the distribution of inverse square localisation lengths is given by

$$p(\xi^{-2}) = (2/\epsilon)\,\pi(2/\epsilon\,\xi^2). \tag{3}$$

Thus, the typical localisation length scales as $\epsilon^{-1/2}$, and the statistical distribution $P_{sc}(\xi/\xi_{av})$ of localisation lengths ξ scaled by the mean value ξ_{av}, is universal for all near-critical crosslink densities, and is predicted by our mean field theory to be given by

$$P_{sc}(y) = (2s/y^3)\,\pi(s/y^2), \tag{4}$$

where the constant $s \simeq 1.224$ is fixed by demanding that $\int_0^\infty dy\,y\,P_{sc}(y) = 1$. This theoretical prediction and its comparison with the data of Barsky and Plischke (unpublished, 1997) are shown in Fig. 5.
6: We have explained the origin of the universality of these results (by which we mean their insensitivity to the specific form and details of the system undergoing the

amorphous solidification transition) by establishing that the amorphous solidification transitions in all these systems are governed by a common Landau theory, as we shall discuss further in the section on universality.

We have also developed a theory of the elastic properties of random networks near the amorphous solidification transition, which we shall discuss further in the section on rigidity. The primary conclusions of this theory are that:

7: The amorphous solid (in the sense that a nonzero fraction of the monomers have become randomly localised) is a solid (in the sense that at the transition the system acquires a nonzero static shear modulus).

8: The shear modulus vanishes continuously as the transition is approached, and does so with the third power of the excess crosslink density ϵ (i.e. the amount by which the crosslink density exceeds its critical value).

9: The shearing of the container associated with elastic deformations does *not* lead to a shearing of the probability clouds associated with the thermal fluctuations of localised particles about their mean positions.

In addition to these results on the mean field structure and response of random networks, we have also obtained a number of further results. Specifically, we have: (i) established the local stability of our mean field theory with respect to arbitrary small fluctuations[22]; (ii) addressed certain correlators that become long ranged at the amorphous solidification transition[9]; (iii) begun to obtain information concerning spatial correlations between localisation lengths[23]; (iv) obtained some results concerning dynamical properties of random networks[16]; and (v) formulated (albeit in terms of a rather formidable "string" theory) the statistical dynamics of randomly crosslinked macromolecular networks. I will not dwell on these developments here.

AMORPHOUS SOLID ORDER PARAMETER: A DIAGNOSTIC TOOL

In the previous section I have presented some of the results that have emerged from our statistical mechanical approach to random network forming media, but I have done so without explaining how these results were obtained. I now give a tour of some of the concepts used to establish the results that I have just catalogued.

The central object through which most of our understanding has come is an order parameter for the amorphous solid state. Unlike more familiar examples from condensed matter physics, such as the magnetisation density and liquid crystallinity order parameters, this one is not a number (or set of numbers) but instead is a function of several variables, and it is via the dependence of the order parameter on these variables that physical information about the amorphous solid state is encoded. This order parameter is quite unusual, and I now discuss it in some detail, my ultimate aim being to show how it encodes this physical information.

How localised is a single particle?

Before encumbering the discussion with issues of internal molecular or macromolecular structure, let us consider a system of N identical structureless point particles, labelled[†] $i = 1, \ldots, N$. By structureless I mean that no internal degrees of freedom play any role. Envisage the system of particles to be enclosed in a cubic container of volume V. By specifying the positions in the container of all of the particles, i.e. $\{\mathbf{R}_1, \mathbf{R}_2, \ldots, \mathbf{R}_N\}$, we specify the microscopic state (i.e. the configuration) of the system. Suppose that the system is in some state of thermal equilibrium. What this means

[†]The issue of indistinguishability is an important and subtle one here: see Sec. 2.4 of Ref. [8].

is that we possess only probabilistic information about the microscopic state [i.e., that the probability density for finding the system to be in the configuration $\{\mathbf{R}_1, \ldots, \mathbf{R}_N\}$ is given by some distribution $\Pi(\{\mathbf{R}_1, \ldots, \mathbf{R}_N\})$, and that the Helmholtz free energy is a minimum with respect to variations in Π].

Let us denote equilibrium averages (i.e. averages taken with respect to Π) by angle brackets $\langle \cdots \rangle$. Then, e.g., the mean position of particle i in this state would be denoted $\langle \mathbf{R}_i \rangle$, and the probability of finding particle i to be at position $\mathbf{r}$ would be denoted $\langle \delta(\mathbf{r} - \mathbf{R}_i) \rangle$, where δ denotes the Dirac delta function. In fact, as the issue of random solidness (and crystalline solidness, for that matter) is intimately connected with the presence or absence of translational symmetry, it proves especially useful to adopt the Fourier perspective when addressing random solids, and we shall find ourselves analysing not the probability of finding particle i to be at position $\mathbf{r}$, viz., $\langle \delta(\mathbf{r} - \mathbf{R}_i) \rangle$, but instead its Fourier transform (ofter referred to as its *characteristic function*)

$$\int_V d^3r \, \langle \delta(\mathbf{r} - \mathbf{R}_i) \rangle \, \exp i\mathbf{k} \cdot \mathbf{r} = \langle \exp i\mathbf{k} \cdot \mathbf{R}_i \rangle, \tag{5}$$

where, owing to the periodic boundary conditions that we have imposed on the volume V, $\mathbf{k}$ is any of the discrete lattice of wave vectors $2\pi(n_x, n_y, n_z)/V^{1/3}$ obtained by allowing each of n_x, n_y and n_z to be any integer (positive, negative or zero).

What are the qualitative properties of the characteristic function $\langle \exp i\mathbf{k} \cdot \mathbf{R}_i \rangle$? The answer to this question depends crucially on the thermodynamic phase of the system, which, in the present context, depends not only on the constituents—atoms, molecules, macromolecules—their interactions and the temperature T, but also on the type of constraints that have been imposed and which degrees of freedom they involve: the *quenched random* information. The necessity of accounting for such information probabilistically, instead of facing the statistical mechanics of a system with a specific realisation of constraints, is what distinguishes random systems from pure ones, and introduces a source of probability beyond that arising in the usual way from statistical mechanics.

So, what is the value of $\langle \exp i\mathbf{k} \cdot \mathbf{R}_i \rangle$ in any fluid state? The essential feature of a fluid state is that system configurations related by any common shift $\mathbf{b}$ of all the particles are equally probable. It is easy to see a consequence of this: for any $\mathbf{b}$

$$\langle \exp i\mathbf{k} \cdot \mathbf{R}_i \rangle_{\substack{\text{fluid} \\ \text{state}}} = \langle \exp i\mathbf{k} \cdot \mathbf{R}_i \rangle_{\substack{\text{fluid} \\ \text{state}}} \exp i\mathbf{k} \cdot \mathbf{b}\,; \tag{6}$$

hence $\langle \exp i\mathbf{k} \cdot \mathbf{R}_i \rangle_{\substack{\text{fluid} \\ \text{state}}} = \delta_{\mathbf{k},\mathbf{0}}$. This simple observation has a useful corollary:

$$\lim_{\mathbf{k} \to \mathbf{0}} \langle \exp i\mathbf{k} \cdot \mathbf{R}_i \rangle_{\substack{\text{fluid} \\ \text{state}}} = 0, \tag{7}$$

$$\langle \exp i\mathbf{k} \cdot \mathbf{R}_i \rangle_{\substack{\text{fluid} \\ \text{state}}} \Big|_{\mathbf{k}=\mathbf{0}} = 1, \tag{8}$$

i.e., the characteristic function is discontinuous at $\mathbf{k} = \mathbf{0}$.

Now, how about $\langle \exp i\mathbf{k} \cdot \mathbf{R}_i \rangle$ in a solid state? In the kind of solid state that we shall be considering, (at least) a nonzero fraction of the particles have become spontaneously localised around certain mean positions, by which we mean that the probability cloud associated with the position of a localised particle is no longer extended over the volume V, instead being more or less concentrated near a mean position $\langle \mathbf{R}_i \rangle$. Whilst we cannot say, in general, precisely what the form of the probability cloud (and its Fourier transform the characteristic function) have, we can make a simple and surprisingly useful caricature via what is known as a *cumulant expansion*. Consider not $\langle \exp i\mathbf{k} \cdot \mathbf{R}_i \rangle_{\substack{\text{solid} \\ \text{state}}}$

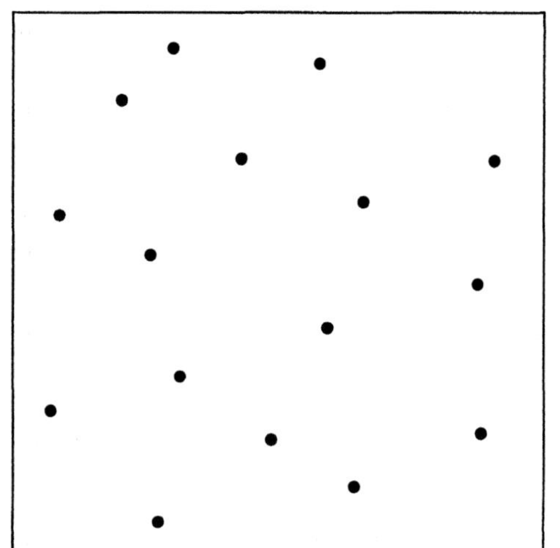

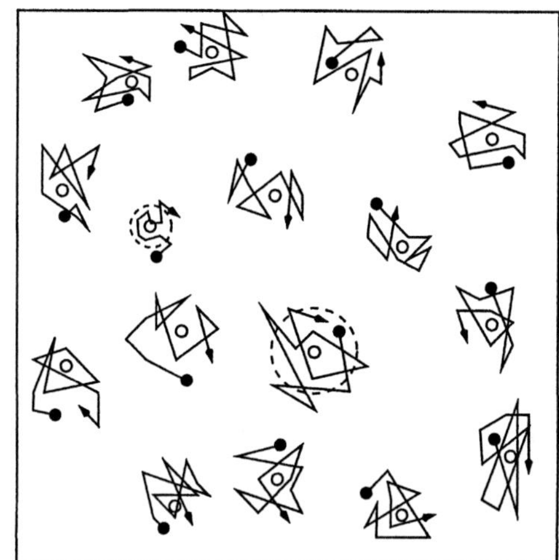

Figure 6. *Left*: atomic positions at some instant (filled circles). This does not reveal the distinction between liquid and amorphous solid states. *Right*: The same atomic positions at some instant (filled circles), together with the subsequent Brownian motion (jagged lines) and the time-averaged positions of the atoms (open circles). The (infinite-time) history distinguishes the liquid and amorphous solid states: beyond a (finite) relaxation time a liquid would exhibit a completely new random set of atomic positions, whereas for infinitely long times the amorphous solid (depicted here) would exhibit a random configuration partially correlated with the earlier one. As the state is amorphous the constituent atoms are localised to differing extents. For two atoms the characteristic volume visited is indicated by broken circles.

but instead its logarithm $\ln \langle \exp i\mathbf{k}\cdot\mathbf{R}_i\rangle_{\substack{\text{solid}\\\text{state}}}$, and expand the exponential function,

$$\begin{aligned}
\ln \langle \exp i\mathbf{k}\cdot\mathbf{R}_i\rangle_{\substack{\text{solid}\\\text{state}}} &= \ln \langle 1 + i\mathbf{k}\cdot\mathbf{R}_i + (i\mathbf{k}\cdot\mathbf{R}_i)^2/2 + \cdots\rangle_{\substack{\text{solid}\\\text{state}}} \\
&= \ln\Big\{1 + i\mathbf{k}\cdot\langle\mathbf{R}_i\rangle_{\substack{\text{solid}\\\text{state}}} - \frac{1}{2}\sum_{\nu,\nu'=x,y,z} k_\nu\, k_{\nu'}\,\langle R_{i,\nu}\, R_{i,\nu'}\rangle_{\substack{\text{solid}\\\text{state}}} + \cdots\Big\} \\
&= i\mathbf{k}\cdot\langle\mathbf{R}_i\rangle_{\substack{\text{solid}\\\text{state}}} - \frac{1}{2}\sum_{\nu,\nu'=x,y,z} k_\nu\, k_{\nu'}\langle(R_{i,\nu} - \langle R_{i,\nu}\rangle_{\substack{\text{solid}\\\text{state}}})(R_{i,\nu'} - \langle R_{i,\nu'}\rangle_{\substack{\text{solid}\\\text{state}}})\rangle_{\substack{\text{solid}\\\text{state}}} + \cdots .
\end{aligned}$$

By retaining only terms to quadratic order, assuming that the r.m.s. displacement $\langle(R_{i,\nu} - \langle R_{i,\nu}\rangle_{\substack{\text{solid}\\\text{state}}})(R_{i,\nu'} - \langle R_{i,\nu'}\rangle_{\substack{\text{solid}\\\text{state}}})\rangle_{\substack{\text{solid}\\\text{state}}}$ is isotropic (and hence given by $\xi_i^2\,\delta_{\nu,\nu'}$ where ξ_i is referred to as the *localisation length* for particle i), and re-exponentiating, one obtains the simple Gaussian result:

$$\langle \exp i\mathbf{k}\cdot\mathbf{R}_i\rangle_{\substack{\text{solid}\\\text{state}}} \approx \exp i\mathbf{k}\cdot\langle\mathbf{R}_i\rangle_{\substack{\text{solid}\\\text{state}}}\ \exp(-\xi_i^2\, k^2/2). \tag{9}$$

The first factor on the right-hand-side is a phase factor that encodes the mean location of the particle; the second factor is a Debye-Waller factor: its variation with wave number k encodes the localisation length. If particle i happens to be delocalised (i.e. $\xi_i = \infty$) then $\langle \exp i\mathbf{k}\cdot\mathbf{R}_i\rangle_{\substack{\text{solid}\\\text{state}}} = \delta_{\mathbf{k},\mathbf{0}}$, and $\lim_{\mathbf{k}\to\mathbf{0}}\langle \exp i\mathbf{k}\cdot\mathbf{R}_i\rangle_{\substack{\text{solid}\\\text{state}}} = 0$. On the other hand, if particle i happens to be localised (i.e. $\xi_i < \infty$) then $\lim_{\mathbf{k}\to\mathbf{0}}\langle \exp i\mathbf{k}\cdot\mathbf{R}_i\rangle_{\substack{\text{solid}\\\text{state}}} = 1$. In other words, via the limit $\mathbf{k}\to\mathbf{0}$ the characteristic function is capable of focusing on the qualitative difference between localisation and delocalisation. The reason for this is that in the limit $\mathbf{k}\to\mathbf{0}$ one is examining longer and longer length scales, and ultimately all localised particles "look" equally (and perfectly) localised, no matter what the value of their (finite) localisation lengths.

Thus we reach the conclusion that inasmuch as it allows one to discriminate between localisation and delocalisation, and its wave number dependence provides access to localisation lengths, the characteristic function can be expected to play a pivotal role in the statistical-mechanical description of the origins and properties of any amorphous solid state. Let us now explore how it does this, by turning from the issue of a single particle i to an assembly of particles $i = 1, \ldots, N$.

Amorphous solid order parameter for many point particles

By examining a single localised particle one cannot tell whether the system of which it is a part is in a crystalline solid state (in which the mean positions of the particles are arranged on some regular, highly symmetrical, lattice, and all localisation lengths are identical, or at least vary regularly between a discrete family of choices) or is in an amorphous solid state (in which the mean positions are arranged at random, with no long-distance regularity, and the localisation lengths also vary at random from one localised atom to another). But we can answer this question by examining a very carefully crafted *order parameter*.

We are now going build and examine the order parameter that is capable of distinguishing between the fluid, crystalline solid and amorphous solid states. (We shall soon find ourselves using this order parameter in our computation of certain properties of the amorphous solid state, including structural and elastic properties.) A little patience is needed at this stage. The object that we are building may, at first sight, look somewhat strange and artificial. But one very quickly gets used to it and, before long, one sees that it is reasonable, natural and, in fact, exceptionally useful.

So, what do we actually do? We take the characteristic function for a *single* particle i, viz., $\langle \exp i\mathbf{k} \cdot \mathbf{R}_i \rangle$, and we form g factors of it at an arbitrary sequence of wave vectors $\mathbf{k}^1, \mathbf{k}^2, \ldots, \mathbf{k}^g$:

$$\langle \exp i\mathbf{k}^1 \cdot \mathbf{R}_i \rangle \langle \exp i\mathbf{k}^2 \cdot \mathbf{R}_i \rangle \cdots \langle \exp i\mathbf{k}^g \cdot \mathbf{R}_i \rangle. \tag{10}$$

Next, we average this quantity over all the particles in the system,

$$\frac{1}{N} \sum_{i=1}^{N} \langle \exp i\mathbf{k}^1 \cdot \mathbf{R}_i \rangle \langle \exp i\mathbf{k}^2 \cdot \mathbf{R}_i \rangle \cdots \langle \exp i\mathbf{k}^g \cdot \mathbf{R}_i \rangle, \tag{11}$$

and finally we average the resulting quantity over realisations of the disorder (a procedure that we denote by square brackets $[\cdots]$):

$$\left[\frac{1}{N} \sum_{i=1}^{N} \langle \exp i\mathbf{k}^1 \cdot \mathbf{R}_i \rangle \langle \exp i\mathbf{k}^2 \cdot \mathbf{R}_i \rangle \cdots \langle \exp i\mathbf{k}^g \cdot \mathbf{R}_i \rangle \right], \tag{12}$$

What is unusual about this quantity is that the ingredient being averaged over all the particles in the system is a *product* of equilibrium averages, in contrast with the the kind of quantities that crop up in conventional statistical mechanics, which typically involve at least as many particle averages as thermal averages. (Examine, e.g., the connected density-density correlator of a simple fluid.)

Now, the chief characteristic of the amorphous solid state is not simply the spatially random arrangement of particle locations that a *snapshot* of the system would reveal—a fluid state would also yield such a random arrangement—but it is the *persistence for all time* of the spatially random arrangement of particle locations; see Fig. 6. (This persistent correlation is not, of course, perfect, owing to thermal fluctuations about the mean locations.) One could view time-persistence as a dynamical statement, and try to capture it by adopting a statistical-dynamical approach, in which the presence of time-persistent correlation would emerge from a *dynamical correlation function* in the limit that all time-differences are large compared with the longest equilibration time, e.g.,

$$\lim_{\substack{t_g - t_{g-1} \to \infty \\ \cdots\cdots \\ t_3 - t_2 \to \infty \\ t_2 - t_1 \to \infty}} \frac{1}{N} \sum_{i=1}^{N} \Big\langle \exp i\mathbf{k}^1 \cdot \mathbf{R}_i(t_1) \, \exp i\mathbf{k}^2 \cdot \mathbf{R}_i(t_2) \cdots \exp i\mathbf{k}^g \cdot \mathbf{R}_i(t_g) \Big\rangle, \tag{13}$$

where $\{t_1, t_2, \ldots, t_g\}$ are an ordered sequence of times, and the angle brackets temporarily refer to *statistical-dynamical* averages. However, in this limit the average *clusters* into factors,

$$\begin{aligned} &\langle \exp i\mathbf{k}^1 \cdot \mathbf{R}_i(t_1) \, \exp i\mathbf{k}^2 \cdot \mathbf{R}_i(t_2) \cdots \exp i\mathbf{k}^g \cdot \mathbf{R}_i(t_g) \rangle \\ &\qquad \longrightarrow \langle \exp i\mathbf{k}^1 \cdot \mathbf{R}_i(t_1) \rangle \, \langle \exp i\mathbf{k}^2 \cdot \mathbf{R}_i(t_1) \rangle \cdots \langle \exp i\mathbf{k}^g \cdot \mathbf{R}_i(t_1) \rangle, \end{aligned} \tag{14}$$

and, bearing in mind the homogeneity in time of the system, the individual factors can be computed via *equilibrium statistical mechanics*, i.e., without having to face the (immense) complexities of statistical dynamics. Thus, in retrospect, we see that an order parameter containing a sequence of factors of characteristic functions is just what is needed, if an approach to the amorphous solid state based on *equilibrium* statistical mechanics is to be successful.

But the order parameter, Eq. (12), depends on the g wave vectors $\{\mathbf{k}^1, \mathbf{k}^2, \ldots, \mathbf{k}^g\}$, i.e., it is a function of several variables, and it would be wise for us to try to get a feeling for what this function looks like, and what information it encodes before attempting to calculate it. (In fact, when we compute the order parameter via replica statistical mechanics we find ourselves facing one further technical hurdle, as g will turn out to be $n+1$ and the limit $n \to 0$ will have to be taken.) Recall our Gaussian approximation to the characteristic function,

$$\langle \exp i\mathbf{k} \cdot \mathbf{R}_i \rangle \approx \exp i\mathbf{k} \cdot \langle \mathbf{R}_i \rangle \, \exp(-\xi_i^2 \, k^2/2), \tag{15}$$

which also contains the result for delocalised particles inasmuch as it becomes $\delta_{\mathbf{k},\mathbf{0}}$ in the $\xi_i \to \infty$ limit. (Higher terms in the cumulant expansion turn out to be irrelevant near the solidification transition.) By inserting this approximation into the order parameter we obtain

$$\begin{aligned} &\left[\frac{1}{N} \sum_{i=1}^{N} \exp i\mathbf{k}^1 \cdot \langle \mathbf{R}_i \rangle \, \exp(-\xi_i^2 \, |\mathbf{k}^1|^2/2) \cdots \exp i\mathbf{k}^g \cdot \langle \mathbf{R}_i \rangle \, \exp(-\xi_i^2 \, |\mathbf{k}^g|^2/2) \right] \\ &= \left[\frac{1}{N} \sum_{i=1}^{N} \exp\left(i\langle \mathbf{R}_i \rangle \cdot \textstyle\sum_{\alpha=1}^{g} \mathbf{k}^\alpha \right) \exp\left(-\xi_i^2 \textstyle\sum_{\alpha=1}^{g} |\mathbf{k}^g|^2/2 \right) \right] \\ &= \int_V d\mathrm{R} \int_0^\infty d\xi \left[\frac{1}{N} \sum_{i=1}^{N} \delta(\mathrm{R} - \langle \mathbf{R}_i \rangle) \, \delta(\xi - \xi_i) \right] \exp\left(i\mathbf{R} \cdot \sum_{\alpha=1}^{g} \mathbf{k}^\alpha \right) \exp\left(-\frac{\xi^2}{2} \sum_{\alpha=1}^{g} |\mathbf{k}^g|^2 \right). \end{aligned}$$

Physical content of the order parameter: percolation and beyond

We are beginning to see the physical content of the order parameter emerging. Accepting, for the sake of discussion, the validity of the Gaussian approximation, the order parameter is seen to depend only on the symmetric combinations of wave vectors $\sum_{\alpha=1}^{g} \mathbf{k}^\alpha$ and $\sum_{\alpha=1}^{g} |\mathbf{k}^g|^2/2$, and is a simple transform of the *normalised joint position-localisation-length histogram* $\left[N^{-1} \sum_{i=1}^{N} \delta(\mathbf{R} - \langle \mathbf{R}_i \rangle) \, \delta(\xi - \xi_i) \right]$, a Fourier transform with respect to $\mathbf{R}$ and an (elementary variation of a) Laplace transform with respect to ξ. So, knowledge of the order parameter as a function of the wave vectors $\{\mathbf{k}^1, \mathbf{k}^2, \ldots, \mathbf{k}^g\}$ yields, via inverse transformations, this histogram (or joint probability distribution) as a useful and unanticipated diagnostic of the equilibrium state of the system. Indeed, once the disorder average has been performed there can no longer be any correlation between the position in the sample and the localisation length of the localised particle found there, so that the histogram becomes simply $V^{-1}p(\xi)$, i.e., proportional to the disorder-averaged distribution of localisation lengths $p(\xi)$.

Let us explore this point a little further. Suppose that the system is in a fluid state (i.e. all particles are delocalised). Then the order parameter takes the value

$$\prod_{\alpha=1}^{g} \delta_{\mathbf{k}^\alpha,\mathbf{0}} \ , \tag{16}$$

i.e., the order parameter is nonzero only if all the wave vectors are zero. This corresponds to a distribution of localisation lengths such that only infinite localisation lengths are found[‡] Henceforth, we shall remove "by hand" from this distribution any delocalised particles (a fraction $1-q$) and rescale the remaining distribution (by dividing by q) so that we arrive at the distribution of localisation lengths pertaining solely to localised particles (if any there be).

Now suppose that the system is in a crystalline solid state (i.e. all particles are localised, their mean particle positions lie on a regular lattice, and their localisation lengths ξ_i take a common value, say ξ_1, at least in the simplest case). Then the order parameter takes the value

$$\exp\Big(-\xi_1^2 \sum_{\alpha=1}^{g} |\mathbf{k}^g|^2/2\Big) \frac{1}{N} \sum_{\mathbf{r}} \exp\Big(i\mathbf{r}\cdot \sum_{\alpha=1}^{g} \mathbf{k}^\alpha\Big) = \exp\Big(-\xi_1^2 \sum_{\alpha=1}^{g} |\mathbf{k}^g|^2/2\Big) \sum_{\mathbf{G}} \delta_{\mathbf{G},\sum_{\alpha=1}^{g}\mathbf{k}^\alpha} , \tag{17}$$

where the sum on the left hand side is taken over all N sites $\mathbf{r}$ of the crystalline lattice inside the volume V (one of which sites lies at the origin), and the sum on the right hand side is taken over all reciprocal lattice vectors of the crystalline lattice. In contrast with the fluid state, the order parameter is no longer nonzero only at the origin (of the space spanned by the g wave vectors), and now has weight on the hypersurfaces in this space defined by $\sum_{\alpha=1}^{g} \mathbf{k}^\alpha = \mathbf{G}$, on which the wave vectors sum to any of the reciprocal lattice vectors $\mathbf{G}$. Moreover, from the way the order parameter varies under the (continuous) variation of the wave vectors over any of these (discrete) surfaces one can extract the (common) localisation length ξ_1.

Now suppose that the system is in the amorphous solid state, i.e., some nonzero fraction q of the particles are localised about random mean positions and with random localisation lengths. A useful way to think of this state is that it is the extremely special case of the crystalline state in which the unit cell is the entire system (i.e. the structure is not built by the periodic repetition of some elemental motif), so that the only reciprocal lattice vector is $\mathbf{G} = \mathbf{0}$. We have already seen the value of the order parameter in the liquid state, Eq. (16). Bearing in mind the notion that a fraction $1-q$ of the particles are delocalised, we expect that the delocalised particles contribute a term $(1-q)\prod_{\alpha=1}^{g} \delta_{\mathbf{k}^\alpha,\mathbf{0}}$ to the sum over particles in the order parameter. As for the remaining contribution, as we are now denoting by $p(\xi)$ the normalised distribution of localisation lengths for the particles in the localised fraction then for the localised part of the position-localisation-length histogram $\big[N^{-1}\sum_{i=1}^{N} \delta(\mathrm{R} - \langle \mathbf{R}_i \rangle)\, \delta(\xi - \xi_i)\big]$ we have

[‡]If you are queasy about having a δ function peak located at $\xi = \infty$ then make the transformation from the variable ξ to a new variable $1/\xi$.

the value $V^{-1} q\, p(\xi)$, and the value of the order parameter becomes

$$(1-q) \prod_{\alpha=1}^{g} \delta_{\mathbf{k}^\alpha,\mathbf{0}} + q\, \delta_{\mathbf{0},\sum_{\alpha=1}^{g} \mathbf{k}^\alpha} \int d\xi\, p(\xi) \exp\Big(-\xi_1^2 \sum_{\alpha=1}^{g} |\mathbf{k}^g|^2/2\Big). \tag{18}$$

The Kronecker δ factor in the second term reflects the complete absence of periodicity, which we refer to as macroscopic translational invariance. Notice that this form of the order parameter tends to the limit q, in the limit that all g wave vectors are small (and add to zero), just as it should given the definition of the order parameter, Eq. (12), and the remarks concerning the (differing) long wave length limits of the characteristic function for localised and delocalised particles.[§] Also notice that the order parameter falls into two pieces: for small q the first piece is simple and large; the second is more interesting and small, and is simply related to the distribution of localisation lengths for the localised particles $p(\xi)$. This decomposition is one of two reasons why a perturbative treatment of the amorphous solidification transition is possible and yields universal results (see the section on universality.). The second reason is that near the transition the characteristic value of the localisation lengths of the localised particles is large.

By now you may well be asking yourself: Why have g copies of the characteristic function $\langle \exp i\mathbf{k}\cdot\mathbf{R}\rangle$ in the order parameter (12) and not just one? Now, as we have seen, for $\mathbf{k} \neq \mathbf{0}$ the characteristic functions differ for localised and delocalised particles, so for $g = 1$ the terms in the order parameter differ at $\mathbf{k} \neq \mathbf{0}$. However, by the time we have summed over all the particles, the phase factors [see Eq. (15)]—which are random, owing to the absence of crystallinity—destructively interfere, ultimately giving complete cancellation for a thermodynamically large system.[¶] Thus we see that for $g = 1$ the order parameter would not distinguish between the two states of interest: the fluid state and the amorphous solid state. By contrast, in a crystalline state the periodicity of the mean particle-positions would guarantee that for certain wave vectors (i.e. for reciprocal lattice vectors) the phase factors would add coherently, leading to a nonzero result and, hence, to a diagnostic capable of distinguishing between the fluid and crystalline states.

Then why not just consider $g = 2$? In spin glass physics,[11, 12, 24] where the difficulty of diagnosing random freezing was first addressed, one commonly finds that the corresponding magnetic order parameter (i.e. the so-called Edwards-Anderson order parameter) can be taken to have $g = 2$. [In fact, this is not always the case for spin glasses: if the fraction of frozen magnetic sites is small but the freezing is strong then one needs to account for the analogues of the $g = 2, 3, 4, \ldots$ order parameters; see Ref. (25).] Similarly, at the amorphous solidification transition only a small fraction of the particles are localised, but for wave vectors smaller than or comparable to the inverse characteristic localisation length their localisation appears strong, so that further factors of characteristic functions do not diminish the contribution of a given localised

§There is an analogy with Bose-Einstein condensation here. In homogeneous systems, Bose-Einstein condensation amounts to a macroscopic population of the single-particle quantum mechanical ground (i.e. $\mathbf{k} = 0$) state. Amorphous solidification amounts to a macroscopic *de*-population of the $\mathbf{k} = \mathbf{0}$ Fourier component of the spatial probability density. In both cases one must be careful to treat the $\mathbf{k} = \mathbf{0}$ contributions to wave vector summations correctly.

¶Consider any nonzero wave vector, even the smallest, given the periodic boundary conditions. Then, as the sample is scanned, the projections of the mean positions of the localised particles on to the direction parallel to the wave vector will cover the line progressively more uniformly, as the sample size is increased. This will be true even if the particles are filtered according to localisation length, and thus the $g = 1$ sector of the order parameter will vanish for the amorphous solid state as well as the liquid state.

particle. Moreover, in the form of the order parameter proposed in Eq. (18) we see that there is no increase in the essential level of complication brought about by including arbitrary g, just more terms in some sums over wave vectors. Furthermore, when we come to construct a self-consistent equation for the order parameter we shall find ourselves needing to consider not integral values of g but *non-integral values*, specifically $g = n+1$ in the $n \to 0$ limit! The fact that the form of the order parameter has such a simple and highly symmetric structure for its dependence on the wave vectors allows us to establish results for integral g [and hence to extract the distribution of localisation lengths for the localised particles $p(\xi)$ from results obtained in the $n \to 0$ limit].

Amorphous solid order parameter for macromolecular networks

If the constituents, prior to the incorporation of constraints, were flexible or semi-flexible linear macromolecules, what would the order parameter look like? At the semi-microscopic level we may index the N distinct macromolecules by $i = 1, \ldots, N$ and locations on the macromolecules by the fractional arclength s (with $0 \leq s \leq 1$). Then the location in space of the monomer a fraction s of the arclength from one end of macromolecule i is denoted $\mathbf{R}_i(s)$. Averaging over the particles then involves two steps: summing over the macromolecules and integrating over their arclengths. This leads to the order parameter having the slightly more elaborate definition:

$$\frac{1}{N}\sum_{i=1}^{N}\int_0^1 ds \, \langle \exp i\mathbf{k}^1 \cdot \mathbf{R}_i(s)\rangle \, \langle \exp i\mathbf{k}^2 \cdot \mathbf{R}_i(s)\rangle \cdots \langle \exp i\mathbf{k}^g \cdot \mathbf{R}_i(s)\rangle. \tag{19}$$

Note that s denotes the arclength location, not the time: the averaging is still over constituents, although they are now labelled in two ways (i.e. which macromolecule and where on it). As for the physical value of the order parameter, it turns out that there is no essential change brought about by the assembling of degrees of freedom into macromolecules, and it is adequate to adopt the form for the order parameter given in Eq. (18), as well as the physical interpretation of its ingredients: q and $p(\xi)$.

In fact, this macromolecular setting for the exploration of the implications of imposing quenched random constraints is important, the constraints being the crosslinks (or endlinks) brought about by vulcanisation, and the resulting solid being rubber [see Ref. (26)].

Amorphous solid order parameter for low-molecular weight networks

Let us now turn to an application of our statistical mechanical technology to the setting of random networks formed from atoms or low-molecular-weight molecules.[7, 27] Let us regard our constituents as point-like particles from which radiate a number A of legs (representing orbitals for possible covalent chemical bonds), as shown in Fig. 3. Network formation then proceeds by the identification of randomly selected pairs of legs that are constrained to form a permanent bond (i.e. to lie on top of one another and to point in opposite directions). Notice that such constraints do not explicitly break either translational or rotational symmetry. However, if present at sufficient densities, they cause a transition to an amorphous solid state possessing a new feature: not only do the particle positions become spontaneously localised but also the orientations of the legs becomes orientationally localised. This feature emerges from the statistical mechanical approach, and leads to a more elaborate order parameter, which we now discuss.

What is the structure of this order parameter? Let us now regard $\mathbf{R}_i$ as the position of particle i (with $i = 1, \ldots, N$), and let us introduce the unit vectors $\mathbf{S}_{i,a}$

(with $a = 1, \ldots, A$) describing the microscopic orientation of leg a on particle i. The the order parameter is found to have the structure[7, 27]

$$\frac{1}{N}\sum_{i=1}^{N}\frac{1}{A}\sum_{a=1}^{A}\langle\exp i\mathbf{k}^1\cdot\mathbf{R}_i\, Y_{\ell^1 m^1}(\mathbf{S}_{i,a})\rangle\,\langle\exp i\mathbf{k}^2\cdot\mathbf{R}_i\, Y_{\ell^2 m^2}(\mathbf{S}_{i,a})\rangle$$
$$\times\langle\exp i\mathbf{k}^3\cdot\mathbf{R}_i\, Y_{\ell^3 m^3}(\mathbf{S}_{i,a})\rangle\cdots\langle\exp i\mathbf{k}^g\cdot\mathbf{R}_i\, Y_{\ell^g m^g}(\mathbf{S}_{i,a})\rangle, \quad (20)$$

where $Y_{\ell m}$ are the usual spherical harmonic functions. (Recall that plane waves are *Euclidian* harmonic functions.) By setting all the orientational indices $\{\ell^1, m^1; \ell^2, m^2; \ldots; \ell^g, m^g\}$ to zero we recover (up to some elementary factors of 4π) the by-now familiar positional amorphous solid order parameter, which is capable of detecting random positional ordering. But we also have two new features. By setting all the wave vectors to zero we arrive at the form

$$\left[\frac{1}{N}\sum_{i=1}^{N}\frac{1}{A}\sum_{a=1}^{A}\langle Y_{\ell^1 m^1}(\mathbf{S}_{i,a})\rangle\,\langle Y_{\ell^2 m^2}(\mathbf{S}_{i,a})\rangle\ldots\langle Y_{\ell^g m^g}(\mathbf{S}_{i,a})\rangle\right], \quad (21)$$

which is an extension (to higher moments and higher angular momenta) of the original Edwards-Anderson spin glass order parameter;[11] and is capable of detecting random orientational ordering of the legs. And in the general case, in which there are thermal averages involving both positionally and orientationally varying harmonic functions, the order parameter is also sensitive to the correlations between orientational and positional thermal fluctuations.

Can we use our physical intuition to develop a hypothesis for the form of this order parameter? Yes, and here is one possible route having an emerging structure that is rather rich and engaging. Let us suppose that we follow our previous strategy of establishing a cumulant expansion for the complex conjugate of the (now more elaborate) characteristic function $\langle\exp(-i\mathbf{k}\cdot\mathbf{R})\, Y^*_{\ell m}(\mathbf{S})\rangle$. The principles are precisely as they were for the purely translational case, but the details are a bit more complicated. Specifically, in addition to the mean position $\langle\mathbf{R}_i\rangle$ we introduce a *most probable orientation* $\overline{\mathbf{S}}$ and a random rotation of it described by the vector $\mathbf{a}$ (such that the axis of the rotation is the unit vector $\mathbf{a}/a$ and the magnitude of the rotation is a). We also parametrise fluctuations and correlators of the positional and orientational degrees of freedom via the

$$\langle(R_\mu - \langle R_\mu\rangle)(R_\nu - \langle R_\nu\rangle)\rangle = \xi^2\,\delta_{\mu\nu}\,, \quad (22)$$
$$\langle a_\mu\, a_\nu\rangle = \eta^2\left(\delta_{\mu\nu} - \overline{S}_\mu\,\overline{S}_\nu\right), \quad (23)$$
$$\langle(R_\mu - \langle R_\mu\rangle)\, a_\nu\rangle = \chi\,\epsilon_{\mu\nu\rho'}\,\overline{S}_\rho. \quad (24)$$

Some comments are in order here. We have assumed that the correlator for the position fluctuations remains anisotropic, despite the presence of a preferred direction $\overline{\mathbf{S}}$. The subtraction of $\overline{S}_\nu\,\overline{S}_{\nu'}$ in the correlator of rotation fluctuations ensures that only rotations that tilt $\overline{\mathbf{S}}$ are included. The ϵ symbol in the fluctuation correlator reflects the polar nature of the position vector and the axial nature of the rotation vector: it ensures a sensible structure for the correlations between the fluctuations in position and orientation. By undertaking a little Wigner $\mathcal{D}$-function technology (to handle the random rotations of the argument of the spherical harmonic function), using these correlators in the cumulant expansion of the characteristic function, and performing a little matrix algebra, one arrives at the following extension of the Gaussian approximation, Eq. (9), that we considered for the case of purely positional degrees of freedom:

$$\langle\exp -i\mathbf{k}\cdot\mathbf{R}\; Y^*_{\ell m}(\mathbf{S})\rangle \approx \mathrm{e}^{-i\mathbf{k}\cdot\langle\mathbf{R}\rangle}\,\mathrm{e}^{-\xi^2 k^2/2}\,\mathrm{e}^{-\eta^2\ell(\ell+1)/2}\,\times$$

$$\sum_{m'=-\ell}^{\ell} \{\ell m| \exp\left(\frac{\eta^2}{2\hbar^2}(\overline{\mathbf{S}}\cdot\hat{\mathbf{L}})^2 + \frac{\chi}{\hbar}(\mathbf{k}\times\overline{\mathbf{S}})\cdot\hat{\mathbf{L}}\right)|\ell m'\}\, Y^*_{\ell m'}(\overline{\mathbf{S}}), \tag{25}$$

where the (nonstandard) notation $\{\cdots|\cdots|\cdots\}$ indicates a quantum mechanical matrix element between the usual angular momentum eigenstates states $|\ell m\}$ and $|\ell m'\}$, and the symbol $\hat{\mathbf{L}}$ denotes a quantum mechanical angular momentum operator. (Of course, there is nothing quantum mechanical about this equation. It is just that the familiar quantum mechanical notation is an economical way of stating results associated with random rotations of the arguments of spherical harmonic functions.)

We now turn to the form of the order parameter that the cumulant expansion for the characteristic function engenders. By applying Eq. (25) to Eq. (20), and introducing the appropriate generalisation of the distribution of localisations lengths, viz., the joint probability distribution $p(\xi,\eta,\chi)$ for finding the particle to have localisation length ξ, orientational localisation angle η and fluctuation correlation strength χ, we find that the order parameter becomes

$$(1-q)\prod_{\alpha=1}^{g}\delta_{\mathbf{k}^\alpha,0} + q\,\delta_{0,\sum_{\alpha=1}^{g}\mathbf{k}^\alpha}\int d\xi\, d\eta\, d\chi\, p(\xi,\eta,\chi)\, \mathrm{e}^{-\xi^2\hat{k}^2/2}\, \mathrm{e}^{-\eta^2\hat{\ell}^2/2}\int\frac{d^2\overline{S}}{4\pi}$$
$$\times\sum_{\hat{m}'}\prod_{\alpha=1}^{g}\left(\{\ell^\alpha m^\alpha|\exp\left(\frac{\eta^2}{2\hbar^2}(\overline{\mathbf{S}}\cdot\hat{\mathbf{L}})^2 + \frac{\chi}{\hbar}(\mathbf{k}\times\overline{\mathbf{S}})\cdot\hat{\mathbf{L}}\right)|\ell^\alpha m'^\alpha\}\, Y^*_{\ell^\alpha m'^\alpha}(\overline{\mathbf{S}})\right), \tag{26}$$

where $\tilde{\mathbf{k}} \equiv \sum_{\alpha=1}^{g}\mathbf{k}^\alpha$, $\hat{k}^2 \equiv \sum_{\alpha=1}^{g}|\mathbf{k}^\alpha|^2$, and $\hat{\ell}^2 \equiv \sum_{\alpha=1}^{g}\ell^\alpha(\ell^\alpha+1)$.[‖]

At first sight this form for the order parameter may seem ungainly, but many of its parts are just associated with the geometry of rotations. Generally, technical trappings such as these fall by the wayside when amorphous solid order parameters are actually computed, and what emerges is information concerning the the physical significant quantity $p(\xi,\eta,\chi)$. At the present time it is not clear what is the ultimate amount of information that can be obtained via this approach. The information tends to emerge in the form of reduced distributions, such as $\int d\eta\, d\chi\, p(\xi,\eta,\chi)$ (i.e. the distribution of ξ, regardless of η and χ), and perhaps $\int d\eta\, d\chi\, \exp(-4c)p(\xi,\eta,\chi)$ (which gives information on the extent of angular localisation). I expect that it will prove possible to extract further information, including some concerning the statistics of the correlation strength χ. Work in progress with K. Shakhnovich is aimed precisely at the problem of determining the physical content of the order parameter discussed in the present section.

GETTING DOWN TO BUSINESS: CALCULATING THE PROPERTIES OF THE AMORPHOUS SOLID STATE

Having spent some time discussing the construction and interpretation of order parameters crafted to allow us to quantify the amorphous solid state, I now briefly and schematically touch upon the primary approach that we have been taking to compute such order parameters, viz., the statistical mechanics of systems having quenched disorder. For the case of randomly crosslinked flexible macromolecules the full technical details (up to and including the level of actually "doing the integrals"!) have been reported in Ref. [8]. The case of endlinked macromolecules has been discussed in Ref. [6]. The beginnings of the picture for the case of low-molecular-weight networks was discussed in Refs. [7, 27].

[‖]M. Makeev and K. Shakhnovich participated in the construction of this hypothetical form for the order parameter.

Obtaining the structure of the state via semi-microscopic models

At the the most basic level, the approach consists of the following sequence of steps (see Ref. [8] for details):

1: Select a semi-microscopic model for the unconstrained fluid system; address kinematic issues (what are the appropriate degrees of freedom) and the energetic issues (what Boltzmann weight should be ascribed to each configuration).

2: Identify the appropriate random constraints, and determine which formerly admissible configurations they rule out.

3: Write a formula for the partition function of the system for a specific realisation of the random constraints. This is the partition function for the unconstrained system, except that the Boltzmann weight is multiplied by a a mathematical *switch* that deletes from the sum over configurations all configurations that fail to satisfy all the constraints.

4: Construct a model for the statistical distribution of realisations of the constraints, which includes a parameter that controls the mean density of constraints. The mathematical form of this weight is determined by the physical procedure selected to introduce the constraints.

5: Avoid the (prohibitive) difficulty of working with a specific realisation of random constraints, and the vast amount of information needed to characterise the realisation, by averaging with respect to the statistical distribution of realisations of the constraints. One must be careful to average in a physically meaningful way, so as to preserve the crucial distinction between the quenched variables and the annealed variables. Specifically, one should average the free energy, the order parameter, correlation functions, etc., but not the partition function. It is this disorder-averaging stage that presents the need to invoke the so-called replica technique, which provides us with a tool for commuting the operations of averaging over the disorder (the quenched variables) and averaging over the configurations (the annealed variables).

6: At this stage we have averaged over the quenched random variables, so no explicit reference to them remains. However, their impact remains, in the form of interactions between the copies of the annealed variables, copies that the replica technique forced us to introduce. And we still must face the task of averaging over the configurations of the (replicas of the) annealed variables, a task that we cannot complete exactly and which we instead address by using a version of mean field theory. From the mean field theory we obtain a self-consistent equation for the order parameter, as well as an expression for the free energy in terms of the order parameter.

Interlude: How to do mean field theory in a simple setting

To illustrate how one can actually implement the mean field approximation in the simplest of settings let us consider the Weiss mean field theory of the magnetism of Ising spins. Put Ising spins $\sigma(\mathbf{x}) = \pm 1$ on the sites $\mathbf{x}$ of a regular lattice of nearest neighbour coordination number z, and let the spins interact ferromagnetically, with an exchange interaction J/z, so that their Hamiltonian is given by

$$\mathcal{H} = -z^{-1} J \sum_{\langle \mathbf{x}\mathbf{x}' \rangle} \sigma(\mathbf{x})\, \sigma(\mathbf{x}'), \tag{27}$$

where the summation is taken over all nearest neighbour pairs of lattice sites. Now aim to compute the equilibrium magnetisation density m, defined by

$$m \equiv \left\langle N^{-1} \sum_{\mathbf{x}} \sigma(\mathbf{x}) \right\rangle = \frac{\mathrm{Tr}\, \mathrm{e}^{-\mathcal{H}/k_{\mathrm{B}}T} N^{-1} \sum_{\mathbf{x}} \sigma(\mathbf{x})}{\mathrm{Tr}\, \mathrm{e}^{-\mathcal{H}/k_{\mathrm{B}}T}}, \tag{28}$$

where k_B is Boltzmann's constant, T is the temperature, and the trace Tr indicates a summation over all 2^N configurations of the Ising spins. To implement the mean field approximation we choose to write $\sigma(\mathbf{x})$ as $m + (\sigma(\mathbf{x}) - m)$, and regard the second term on the right hand side, the fluctuation $\sigma(\mathbf{x}) - m$, as a small quantity. We then insert this rewriting of $\sigma(\mathbf{x})$ into $\mathcal{H}$, expand in powers of the fluctuation, and retain terms only to *linear* order in the fluctuation. In this way, we neglect interactions that tend to correlate the fluctuations, and we obtain for m the approximation

$$m \approx \frac{\mathrm{Tr}\, e^{-\mathcal{H}_{\mathrm{MF}}/k_B T} N^{-1} \sum_{\mathbf{x}} \sigma(\mathbf{x})}{\mathrm{Tr}\, e^{-\mathcal{H}_{\mathrm{MF}}/k_B T}}, \tag{29}$$

where

$$\mathcal{H}_{\mathrm{MF}} \equiv \frac{1}{2} N J m^2 - J m \sum_{\mathbf{x}} \sigma(\mathbf{x}). \tag{30}$$

The Ising spins no longer interact directly with one another, so the trace can be performed easily, and we arrive at the following self-consistent equation:

$$m = \tanh\left(mJ/k_B T\right). \tag{31}$$

It is straightforward to ascertain (e.g. graphically or numerically) that for small values of the control parameter $J/k_B T$ the equilibrium state of the system is paramagnetic (i.e. not ordered, $m = 0$) whereas for large values it is ferromagnetic (i.e. ordered, $m \neq 0$). Moreover, there is an analytically locatable critical value of the control parameter $J/k_B T_c = 1$ at which the continuous transition between the paramagnetic and ferromagnetic states occurs, and in the vicinity of this regime one finds $m \approx \varepsilon^{1/2}$, where $J/k_B T \equiv 1 + \varepsilon/3$.

Although the technical implementation of the mean field approximation for amorphous solids is substantially more elaborate, the spirit is precisely as it is here, the essence being the neglect of correlations between *fluctuations of order parameter variables*. Moreover, the order parameters that we have described in detail (in the section on order parameters as diagnostic tools) turn out to be precisely the right objects for undertaking this approximation.

Mean field theory for vulcanised macromolecular networks

By applying the method sketched in the previous subsection to the amorphous solid order parameter for randomly crosslinked macromolecular networks, Eq. (12), which we denote by $\Omega(\mathbf{k}^0, \ldots, \mathbf{k}^n)$, we arrive at the self-consistency condition

$$\Omega(\mathbf{k}^0, \ldots, \mathbf{k}^n) = \frac{\left\langle \int_0^1 ds \prod_{\alpha=0}^{n} \exp\left(i\mathbf{k}^\alpha \cdot \mathbf{c}^\alpha(s)\right) \times \exp\left(\frac{\mu^2}{V^n} \mathrm{Re} \overline{\sum_{\mathbf{p}^0,\ldots,\mathbf{p}^n}} \Omega(\mathbf{p}^0, \ldots, \mathbf{p}^n)^* \int_0^1 d\sigma \prod_{\alpha=0}^{n} \exp\left(i\mathbf{p}^\alpha \cdot \mathbf{c}^\alpha(\sigma)\right)\right) \right\rangle_{n+1}^{\mathrm{W}}}{\left\langle \exp\left(\frac{\mu^2}{V^n} \mathrm{Re} \overline{\sum_{\mathbf{p}^0,\ldots,\mathbf{p}^n}} \Omega(\mathbf{p}^0, \ldots, \mathbf{p}^n)^* \int_0^1 d\sigma \prod_{\alpha=0}^{n} \exp\left(i\mathbf{p}^\alpha \cdot \mathbf{c}^\alpha(\sigma)\right)\right) \right\rangle_{n+1}^{\mathrm{W}}}, \tag{32}$$

where the angle brackets $\langle \cdots \rangle_{n+1}^{\mathrm{W}}$ indicate an average taken with respect to a system of $n + 1$ replicas of a single free macromolecule, the overbar on the summation indicates that only terms with at least two nonzero wave vectors should be included, the limit $n \to 0$ must be taken and, once again, μ^2 is the crosslink density control parameter.

[Equivalently, one can obtain this equation for the order parameter via a field theoretic approach. This has the virtue of being readily extendable to more detailed questions involving, e.g., the role of fluctuations; see Ref. (8).] The exclusion of terms with only one nonzero wave vector amounts to the incorporation of the fact that steric repulsions between the constituents, which are assumed to be present, have the effect of maintaining the stability of the system with respect to macroscopically inhomogeneous states, such as crystalline states. One can see a kind of frustration at work here: the crosslinks create a tendency for the the monomers to coalesce (simultaneously in all replicas); the monomer repulsions create a tendency for the monomers to remain homogeneous (i.e. not to coalesce) but act separately in each replica. The system accommodates these competing tendencies by adopting an amorphous solid state, which is macroscopically homogeneous but microscopically inhomogeneous.

In the simple case of the Weiss mean field theory of magnetism, as we have just reminded ourselves, the self-consistent equation for the order parameter can be handled easily: it is a single equation (albeit transcendental) for a single number (the magnetisation density), the equation can be determined exactly, and the physical content of the mean field theory can readily be extracted at all temperatures. In the case of amorphous solidification, the situation is not so simple. Instead of having to seek the single number that solves a single equation, one must fact the task of determining the order parameter $\Omega(\mathbf{k}^0, \ldots, \mathbf{k}^n)$, a complex-valued function of $3(n+1)$ variables in the limit $n \to 0$, this function satisfying a complicated functional equation, an equation that we are not able to compute in closed form. Even if we restrict ourselves to the vicinity of the transition, where we can at least construct explicitly (via perturbation theory) the equation that the order parameter satisfies, we do not know how to find the general solution.

However, despite the complexity of the situation we are able to make progress because the form of the order parameter constructed on physical grounds, Eq. (18), turns out to be an *exact solution of the order parameter self-consistency condition*, and yields relatively simple equations for the physical ingredients: the fraction of localised particles q and the distribution of localisation lengths for the localised particles $p(\xi)$. These results were given in Eqs. (1) and (2). By solving these equations we arrive at the results reviewed in the section on goals and results, and in Fig. 5.

When one considers the statistical mechanics of randomly constrained atomic and (low molecular weight) molecular networks one finds a self-consistent equation similar in form to Eq. (32) but for the order parameter (20) rather than (19). We have considered various variational schemes for obtaining approximate solutions for the order parameter for randomly constrained atomic and molecular networks, the least crude being described in Ref. [27]. Work in progress with K. Shakhnovich aims at constructing an *exact* value for this order parameter, valid at least in the vicinity of the solidification transition.

Why statistical field theory?

One can develop mean field theories just as we did in the interlude on mean field theory, i.e., via an essentially *ad hoc* neglect of the correlations between order parameter fluctuations. Alternatively, one can proceed more systematically, via a method from statistical field theory known as the collective field (or auxilliary field or Hubbard-Stratonovich) method. The virtue of the collective field method is that it allows one to set up a framework for going beyond mean field theory in a systematic way, although physical insight remains a vital ingredient for the initial introduction of the appropriate fields. The development presented in Ref. [8] proceeds via the collective field method.

Transforming from a picture in which the coordinates of the atoms (or molecules or macromolecules) is at the forefront to one in which the order parameter fields are at the forefront is also useful, in that it yields a continuum field description, which is readily addressed via the calculus of functional derivatives.

UNIVERSALITY OF THE EQUILIBRIUM AMORPHOUS SOLIDIFICATION TRANSITION: A LANDAU THEORY

Having obtained a picture of the vulcanisation transition and the amorphous solid state by analysing a particular semi-microscopic model of vulcanised macromolecular matter, it seems reasonable to examine the extent to which the results that emerged are sensitive to the details of the model: the constituents are flexible macromolecules, the constraints correspond to vulcanisation, etc. By applying our statistical mechanical approach to other systems, such as vulcanised stiff and semi-flexible macromolecules, endlinked macromolecules[6] (see Fig. 2), and manifolds,[28] we have arrived at the observation that the chief characteristics of the transition are *insensitive* to the specifics of the system giving rise to the amorphous solid state. Specifically, the behaviour (with crosslink density) of the fraction of localised particles, the scaling of the distribution of localisation lengths, and the precise form of the scaled distribution turned out to be identical across a broad range of systems, indicating a form of universality at the amorphous solidification transition.

Further evidence for universality comes from extensive computer simulations, undertaken by Michael Plischke and Sandra Barsky at Simon Fraser University.[19] These simulations addressed the vulcanisation transition for fairly realistic versions of vulcanised macromolecular systems, including Lennard-Jones inter-particle interactions. The analytical predictions were rather accurately verified by these simulations (see Fig. 5).

Encouraged by the evidence that several basic aspects of the amorphous solidification transition are universal and independent of the details of the system undergoing the transition, we sought an explanation for this universality in terms of the Landau approach to phase transitions.[4] Our semi-microscopic approach had led to specific forms of Landau theories that had a common structure and, hence, to common predictions, and we thus asked the question: To what extent is it inevitable that systems undergoing amorphous solidification should exhibit common universal properties at this transition?

To answer this question we made the following assumptions:

1: The order parameter is of the form given in Eq. (12), with $g = n + 1$ (and the $n \to 0$ limit implied).

2: Near to the transition only a small fraction of the particles are localised.

3: The typical localisation length of localised particles is substantially larger than the objects (macromolecules, manifolds,...) becoming localised.

Quite strikingly, it turns out that under these assumptions the part of the Landau free energy that controls the amorphous solidification transition in the transition regime is *essentially unique* and, consequently, so are the properties of the resulting amorphous solid state. Given this perspective, the collection of results indicating universality across a spectrum of amorphous solid forming systems becomes even more convincing.

What form does the Landau free energy take? Let us denote by $\hat{k}$ the collection of $n+1$ (3-component) wave vectors $\{\mathbf{k}^0, \mathbf{k}^1, \ldots, \mathbf{k}^n\}$. Then the Landau free energy (per object being linked) is given by

$$n\mathcal{F}_n(\{\Omega_{\hat{k}}\}) = \overline{\sum}_{\hat{k}}\Big(-\epsilon + \frac{1}{2}|\hat{k}|^2\Big)|\Omega_{\hat{k}}|^2 - \overline{\sum}_{\hat{k}_1 \hat{k}_2} \Omega_{\hat{k}_1}\, \Omega_{\hat{k}_2}\, \Omega_{-\hat{k}_1-\hat{k}_2}\,. \tag{33}$$

in terms of which the disorder-averaged free energy (per particle) f is given by

$$f = \lim_{n \to 0} \min_{\{\Omega_{\hat{k}}\}} \mathcal{F}_n(\{\Omega_{\hat{k}}\}). \tag{34}$$

By taking the first variation of $\mathcal{F}_n$ with respect to $\Omega_{-\hat{k}}$ we obtain the stationarity condition (i.e. the self-consistent equation for the order parameter):

$$0 = nd \frac{\delta \mathcal{F}_n}{\delta \Omega_{-\hat{k}}} = 2\left(-\epsilon + \frac{1}{2}|\hat{k}|^2 \right)\Omega_{\hat{k}} - 3\overline{\sum}_{\hat{k}_1} \Omega_{\hat{k}_1} \Omega_{\hat{k}-\hat{k}_1} \,. \tag{35}$$

By recalling the meaning of the overbar on the wave vector summations mentioned shortly after Eq. (32), we see that this self-consistent equation for the order parameter only applies to values of $\hat{k}$ containing more than one nonzero 3-vector.

The hypothesis for the form of the order parameter specified in Eq. (18) exactly solves this self-consistency condition. Given the very general basis on which the Landau free energy was built, we now see that it is quite reasonable for various systems having significant distinctions at the microscopic level to give rise to common critical behaviour.

One might go so far as to say that the solidification of amorphous solids is simpler than that of their crystalline counterparts, at least insofar as the existence of universal features is concerned. The reason for this curious state of affairs can perhaps be traced to the presence of small parameters (the localised fraction and the inverse localisation length) near the amorphous solidification transition. By contrast, a small localised fraction seems incompatible with conventional crystallisation; and consequently the interparticle spacing near the crystallisation transition seems to be extremely significant. As the crystallisation transition is approached and the position fluctuations begin to grow, a discontinuous transition intercedes.

RIGIDITY: AN EMERGENT PROPERTY OF RANDOM NETWORKS

Elastic properties of the amorphous solid state

So far in this article I have emphasised the *structural* aspects of the amorphous solid state, especially the spatially random localisation of particles (i.e. the spontaneous breaking of translational symmetry that is diagnosed by the order parameter). It seems intuitively clear (and this intuition is borne out by an explicit calculation[5] that we shall discuss shortly) that once a nonzero fraction of particles is localised then the system will be *rigid*, in the sense that a (volume preserving) change in the *shape* (away from the shape the sample had during the constraint-imposing process) leads to an increase in the free energy per unit volume. Such shape deformations are known as *shear* deformations.

When systems display an increase in free energy density upon shear deformation, we say that they are *rigid*.** More precisely, we say that they are equilibrium (or zero-frequency) elastic media. (This notion of rigidity is precisely the one used throughout the physics and elasticity literature.) Just as it does in the context of crystalline solids,

**Equivalently, one can view rigidity as the phenomenon in which a static shear deformation causes an indefinitely persisting shear stress. In the (predominant) situation in which the rigidity is linear (i.e. the stress is proportional to the strain in the limit of small-amplitude strains) rigidity can be quantified by the linear shear modulus, which is the stress required to produce a unit strain. This shear modulus can be extracted (either in experiment or theory or computer simulations) by measuring the stress and strain, or by measuring the increase in the free energy caused by a small deformation.

rigidity in amorphous solids emerges as a consequence of the spontaneous breaking of translational symmetry that is occurring at the phase transition. Familiar extensions of this notion of rigidity arise in the context of, e.g., nematic liquid crystals (where Frank elasticity emerges as a consequence of the spontaneous breaking of rotational invariance) and superfluids (where phase rigidity emerges as a consequence of the the breaking of gauge symmetry).

So, how have we attacked this issue of rigidity to shear deformations? In fact, after making two indirect and *ad hoc* approaches, we finally tackled the issue directly[5] by simply asking the question: by how much does the (bulk part of the) free energy of the system change under a volume preserving deformation of the state (that is imposed after the random constraints have been incorporated)? Constructing the appropriate Landau free energy is straightforward (see Refs. [10] and [5]): the replica associated with the statistical distribution of crosslinks remains undistorted, whereas the remaining n replicas are distorted, and this is accomplished by altering the shape of the (formerly cubic) box on which periodic boundary conditions are imposed. As a consequence, the (fine but discrete) lattice of wave vectors of plane waves available to express any function of position is altered, and this is how the shape change is fed into the calculation of the free energy. Now, as the form of the free energy (functional of the order parameter) changes under deformation, so does the form of the self-consistent equation obeyed by the order parameter. Thus there are two sources to the change in the free energy: the change in the functional and the change in the (equilibrium value of its) argument.

Now, just as it does in the case of the undistorted system, the task of determining the value of the order parameter that solves the stationarity condition requires physical input, in the form of a parametrised hypothesis for its value. Specifically, one must make a hypothesis about the manner in which the particle positions and their probability clouds respond to the deformation of the sample. Our strategy for doing this goes as follows. Focus on a localised particle. Prior to distortion the motion of this particle is characterised by a spherical probability cloud of spatial extent ξ. Next, ask: What is the response of the structure due to a shear deformation? We answer by considering four physical elements: (i) the mean position of the particle moves affinely (i.e. according to the geometry of the deformation of the container); (ii) its probability cloud moves affinely (i.e. is sheared); (iii) there is a random shift in the location of the mean position; and (iv) there is a random shift in the shape of the probability cloud. In accordance with our probabilistic approach to heterogeneous states, we then propose a form for the order parameter in which the statistical distribution of localisation lengths is extended to a joint probability distribution for the extent of the cloud prior to deformation, together with the random shift and shape-change of the cloud upon deformation. It turns out that the resulting form for the order parameter exactly solves the self-consistency condition, and leads to a partial integro-differential equation for the joint probability distribution, the form of the deformation appearing as a parameter in this equation.

What do we find for the joint distribution characterising the response of the structure to shear deformations? The situation turns out to be simpler than we had anticipated. There is no random shift in the location of the mean position; and the anticipated random shift in the shape of the probability cloud is not in fact random, but serves to undo the presumed affine distortion. In other words, our physical picture that the entire probability cloud is distorted affinely is wrong: the spherical cloud is, on average, simply deterministically relocated during the distortion.

As for the macroscopic implications of the transition to the amorphous solid state, they are that the medium becomes a homogeneous, isotropic elastic solid, the shear modulus of which vanishes, as the transition is approached from the amorphous solid state, with the third power of the excess constraint density. This result is consistent with a simple power-counting argument in which one argues that the elastic free energy density should be built from two powers of the gradient of the order parameter (as it is in other broken-symmetry settings). Then each power of the order parameter brings one power of the excess constraint density, and the two gradients bring two inverse powers of the typical localisation length (i.e. one further power of the excess constraint density, making a total of three powers in all. This should be contrasted with the conventional liquid-to-crystal first-order solidification transition, in which there are no genuinely small parameters.

Thermodynamic versus mechanical rigidity

Especially given the extensive attention paid to the issue of *rigidity percolation* at this workshop (see several other articles in this volume), it is certainly worth remarking that our work shows no evidence whatsoever that the constraint density necessary for the onset of macroscopic connectivity and that necessary for the onset of rigidity are anything but identical. Indeed, it seems rather plausible that a sharp distinction between the onset of macroscopic connectivity and of rigidity can only arise at zero temperature, i.e., when entropic effects play no role. I do not see how one can rule out entirely the idea of an additional phase transition at nonzero temperatures, occurring at a constraint density larger than that at which rigidity sets in, presumably signalling some underlying geometrical phenomenon. However, given that the system is already rigid, in the sense that it has a nonzero shear modulus, it seems—at least to me—rather more plausible that one will find only a quantitative and not a fundamental qualitative change in the properties of the system at constraint densities larger than the one at which rigidity first sets in. Moreover, if one is considering atomic systems at zero temperature, so that entropic effects play no role, one may be in a position in which it is necessary to consider quantal fluctuations. It seems a plausible (and indeed attractive) notion that once the structure is extended (in the sense of connectivity) then the value of the zero-point energy (arising from quantal fluctuations in the quantal ground state) will be dependent on the shape of the container, and will lead to a quantal (rather than thermal) version of rigidity. These and other interesting issues remain to be investigated.

CURRENT AND FUTURE DIRECTIONS

The theoretical developments that I have surveyed have led to a fairly detailed description of the equilibrium properties of the amorphous solid state in the vicinity of the transition from the liquid to this new state. At the semi-microscopic level there is now a picture of the structure of the state in terms of the heterogeneity of the particle localisation lengths. At this level there is also a picture of the response of the structure to an externally applied shear stress, which leads to an understanding of the emergent macroscopic rigidity of the amorphous solid state.

Experimental probes of the amorphous solid state

Given this detailed theoretical picture, which has received encouraging support from computer simulations, it would be very useful and instructive for the picture to be confronted with experiments aimed at determining the extent to which it is valid.

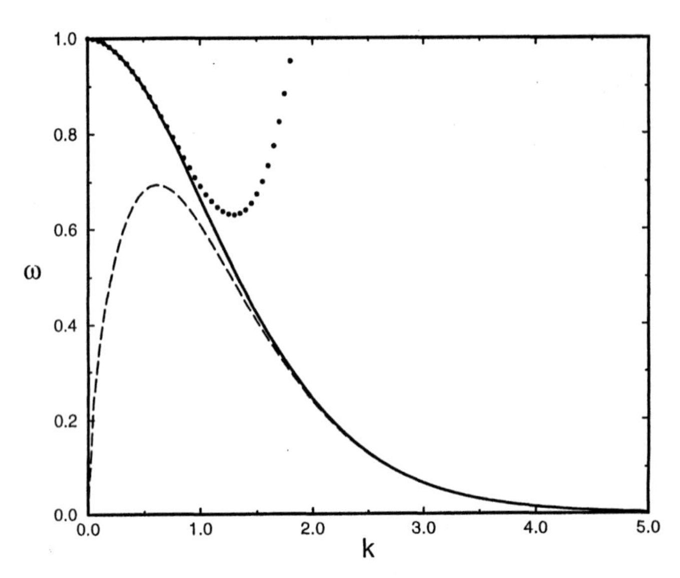

Figure 7. Scaling function $\omega(k)$ for the order parameter (full line), which should be accessible via neutron scattering experiments. Asymptotic form for $k \rightarrow 0$ (dotted line); Asymptotic form for $k \rightarrow \infty$ (broken line).

The are a number of possible experimental settings, some of which are:
1: Neutron scattering (of the quasi-elastic incoherent type), and possibly x-ray scattering (see Fig. 7), which would provide a rather direct examination of the order parameter, and thereby the distribution of localisation lengths, both for unsheared and sheared samples.
2: Fluorescent labelling of crosslinked biological polymers (e.g., of actin fibres), and the subsequent video imaging of their motions, which would provide access to the distribution of localisation lengths, as well as to dynamical issues. (I thank Josef Käs for sharing his expertise on this technique with me.)
3: Accurate macroscopic elasticity measurements, which would test the scaling of the shear modulus with crosslink density. [For further details see Ref. (5).]

Open theoretical issues

Many issues on the theoretical side also remain open. Some of these are:
1: How to access the regime of constraint densities well beyond the regime of the phase transition. Current theoretical results are limited, for technical reasons, to the transition regime.
2: How to incorporate the effects of order parameter fluctuations. In fact, the critical regime is believed to be rather narrow[29]; nevertheless, as the system is one having a random equilibrium state it would be instructive to take the theory beyond the mean field level, even if the observation of corrections to mean field theory should turn out to be difficult.
3: How to calculate the next level of physical diagnostics, such as the probability that two localised particles whose mean positions are separated by a distance R have

localisation lengths ξ and ξ'. Some results have already been obtained in this direction.[23]
4: Dynamical properties of random networks, including the divergence of the viscosity (as the transition is approached from the liquid state).
5: The implementation of this technology to the glass transition. A widely considered physical picture of glass is that it is a state arrived at by the "freezing in" of correlations in the liquid state. Our methods for dealing with the consequences of "frozen in" correlations provide a natural framework for exploring this picture.
6: The development of probabilistic methods for addressing the typical properties of random networks made from various ingredients, such as atoms and molecules, but also macroscopic structural units (struts, ropes, joints, etc.) of interest in engineering settings.

I conclude with a brief comment on the philosophy of the work that I have surveyed. Our aim has been to construct a fundamental theory of the formation and properties of the amorphous solid state. To be fundamental, such a theory must capture the event in which particles become spontaneously localised. That is, the theory must feature, *inter alia*, regimes in which the particles roam throughout the container as well as regimes in which (at least some of) the particles become trapped at random positions about which they fluctuate.

The construction of this theory has required the introduction and development of new theoretical concepts and calculational tools.[††] This has, from time to time, elicited the response: Can't you just simulate? Well, of course, one can simulate random systems such as vulcanised macromolecular networks. And, done well, interesting and useful results are likely to emerge. However, I do not believe that simulation results, by themselves, constitute a theory. Instead, I believe that the deepest levels of understanding (not just in terms of bald numbers but also in terms of the emergence of insight, conceptual advances, connections and intellectual coherence) are more likely to emerge when a full theory of the type we have been seeking is sought, with information from simulations and experiments being regarded more as signposts than as destinations. Of course, we may find ourselves unable to produce sharp enough tools, so that certain highly desirable issues may remain beyond our reach. But that possibility should not deter us from the aim of establishing the deeper level of understanding that a real theory brings.

Acknowledgments

I would like to thank Michael Thorpe and Philip Duxbury for organising this stimulating and enjoyable workshop. I would also like to thank Michael Plischke and Sandra Barsky for making available to us unpublished results from their computer simulations of macromolecular networks. The research surveyed in this article has been done in collaboration with several people, most notably Nigel Goldenfeld, Annette Zippelius, Horacio Castillo, Weiqun Peng and Konstantin Shakhnovich. The support of this research by the U.S. National Science Foundation, NATO and the University of Illinois at Urbana-Champaign is gratefully acknowledged.

[††] "If one finds this [...] discouraging," as K. G. Wilson has written,[30] "one should remember that the successful tricks of one generation become the more formal and more easily learned [...] methods of the next generation."

REFERENCES

1. P. M. Goldbart and N. Goldenfeld, Phys. Rev. Lett. **58**, 2676 (1987); Phys. Rev. A **39**, 1402 (1989); *ibid.*, **39**, 1412 (1989).
2. P. M. Goldbart and A. Zippelius, Phys. Rev. Lett. **71**, 2256 (1993).
3. H. Castillo, P. M. Goldbart and A. Zippelius, Europhys. Lett. **28**, 519 (1994).
4. W. Peng, H. E. Castillo, P. M. Goldbart and A. Zippelius, Phys. Rev. B **57**, 839 (1998).
5. H. E. Castillo and P. M. Goldbart, *Elasticity near the vulcanization transition*, Phys. Rev. B (in press, 1998) [cond-mat/9712050]; a more detailed account is in preparation.
6. M. Huthmann, M. Rehkopf, A. Zippelius and P. M. Goldbart, Phys. Rev. E **54**, 3943 (1996).
7. P. M. Goldbart and A. Zippelius, Europhys. Lett. **27**, 599 (1994).
8. P. M. Goldbart, H. Castillo and A. Zippelius, Adv. in Phys. **45**, 393 (1996).
9. A. Zippelius and P. M. Goldbart, *Vulcanised matter: A model glass?* in *Spin Glasses and Random Fields*, p. 357, edited by A. P. Young (World Scientific, Singapore, 1997).
10. R. T. Deam and S. F. Edwards, Phil. Trans. R. Soc. **280A**, 317 (1976).
11. S. F. Edwards and P. W. Anderson, J. Phys. F **5**, 965 (1975).
12. For a discussion quenched randomness see, e.g., *Spin Glasses*, by K. H. Fischer and J. A. Hertz (Cambridge University Press, 1991), Sec. 2.3.
13. See, e.g., the experiments on silica networks by B. Gauthier-Manuel, E. Guyon, S. Roux, S. Gits and F. Lefaucheux, J. Phys. (Paris) **48**, 869 (1987).
14. E. Marinari, G. Parisi and F. Ritort, J. Phys. A **27**, 7615 (1994); *ibid.*, 7647.
15. J.-P. Bouchaud and M. Mézard, J. Phys. I (France) **4**, 1109 (1994).
16. K. Broderix, P. M. Goldbart and A. Zippelius, Phys. Rev. Lett. **79**, 3688 (1997).
17. S. Seung, H. Sompolinsky and N. Tishby, Phys. Rev. A **45**, 6065 (1992), especially Sec. V.D.
18. D. Stauffer, *Introduction to Percolation Theory* (Taylor and Francis, Philadelphia, 1985).
19. S. J. Barsky and M. Plischke, Phys. Rev. E **53**, 871 (1996); S. J. Barsky, Ph.D. thesis, Simon Fraser University, Canada (unpublished, 1996); S. J. Barsky and M. Plischke (unpublished, 1997). These authors have kindly made unpublished data of theirs available to us.
20. P. Erdős and A. Rényi, Magyar Tud. Akad. Mat. Kut. Int. Közl. **5**, 17 (1960), especially Theorem 9b; reprinted in Ref. 21, Chap. 14, article [324]. For an informal discussion, see P. Erdős and A. Rényi, reprinted in Ref. 21, Chap. 14, article [v].
21. *Paul Erdős: The Art of Counting*, edited by J. Spencer (MIT Press, 1973).
22. H. E. Castillo, P. M. Goldbart and A. Zippelius, *Randomly crosslinked macromolecular systems: Linear stability of the amorphous solid state* (in preparation, 1998).
23. W. Peng and P. M. Goldbart, work in progress (1998).
24. See, e.g., M. Mézard, G. Parisi and M. A. Virasoro, *Spin Glass Theory and Beyond* (World Scientific, Singapore, 1987); and K. Binder and A. P. Young, Rev. Mod. Phys. **58**, 801 (1986).
25. L. Viana and A. J. Bray, J. Phys. C **18**, 3037 (1985).
26. C. Goodyear, *Gum-Elastic and its Varieties, with a Detailed Account of its Applications and Uses, and of the Discovery of Vulcanization* (New Haven, 1855).
27. O. Theissen, A. Zippelius and P. M. Goldbart, Int. J. Mod. Phys. B **11**, 1945 (1996).
28. C. Roos, A. Zippelius and P. M. Goldbart, J. Phys. A **30**, 1967 (1997).
29. P. G. de Gennes, J. Physique Lett. **38**, L355 (1977).
30. K. G. Wilson, Rev. Mod. Phys. **47**, 773 (1975), specifically p. 805.

GRANULAR MATTER INSTABILITY: A STRUCTURAL RIGIDITY POINT OF VIEW

Cristian F. Moukarzel

Instituto de Física,
Universidade Federal Fluminense,
24210-340 Niterói RJ, Brazil.
email: **cristian@if.uff.br**

1 INTRODUCTION

Granular materials are ubiquitous in nature and very common in industrial processes, but it is only recently that their unusual properties have begun to receive detailed attention from the physicists community[1,2]. The earliest documented studies of granular matter date back to Faraday[3], who discovered the convective behavior of vibrated sand, and Reynolds [4], who noted that compactified granular matter cannot undergo shear without increasing its volume.

The behavior of vibrated granular matter in some aspects resembles that of a fluid, although there are crucial differences. Size segregation[5], for example, at first sight defies intuition. When a mixture of particles of different sizes is subject to vibration, the larger ones migrate to the top, irrespective of density. Also interesting is the layering instability [6] of a binary mixture under pouring. Instead of a homogeneously mixed pile, under certain conditions an alternation of layers of both kinds of particles can be obtained.

Similar demixing phenomena occur in granular materials subject to various kinds of external excitation. These seem to contradict the naive expectation that shaking should favor mixing, or take the system to a low-energy state. Many of the unusual properties of vibrated granular matter are in fact due to the dissipative character of interparticle collisions. An interesting example of the consequences of dissipation is inelastic clustering[7], by which particles tend to cluster together as their relative kinetic energy is completely lost during collisions.

The compactification of vibrated sand has been recently found to be logarithmically slow[8], resembling glassy behavior, and a spin model with frustration has been proposed to model this process [9]. This provides a bridge between the dynamics of spin glasses and vibrated granular matter.

It is thus clear that granular materials present extremely interesting dynamic phe-

Rigidity Theory and Applications
Edited by Thorpe and Duxbury, Kluwer Academic/Plenum Publishers, 1999

nomena, but it is already at the much simpler level of static properties that unusual behaviors show up. Stress propagation in piles or packings of granulate matter has many uncommon features. When grains are held in a tall vertical silo, for example, the pressure at the bottom does not indefinitely increase with height but saturates after a certain value[10]. The excess weight is deviated towards the walls and equilibrated by friction forces. A related phenomenon is the formation of a pressure "dip" right below the apex of a conical pile of granular matter[11], instead of the expected pressure maximum. These phenomena indicate that gravity-induced stresses do not propagate vertically but often deviate laterally. Pressure saturation in silos and the pressure dip under piles are both due to the formation of "arches"[2,12,13]. Many proposals to explain arching[12,13] rest on the idea that friction plays an essential role, but recent studies[14,15,16,17] show that friction is not necessary.

Photoelastic visualization experiments[18,19,20,21] show that stresses in granular matter concentrate along well defined paths. It is not clear whether the characteristic size of these patterns is finite, or limited by system size only. Stress-concentration paths are observable even on regular packings of monodisperse particles[18,19], their exact location being sensitively dependent on very weak particle irregularities. Stress-paths often suffer sudden rearrangement on a global scale when the load conditions are slightly changed[18,20,22]. For similar reasons, the fraction of the total weight that reaches the base of a silo can vary by large amounts under very weak perturbations, or when repeating the filling procedure with exactly the same amount of grain[13]. These phenomena demonstrate that slight perturbations can produce macroscopic internal rearrangements in granular matter. In other words, granular matter are internally *unstable.*

In part because of the technological importance of the problem, and also because of its interest from the point of view of basic science, much work has been done in recent years to understand the propagation of stresses in granular systems. On the numerical side, several methods have been implemented. Classical Molecular Dynamics simulations[16,21,23,24], which usually include a fictitious damping term in order to allow the system to come to rest, are normally very cpu-intensive and thus limited to relatively small sizes. Alternatively, the elastic equations can be solved using symbolic software in order to obtain stress values which are free of numerical error[25]. Lattice automata based on random contact disorder[26], are able to reproduce the observed dip under granular piles. Contact Dynamics simulations[22] provide an efficient way to include friction forces, and allowed numerical visualizations of stress concentration on relatively large systems. Lubricated Granular Dynamics[14] is a method to obtain the equilibrium contact network of infinitely stiff networks and is based on the use of a fictitious damping with a singularity at zero distance.

A large number of theoretical approaches to this problem are formulated on a continuum[12,27,28,29,30] and thus rest on the assumption that a length scale exists, below which fluctuations are negligible when compared to averages. It is not clear whether this assumption is easily justified for granular matter, where stress fluctuations seem to be at least as large as average stresses[20,21].

A different type of modeling strategy starts by formulating a stochastic rule for stress propagation on a lattice, which is thereafter solved by various methods[21,24,31,32,33], or taken as the starting point for a continuum description[31]. In the simplest version of this approach[21,24], only the vertical component of the transmitted force is considered, i.e. the problem is reduced to a scalar one. Despite the roughness of the approximation, this procedure gives good results for the average distribution of stresses $P(w)$, in particular the observed exponential decay for large stresses[20,21,22]. The occurrence of small stress is though strongly underestimated within this simple scalar model[21,31]. This is due to the fact that scalar "stresses" propagate vertically, with at most a diffusive width due to disorder, and therefore arching is not possible [31]. In order to correctly reproduce the small-stress part of $P(w)$, which is arch-dominated, the vectorial nature of stresses has to be taken into account [31,32,33]. This brings in the problem of stress signs, since now negative (traction) stresses, which do not exist on non-cohesive granular matter, cannot be easily avoided [31].

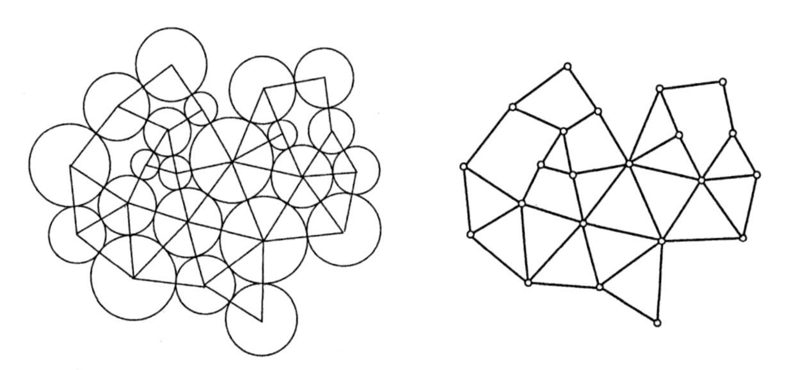

Figure 1: The contact network (right) associated to a granular pile is a graph in which nodes represent particles and are connected by an edge or bond whenever there is a nonzero (compressive) force between the corresponding particles.

It is thus clear that granular matter has, from the point of view of its static properties, two noteworthy characteristics:

- Stresses are not homogeneously distributed over the system but concentrate on paths that form a sparse network.
- The exact location of stress paths is susceptible of change under very weak perturbations, showing that granular matter is extremely unstable.

Although there have been many proposals to describe stresses in granular matter, most of these models are largely phenomenological in nature and sometimes contain unclear ad hoc assumptions. A deeper understanding of the above described particularities of granular matter has remained elusive. It is the purpose of this work to show that structural rigidity concepts can help us advance in this direction.

We first review some structural rigidity notions in Section 2, and demonstrate in Section 2.1 that the contact network of a granular system becomes *exactly isostatic* in the limit of large stiffness. The consequences of this are discussed in Section 3. An immediate consequence of isostaticity is the possibility of stress concentration, as briefly discussed in the beginning of Section 3. Most of the previous theoretical and numerical effort has concentrated on the description of stresses. There is though a complementary aspect to this problem, which has not been explored. This is the study of how *displacements* induced by a perturbation propagate in a granular system. We thus leave further discussion of stresses for future publications [17], and concentrate in understanding the behavior of induced displacements upon perturbation. This will lead us to the central results of this work, respectively:
a) in Section 3.1 it is shown that an *isostatic phase transition* takes place in the limit of infinite stiffness, and that the isostatic phase is characterized by an anomalously large susceptibility to perturbation.
b) Section 3.2 contains a discussion of the load-stress response function of the system in

the light of **a**), which shows that isostaticity is responsible for the observed instability of granular matter.

We will furthermore find that very large displacements are produced on isostatic networks when a site is perturbed. Section 3.3 clarifies the origin of these anomalously large displacements, while Section 4 contains our conclusions.

2 STRUCTURAL RIGIDITY AND GRANULAR NETWORKS

The contact network of a frictionless packing of spherical particles can be defined in the following way (Fig. 1): we let each particle center be represented by a point in space, and connect two of these points by a line (bond, or edge) whenever there is a nonzero compression force between the corresponding particles. The networks so generated can be seen as particular cases of what is usually called *frameworks* in rigidity theory, i.e. structures made of points connected by rotatable bars.
Structural rigidity [34,35] studies the conditions that a network of points connected by central forces has to fulfill in order to support applied loads, i.e. be *rigid*. The first studies of rigidity of structures from a topological point of view date back to Maxwell[36]. Structural rigidity concepts were first introduced in the study of granular media by Guyon et al [37], who stressed that granular contact networks differ from linear elastic networks in an important aspect: the first are only able to sustain compressive forces between grains. Technically speaking, force networks with a sign-constraint on stresses are called *struts*. Another typical example of sign-constrained networks are spider webs, or cable structures, the elements of which (strings) can only sustain traction forces [38]. Structures with interesting properties can be obtained by combining elements of both types, in which case the resulting framework is called a *tensegrity structure* [39]. Several

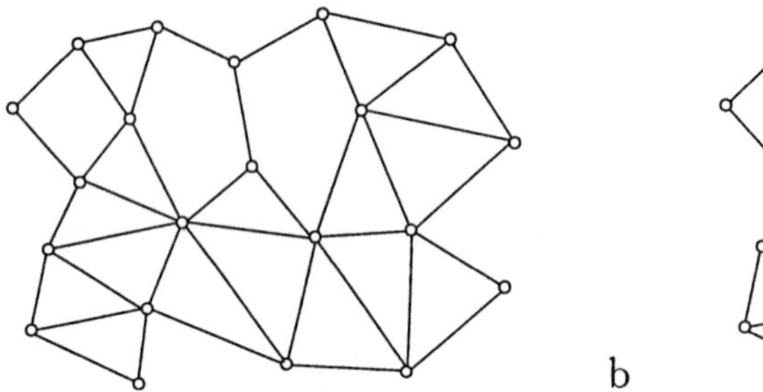

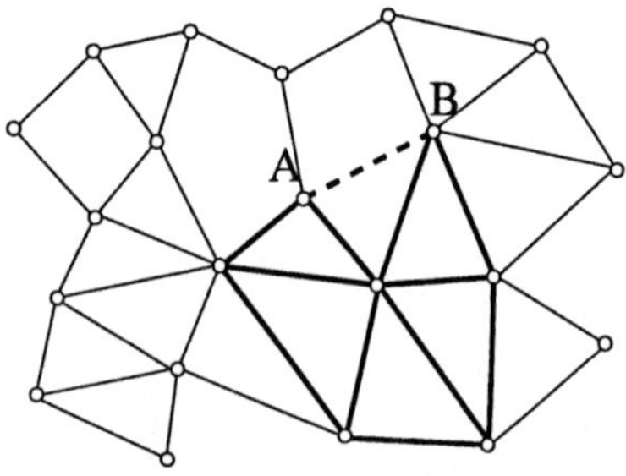

Figure 2: **a)** A minimally rigid network in two dimensional space, composed of n points and $2n-3$ bars. If any bar in this network is removed, some of the points would cease to be rigidly connected to the rest. All bars in this network are therefore essential for rigidity. **b)** A *redundant* bar (dashed line) between points A and B can induce stresses ("self-stresses") in some part of the network (dark lines). The locus of self-stresses is the *overconstrained* subgraph associated with bar AB. This is exactly the set of bars which already provides a rigid connection between points A and B. If the length of bar AB is exactly equal to the distance d(A,B), i.e. if there is no *length-mismatch*, self-stresses will be zero.

applications of rigidity related ideas and tools have been already presented in this book. Let us here only briefly refresh some concepts which we need for our discussion.

A point in d dimensions has d degrees of freedom, while a rigid cluster has $d(d+1)/2$. Therefore if a set of n points forms a rigid cluster, it must be connected by at least $dn - d(d+1)/2$ bars. If a rigid cluster of n points has exactly $dn - d(d+1)/2$ bars, it is said to be (generically) *isostatic*, or minimally rigid (Fig. 2a). A framework with more bars than necessary to be rigid is *hyperstatic* or *overconstrained* (Fig. 2b). Excess bars, which can be removed without introducing new degrees of freedom, are called *redundant.* A bar is redundant when the two points it connects are already rigidly connected. Unless the length of the redundant bar is exactly equal to the distance between the points it connects, self-stresses will be generated in some parts of the framework. Thus self-stresses are non-zero only within overconstrained subgraphs, and can be absent if there are no length-mismatches.

We will discuss granular piles under the action of gravity, in which forces are transmitted to a supporting substrate. We could as well consider any other load condition, provided that an infinitesimal gravity field is added in order to remove indeterminacies in the positions of the particles. Because of the action of gravity, contact networks associated to static granular piles must be rigid. Otherwise some of the grains would be set in motion by gravitational forces. Because of the fact that grains are rigidly connected to a lower boundary, a redundant contact anywhere on the system will usually produce an overconstrained subgraph that extends all the way down to the rigid boundary.

Some early attempts to numerically study the static properties of granular materials

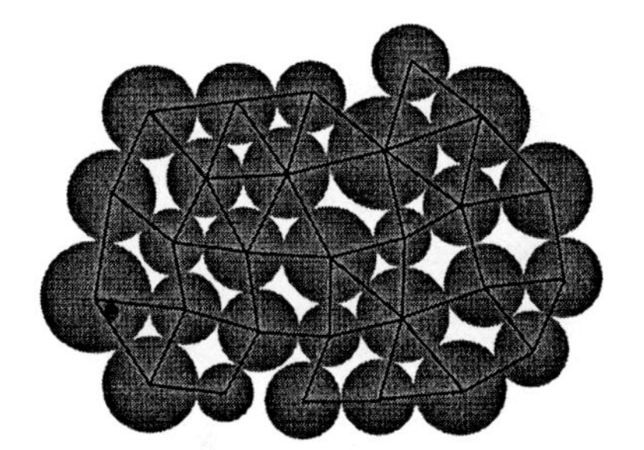

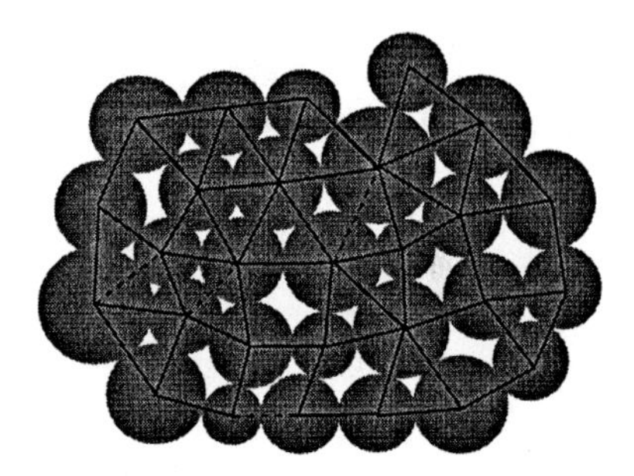

Figure 3: In the limit of large stiffness-to-load ratio, i.e. when the compressive forces are small, or the rigidity large (left), the contact network of a granular packing is sparse and, as discussed in the text, becomes *minimally rigid.* If the compressive forces are increased, or the stiffness decreased (right), excess contacts (dashed lines) are created.

have ignored the sign constraint, thus modeling granular piles as randomly diluted linear frameworks [40]. Due to this, it has been sometimes suggested that rigidity percolation concepts might be applied to granular networks [37,41]. But this would require forces with power-law distribution, since the elastic percolation backbone is a fractal object at the critical point [40,42], while experimental and theoretical studies [21,20,22] suggest that force distributions display exponential decay on granular systems.

This suggests that the sign-constraint associated with non-cohesive granular matter cannot be neglected. As we shall soon show, it is possible to see from a topological point of view that the sign-constraint has far-reaching consequences for the static behavior of granular aggregates. We demonstrate next that this restriction forces a stiff granular system to choose, among all possible equilibrated contact networks, only those with the specific topological property of minimal rigidity.

2.1 Isostaticity of stiff networks with a sign constraint

Consider now a d-dimensional frictionless granular pile in equilibrium under the action of external forces $\vec{F}_i$ (gravitational, etc) on its particles. We represent the contact network of this pile by means of a linear-elastic central-force network in which two sites are connected by a bond if and only if there is a nonzero compression force between the two corresponding particles in the pile (Fig. 1).

If the external compression acting on the pile is increased, particles will suffer a larger deformation, and therefore the number of interparticle contacts will increase (see

Fig. 3). Equivalently, if compression forces are weakened, or the stiffness of the particles made larger, the number of interparticle contacts, and thus the number of bonds on the equivalent contact network, will be reduced because there are no cohesive forces between particles. But there is a lower limit for the number of remaining contacts, given by the condition that all particles be rigidly connected, otherwise they would move until new contacts are established. Therefore one may expect that in the limit of infinite stiffness the resulting contact network will be minimally rigid. Let us now formalize this observation.

Because of linearity, stresses f_{ij} on the bonds of the linear-elastic equivalent network can be uniquely decomposed as

$$f_{ij} = f_{ij}^{\mathbf{self}} + f_{ij}^{\mathbf{load}} \tag{1}$$

, where $f_{ij}^{\mathbf{self}}$ are self-stresses, and $f_{ij}^{\mathbf{load}}$ are load-dependent stresses.

Load-dependent stresses are linear in the applied load: if all loads are rescaled, $f_{ij}^{\mathbf{load}}$ are rescaled by the same factor. But $f_{ij}^{\mathbf{load}}$ are not changed if *all* elastic constants, or *stiffnesses*, are multiplied by a factor.

Self stresses in turn do not depend on the applied load. They are in general linear combinations of terms of the form $k_{ij}\epsilon_{ij}$, where k_{ij} is the stiffness of bond ij and ϵ_{ij} its length-mismatch (See Fig. 2). The length mismatch ϵ_{ij} of a bond is defined as the difference between its repose length and its length in an equilibrium configuration under zero external loads.

As already discussed, length mismatches, and thus self-stresses, only arise within over-

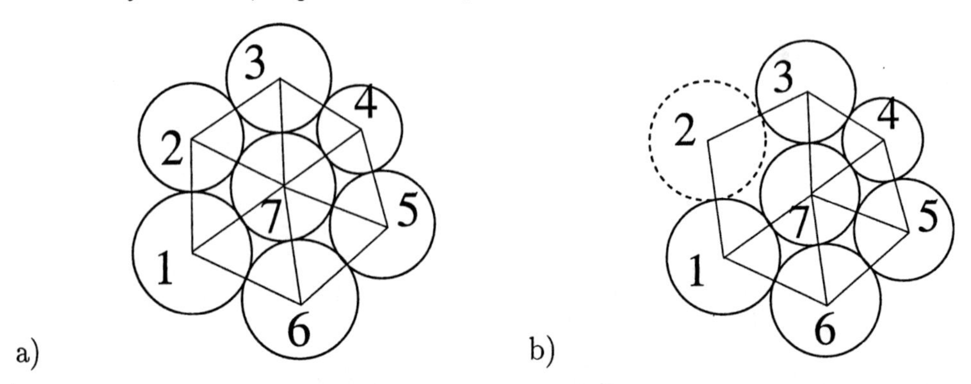

Figure 4: **a)** In order for seven particles to be in contact forming an overconstrained graph as shown here, *one* condition has to be satisfied by the radii. One is free to choose the radii of 6 of them, but the seventh will be uniquely determined by this choice. **b** If one of the particles is for example slightly too large to satisfy the required condition, one contact will be opened, restoring isostaticity. In this example the contact between particles 2 and 7 is open, but any other bond between the central particle and its neighbors could have been chosen. If on the other hand particle 2 were too small, one of the contacts between the external particles would be lost.

constrained subgraphs [34,37]: those with more contacts, or bonds, than strictly necessary to be rigid. For if a graph is not overconstrained, then all its bonds, regarded as linear springs, can *simultaneously* attain their repose lengths while still being in equilibrium under zero external load.

It is easy to see that a bounded overconstrained subgraph with nonzero self-stresses must have at least one bond subject to a negative (traction) stress: As discussed, self-stresses are equilibrated without the intervention of external forces. It then suffices to consider a joint belonging to the envelope of the overconstrained subgraph. Since bonds can only reach it from one side of the frontier, self-stresses of *both signs* must necessarily exist in order for this joint to be equilibrated.

Now imagine rescaling all stiffnesses according to $k \to \lambda k$ (both in the granular pile and in our equivalent elastic system). In doing so, all self-stresses are rescaled by λ, but the load-dependent stresses remain constant. Therefore, if self-stresses were non-zero, in the limit $\lambda \to \infty$ at least one bond of the network would have negative *total* stress (eq. (1)). This is not possible since traction forces are not allowed on our granular pile by hypothesis. Thus the existence of self-stresses is not possible in the limit of large stiffness if there is a sign-constraint on total stresses.

For reasons already reviewed, in order for self-stresses to be zero, one or more of the following conditions will have to be satisfied by granular contact networks in the limit of large stiffness:

a) to have all length mismatches equal zero on overconstrained graphs.
b) to have no overconstrained graphs at all (the network is isostatic).

Condition a) requires that, even when overconstrained graphs exist, particle radii satisfy certain conditions in order to exactly fit in the holes left by their neighbors (see Fig. 4). But as soon as particles have imperfections or polydispersity (no matter how small if their rigidity is large enough) self-stresses would appear if the contact network is hyperstatic. In other words condition a) cannot be *generically* satisfied, if by generic we understand for a "randomly chosen" set of radii. Therefore condition b) must generically hold, i.e. there will be no overconstrained subgraphs in the limit of large stiffness-to-load ratio.

From the point of view of this work, most experimentally realizable packings fall under the category "generic", since small imperfections in radius are unavoidable in practice.

In view of the above, we can now conclude that:

> the contact network of a polydisperse granular pile becomes *isostatic* when the stiffness is so large that the typical self-stress, which is of order $k\epsilon$ would be much larger than the typical load-induced stress.

Exceptions to this rule are packings with periodic boundary conditions, because they are not bounded, and packings in which the radii satisfy exact conditions in order to have zero length-mismatches in overconstrained graphs, because they have no self-stresses [43]. One can for example consider a regular triangular packing of exactly monodisperse particles, in which case the associated contact network will be the full triangular lattice [25], i.e. hyperstatic. But the contact network (and the properties of the system as we shall soon see) will be drastically modified as soon as a slight polydispersity is present [23,16,19] if the stiffness is large enough. Therefore, while one is free to consider specific packings which are not isostatic, in practice these cannot be realized for hard particles, since any amount of polydispersity, no matter how small, will force some contacts to be opened so as to have an isostatic contact network. We now discuss how this affects the properties of granular systems.

3 CONSEQUENCES OF ISOSTATICITY

Isostaticity has been sometimes imposed in numerical models [26], as a condition allowing one to calculate stresses by simple propagation of forces. Recently isostaticity was reported by Ouaguenouni and Roux [14], who use an iterative numerical algorithm to find the stable contact network of a set of rigid disks. Our discussion in the previous section shows that isostaticity is a generic property of stiff packings, and appears because negative stresses are forbidden (an equivalent conclusion would be reached if only traction stresses were allowed[38]). We will now show that isostaticity has important consequences on the way the system reacts when it is perturbed, but before starting a more rigorous analysis, let us first discuss some of the most important differences between isostatic and hyperstatic systems, on an intuitive level.

Imagine perturbing an elastic network by letting an equilibrated pair of collinear forces act on the ends of a given bond, and consider how *stresses* and *equilibrium positions* are modified in the whole system. A properly chosen change in the repose length of this bond would have exactly the same effect, so we can alternatively consider the perturbation to be a change in length or a couple of forces.

- On overconstrained rigid networks, stresses are *correlated* over long distances: if a bond is stretched as described above, stresses will be modified on other bonds far away from the perturbation. This is so because self-stresses *percolate* through the system.

  But for the same reason the *displacements* of the sites from their original positions, induced by this local perturbation, decay very fast with distance. The reason for this is that self-stresses oppose the perturbation and thus tend to "quench" its effect.

- On isostatic systems, *stresses are uncorrelated.* If we change the length of (or apply a equilibrated pair of forces to) an arbitrary bond, stresses on all other bonds remain unchanged, because they only depend on external loads, and not on bond lengths. This is a trivial property of isostatic systems.

  But the equilibrium position of many sites will be in general modified if one of the lengths is changed, and therefore displacements induced by a perturbation may be felt far away from its origin.

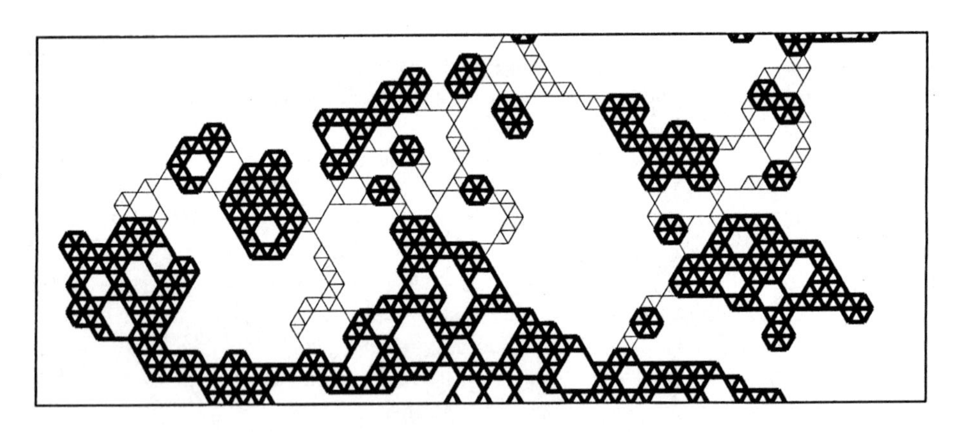

Figure 5: Part of the stress-carrying backbone for rigidity percolation on a randomly diluted triangular lattice. Overconstrained rigid clusters (thick lines) are isostatically connected to each other by *cutting bonds* (thin lines). Cutting bonds are critical in the sense that they provide a minimally rigid connection. All of them are essential for rigidity.

Therefore on a hyperstatic system, perturbations of stresses propagate over long distances, while on isostatic systems it is the displacement field which may display long-range correlations. Stresses are uncorrelated on isostatic systems, and thus arbitrarily large stress gradients are possible. Hyperstatic systems, on the opposite hand, have smoothly varying stresses because of the strong correlations introduced by overconstraints. This provides some indication that stress concentration is possible *because* of isostaticity.

We can get further insight into the meaning and potential consequences of isostaticity from recent studies of central-force rigidity percolation [42]. Rigid backbones, the stress-carrying components of rigidly connected clusters, are found to be composed of large overconstrained clusters, *isostatically* connected to each other by *critical bonds* (also called red bonds, or cutting bonds – See Fig. 5). Overconstrained clusters have more bonds than necessary to be rigid, so any one of them can be removed without compromising the stability of the system. But the rigid connection among these clusters, provided by critical bonds, is isostatic, or minimally rigid. In other words, cutting one critical bond is enough to produce the collapse of the entire system, because each isostatic bond is by definition *essential* for rigidity. Thus we may expect that stretching a critical bond will have a measurable effect on a large number of sites.

But in percolation backbones, critical bonds only exist very close to the rigidity percolation density p_c. Above p_c there is percolation of self-stresses [42] and thus the rigid backbone is hyperstatic. Even exactly at p_c the number of critical bonds is not extensive, but scales as $L^{1/\nu}$ where ν is the correlation-length exponent [42,44]. Consequently critical bonds are relatively few at p_c, and virtually absent far from p_c. Thus if we perturb (cut or stretch) a randomly chosen bond in a percolation backbone, most of the times the effect will be only be local since no critical bond will be hit.

The important new element in stiff granular contact-networks is the fact that *all* contacts are isostatic, i.e. there is *extended isostaticity*. In this case, if any of the bonds (contacts) where removed, the pile would cease to be rigidly connected to the supporting boundary below it. Because of this, if the length of *any* of the network's bonds is changed (which corresponds to a variation in one of the particles radii) the equilibrium position of a finite fraction of the particles will also be changed. For these reasons, one may expect that isostaticity will produce a large sensitivity to perturbation in granular networks.

3.1 Susceptibility to perturbation

We now quantify the degree of susceptibility to perturbation, and then see whether the intuitively appealing ideas we have just discussed are in fact verified on specific models. In order to provide a formal definition of susceptibility, we introduce an infinitesimal change in the length l_{ij} of a randomly chosen bond of the network, and record the *induced displacement* $\vec{\delta}_i$ suffered by all particle centers. We then define the system's susceptibility D as

$$D = \sum_{i=1}^{N} \vec{\delta}_i^2 \qquad (2)$$

, where N is the total number of particles on the system. These measurements are done for variable amounts of *overconstraints* (excess contacts) randomly located on an elastic network, and averages are performed over disorder. In this way $D(O_v)$ is obtained, where O_v is the *density of overconstraints*. According to our previous discussions, we expect D to increase as $O_v \to 0$, which is the isostatic limit.

We start by discussing a toy model shown in Fig. 6: a quasi one-dimensional system composed of linear elastic bonds, in which diagonals are present with probability O_v. The system with no diagonals ($O_v = 0$) is exactly isostatic, therefore each diagonal is an overconstraint, or redundant bond. We assume for simplicity that all bonds have the same stiffness k and length l (diagonals have a length $\sqrt{2}l$). We are interested in calcu-

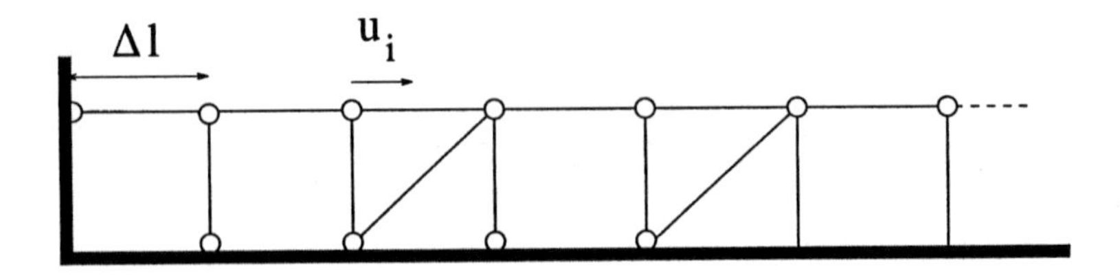

Figure 6: A linear chain of springs in which overconstraints (diagonals) are present with probability O_v is useful as a toy model to understand the influence of isostaticity on the propagation of perturbation. The characteristic distance for the decay of displacements induced by a perturbation (a bond-stretching) diverges in the limit $O_v \to 0$ and thus the susceptibility to perturbation also diverges (see text).

lating the average horizontal displacements $u(x)$ induced by a length change Δl in the left-most horizontal bond, as a function of the density O_v of overconstraints (diagonals). A simple calculation [17] shows that, after averaging over disorder, the displacement field $u(x)$ satisfies

$$\frac{\partial^2 u}{\partial x^2} = \kappa^2 O_v u \tag{3}$$

, where κ is some constant. Therefore $u(x) = u(0)\exp\{-\kappa O_v^{1/2} x\}$ and we see that there is exponential decay with distance, with a characteristic length $\xi(O_v) \sim O_v^{-1/2}$. This "persistence length" diverges at $O_v = 0$, which corresponds to the isostatic limit. Consequently $O_v = 0$ is a critical point, and D as defined above is divergent there. The divergence of D in this model is linear with system size.

Now let us see whether isostaticity has comparable effects in two dimensions. In

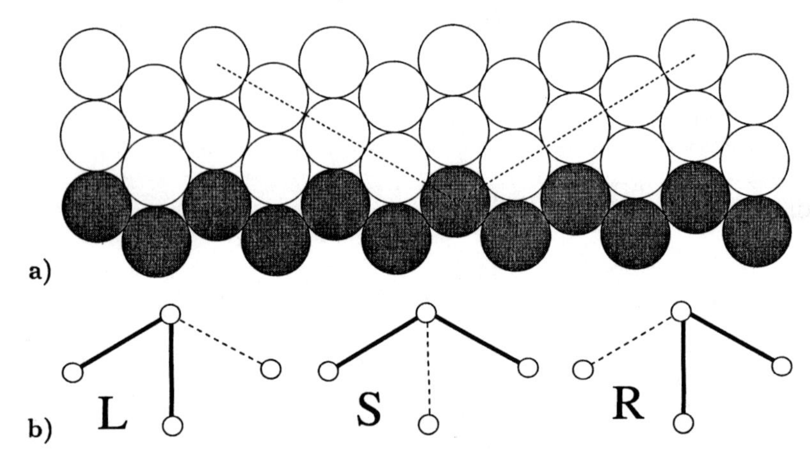

Figure 7: **a)** In two dimensions, a triangular packing is used in order to numerically measure susceptibility to perturbation. The two first layers of particles (shaded) is regarded as a fixed rigid boundary supporting the load of the upper ones. If a site is shifted, only particles within its "cone of influence" (dashed line) can be displaced. In this example, 6 layers of 6 particles each are displayed. **b)** Appropriately choosing among these three isostatic configurations for each site, only compressive stresses are produced. First S is chosen with probability 1/2. If S is not chosen then either R or L are, depending on the sign of the horizontal force acting on the site (see text).

the spirit of previously proposed models [41,26] we consider a triangular packing oriented as in Fig.7a, made of very stiff disks with small polydispersity, under the action of gravitational forces. The polydispersity is assumed to be small enough such that disk centers are approximately located on the sites of a regular triangular lattice, and the stiffness to load ratio large enough such that the contact network is isostatic, according to our discussion in the previous section.

The full problem of generating realistic contact networks that respect the constraint of no traction forces is a difficult one. Several approaches have been proposed [41,26], all involving some degree of approximation even for geometrically simple settings. In all these models, the triangular lattice was oriented with one of its principal axis horizontal, i.e. normal to gravity. There is though some advantage in considering a different orientation, such that one of the principal axis of the lattice is parallel to gravity (Fig. 7). In this case there is no need for recursive checks of positiveness of stresses since the disordered contact network can be built in a fashion that guarantees positive stresses. Our model is defined as follows: We ask that each site be supported by exactly two out of its three lower neighbors, thereby ensuring that only isostatic contact networks are generated. This condition gives three possible local configurations which we call left (L) symmetric (S) and right (R) respectively and are depicted in Fig.7b. Choosing a local configuration of bonds on each site produces a sample, or one realization of the disorder. Clearly not any choice of bond configurations give rise to a contact network with positive stresses. But it is possible to satisfy the positivity constraint and still have disorder in the following way:

1. For each site, starting from the uppermost layers and proceeding downwards, we choose configuration S with probability 1/2 [45].

2. If S was not chosen, then either R or L are, according to the sign of the horizontal force-component F_x acting on that particle: if $F_x > 0$ (F points rightwards), R is chosen. If $F_x < 0$, then L is chosen. If the horizontal component is zero then R or L are chosen at random.

Our model has no geometrical disorder, and this is justified by our assumption of small polydispersity, but we keep contact disorder and isostaticity which are the important characteristics of real granular networks in the limit of large stiffness.

There is no reason to think that the method we have chosen generates all possible equilibrated contact networks that satisfy isostaticity and positiveness of stresses. It seems in principle possible to have some sites making contact to all three downward neighbors and still have isostaticity, by simultaneously opening some other contacts. But our aim here is not to provide a realistic model for granular contact networks but to test whether isostaticity has important effects on the properties of a two-dimensional network.

In order to accomplish this we will measure, on the networks so generated, the susceptibility defined above, and compare the results with those obtained on systems with a finite density O_v of overconstraints randomly located on the network. A non-zero density of overconstraints O_v mimics, as discussed previously, the effect of increasing the mean pressure on the system (or reducing the stiffness), since this would produce a larger number of contacts, in excess of isostaticity, to be established between particles.

A finite density O_v of overconstraints is introduced in this model by letting *all three bonds* be connected below a given site, with probability O_v. Each third bond introduced in this way creates an overconstrained subgraph that extends all the way down to the rigid boundary. The limit $O_v = 1$ gives the fully connected triangular lattice, which of course has no disorder. After building a contact network with a specified density of overconstraints as described above, an infinitesimal upwards shift is introduced in a randomly chosen site on the lowest layer, and the induced displacement field $\vec{\delta}_i$ is measured.

If the network is isostatic ($O_v = 0$) one can calculate all stresses [41,26] and displacements [17] in a *numerically exact fashion* so that systems of 2000×2000 particles may be

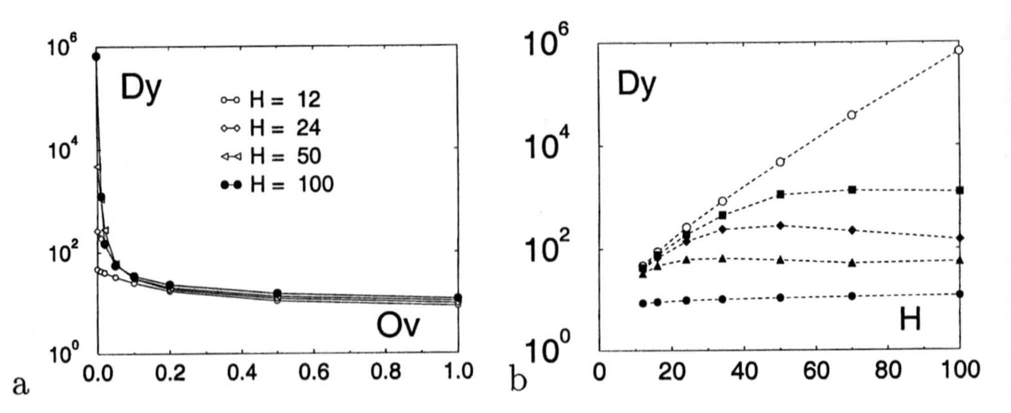

Figure 8: **a)** Susceptibility D_y (defined in the text), versus density of overconstraints O_v, as measured on two-dimensional triangular packings of total height $H = 12, 24, 50$ and 100 layers. $O_v = 0$ is the isostatic limit, and corresponds to granular packings with infinite stiffness, as demonstrated in the text. **b)** Susceptibility D_y versus system height H, for fixed fractions O_v of overconstraints: 0.00 (empty circles), 0.01 (squares), 0.02 (diamonds), 0.05 (triangles), 1.00 (full circles). It is clear in this figure that, for $O_v > 0$, D_y *saturates* to a finite value in the $H \to \infty$ limit. For isostatic systems, on the other hand, D_y diverges *exponentially fast* with H. See also Fig 9a).

simulated on a workstation. The idea is that stresses are propagated downwards and displacements upwards. The way in which the induced displacements are propagated upwards is easily calculated [17] by noting that, when the network is isostatic, all bond lengths (except the perturbed one) must remain *constant*. On the other hand when the network is overconstrained ($O_v > 0$), stresses and displacements can no longer be exactly calculated. In this case one has to solve the elastic equations in order to find the new equilibrium positions after the perturbation. This is done in the limit of linear elasticity (since the perturbation is infinitesimally small) by means of a conjugate gradient solver. In this case the calculations are much more time-consuming so that only relatively small systems, or order 100×100 particles can be studied if $O_v \neq 0$. Supercomputers are required for this part of the calculation [46]. A cross-check of the computer programs was done by comparing the results obtained with the direct solver for isostatic systems, with those produced by the conjugate gradient solver with no overconstraints, on systems of up to 100×100 particles. Excellent agreement was found in all cases, for stresses as well as as for displacements.

In this way the susceptibility $D_y(H, O_v) =< \sum_{i=1}^{N} \delta y_i{}^2 >$ is measured, where δy_i is the vertical displacement of site i due to the perturbation, and $<>$ stands for average over disorder realizations. The system consists of H layers of H grains each, so that $N = H \times H$. Figure 8a shows the susceptibility D_y as a function of the density of overconstraints O_v, for several system heights H. We see that D_y increases rapidly on approach to the isostatic limit $O_v = 0$, and that this increase is faster for larger systems, meaning that D_y diverges at $O_v = 0$ in the $H \to \infty$ limit. Figure 8b shows the same data now plotted as a function of system size, for several values of the density of overconstraints. For any $O_v \neq 0$, D_y goes to a *finite limit* for large sizes, while it diverges with system size if $O_v = 0$. Data for much larger systems can be obtained in the isostatic case using the direct solver program, and are displayed in Fig. 9a. This plot shows that $D_y(O_v = 0)$ is of the form $\log D_y \sim H$, that is, D_y diverges *exponentially fast* with system size.

These numerical results demonstrate that a *phase transition* occurs at $O_v = 0$, where

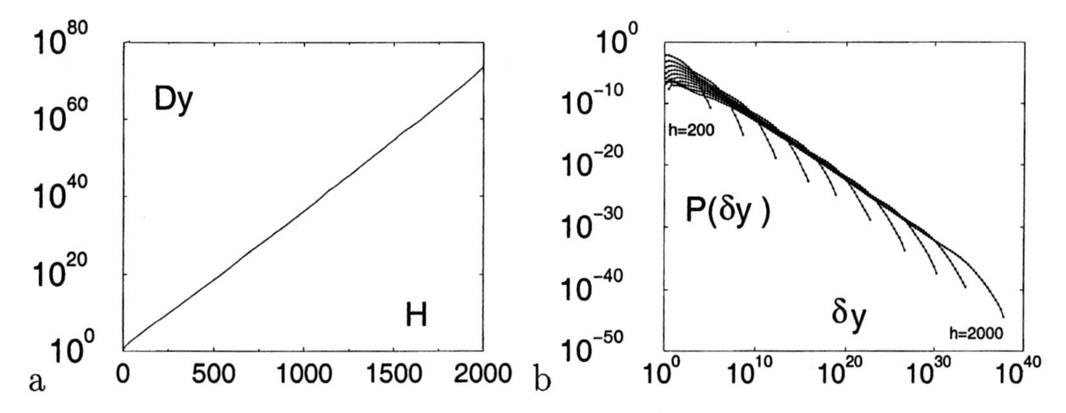

Figure 9: **a)** D_y grows *exponentially* with height when $O_v = 0$. **b)** The probability $P_h(\delta y)$ to have an induced displacement δy is power-law distributed on isostatic networks. Results are shown at $h = 200, 400, 600, \ldots, 2000$ layers above the perturbation, for positive values of δy only. $P_h(\delta y)$ is approximately symmetric, and has a finite peak of weight 0.47 at $\delta y = 0$. Thus at any height above the perturbation, approximately one half of the sites within the influence cone are not vertically shifted.

anomalously large susceptibility sets in. In a way which is consistent with our intuitive expectations in the previous section, and with the one dimensional toy model, isostaticity is also in two dimensions responsible for a large susceptibility to perturbation. An important and surprising difference is the fact that, at the isostatic critical point $O_v = 0$, the susceptibility D increases exponentially fast with system size, whereas it grows only linearly with size in one dimension.

Seeking to understand the surprisingly fast growth of D with system size, the probability distribution $P_h(\delta y)$ to have a vertical displacement δy, h layers above the perturbation has been measured on isostatic systems of 2000×2000 particles. Only sites within a 120 degree cone whose apex corresponds to the perturbed bond may feel the effect of the perturbation (Fig. 7a). $P_h(\delta y)$ thus gives the probability for a randomly chosen site inside the influence cone of the perturbed bond and h layers above it, to have a vertical displacement δy. Sites outside this cone have $\delta = 0$, so our measurements essentially correspond to a system of infinite width.

Figure 9b shows the result of these measurements at $h = 200, 400, 600, \ldots, 2000$ layers above the perturbation. Only positive values of δ are displayed in this figure since $P_h(\delta y)$ is approximately symmetric. $P_h(\delta y)$ is found to be consistent with a power-law behavior with an h-dependent cutoff:

$$P_h(\delta y) \sim h^{-\rho} |\delta y|^{-\theta} \tag{4}$$

for $\delta y < \delta_M(h)$.

It is also evident from Fig. 9d, that the cutoff $\delta_M(h)$ grows exponentially with increasing distance h from the perturbation. Fitting the curve corresponding to $h = 2000$ in the interval $10^{10} < \delta y < 10^{30}$, an estimate $\theta = 0.98$ is obtained, suggesting that $\theta \to 1$ asymptotically. Normalization then requires $\rho = 1$, since the cutoff $\delta_M(h)$ increases exponentially with h.

Similar measurements where done for (smaller) systems with a finite density of overconstraints O_v, in which case the distribution of displacements presents a size-independent bound (Fig. 11b).

Thus, in two dimensions an isostatic phase transition takes place at $O_v = 0$, and the resulting isostatic phase is characterized by a susceptibility to perturbation that grows exponentially fast with system size. The distribution of induced displacements is power law, with a cutoff that grows exponentially with distance from the perturbation. Of course one does not expect to be able to really measure exponentially large values of displacements on granular systems. The calculations reported here are valid for infinitesimal displacements, and in with this in mind the contact network is considered to remain unchanged during the perturbation. In practice, internal rearrangements would occur before we could detect very large values of displacement on a real pile. So how can we know if the huge susceptibility to perturbation that our calculations predict have any observable effect? This is is discussed in the next section.

3.2 Isostaticity implies instability

In order to clarify the relevance of the findings described in section 3.1 in relation with the observed unstable character of granular packings[18,19,21,20], we must first demonstrate the equivalence between induced displacements on site i and the load-stress response function $\mathcal{G}(i,b)$ of the stretched bond b with respect to a load on site i.

The network's total energy can be written as

$$E = \sum_{i=1}^{N} W_i y_i + 1/2 \sum_{b} k_b (l_b - l_b^0)^2 \tag{5}$$

, where the first term is the potential energy (W_i are particle's weights) and the second one is a sum over all bonds and accounts for the elastic energy. l_b are the bond lengths in equilibrium and l_b^0 their repose lengths (under zero force). Upon infinitesimally stretching bond b', equilibrium requires that

$$\sum_i W_i \frac{\partial y_i}{\partial l_{b'}} + \sum_{ov} k_{ov}(l_{ov} - l_{ov}^0) \frac{\partial l_{ov}}{\partial l_{b'}} = 0 \tag{6}$$

where the second sum goes over bonds ov that belong to the same *overconstrained* graph as b' does. This is so since bonds not overconstrained with respect to b' *do not change their lengths* as a result of stretching b'. Since stress f_b on bond b is $f_b = k_b(l_b^0 - l_b)$ this may be rewritten as

$$\sum_{ov} f_{ov} \frac{\partial l_{ov}}{\partial l_b} = \sum_i W_i \frac{\partial y_i}{\partial l_b} \tag{7}$$

If there are no overconstrained graphs the left hand sum only contains bond b itself, therefore,

$$f_b = \sum_i W_i \frac{\partial y_i}{\partial l_b} \tag{8}$$

showing that, in the isostatic case, the displacement $\delta y_i^{(b)} = \frac{\partial y_i}{\partial l_b}$ induced on site i by a stretching of bond b is the *response function* $\mathcal{G}(i,b)$ of stress f_b with respect to an overload on site i.

Taking averages with respect to disorder on equation (8), we obtain

$$< f_b > = \sum_i W_i < \delta y_i^{(b)} > \tag{9}$$

, and since average stresses on a given layer grow linearly with depth, we must have

$$< \delta y_i^{(b)} >_H \sim H^{-1} \tag{10}$$

We have seen that the second moment of $P_H(\delta y)$ diverges as $\exp\{H\}$, while (10) shows that its first moment goes to zero with increasing H. This can only happen if $P_H(\delta y)$ is approximately symmetric (this is numerically verified), which demonstrates that large positive and negative values of δy appear with similar probability. Given now the equivalence between induced displacements and the load-stress response function, the existence of large negative induced displacements means that a positive overload at a random site i, would often produce a (very large) negative stress on any arbitrarily chosen bond b. This in turn indicates that the system will have to rearrange itself in order to restore compressive forces, since negative stresses are not possible. In other words, isostatic packings are *unstable* to small perturbations, and will reorganize themselves on the slightest change in load, in order to find a new stable (compression only) contact network.

In order to finish the demonstration that *instability is a consequence of isostaticity*, we still have to show that a finite density of overconstraints would make the response function bounded again.

When there are overconstraints, $P_H(\delta y)$ is no longer critical but bounded as our numerical simulations show (see Fig. 11). But in this case δy is no longer identically equal to the load-stress response function, i.e. (8) no longer holds. One can nevertheless see, by looking at formula (7), that the weight-stress response function $\mathcal{G}(i,b)$ must be bounded if $P_H(\delta y)$ is. Therefore in the overconstrained case a finite overload of order $< f >$ is necessary in order to produce rearrangements, and the system is thus no longer unstable.

3.3 Pantographs

The exponential growth of $\delta_M(H)$ is responsible for the observed exponential behavior of the total susceptibility in the isostatic case. But we have yet to understand for which reason exponentially large values of displacements do exist. Surprisingly this

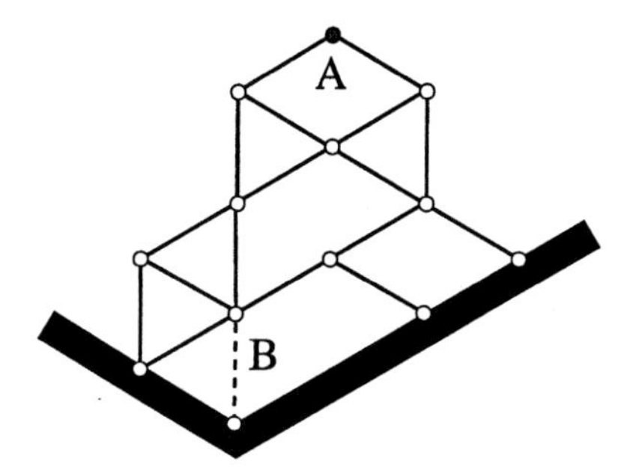

Figure 10: The observed exponential growth of induced displacements with distance to the perturbation is due to the existence of "pantographs" as the one shown in this figure. Upon stretching bond B by a small amount δ, site A moves vertically by an amount 2δ. Conversely a unitary weight at A produces a stress of magnitude 2 on bond B. This multiplicative effect only exists on isostatic systems, and is lost if the network is overconstrained. In the overconstrained case, a stretching of bond B would generate internal stresses on other bonds that oppose the deformation, and the displacement of site A would be much smaller.

can be explained in very simple terms. The appearance of exponentially large values of displacements is due to the existence of "lever configurations" or "pantographs", which amplify displacements.

Fig. 10 shows an example of a pantograph with amplification factor 2. When the dashed bond is stretched by ϵ, site A is vertically shifted by 2ϵ. Given that there is a finite density of similar pantographs on the system, it is clear that displacements will grow exponentially with system height.

This amplification effect only exists in the *isostatic limit*: Pantographs as the one in Fig. 10 are no longer effective if blocked by overconstraints. For example, an additional (redundant) bond between site A and the site below it would "block" the amplification effect of the pantograph. Then, a unitary stretching of bond B would induce stresses in the whole pantograph, but only a small displacement of site A.

In order to understand why the transition occurs at zero density of overconstraints

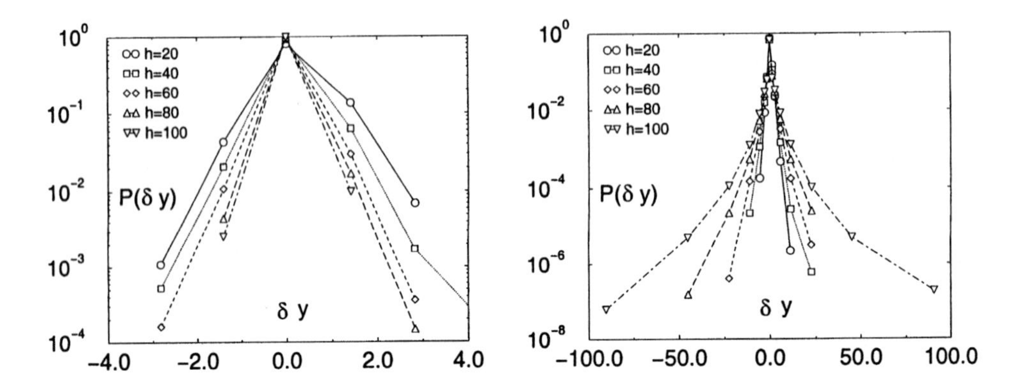

Figure 11: The effect of isostaticity is dramatically illustrated by a comparison of $P_h(\delta y)$ with and without overconstraints. Here δy the displacement of a site, induced by a bond-length perturbation h layers below it. For any nonzero density of overconstraints (left, $O_v = 0.02$ in this case), induced displacements *decrease* with distance. This is the usual behavior on an elastic continuum. On the other hand if $O_v = 0$, i.e. when the system is *isostatic* (right), the distribution of induced displacements gets *broader* when the distance h to the perturbation increases. This is due to the multiplicative effect of pantographs (see Fig. 10). These results where obtained on systems of total height 100 layers.

and not at any finite density, it is extremely important to notice that pantographs are composed of all sites suffering displacement when the perturbed bond is stretched. Thus a typical pantograph covers a finite fraction of the system, and any non-zero fraction of redundant bonds is enough to place *at least* one excess bond on it, eliminating the lever effect. This explains why anomalously large induced displacements only exist in the isostatic limit $O_v = 0$.

4 CONCLUSIONS

We have shown that the contact network of granular packings becomes *exactly isostatic* in the limit of large stiffness-to-load ratio, i.e. when the stiffness is large or the mean compressive load is small. We have furthermore provided analytical (in 1d) and numerical (in 2d) evidence that isostaticity is responsible for the appearance of a large susceptibility to perturbation, defined as the sum of the square displacements induced by a small bond-stretching. When an arbitrary bond is stretched on an overconstrained system, the effect of this perturbation is only felt locally. On the contrary, on an isostatic system the induced displacements *grow* with distance. This surprising phenomenon makes the susceptibility diverge *exponentially* fast with system size, and is produced by the existence of "pantographs": network mechanisms that amplify displacements in the same way a lever does.

We have also clarified the relationship between the susceptibility to perturbation defined in this work and the experimentally observed instability of granular networks. This was done using an equivalence between induced displacements and the weight-stress response function. The existence of negative values for the response function and the relation of this fact with instability were first discussed in the context of a phenomenological model for stress propagation [31]. In that model, the appearance of large negative values for the response function (correctly identified as a signature of instability by the authors) is a consequence of ad-hoc assumptions about the way in which stresses propagate downwards. That work though does not correctly identify the physical origin of instabilities. The vectorial character of the transmitted quantity is **not** the reason (while it is a necessary condition), as easily illustrated by an overconstrained network. The reason by which granular contact networks are unstable is that they are isostatic.

Thus stiffness produces isostaticity. Isostaticity is responsible for the tendency to global rearrangement upon slight perturbation of stiff granular materials. Any non-zero density of overconstraints is enough to destroy criticality and therefore drastically reduce instabilities. Therefore "soft" granular packings are stable.

Acknowledgments

The author is supported by a PVE fellowship granted by CNPq, Brazil.

REFERENCES

1. H. M. Jaeger, S. R. Nagel and R. P. Behringer, Rev. Mod. Phys. **68**, 1259 (1996); Physics Today **49**, 32 (1996).
2. "Friction, Arching, Contact Dynamics", Edited by D. E. Wolf and P. Grassberger, World Scientific (Singapore), 1997.
3. M. Faraday, Phil. Trans. R. Soc. London **52**, 299 (1831).
4. O. Reynolds, Philos. Mag. **20**, 469 (1885).
5. Rosato, A. *et al*, Phys. Rev. Lett. **58**, 1038 (1987); See also [1].
6. H. A. Makse, S. Havlin, P. R. King, and H. E. Stanley, , Nature **386**, 379 (1997); H. A. Makse, P. Cizeau, and H. E. Stanley, Phys. Rev. Lett. **78**, 3298 (1997); H. A. Makse, , Phys. Rev. E**56**, 7008-7016 (1997).
7. I. Goldhirsch and G. Zanetti, Phys. Rev. Lett. **70**, 1619 (1993).
8. Knight J. B.*et al*, Phys. Rev. E**51**, 3957 (1995).
9. M. Nicodemi, A. Coniglio and H. Herrmann, Phys. Rev. E **55**, 3962 (1997); see also [2].
10. Janssen, Z. Verein Deutsch. Ing. **39**, 1045 (1895).
11. J. Smid and J. Novosad, Ind. Chem. Eng. Symp. **63**, D3V 1-12 (1981).
12. S. F. Edwards and C. C. Mounfield, Physica A **226**, 1,12,25 (1996).
13. P. Claudin and J. P. Bouchaud, Phys. Rev. Lett. **78**, 231 (1997); see also [2]
14. S. Ouaguenouni and J.-N. Roux, Europhys. Lett. **35**, 449 (1995); Europhys. Lett. **39**, 117 (1997); see also [2]
15. U. Bastolla, F. Radjai and D. Wolf, in ref. [2].
16. S. Luding, Phys. Rev. E **55**, 4720 (1997).
17. C. Moukarzel, to be published.
18. P. Dantu, in *Proc. of the 4th. Int. Conf. on Soil Mech and Fund. Eng. (Butterworths, London, 1957).* P. Dantu, Ann. Ponts Chaussees **4**, 144 (1967).
19. T. Travers *et al*, J. Phys. A **19**, L1033 (1986); Europhys. Lett. **4**, 329 (1987).
20. B. Miller, C. O'Hern and R. Behringer, Phys. Rev. Lett. **77**, 3110 (1996); see also [2]
21. C. Liu et al, Science **269**, 513 (1995); see also [2]

22. F. Radjai, M. Jean, J. J. Moreau and S. Roux, Phys. Rev. Lett. **77**, 3110 (1996); see also [2].
23. D. Stauffer, H. J. Herrmann and S. Roux, J. Physique **48**, 347 (1987).
24. S. N. Coppersmith et al,Phys. Rev. E **53**, 4673 (1996).
25. G. Oron and H. J. Herrmann, 1997, PRE to appear; see `cond-mat/9707243`.
26. J. Hemmingsson, Physica A **230**, 329 (1996); J. Hemmingsson, H. J. Herrmann and S. Roux, J. Physique I**7**, 291 (1997).
27. R. M Nedderman, "Statics and Kinematics of Granular Materials" (Cambridge University Press, 1992).
28. F. Cantelaube and J. D. Goddard, in "Powder and Grains 97", edited by R. P. Behringer and J. T. Jenkins, Rotterdam, 1997, Balkema.
29. J. P. Wittmer, P. Claudin, M. E. Cates and J.-P. Bouchaud, in [2].
30. M. E. Cates and J. P. Wittmer, Physica A **249**, 276 (1998).
31. P. Claudin, J.-P. Bouchaud, M. E. Cates and J. P. Wittmer `cond-mat/9710100`, submitted to Phys. Rev. E; `cond-mat/9711135`.
32. J. E. S. Socolar, 1997, PRE to appear; see `cond-mat/9710089`.
33. C. Eloy and E. Clément, J. Physique I **7**, 1541 (1997).
34. Henry Crapo, Structural Topology **1** (1979), 26-45; T. S. Tay and W. Whiteley, Structural Topology **11** (1985), 21-69.
35. B. Servatius, this conference.
36. J. C. Maxwell, Phil. Mag. Series 4 **27**, 250 (1864).
37. E. Guyon, S. Roux, A. Hansen, D. Bideau, J.-P. Troadec and H. Crapo, Rep. Prog. Phys. **53**, 373 (1990).
38. W. Tang and M. F. Thorpe, Phys. Rev. B **37**, 5539 (1988); Phys. Rev. B **36**, 3798 (1987).
39. R. Connelly, this conference.
40. S. Feng and P. Sen, Phys. Rev. Lett.**52**, 216 (1984).
41. S. Roux, D. Stauffer and H. J. Herrmann, J. Physique **48**, 341 (1987).
42. C. Moukarzel and P. M. Duxbury, Phys. Rev. Lett. **75**, 4055 (1995); see also: C. Moukarzel, P. M. Duxbury and P. L. Leath, Phys. Rev. Lett. **78**, 1480 (1997).
43. I am indebted to H. Herrmann and J. Goddard for helpful remarks about this point.
44. A. Coniglio, J. Phys. **A15**, 3829 (1982).
45. This value can be justified by simple geometrical reasoning, assuming that the radii are random variables with a small dispersion, and that the particles are centered on the sites of a regular triangular lattice.
46. I thank HLRZ Jüelich for providing access to its supercomputing facilities.

RIGIDITY AND MEMORY IN A SIMPLE GLASS

P. Chandra[1], L. B. Ioffe[2,3]

[1]NEC Research Institute
4 Independence Way
Princeton, NJ 08540

[2]Department of Physics, & Astronomy
Rutgers University
Piscataway, NJ 08865-8019

[3]Landau Institute for Theoretical Physics
Moscow, RUSSIA

MOTIVATION

Why is a glass rigid despite its amorphous structure? In a rigid system, a local disturbance leads to a bulk response. For example, application of a small force to one end of a crystal will lead to its collective motion; energetically it is favorable for its constituent atoms to maintain their relative positions. By contrast a similar force applied to a liquid will lead to local rearrangements, and all memory of its prior configuration will be soon forgotten. Rigidity implies slow relaxation processes, long time-scales and thus long-range temporal correlations which are usually associated with long-range spatial ordering. So why does a glass have mechanical properties at all similar to those of a crystal?

As a point of reference, let us consider the response of an ideal crystal to application of shear. We note that its low-temperature state is uniquely defined (modulo transformations of the crystal group). The system is cooled into its crystalline phase and after a waiting time, t_w, a shear deformation is applied. For all practical measurement times, $t_m = t_w + t$, it recovers its inital state; it displays perfect memory of its original configuration independent of specifics associated with the shearing and measurement processes.

At low temperatures a crystal has long-range spatial order. By contrast the low-temperature structure of glasses is determined by the incompatibility between local and global ordering, and is *not* unique on any spatial scale.[1, 2] Though such frustration certainly plays a key role in glass formation, purely structural approaches do not address the thermal or dynamical aspects of glassy behavior. For example, unlike a liquid or a crystal, the low-temperature state of a glass is dependent on its sample history. Therefore the response of a glass to shear is not simple. First of all, its inital low-temperature state is a function of sample history; the number of such configurations

Table I.

	Liquid → Solid	Metal → Superconductor
Conductivity Equation	$\Pi = \eta \dot{u}$	$J = \sigma E = \sigma \dot{A}$
Drude Relation	$\eta = \eta_0 \frac{1}{1 - i\omega\tau}$	$\sigma = \frac{ne^2\tau}{m} \frac{1}{1 - i\omega\tau}$
London Kernel	$\Pi = Gu$	$J = -QA$
Kubo Relation	$G = -i\omega\eta$	$Q = -i\omega\sigma$

scales exponentially with the number of sites. Furthermore the response is dependent on both t_w and $t_w + t$, and not simply on their difference, t. On short time-scales $(t < t_w)$, a glass exhibits perfect memory of its original configuration; by contrast on longer ones $(t \gg t_w)$ a glass "flows" and thus forgets its past, a phenomenon often referred to as "ageing".

Maxwell was one of the first physicists to address the crossover between viscous and elastic behavior in supercooled liquids.[3] Conceptually his treatment is amusingly similar to later studies of the metal-superconductor transition;[4] it also provides a nice study of "generalized rigidity", an idea that has been emphasized in conjunction with broken symmetry by Anderson.[5] The analogy is presented in Table I. In a metal the flow equation is Ohm's Law: the current (J) is proportional to the field (E); alternatively the flow of charge is proportional to the time derivative of the vector potential $(\dot{A})$. Analogously in a liquid the shear (Π) is proportional to the time derivative of the strain $(\dot{u})$; the shear can be viewed as the flow of momentum. At finite frequency charge currents in a simple metal relax on a characteristic time-scale τ leading to a Drude relation for the conductivity σ; momentum currents in a liquid decay in a similar fashion with an analogous relation for the viscosity η. Then, for example at a particular temperature, this characteristic time-scale τ diverges and the system displays rigidity. In the resulting solid, the shear is proportional to the strain. In the superconductor the supercurrents are proportional to the vector potentional as described by the London equation. In the superconductors and the solids, the rigidity is associated with the development of supercurrents of charge and momentum respectively. The analogue of the London kernel in the superconductor is the shear modulus, and there is a Kubo relation for both of these quanties.

At this point we have only discussed the development of a solid, but have not distinguished between a crystal and a glass. This distinction is summarized in Table II. As we have already mentioned above, a crystal breaks translational symmetry and

Table II.

	Glass	Crystal
Long-Range Spatial Ordering	No	Yes
Number of States	$\sim \exp N$	$O(1)$
Fluctuation-Dissipation Theorem	Violated	Obeyed
Memory Effects	$\lim_{t\to\infty} \lim_{t_w\to\infty} G_{t_w+t,t_w} \neq 0$	
Ageing	$\lim_{t\to\infty} G_{t_w+t,t_w} = 0$	$\lim_{t\to\infty} G_{t_w+t,t_w} \neq 0$ (fixed t_w)

has long-range spatial ordering; by contrast there is *not* any such simple long length-scale associated with a glass. A crystalline state is unique up to transformations of the crystal group. In a glass, there is no preferred order so that local changes can lead to new metastable states. As a result, the number of states in a glass scales exponentially with the number of sites in the system. Characterization of the low-temperature glassy state then becomes rather difficult. Conventionally, for example in a crystal, one applies a small symmetry-breaking field to restrict one's statistical trace to the states in a single well. Then, using this restricted trace, one can proceed with equilibrium statistical mechanics as usual in the low-temperature phase. However because there are an exponential number of states in a glass, it is not possible to distinguish different metastable configurations by application of a simple field. As a result one cannot determine properties of the glass using a conventional Gibbs approach; more specifically, one cannot employ the fluctuation-dissipation theorem.

As previously mentioned, the similarities and differences between a glass and a crystal are best displayed by considering the behavior of the shear modulus (e.g. Table II). The memory of a glass improves with increasing waiting time (t_w); indeed if one takes the waiting time to infinite first and then lets the measurement time ($t_w + t$) diverge it has perfect memory just like a crystal. Here the order of limits is crucial. By contrast for fixed waiting time in the limit of infinite measurement time, a glass loses its rigidity; by contrast the response of a crystal is invariant under time-translation, and thus its memory is perfect and independent of the order of limits associated with t_w and t.

In a nutshell, glassiness is solidification without crystallization. It occurs in a large number of materials of diverse microscopic character; therefore one feels that there should exist an underlying reductionist picture, the analogue of the Ising model for second-order phase transitions, which has not yet been found. In particular, how is a system simultaneously out-of-equilibrium and stable on very long (sometime millenium) time-scales? How does one keep an intrinsically non-random system away from its thermodynamic ground-state? There has been tremendous technical progress in our theoretical understanding of disordered models; is it possible to link "dirty" and "clean" glasses? There have been several studies of such mappings, which seem to exist, and we refer the interested reader to an excellent review on this subject.[6] Here the strategy is similar to that associated with energy level spectra in complex nuclei, where progress was made through studies of random Hamiltonians. However in what follows, we shall discuss a complementary approach: a direct study of a specific non-random system that displays glassiness. We have taken a somewhat "Galilean" approach to the glass problem: we have abstracted it to a point that we can manage, hoping to have gleaned its essential features. Then we have characterized it with the available tools at our disposal to check whether it has the desired properties. Finally we have tried to make contact with experiment by making predictions for measurements to check whether we are on the right track! In the next section we hope to give the reader a flavor of the model of study and for our main analytic results. Next we'll discuss some early experiments on this "simple glass" and some predictions for measurements that we'd like to see. We'll mention briefly how this "Galilean" system, with both memory and a large number of states, can be used for information storage. Finally we'll end with a summary and everpresent open questions which should be addressed in the future.

A "GALILEAN" MODEL FOR A SIMPLE GLASS

In the spirit of the Maxwellian analogy between liquids and metals discussed above, we have studied the glass transition in a long-range superconducting array. The physical system of study is a stack of two mutually perpendicular sets of N parallel thin wires with Josephson junctions at each node (Figure 1) that is placed in an external tranverse field H. The thermodynamic variables of this array are the $2N$ quantum mechanical phases of the superconducting order parameter associated with each wire; these play an analogous role to that of the atomic positions or density waves in a real glass. We note that in this network each horizontal/vertical wire is coupled to each vertical/horizontal one by a Josephson junction, so that in the thermodynamic limit ($N \to \infty$ for fixed array area) the number of neighbors diverges.

The energy scales associated with this long-range array are associated with the individual superconducting wires and with the Josephson junctions. In the absence of a field the macroscopic phase of a superconductor is constant in equilibrium, assuming that its thickness l is larger than the coherence length ξ_0 so that phase slips are energetically unfavorable. Application of a field results in the procession of this macroscopic phase, where the rate of procession is determined by the amplitude of the applied field. In the absence of a field the phase differences would be zero at each Josephson junction, but this is no longer possible at finite H. Thus frustration in this network can be "tuned" by application of a transverse field; the system is then overconstrained since there are $2N$ phases and N^2 Josephson junctions with competing energetic requirements.

Here we assume that the Josephson couplings in the array are sufficiently small so that the induced fields are negligible in comparison with H; we shall return to this issue when discussing the experimental realization of this network. We can therefore describe the arary by the Hamiltonian

$$\mathcal{H} = -J \sum_{jk} \cos\left(\phi_j - \phi_k - \frac{2\pi\Phi_{jk}}{\Phi_0}\right) = \mathrm{Re} \sum_{jk} s_j^* J_{jk} s_k \tag{1}$$

with $1 \leq (j,k) \leq N$ where $j(k)$ is the index of the horizontal(vertical) wires and Φ_0 is the flux quantum; the effective spins $s_k = e^{i\phi_k}$ where the ϕ_k are the superconducting phases of the $2N$ wires. The couplings are site-dependent and are related to the enclosed flux at a given node such that

$$J_{jk} = J \exp \frac{2\pi i \Phi_{jk}}{\Phi_0} = J \exp \frac{2\pi \alpha k}{N} \tag{2}$$

where we have introduced the flux per unit strip $\alpha = \frac{NHl^2}{\Phi_0}$, with l is the inter-node spacing; here we note that the sign of the coupling depends on the enclosed flux and can be both positive and negative.

Because every horizontal (vertical) wire is linked to every vertical (horizontal) wire, the number of nearest neighbors in this model is N and it is accessible to an analytic mean-field treatment. For the same reason, the free energy barriers separating its low-temperature solutions scale with N. This situation is in marked contrast to that in conventional 2D arrays where the coordination and hence the barriers are low.[7] A similar long-range network with positional disorder was studied previously.[8] For $\alpha \gg 1/N$ that system displays a spin glass transition which was mapped onto the Sherrington-Kirkpatrick model[9] for $\alpha \gg 1$; in this field regime there is no residual "ferromagnetic" phase coherence between wires. Physically this glassy behavior occurs because the phase differences associated with the couplings J_{jk} acquire random values and fill the interval $(0, 2\pi)$ uniformly for $\alpha \gg 1/N$. More specifically, there will be no commensurability if the sum

$$\sum_{k=1}^{N} J_{jk} J_{kl}^{\dagger} = J_0^2 \sum_{k=1}^{N} \exp 2\pi i \left\{ \frac{\alpha k(j-l)}{N} \right\} \tag{3}$$

is smooth where $\{j,l\}$ (k) are indices labelling horizontal (vertical) wires; this will occur only if the expression in curly brackets on the r.h.s. of (3) is *not* an integer, a condition always satisfied for the disordered array. For the periodic case, this situation is realized in the "incommensurate window" $1/N \ll \alpha \leq 1$; here the phase-ordering unit cell is larger than the system size so that the "crystalline" phase is inaccessible. There are thus no special field values where the number of low-temperature solutions are finite, in contrast to the situation for $\alpha > 1$. In the thermodynamic limit of $N \to \infty$ (with fixed array area), the high-temperature approach to the glass transition in this system has been studied[10, 11] using a modified Thouless-Anderson-Palmer (TAP) method.[12] Here we discuss the qualitative picture that emerges from these results (cf. Fig. 1), referring the interested reader elsewhere for a more detailed quantitative treatment. As T approaches T_m^+, where $T_m^+ \sim T_0 \approx \frac{J\sqrt{N}}{2\sqrt{\alpha}}$, there appear a number of metastable states in addition to the paramagnetic free-energy minimum; most likely they are energetically unfavorable and thus do not "trap" the system upon cooling from high temperatures. As $T \to T_G^+$, the paramagnetic minimum is "subdivided" into an extensive number of degenerate metastable states separated by effectively infinite barriers, and the system is dynamically localized into one of them. Qualitatively, in the interval $T_m > T > T_G$ there appear many local minima in the vicinity of the paramagnetic state separated by *finite* barriers; these barriers increase continuously and become infinite at $T = T_G$. Each of these minima is characterized by a finite "site magnetization" $m_i = \langle s_i \rangle_T$ where "site"

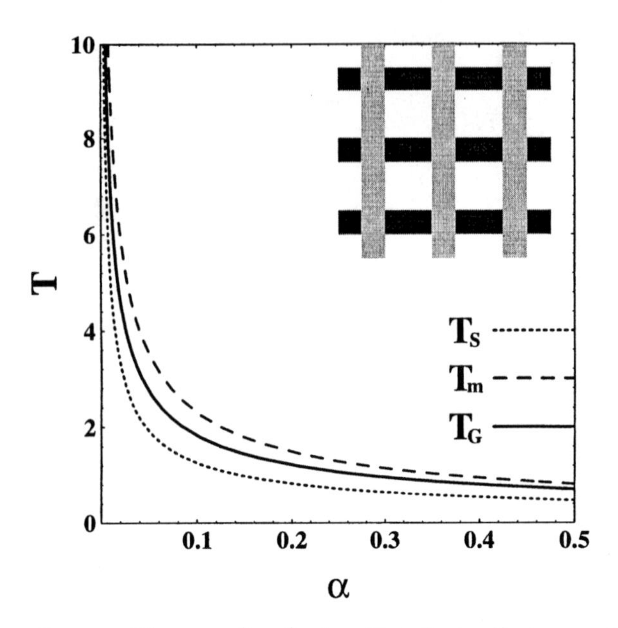

Figure 1. The phase diagram of the array (inset) where T_G indicates the temperature asosciated with the dynamical instability discussed in the text, T_S is the speculated equilibrium transition temperature and T_m is the "superheating" temperature where the low-temperature metastable states cease to exist.

refers to a wire. When $T > T_G$ thermal fluctuations average over many states so that $\langle m_i \rangle \equiv 0$. At $T = T_G$ the system is localized in one metastable state and there is an associated jump in the Edwards-Anderson order parameter, $\left(q = \frac{1}{N}\sum_i \langle m_i \rangle^2\right)$. The low-temperature phase is characterized by a finite q and by the presence of a memory, $\lim_{t' \to \infty} \Delta(t, t') \neq 0$ where $\Delta(t, t')$ is the anomalous response, defined as $\Delta(t, t') \equiv \langle \frac{\partial m(t)}{\partial h(t')} \rangle$. At $T = T_G$, the metastable states are degenerate and thus there can be no thermodynamic selection. However at lower temperatures interactions will probably break this degeneracy and select a subset of this manifold; then we expect an $(t \to \infty)$ equilibrium first-order transition (T_S) which should be accompanied by a jump in the local magnetization. In order to observe this transition at T_S the array must be equilibrated on a time-scale (t_E) longer than that (t_A) necessary to overcome the barriers separating its metastable states; t_A scales exponentially with the number of wires in the array. Thus the equilibrium transition at T_S is observable *only* if $t_E \to \infty$ *before* the thermodynamic limit $(N \to \infty)$ is taken; in the opposite order of limits only the dynamical transition occurs.

The periodic array thus exhibits a first-order thermodynamic transition *preceeded* by a dynamical instability; the glass transition at T_G is characterized by a a diverging relaxation time and an accompanying jump in the Edwards-Anderson order parameter. In general the dynamical behavior of this network is described by coupled integral-differential equations for the correlation $(D_{tt'} = \langle s(t)s(t') \rangle)$ and the response $(G_{tt'} = \langle \frac{\partial s(t)}{\partial h(t')} \rangle$ where $t > t')$ functions. At temperatures above the glass transition $T > T_G$ the response and the correlation functions are related by the fluctuation-dissipation theorem and there is time-translation invariance; the resulting equation

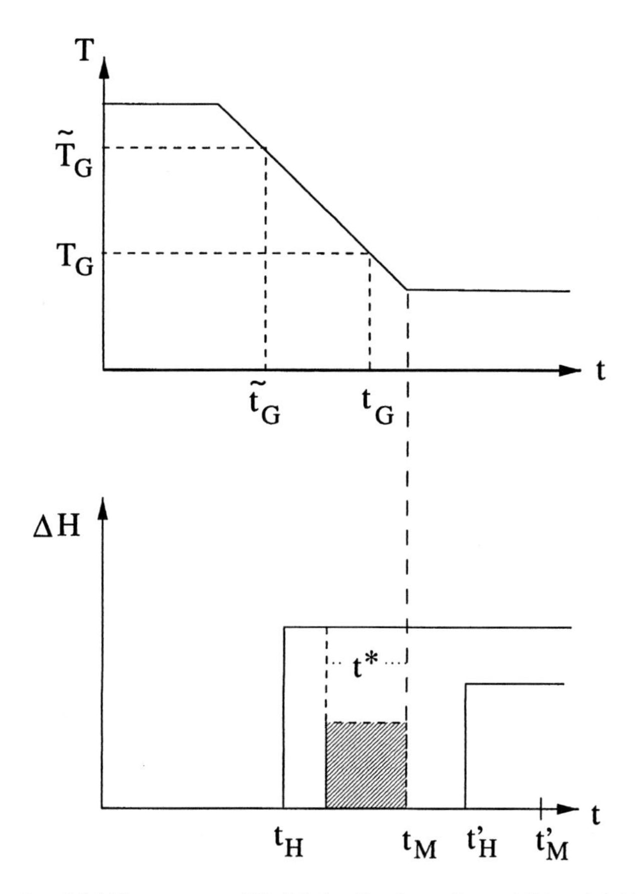

Figure 2. Schematics of (a) Temperature (T) (b) Applications of an additional field (ΔH) vs. time (t) indicates the time-scales involved in a finite-cooling experiment as discussed in the text; $\tilde{T}_G$ and $\tilde{t}_G$ refer to the "effective glass transition" when the system goes out of equilibrium, and t_H, t_M and t^* are the time-scales associated with the onset of an additional field, a subsequent measurement and ageing.

for the system's dynamical evolution is structurally similar to that derived for density-density correlations in the mode-coupling approach to the liquid-glass transition.[13] Furthermore the general coupled equations describing the dynamical behavior of this *non-random* network in the regime ($1/N \ll \alpha < 1$) are *identical* to those obtained for the $p = 4$ (disordered) spherical Potts model.[14] The possible connection between non-random glasses and $p \geq 3$ (disordered) spherical models has been previously suggested in the literature,[15, 16, 17] and these results give support to those conjectures. Furthermore these long-range Josephson arrays can be built in the laboratory, allowing for parallel theoretical and experimental studies of the "simplest spin glasses" that have previously been an abstraction. Several physical properties of the array have been characterized[18] in its non-ergodic regime ($T < T_G$). We note that the glass transition temperature T_G described above corresponds to the system going out of equilibrium as it is cooled infinitesmally slowly from high temperatures. In practice any physical cooling process occurs at a finite rate, and the effective glass transition occurs when the system drops out of equilibrium; thus the observed glass transition temperature is a function of the time-scale of the experimental probe. If we define the reduced temperature $\Theta \equiv \left(\frac{T-T_G)}{T_G}\right)$ then the effective glass transition will occur when the time associated with cooling, $t_c \approx \frac{\Theta}{d\Theta/dt}$ is equal to the relaxation time at that tempera-

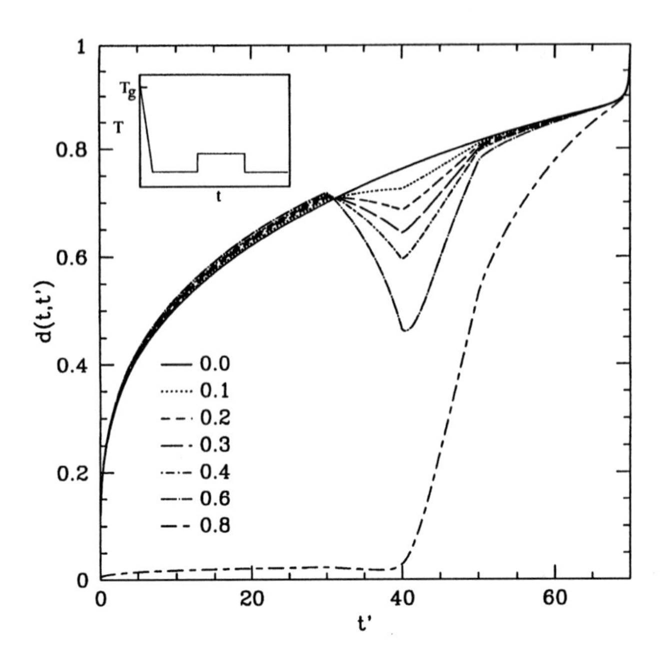

Figure 3. The rescaled correlation function $d(t, t')$ for the cooling-heating regime. Here the temperature was reduced infinitely fast to a temperature $T_i < T_g$ at $t' = 0$; the system was equilibrated for $0 < t' < 30$, then heated linearly for $30 < t' < 40$, then cooled linearly back to its original temperature T_i for $40 < t' < 50$, and then measured at $t = 70$. Different curves correspond to different amplitudes of the heat pulse as shown. We see that for heat pulses of amplitude less than 0.6 the system recovers its original state; however, for larger heat pulses all memory is lost.

ture, $\tau_R \approx t_0\theta^{-\nu}$ where t_0 and ν are determined from the high-temperature analysis elsewhere.[11] There is evidence for history-dependence; the system's response is very different if an additional field is turned on during or after the cooling process, reminiscent of the zero-field cooled vs. field-cooled susceptibility observed in spin glasses. Furthermore there exists an "ageing" time-scale t* this system (cf. Fig. 2); if the field is turned on during the cooling process at time t_H and a measurement is taken at time t_M such that $t_H - t_M < t*$ the system "remembers" the cooling process; otherwise $(t_H - t_M > t*)$ it "forgets" it completely.

The low-temperature "memory" of the superconducting array is perhaps best exhibited by its response to an applied heat pulse [18]. The correlation function $D_{tt'}$ as a function of t', displayed in Figure 3, exhibits the system's overlap between its state at time t and time t' after a fast temperature quench. Heat pulses of varying amplitudes (c.f. Fig. 3) are applied after the quench, and the system's memory is probed. For all but the largest amplitude pulse, the system recovers its original dynamical trajectory indicating the presence of long-range temporal correlations. Such measurements have been performed experimentally on real spin glasses to determine the structure of their low-temperature states.[19] An analogous numerical analysis of the superconducting array indicates that *all* metastable states appear at the glass transition; this absence of further subdivision of states is consistent with analytic work on the structure of states in the $p = 4$ (disordered) spherical model.[20]

DISCUSSION

Since the experimental realization of these arrays is important, we would like to identify several assumptions in the theoretical treatment; in order to test its predictions and probe beyond its realm (e.g. finite-size effects) the fabricated system must satisfy certain physical requirements[21, 22] In particular the effects of everpresent screening currents must be minimized to ensure that the external field frustrates the superconducting phases effectively. More specifically the induced flux in the array should be less than a flux quantum so that all fields and phase gradients produced by diamagnetic currents are negligible. The resulting condition on the array limits the number of wires; there is thus a delicate balance between the need for many neighbors (and high free-energy barriers) and an external field that effectively frustrates the phases. Furthermore the model has only been considered in the classical limit where quantum mechanical fluctuations of the phases are negligible. These constraints put some strong restrictions on the choice of array parameters, which have been elaborated elsewhere.[21, 22]

Once the arrays are fabricated to specification, what experiments should be performed? Recently there have been some measurements probing the static properties of such long-range arrays.[22] Commensurability effects were studied in a small network as a function of field; the results were in good agreement with mean-field theory and in particular indicated the absence of long-range spatial phase ordering for field strengths $\alpha < 1$. What other experimental measurements would we like to see? History-dependence of the critical current j_c should be studied to establish the presence of large barriers crucial for the development of glassiness; more specifically the dependence of j_c on path in the $T - H$ plane should be investigated. The critical current j_c should be measured for a large number of thermal cyclings; if the system is glassy it should display a well-defined distribution $P(j_c)$. Finally the diverging relaxation time at T_g should be accessible via the a.c. response to a time-varying field $H(t)$; the associated a.c. susceptibility is

$$\chi_\omega = \frac{\partial M_\omega}{\partial H_\omega} = \frac{C(\alpha)\omega}{\omega + i/\tau_R(T)} \tag{4}$$

where τ_R is the longest response of the system that diverges at $T = T_g$ and $C(\alpha)$ is to be found elsewhere.[11] The $\omega \to 0$ limit of the a.c. susceptibility jumps to a finite value at $T = T_g$, indicating the development of a finite superconducting stiffness at the transition. Therefore measurement of this a.c. response in a fabricated array would be a direct probe its predicted glassiness.

This array has long-range temporal correlations (memory) and has an extensive number of metastable states;[23] can it be used for information storage? Indeed high connectivity and nonlinear elements (here the Josephson junctions) are key features required for the construction of associative memories.[24] Here one would like to store p patterns in such a way that if the memory is exposed to a slightly different one, it produces the stored pattern most similar to it. A simple model for such a memory is based on an array of McCulloch-Pitts neurons.[24] The patterns are stored in couplings which are chosed to minimize an energy function. Each nonlinear element has multiple inputs $n_i = 0, 1$; in the simplest model the couplings can have arbitrary sign and the ouput is "computed" by each element using the simple formula $n_i = \Theta\left(\sum_j J_{ij} n_j\right)$. Clearly the output is robust to errors in the input due to the multiple connections present. An array of such artificial neurons is thus content-addressable and fault-tolerant.

It is quite straightforward to adapt the long-range array described above to become a superconducting analog of a McCulloch-Pitts networks.[25] The couplings at each

Josephson node can be "written" by a superimposed array of superconducting quantum interference devices (SQUIDs). The inputs and outputs are rapid single flux quantum (RSFQ) voltage pulses.[26] For each stored pattern there is an input-output dictionnary associated with the pulses, and the system would be tolerant to input errors. For an array of $N = 1000$ wires with internode spacing $l = 0.5\mu$, the capacity would be $C = 0.1N^2 = 10^5$ bits with an access time per bit of $\tau_a \sim 10^{-12}$ sec. Because the $N = 1000$ bits in a given pattern could be accessed in parallel, the access time per image could be $\tau_{DT} = \frac{\tau_A}{N} = 10^{-15}$ seconds. Thus it is a candidate for nonlocal information storage; it is faster but has lower capacity per unit area than the best optical holographic memories.[28]

In conclusion, we have presented a summary of recent work on a periodic "Galilean" glass that displays rigidity in the absence of long-range spatial order. This glass transition is characterized by a diverging relaxation time and an accompanying jump in the Edwards-Anderson order parameter. At temperatures above the glass transition its dynamical equation is similar in structure to that studied in mode-coupling approaches to the liquid-glass transition. More generally, its evolution is described by coupled integral-differential equations which are identical to those of a well-studied disordered model. Preliminary experiments have probed the static structure of the array, indicating the absence of commensurability except for special field values. Predictions for further dynamical measurements have been made, including a suggestion for probing the diverging relaxation time at the glass transition.

Naturally there remain many open questions. How many of the extensive number of metastable states are actually physically accessible? Recent work[27] suggests that a small number of these states have basins of attraction far exceeding the average. If true, this result could have more general implications for other complex systems, particularly for the problem of protein-folding. Quantum effects in this array remain to be studied and could have interesting consequences. The memory discussed above is completely passive; in principle one could construct an adaptation of this array that both remembers and learns. Finally how robust are the properties of this long-range glass to spatial inhomogeneities (finite-size effects)? There is still much to be learned in the study of rigidity, even in the simplest systems.

REFERENCES

1. For example, many chalcogenide glasses are well described by continuous random networks; here the rigidity has been studied using a stability analysis involving the geometrical constraints. This structural treatment has led to predictions or optimal glass compositions in a number of covalnet systems consistent with experiment. For a general review of this constraint-counting approach and its interplay with experiments see M.F. Thorpe, *J. Non-Crys. Sol.* **182**, 135 (1995).
2. For a review of different conceptual approaches to the glass problem see J. Jackle, *Rep. Prog. Phys.* **49**, 171 (1986).
3. For a discussion of Maxwell's approach in contemporary language see J. Zarzycki, *Glasses and the Vitreous State*, Cambridge University Press, Cambridge, 1982).
4. We thank P. Coleman for several discussions of this analogy.
5. P.W. Anderson, *Basic Notions of Condensed Matter Physics*, (Benjamin/Cummings, Menlo Park, 1984).
6. J-P. Bouchaud, L. Cugliandolo, J. Kurchan and M. Mezard in *Spin Glasses and Random Fields*, A.P. Young ed., (World Scientific, Singapore, 1997).
7. e.g. J.E. Mooj and G.B.J. Schon ed., "Coherence in Superconducting Networks," Physica **152B**, 1 (1988).
8. V. M. Vinokur, L. B. Ioffe, A. I. Larkin, M. V. Feigelman, Sov. Phys. JETP **66**, 198 (1987).
9. D. Sherrington and S. Kirkpatrick, Phys. Rev. B **35**, 1792 (1975).
10. P. Chandra, L.B. Ioffe and D. Sherrington, Phys. Rev. Lett. 75, 713, (1995).

11. P. Chandra, M.V. Feigelman and L.B. Ioffe, Phys. Rev. Lett. 76, 4805, (1996).
12. D.J. Thouless, P.W. Anderson and R.G. Palmer, *Phil. Mag.* **35**, 593 (1977).
13. W. Gotze, *Z. Phys.* *B***56**, 139 (1984); E. Leutheusser, Phys. Rev. A 29, 2765, (1984).
14. A. Cristanti, H. Horner, J.-J. Sommers, *Z. Phys. B*, **92**, 257 (1993).
15. The mapping between periodic and disordered systems was first suggested by T.R. Kirkpatrick and D. Thirmulai, Phys. Rev. Lett. 58, 2091 (1987); T.R. Kirkpatrick and D. Thirmulai, Phys. Rev. B 36, 5388, (1987); T.R. Kirkpatrick, D. Thirumalai and P.G. Wolynes, Phys. Rev. B 40, 104, (1989).
16. G. Parisi in J.J. Brey, J. Marro, J.M. Rubi and M. San Migual, eds. *Twenty Five Years of Non-Equilibrium Statistical Mechanics; Proc. of the Thirteenth Sitges Conference*, (Springer-Verlag, Berlin 1995), pp. 135-42.
17. S. Franz and J. Herz, Phys. Rev. Lett. 74, 2115, (1995).
18. P. Chandra, M.V. Feigelman, L.B. Ioffe and D.M. Kagan, Phys. Rev. B 56, 11553 (1997).
19. E. Vincent, J.Hamman, M. Ocio, J.-P. Bouchaud and L.F. Cougliandolo in *Complex Behavior of Glassy Systems* eds. M. Rubi and C. Perez-Vicente (Springer-Verlag, Berlin, 1997)pp. 184 - 219.
20. J. Kurchan, G. Parisi and M.A. Virasoro, *J. Phys.* *I***3**, 1819 (1993).
21. P. Chandra, M.V. Feigelman, M.E. Gershenson and L.B. Ioffe in *Complex Behavior of Glassy Systems* eds. M. Rubi and C. Perez-Vicente (Springer-Verlag, Berlin, 1997) pp. 376 - 384.
22. H.R. Shea and M. Tinkham, Phys. Rev. Lett. 79, 2324 (1997).
23. P. Chandra, L.B. Ioffe and D. Sherrington, to be published.
24. J. Hertz, A. Krogh, R.G. Palmer, *Introduction to the Theory of Neural Computation* (Addison-Wesley, Redwood City, 1991).
25. P. Chandra and L.B. Ioffe, *U.S. Patent No. 5,629,889* (1997); P. Chandra and L.B. Ioffe, to be published.
26. K. Likharev in *The New Superconducting Electronics*, H. Weinstock and R.W. Ralston eds., (Kluwer, Dordrecht, 1992).
27. P. Chandra and L.B. Ioffe, to be published.
28. R. Linke, Private Communication.

CONSTRAINT THEORY, STIFFNESS PERCOLATION AND THE RIGIDITY TRANSITION IN NETWORK GLASSES

J. C. Phillips*

Bell Laboratories, Lucent Technologies
Murray Hill, N. J. 07974-0636

INTRODUCTION

The geometrical ideas underlying percolation theory are so simple as to be understandable at kindergarden level, yet among theoretical physicists the overwhelming preference for treating disordered physical systems was, is, and is likely for the foreseeable future to remain, the effective medium approximation (EMA), in which one first averages over the disorder and then calculates. This is a truly dreadful model, especially near phase transitions, as was stressed quite forcefully by Landauer in his classic review article[1]. In fact, near phase transitions, there is generally some kind of change in effective dimensionality, upon which all analytic properties, including critical exponents, are strongly dependent.

One must now ask, if the EMA is so bad, why does (almost) everyone use it? It seems that most physicists are uncomfortable with percolation theory. While there is an extensive mathematical literature on geometrical (connectivity) percolation, this model is too simple to describe the wide differences in physical properties observed in various classes of disordered materials. Moreover, it is a highly non-trivial matter to extend geometrical percolation to include physical interactions. During the 60's and 70's some of the world's most outstanding theoretical and experimental physicists focused their efforts on the study of magnetic critical transitions, largely in crystals, with a small but important effort on dilute antiferromagnets[2]. These are very simple materials, in which the critical behavior is dominated by clusters formed by short-range exchange interactions between spins, which bear a close resemblance to blobs studied by percolation theory. Yet even in this simple case a tremendous effort was required to obtain theoretical and experimental results which could be compared and pronounced to be scientifically successful.

This example warns us just how complex the physics of disordered materials can be, once we treat interactions realistically, rather than by naïve EMA concepts. However, each time we succeed in analyzing a new class of disordered systems realistically, we contribute to the development of that elusive but extremely valuable quality which scientists call physical intuition. Here one must distinguish between mathematical models, which, however interesting, are merely logical, and physical systems, whose forces are given by nature. For example, network glasses with harmonic interatomic forces seem to be an obvious example of a simple physical system in which percolative effects may be present. Yet the most sophisticated phenomenon which can be treated within the EMA is apparently superconductivity, for which a microscopic theory was produced in 1957, in other words, long, long ago, when the annual output of physics papers was ten times smaller than it is now. The first glimmerings of a microscopic theory of network glasses did not appear until 1979, and this conference is the first one devoted entirely to addressing this problem at a microscopic level comparable to that of the many conferences in the 60's and 70's devoted to effective medium theories of magnetic critical phenomena.

THE PROBLEM OF NETWORK GLASSES

Whereas the phenomenon of superconductivity was first discovered in 1908, the technology of glasses is more than 2000 years old. The science of network glasses developed in this century, first with limited thermochemical studies, often on complex organic glasses (polymers), where steric and network effects are both present. Steric effects are difficult to quantify, either experimentally or theoretically. There is only one universal phenomenon that embraces network, polymeric and other organic glasses, as well as fused salts, and this is the relaxation properties, which are generally described by stretched exponentials. The most successful theory of such relaxation is apparently the one based on dispersive transport to relaxation centers, but even this theory, although almost universally applied in practice, is still regarded as somewhat mysterious, although quantitative understanding of "magic" relaxation fractions is emerging[3]. This description is essentially a modified effective medium theory which does not have much to say about phase transitions, or the origin of the glass-forming tendency, although its apparently universal success is attractive.

The existence of network glasses poses a very fundamental problem for microscopic theory, which is the same as that posed by superconductivity, namely, why? The prototypical network glass is SiO_2, which forms the chemical basis, or percolative backbone, for almost all technological applications of network glass, up to and including optical fibers. However, it is difficult to build a theory on only one example, which cannot be substitutionally alloyed, but only modified with additives. Indeed, from the point of view of materials science, the network glass-forming ability seems to be rarer than that of superconductivity. (For example, aluminum silicates (clays) never form glasses, although the crystalline compounds have network structures even more complicated than that of quartz.) And because the crystal-glass transition is essentially first-order (to the extent that one can describe freezing from an equilibrium to a non-equilibrium state in equilibrium thermodynamic terms), one expects to gain the most insight by varying interactions continuously in substitutional alloys.

CHALCOGENIDE ALLOY GLASSES

For the foregoing reasons the glasses based on S, Se and to a lesser extent, Te (the chalcogenides), alloyed with Ge, As and to a lesser extent, Si, Sn, P, Sb[4], Bi, C, and even H and I, have been most studied and analyzed in terms of their network properties. Great care is needed to be certain that phase separation and/or crystallization have not occurred in preparation of homogeneous samples by quenching from the melt, but with such care it is possible (and even easy for compositions near $As_2(S,Se)_3$) to make excellent glasses. It was on reflecting on chemical trends in the glass-forming tendency (GFT) that I first discovered the earliest version of constraint theory[5]. (As an aside, at this point I apologize to the many workers who have recently done excellent work in this field. It was my intention, when I began this article, to review the most outstanding recent contributions to this subject. The results of a computer print-out showed that ref. 5 has been cited more than 500 times, and upon reading 30 or 40 of the 250 most recent references (since 1990), I realized that the level of research in this field has advanced to such a level that this would require the discussion of several hundred papers, in other words, a full-scale review about five-ten times longer than this contribution.)

What happens, I asked myself, as we vary the coordination number N_c in an alloy series from Ge ($N_c = 4$) to Br ($N_c = 1$)? Physically we go from an overconstrained material (which crystallizes explosively when it is prepared as an evaporated amorphous film and heated) to an underconstrained molecular glass, Br_2. Neither of these systems is a network glass, but As_2Se_3 ($N_c = 2.4$) is such a good glass that it is very difficult to grow crystals of this material, although it satisfies all the simple chemical rules (including even single bonding) for good charge- and valence-alternation coordination compounds. Thus it is clear that the GFT is maximized at $N_c = 2.4$. Moreover, it seems that this must occur for a good reason, namely that at this composition the material is neither overconstrained (with excessive strain energy which causes explosively exothermic crystallization) nor underconstrained (and so effectively disconnected, leading to the formation of a molecular crystal). It is thus trapped in a kind of configurational limbo, where fluctuations do not provide an effective pathway to the crystalline phase.

DEFINING BONDING CONSTRAINTS

The problem with this idea is that, stated in this form, it is uncontrolled. Lagrange knew what he meant when he talked about mechanical constraints and degrees of freedom, whether for a ball rolling down an inclined plane, or for a bicycle (an old Cambridge Tripos problem; you might like to try it on your students, but first work it out yourself). But what do we mean when we talk about interatomic force-field constraints? Even in an insulator there are a great many such forces, first neighbor, second, and so on. Non-central semi-empirical force fields have been developed for many covalent systems, especially hydrocarbon molecules, and from studying these one can see that for light atoms there is a well-defined hierarchy of forces, bond-stretching, -bending, -rotating (relative to a common axis), and so on. (For heavier atoms central Van der Waals forces also become significant.) However, even after we know all the bond forces, the question of whether they are acting effectively as constraints for temperatures near the glass transition temperature, that is, somewhat below the equilibrium melting point of the crystal, or near the solidus and liquidus in alloys, seems quite complex, as it involves questions of geometry (atoms have volumes, they are not points!).

The simple way around this difficulty is provided by studying relevant crystal structures and radial distribution functions measured by diffraction in the glass. (In binary alloys a combination of X-ray and neutron diffraction is usually enough to determine the separate pair distribution functions.) In other words, what seems to be an ambiguous problem mathematically can actually be an overdetermined one in practice, when use is made of all available experimental data. This requires some patience and skill in bibliographical work, but it is really not nearly so much of a problem as it might have appeared to be *a priori*, because the modern data base is enormous. For the chalcogenides many examples are given[5,6]. Here we will discuss SiO_2, and why its GFT is so different from GeS_2, even though both are $(N_c = 4)(N_c = 2)_2$ alloys.

LIMBO: THE GLASS-FORMING CONDITION

As we saw above, network glasses are trapped in a kind of configurational limbo, where fluctuations do not provide an effective pathway to the crystalline phase. Mathematically this is expressed by the condition

$$N_{con} = N_d \tag{1}$$

that is, the number of constraints N_{con} is equal to the number of degrees of freedom N_d; the latter, per atom, is the dimensionality d. For the simplest chalcogenide elements, Ge, As, S and Se, the bonding constraints/atom are both stretching (α) and bending (β) and are given for $d > 2$ ($d = 2$ is a special case) by

$$N_{con}{}^{\alpha} = N_c/2 \quad \text{and} \quad N_{con}{}^{\beta} = (d-1)N_c - d \tag{2}$$

where the bond angles are oriented by $(d-1)$ polar angles relative to a polar axis whose absolute orientation is specified relative to d Cartesian unit vectors; for $d = 3$ this gives $N_c = 2.4$. Not only does this agree with experiment for the (III,II) alloys $As_x(S,Se)_{(1-x)}$, but it also correctly predicts the GFT for the (IV,II) $Ge_xSe_{(1-x)}$ alloys[5,7], verifying that N_{con} is linear in N_c, as shown in Fig. 1.

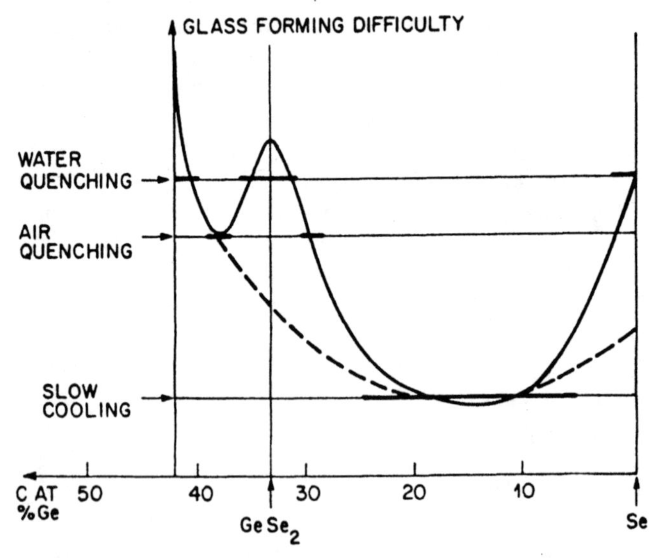

Fig. 1. The glass-forming difficulty, with(out) solid (dashed) curve, cluster effects[5,7].

Beyond the almost unique kinetic data shown in Fig. 1, there have been many microscopic studies which show a critical composition x_c near the mean-field value 0.2 for $Ge_xSe_{(1-x)}$ alloys, as discussed at this conference by Boolchand.

"EXCEPTIONS" WHICH CONFIRM THE CONSTRAINT CONCEPT

As can be seen from Fig. 1, $GeSe_2$ is a very poor glass-former; it readily crystallizes into one of two structures; the high-T phase has 48 atoms/cell, and the low-T phase has 72. (One of the early "explanations" for why it is so hard to grow crystals of As_2Se_3 or SiO_2 is that they have a large (~ 10) number of atoms /cell. However, the cells of $GeSe_2$ are far larger, and they pose few problems for modern crystal-growing techniques.) By contrast, SiO_2 is a superb glass-former; how can this be explained by constraint theory?

The answer is contained in the difference in chemical bonding. The difference in atomic electronegativity in $GeSe_2$ is small, while that in SiO_2 is large; the bonding in the former, where the atoms are of nearly equal size, is almost entirely covalent, while that in the latter is largely ionic, with a weak directional component. Indeed the O electronic 2s-2p shell is nearly full, which makes the bond-bending energies of the small O atoms negligible, so that there are bending forces only at the Si atoms, and only the smaller O-O spacings are constrained, not the larger Si-Si spacings, as shown by the radial distribution function in Fig. 2. Note how broad the peak for the larger Si-Si spacings is, and how narrow that of the smaller O-O spacings is. Constraint theory succeeds because distinctions between broken and intact constraints are so easily made[8]. Incidentally, there are many older molecular dynamics simulations of the glassy SiO_2 radial distribution function, and almost all of them failed to reproduce this observed qualitative difference, but instead yielded two peaks of similar widths, "in good agreement with experiment" according to their authors. (What they meant was that the force parameters had been adjusted to center the peaks correctly; the failures to reproduce the large difference in widths meant that the calculations were not really describing a glass-forming material.) Recent simulations, which use more sophisticated potentials with more adjustable parameters, larger sample sizes, and longer annealing times, are now in excellent agreement with experiment.[9]

My favorite exception which proves the theory to be correct is the pseudobinary alloy $Ge_{(1-x)}Sn_xSe_\mu$ ($\mu = 2, 2.5$), which has been studied by both Raman (bond) and Mossbauer (site) spectroscopy[10]. For these values of μ and the binary system with x = 0 the network is overconstrained and is a poor glass-former. In a partially ionic chalcogenide environment one expects Sn atoms to see a mixture of tetrahedral and octahedral coordinations, as confirmed by Mossbauer spectroscopy[10]. Because the Sn bond-bending forces at the glass transition temperature (~ 600K) do not act as constraints (compare elemental Sn, where the $4 \rightarrow 6$ coordination conversion occurs at 300K), the total number of constraints per formula unit of three atoms ($\mu = 2$) for Sn in tetrahedral sites is[11] $N_{con}{}^{\alpha}(4) + (1-x)N_{con}{}^{\beta}(4) + 2(N_{con}{}^{\alpha}(2) + N_{con}{}^{\beta}(2))$. By tuning x we bring this expression to 3d = 9 when x = 0.40, which is in excellent agreement with experiment[11].

BEYOND EFFECTIVE MEDIUM THEORY

The simple constraint theory which we have just described is a considerable oversimplification of the true situation in most glasses; especially near and below the optimal glass-forming composition in binary alloys there is a very pronounced tendency to form clusters with compositions above and below the average composition, as such clusters can adopt special low-energy surface configurations, and also may reflect inhomogeneities in the melt which have been quenched into the glass. I discussed several examples of such clusters[5], but the most sophisticated comment on them has probably been made in conjunction with the analysis of the composition dependence of anomalously large low-energy Raman scattering intensities in $Ge_xSe_{(1-x)}$ glass alloys[12], where a percolative transition from two-to-three-dimensional fractal morphology is identified near average coordination $N_c = 2.7$ (the Tanaka transition[13]).

Both the continuous Phillips-Thorpe stiffness transition at $N_c = 2.4$ and the nearly first-order Tanaka transition at $N_c = 2.7$ are evident in the thermal (T_g and ΔC_p) and kinetic (viscosity) properties[14]. However, only the Tanaka transition is evident in the optical gap and the Urbach tail parameter[15], and it is also associated with the free-volume maximum, all of which are sensitive to internal surfaces, present with chains and surfaces. This suggests that if the Tanaka transition[13] is fractal two-to-three-dimensional, then the P-T stiffness transition might be one-to-two dimensional in real glasses[16], which, unlike those studied in computer simulations, are not derived by disordering a three-dimensional lattice. It seems to me that this intriguing possibility deserves further study, for example, by computer simulations which start from assemblies of disordered chains or sheets, which are then annealed at just the right temperature for massive and facile cross-linking. Note that it is this annealing temperature itself which can be used as the observable which monitors the two topological transitions; as it is already known that the valence force fields are linear functions of $\langle N_c \rangle$, the two critical compositions where the topological transitions occur should be easily identified. Such models may, for example, also explain the small effects of stoichiometric As_2Se_5 and As_2Se_{10} clusters (which could form chains) on the reversible glass transition temperature which have been reported[14]. Large clusters can displace ideal compositions by several % in careful experimental studies, as discussed here by Boolchand.

One can greatly reduce such cluster effects in the binary Ge-Se system by forming ternary alloys with As as well. In general it appears that formation of such ternary alloys, because it erases effects associated with medium-range order, enhances effects associated with mean-field constraint theory, thus making $\langle N_c \rangle$ a good coordinate and producing more ideal behavior near the stiffness transition at $\langle N_c \rangle = 2.4$. Effects associated with photoinduced stress are reversible and minimized[17] at $\langle N_c \rangle = 2.4$ and become irreversible and are maximized up to $N_c = 2.67$. It has been suggested[18] that the cause of the latter is chemically specific motion of a chalcogen around Ge or As made possible by the free-volume maximum. A more general explanation is the following. At the optimal composition $\langle N_c \rangle = 2.4$ the glass has a kind of global memory because at this point the network is optimally constrained by stretching and bending forces and dihedral forces have not come into play. As $\langle N_c \rangle$ increases above 2.4, some of the weaker bending forces must be replaced by stronger stretching ones, in other words, some of the bending constraints are broken, while some remain intact. The photostructural changes utilize the free-volume maximum to relax the metastable evaporated film network by non-chemically-specific exchanges between broken and intact bending force constraints, as discussed elsewhere[19] in the context of more revealing hole-burning experiments (see below).

THE GLASS TRANSITION TEMPERATURE

Constraint theory has provided a powerful means for studying extensive backbone mechanical percolation effects. I mention here that the thermal effects of network additives or modifiers, for small concentrations of the latter, are surprisingly well understood in terms of a simple local theory[20] which also depend only on the average coordination numbers of the additives and backbone.

INFRARED HOLE BURNING

The very small mass of H causes vibrational modes of H or H_2O to split off from the main vibrational bands of Ge-As-Se chalcogenide glasses and makes such excitations have long lifetimes which can be measured with infrared hole-burning methods[21,22,23]. In such experiments one must, of course, be careful to distinguish isolated - H or H_2O from condensed species in pores, but this is easily done[22]. The results are quite spectacular: linear dependencies of relaxation rates on $< N_c >$ are generally observed, with an unambiguous break in slope at $N_c = 2.4(1)$, which is accompanied by a small but unambiguous cusp[19]. Considering the complexity of the relaxation process, and the fact that the data exclude any significant chemical localization[18] of the additives which would produce dependencies on Ge or As content, this is a remarkable experiment which shows that such ternary glasses composed of covalent atoms of nearly equal sizes behave almost ideally as globally constrained Zachariasen continuous random networks.

DEFECTS AND CHEMICAL ORDERING

Ordinarily one thinks of the glass transition as a high-temperature phenomenon, and so perhaps one would not expect to learn much about glass structure from low-temperature measurements. This expectation is wrong, as was demonstrated most elegantly by Lohneysen et al.[24,25] They measured the low-temperature specific heat of g-As_xSe_{1-x}, and fitted their data with $a(x)T + b(x,T)T^3$. They found that both the Phillips-Thorpe stiffness transition at $\langle N_c \rangle = 2.4$ and the Tanaka $d = 2 \rightarrow 3$ transition at $\langle N_c \rangle = 2.7$ contributed to their observed function dependence, with a double maximum in a(x); subtracting the contribution of the latter, they found that a(x), which measures the concentration of two-level point defects, goes to zero at $\langle N_c \rangle = 2.4$, thus proving that in the absence of interlayer cross-linking one would have a crystal, which confirms the known layered structure of c-As_2Se_3. An alternative, and more elegant interpretation, is that the defects are all associated with chemical disorder (homopolar bonds), and that the concentration of these is, of course, minimized at $x = 0.4$.

Even more interesting is the function b(x,T), which exhibits a maximum at $T = T_m(x) \sim$ 5K. Between $x = 0$ (Se) and $x = 1$ (As), b increases by 50%, as the overall temperature scale increases with $\langle N_c \rangle$. However, the increase is not monotonic; starting from $x = 0$, it increases up to $x = 0.4$, and then drops abruptly to a deep minimum at $x = 0.5$, where the competition between the two transitions is strongest. After excluding the increase of the temperature scale with $\langle N_c \rangle$ one can interpret this result as indicating the fractional volume which is associated with the largest transverse acoustic phonon dispersion; for the ideal composition $\langle N_c \rangle = 2.4$, where a layered, chemically ordered network exists, this volume is maximized, as is the disorder-induced broadening of the transverse acoustic phonons, which is maximized at $x = 0.5$. It would certainly be of great interest to extend these measurements to the study of $Ge_{(1-x)}Sn_xSe_\mu$ ($\mu = 2$) glasses, where chemical ordering should not be an important factor.

PHONON LOCALIZATION

If the Feltz conjecture[16] (that the fractal dimensionality d* of the network is given by $d^*(\langle N_c \rangle) = 1, 2, 3$ for $\langle N_c \rangle = 2, 2.4, 2.67$) is correct, then one should expect to be able to observe the effects of phonon localization in thermal diffusivity as measured by the photoacoustic effect. This indeed appears to be the case, as data on three alloy systems (Ge-Sb-Se, Ge-As-Te and Si-As-Te) all show[26] a thermal diffusivity which peaks sharply at $\langle N_c \rangle = 2.60$, which is close enough to 2.67 to be consistent with the Feltz conjecture; see Fig. 2. In other words, for $\langle N_c \rangle$ decreasing below 2.60, there is increasing localization due to scattering at internal glassy network surfaces. Above $\langle N_c \rangle = 2.60$ presumably nanocrystallites are formed, with even larger scattering at grain boundaries.

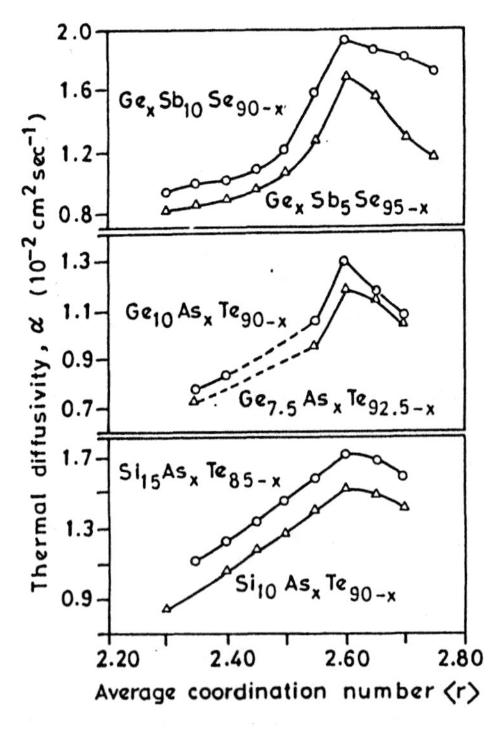

Fig. 2. Minimum phonon localization (as measured by the maximum in the thermal diffusivity) for several ternary alloy systems, as a function of $\langle r \rangle = \langle N_c \rangle$, from [26].

There are a number of subtle and informative aspects to the data shown in Fig. 2. Obviously, $<N_c>$ is a good coordinate. The diffusivity peaks more sharply for the Ge than for the Si alloys, in accordance with the general trends in the chalcogenide glass-forming tendencies of these elements. The peak with 5% Sb is sharper than that with 10% Sb; regrettably, no data on diffusivity were obtained for the best ternary isotropic Zachariasen network glass-former, Ge-As-Se, as these would have provided evidence on the degree to which the locally layered (or fractal) character is necessary to minimize phonon localization.

STRETCHED EXPONENTIAL RELAXATION

In a recent lengthy review[3] of this phenomenon, which covered a very side range of molecular and inorganic glasses, including polymers, electrolytes, alcohols, Van der Waals supercooled liquids and glasses, orientational glasses, water, fused salts, heme proteins and a number of electronic glasses, the network glasses provided some of the most reproducible and reliable data. Moreover, among the network glasses the chalcogenide alloys were the most interesting, because the exponential stretching fraction β (relaxation described by $\exp[-(t/\tau)^{\beta}]$) was studied as a function of composition[27]. The method used was the mechanical relaxation of bent bars, and it showed stretched exponential behavior over *six* decades in time, enough to convince skeptics that Kohlrausch (1854!) had a pretty good idea; six-decade stretched exponential relaxation in the nanosecond regime has also been observed by spin-polarized neutron scattering[3].

The composition dependencies of the stress activation energy barrier E for τ and of β itself are very interesting, and are shown in Figs. 3 and 4. At $<N_c> = 2$, the Se chains are entangled, just as in organic polymers like synthetic rubber, and these entanglements lead to

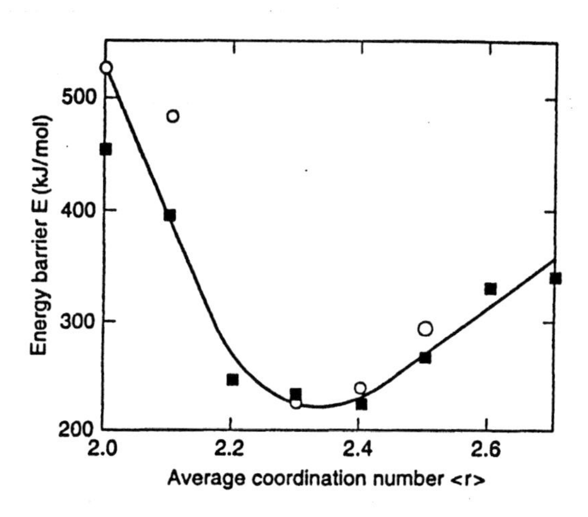

Fig. 3. The Arrhenius activation energy E as determined from the Kohlrausch relaxation time $\tau(T)$ as a function of average coordination number $<r> = <N_c>$.

a large value for E. As $\langle N_c \rangle$ increases above 2, E drops rapidly, reflecting the fact that fairly widely spaced entanglements are being replaced by some cross-linkages, and that, as all the bond-stretching and -bending constraints are satisfied, the remaining constraints near the cross-linkages are essentially those associated with dihedral angle rotations, which are weaker than entanglement forces. Thus E reaches a minimum near 2.35 ~ 2.4, and then climbs rapidly as bond-stretching constraints replace bending ones.

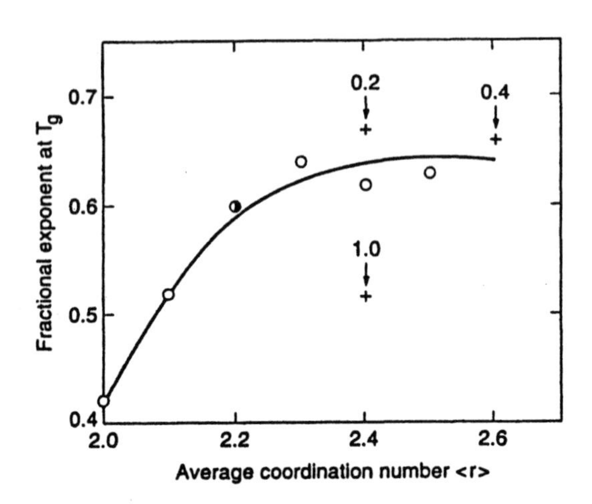

Fig. 4. The dependence of the Kohlrausch stretching fraction for stress relaxation in binary (Ge-Se and As-Se, crosses) and ternary (circles) chalcogenide alloys[27].

As for β, this increases from 0.43 = 3/7 at $\langle N_c \rangle = 2$, to ~ 0.60 = 3/5 (the same value as in oxide glasses) above $\langle N_c \rangle = 2.4$. The general theory of stretched exponential relaxation is discussed at length[3]; it assumes that excitations diffuse to static traps, where they disappear (as into black holes), and it gives $\beta = d/(d + 2)$, where d is the effective dimensionality of diffusion in configuration space. Of course, $\beta(d = 3)$ is 3/5, which is the value obtained in many MDS of molecular relaxation with short-range forces[3], and is also the value obtained[28] in Se at very short times due to the relaxation of Se_8 rings. The value $\beta = 3/7$ corresponds to $d = 3/2$, and is probably associated with the restricted relaxation of chain kinks[29]. Note that, as one would expect from a trap-based model, it is just at $\langle N_c \rangle =$ 2.4 that $\langle N_c \rangle$ is least effective as a measure of configuration space dimensionality; the disagreement between the + 's [the two binaries, x or y = 0] and the o's is largest there because of differences in effective dimensionality, while the ternary o's give the smoothest results. In fact, β is extremely sensitive to differences in local network topology, much more so than any extensive property.

IS THERE A STIFFNESS TRANSITION? THERE ARE ACTUALLY TWO!

Although the stiffness transition which is the subject of this conference was predicted by Thorpe long ago (see his paper), its observation experimentally in the elastic constants proved to be extremely difficult, and "non-observation" of the transition has been reported in many papers. Nevertheless, it seems to me that the transition was finally observed quite well in very careful measurements[30] of longitudinal and shear sound velocities of flat discs; the authors, in stating that their "elastic moduli, however, failed to show any dramatic changes at this threshold" grossly *underestimated* (this is indeed rare!) the importance and value of their results, which are shown in Fig. 5.

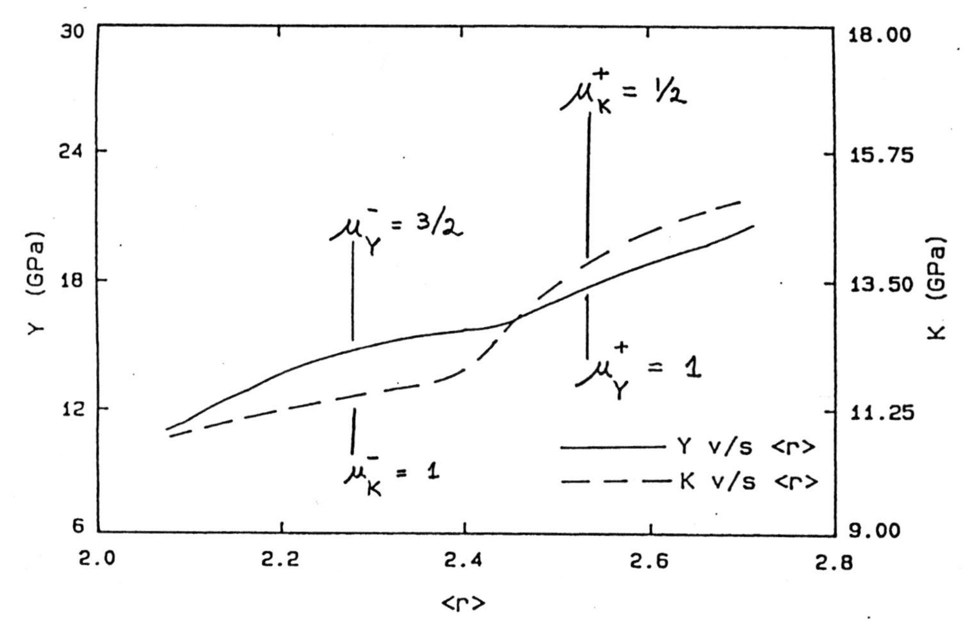

Fig. 5. The bulk modulus K and the Young's modulus Y in ternary Ge-Sb-Se alloys as a function of <r> = $\langle N_c \rangle$ as measured macroscopically on elastic disks[30]. The exponent estimates are those of the present author.

It seems that they were looking for very sharp kinks at the threshold, such as were shown in the idealized theoretical models, and did not make enough allowance for the many factors, both at the sample level (chemical and mechanical degradation of surfaces) and at the atomic level (the effects, chiefly, of breakdown of the effective medium approximation, discussed here for the first time specifically for elastic properties) which smooth out such kinks. Probably the sample problems have already been treated quite well (three systems were studied, and the sample problems were minimized empirically in the Ge-Sb-Se system at 8% Sb), so we turn to the problems associated with the effective medium approximation.

As the authors remark[30], a minimum in the molar volume, a scalar property, at $\langle N_c \rangle$ = 2.4 has been observed in more than 20 cases; this is not surprising, as it is a bulk property and is little affected by chemical and mechanical degradation of sample surfaces. So far there has been no explanation for this observation, which one can take as the starting point for a simple nano-domain model. In the underconstrained regime, $2.10 \leq \langle N_c \rangle \leq 2.40$, the quenched glass is supposed to resemble a jigsaw puzzle with loosely fitting pieces; the spacing between the pieces decreases and their average size increases, as $\langle N_c \rangle$ approaches 2.40. The spacing is not constant, and there are a few contact points at the surfaces where the nanodomains touch. The decrease of the molar volume is associated primarily with the decrease of the filling factor associated with these spaces. When $\langle N_c \rangle$ reaches 2.4 that filling factor is minimized. Above $\langle N_c \rangle = 2.4$, the nanodomains lose their deformability, as there are no longer any soft or "floppy" deformations available which preserve the bond-stretching and -bending constraints. Each nanodomain becomes rigid and is no longer able to deform to fill spaces; the nanodomain average size therefore decreases to reduce the contact stresses, and the filling factor of the interdomain spaces increases, which increases the molar volume.

Next consider the elastic constants shown in Fig. 5. The bulk modulus K is essentially a scalar quantity; hydrostatic compression should not deform the nanodomain jigsaw pieces much, and a nice threshold with a sublinear exponent is already observed in K just at $\langle N_c \rangle_K = 2.40$. This threshold exponent could easily be 0.5 or smaller in ideal samples without chemical and mechanical degradation; admittedly we may never know how small that exponent is, but these data certainly deserve a positive assessment as regards observation of the stiffness transition. Moreover, the large difference between K and the Young's (close to the shear) modulus Y is extremely instructive. We recall that Y is defined as the longitudinal stress σ_1 per unit strain ε_1 when the transverse strains ε_2 and ε_3 are free to change; in an isotropic or cubic medium, $\varepsilon_1 + \varepsilon_2 + \varepsilon_3 = 0$. Bond-stretching ($\alpha$) and -bending ($\beta$) force-constant models of cubic diamond give K as the sum of α and β terms, with $Y \sim \beta$ terms only. In Fig. 12 we see that Y is linear in $\langle N_c \rangle$ for $\langle N_c \rangle > 2.44 = \langle N_c \rangle_Y$ and saturates as $\langle N_c \rangle$ approaches 2.44 from below. If further studies confirm that $\langle N_c \rangle_K \neq \langle N_c \rangle_Y$ this would support the jigsaw model, and would even provide an estimate for the average size of the jigsaw clusters, assuming that the bending force constants at the cluster interfaces are ineffective against shear. Note that recent Raman data[31] on $Ge_x(S,Se)_{1-x}$ glass alloys found $\langle N_c \rangle = 2.46$ for the transverse (shear) optic modes. In general, it seems that the critical hydrostatic value for $\langle N_c \rangle$ is 2.40, while the critical shear value is around 2.45(1).

The following critical exponents μ are suggested quantitatively for K and Y above (+) and below the P-T transition: $\mu_K^- = 1$, $\mu_K^+ = \frac{1}{2}$, $\mu_Y^- = 3/2$, and $\mu_Y^+ = 1$. A theoretical model of these exponents would be of great interest; I hope to provide one at a later time, and other theorists are strongly encouraged to analyze the problem. By the way, the exponents have been computed within an effective medium model for μ_K^+ and μ_Y^+, where they have both been found to be ~ 3/2, and this value seems to work well for the Raman optic mode frequency composition dependence (Thorpe, Boolchand, this conference). It is evident that the effective medium approximation is more successful for the optic than for the acoustic modes, with the latter showing more singular behavior. This pattern cannot be explained by conventional continuum theories, such as scaling or renormalization group treatments of continuous phase transitions in equilibrium systems, but it may be explicable by the jigsaw model discussed above.

Another example of the limitations of continuum methods occurs when one attempts to apply long-wave length fractal concepts to analyze Raman scattering intensities below 80 cm^{-1} in Ge_xSe_{1-x} glasses[32]. The experimental results (see Fig. 6) are very interesting: for $\langle N_c \rangle$ below 2.44, the ln(intensity) vs. $\ln\omega$ plots contain three linear regions, which the authors assign (in order of increasing frequency) to (1) Debye phonons, (2) bending fractons, and (3) stretching fractons. For $\langle N_c \rangle$ above 2.44, (2) disappears, and the fractal dimension (ostensibly a measure of fracton localization) of the bending modes is always larger than that of the stretching modes, which is contrary to the predictions of fracton theory. All these difficulties are removed if one abandons the continuum model, and replaces it by the jigsaw model. In the jigsaw model the three regions correspond to wave lengths (1) longer than domain dimensions (Debye phonons, or interdomain excitations, as before[32]); (3) intradomain excitations, both bending and stretching, and (2) domain interface bending excitations, corresponding to shear modes in the supercooled liquid. As noted above, shear "locks" at $<N_c> = 2.44$, and it is no longer possible to separate and localize bending modes on the domain interfaces. The jigsaw or domain concept reconciles a wide range of quantitative experimental results, and it extends the consequences of the "floppy" or cyclical mode concept in an elegant way.

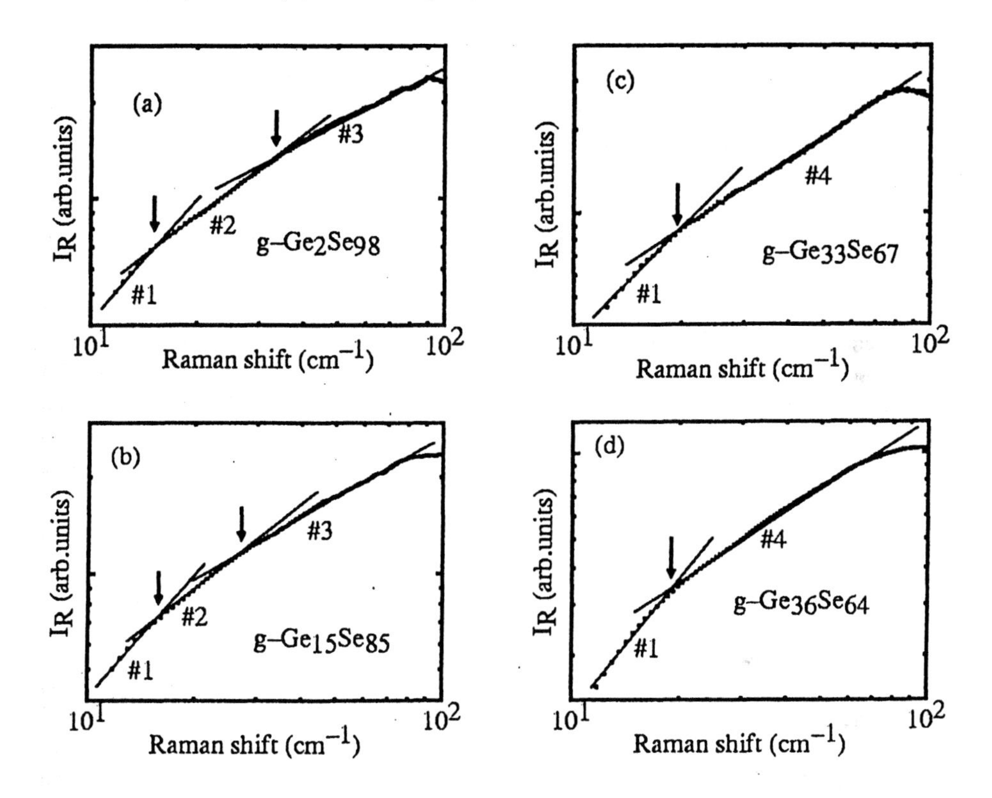

Fig. 6. Log-log plots of a few of the Raman spectra of Ge_xSe_{1-x} glasses reported in ref. 32. The length of segment #2 extrapolates to 0 at $x = 0.22$.

Among all the glasses the candidate most likely to exhibit these "jigsaw" effects on the largest possible scale is obviously SiO_2 grown "epitaxially" on Si. Direct evidence for 130A oxide grains in oxide films grown on Si(001) with thicknesses ranging from 80 to 1000A has recently been obtained by X-ray diffraction[33,34]. These data show that these large oxide grains form a layered structure which is distributed throughout the oxide film even up to the 1000A limit, and constitute convincing evidence of the strength of the cluster-forming tendency in a naturally occurring (and technologically absolutely essential!) context. Some very pretty electron microscope pictures of such cluster formation in chalcogenide alloy films, on a scale as large as 1000 A, have also been reported.[35]

Of course, the essence of structural arrest and the nature of the glassy state should ultimately be obtained from studies of the structural behavior through the glass transition itself. Remarkably thorough and precise studies[36] of the thermal expansions α_g and α_L of both the glass and the supercooled liquid Ge-Sb-Se systems as a function of $\langle N_c \rangle$ showed spectacular behavior: a break in slope for α_g and a steep minimum for α_L, as well as a broader, but still quite clear, minimum in the heat-capacity jump ΔC_p. The fact that the concepts of constraint theory remain valid in the supercooled liquid is quite important and can be explained in terms of "minimum configurational reordering" (see the discussion of "limbo" above). The effects are larger and more symmetrical about $\langle N_c \rangle = 2.4$ in the supercooled liquid than in the glass because the latter is not fully relaxed, and because the longer-range forces are more important in the arrested structure.

The degree to which the non-equilibrium glass transition should be regarded as analogous to a first-order or second-order phase transition is one of the questions which has been most discussed, especially in the older thermochemical literature. If the transition is second-order, so that differences can be replaced by derivatives, the Prigogine-Defay ratio $\Pi = \Delta C_p \Delta \kappa_T / [TV(\Delta\beta)^2] = 1$, where κ_T and β are the isothermal compressibility and volume expansion coefficients. If there are many nearly equivalent internal degrees of freedom which can be well-described by a single order parameter, then Π will be close to 1. On the other hand, if there are large clusters which are easily distinguished, Π may be much larger than 1. This is illustrated in Fig.7, which shows a very large maximum value for Π at $\langle N_c \rangle = 2.4$ in the lightly doped binary glass, compared to a much smaller maximum in the undoped binary.[37]

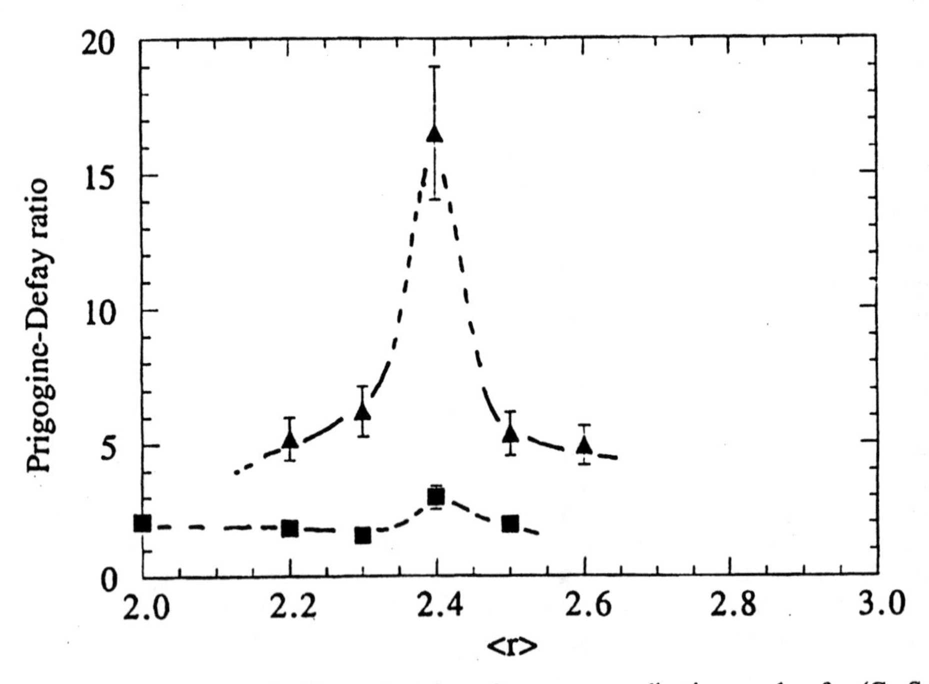

Fig. 7. Prigogine-Defay ratio Π as a function of average coordination number for (Ge-Se) (squares) systems, and $(Ge\text{-}Se)_{0.9}Sb_{0.1}$ (triangles) systems, from ref. 37.

One of the interesting conclusions which has emerged from the foregoing discussion concerns the nature of phase transitions in non-equilibrium systems. One would expect that departures from equilibrium would broaden phase transitions, and of course this does happen. However, in network glasses the entropy at the glass transition is often not much higher than that of the crystal at the same temperature, as the short-range orders are rather similar. Thus the broadening can be small, and we have seen that when macroscopic quenching stresses are reduced, what can happen is that, rather than only broadening, the glass transition can split as well, as a result of broken symmetry. This is a new and quite unexpected application of the concept of broken symmetry, which has previously been used in soft condensed matter physics chiefly to discuss equilibrium behavior of liquid crystals[38]. In addition to splitting, the transitions may also develop more first-order character, but it is the splitting itself which is most easily observed.

Inasmuch as Boolchand and Thorpe will discuss their work in other papers at this conference, little mention of it has been made here, although the quantitative matching of theory and experiment in their work is almost unique in the history of the physics of glasses.

OTHER APPLICATIONS

Although the following topics lie outside the range of this conference, I mention them here to show how similar the physics of percolative phase transitions of disordered solids can be. In fact, it is this generic similarity, together with the relative simplicity of classical mechanical interactions, that makes this conference of special interest to theorists. (One of Shockley's favorite rules was to "do the simplest case first"; this rule certainly worked for him.)

The metal-insulator transition in semiconductor impurity bands is surely percolative, and this aspect was recognized in early modeling of the conductive network in terms of random resistors. However, it is only at low temperatures, with very nearly randomly distributed dopants, that the transitions become sharp enough to determine precise values of critical exponents, as has been done in two classic experiments[39,40]. In this limit quantum effects dominate, as is now generally accepted in the context of an effective medium Coulomb pseudogap which describes the temperature dependence of the dielectric properties on the insulating side of the transition; however, this effective medium model does not describe the insulating composition dependence. That dependence, as well as all the power-law exponents on the metallic side, is described for the first time by a non- effective medium percolative model[41] which includes a new 3-d filamentary electronic phase - the first such new phase since the discovery of superconductivity. Associated with this new phase are multiple transitions, which are in some respects similar to the multiple phases which may be present in the data Boolchand has discussed at this conference.

Of special interest are the critical exponents associated with these transitions, especially the smaller exponents. By finite-size scaling arguments Josephson long ago derived[42,2] the inequality $\mu \geq 2/d$ for the composition exponent μ. He showed that this inequality explains the superfluid density dependence near the Λ transition of liquid He, where the experimental value of $\mu = 0.66$ attains the theoretical lower limit with $d = 3$. For 15 years the major unsolved problem in the metal-insulator transition has been the observation of the composition dependence of the metallic conductivity which is described by $\mu = ½$, apparently in contradiction of the Josephson inequality. However, in the filamentary model[43] this experimental value of μ is explained by the observation that, in effect, one should replace d by d + 1 in systems where the response is affected not only by spatial fluctuations, but also by fluctuations in the component of the internal electric field which is locally tangential to the current-carrying filament. Thus once again the lower limit of the Josephson inequality is attained in a disordered system.

Returning now to Fig. 5, we see that if our present estimate of $\mu_K^+ = ½$ stands up in further experiments, then once again we have a major anomaly which theory should attempt to explain. Perhaps the explanation is the same as in the electronic case, and the anomaly is connected with fluctuations in the longitudinal internal electric field. However, there is more to this problem than just an anomalously small critical exponent, as the strength of the effect varies with the alloy composition. In this respect, there is an interesting parallel between the data shown in Figs. 2 and 5, namely, the largest singularities are found in the $Ge_xSb_ySe_{1-x-y}$ system with y = 0.05. Obviously these alloys should receive further detailed studies. The significance of the small Sb content may be that Sb, being larger than Ge or Se, is a convenient nanodomain network nucleation center which obstructs nanoscopic phase separation. The Sb dopants can also be viewed as prototypes of the relaxation centers (or "black holes") discussed[3] in connection with stretched exponential relaxation.

The metal-insulator transition plays a key role in understanding the extraordinary superconductive properties of the layered cuprates, particularly their amazingly high transition temperatures, which coincide with maximally anomalous normal-state transport properties. A strong case has been made that those anomalies are also associated with a percolative filamentary phase which arises as a result of impurity-assisted resonant interlayer tunneling[43]. The synergies between these models are not accidental; they arise as a result of a common theme, filamentary percolation, which is generally either absent from, or treated incorrectly by, effective medium or continuum theories, which so far have not been able to discuss the fractal aspects of intermediate filamentary phases.

Many theorists find even the existence of filamentary phases counter-intuitive. One can compare them, for example, to Abrikosov's type II filamentary superconductive phase, which was a great surprise to many theorists, including Landau. Abrikosov's filaments surround magnetic flux lines in superconductors, which abhor magnetic fields. Isn't it therefore impossible to have filamentary structures in systems where there is neither superconductivity nor a magnetic field? This is what almost everyone believes. However, I believe that the combination of randomness and Coulomb interactions does generate filamentary topologies, and that these are needed to explain data such as that shown in Fig. 5. It will be interesting to see if alternative explanations can be found for the various observations supporting filamentary topologies.

I am grateful to Prof. P. Boolchand for drawing my attention to ref. 32 and to K. Evans-Lutterodt for drawing my attention to refs. 33, 34.

REFERENCES

*Retired.

1. R. Landauer, in: *Electrical Transport and Optical Properties of Inhomogeneous Media*, J. C. Garland and D. B. Tanner, eds., American Institue of Physics Conference Proceedings, (Am. Inst. Phys., New York), Vol. 40, pp.2-43 (1978).
2. M. F. Collins, *Magnetic Critical Scattering,* Oxford Univ. Press, Oxford (1989) pp. 14, 23, 29, 37, 60, 167.
3. J. C. Phillips, *Rep. Prog. Phys.* 59:1133 (1996).
4. U. Senapati and A. K. Varshneya, *J. Non-Cryst. Sol.* 185:289 (1995).
5. J. C. Phillips, *J. Non-Cryst. Sol.* 34:153 (1979).
6. J. C. Phillips, *J. Non-Cryst. Sol* 43:37 (1981).
7. R. Azoulay, H. Thibierge and A. Brenac, *J. Non-Cryst. Sol.* 18:33 (1975)
8. M. Zhang and P. Boolchand, *Science* 266:1355 (1994).
9. K. Vollmayr, W. Kob and K. Binder, *Phys. Rev. B*54:15808 (1996).
10. M. Stevens, J. Grothaus, P. Boolchand, and J. G. Hernandez, *Sol. State Comm.* 47:199 (1983).
11. J. C. Phillips, *Sol. State Comm.* 47:203 (1983).
12. A.. Boukenter and E. Duval, *Phil. Mag. B* 77:557 (1998).
13. K. Tanaka, *Phys. Rev. B* 39:1270 (1989).
14. T. Wagner and S. O. Kasap, *Phil. Mag. B* 74:667 (1996).
15. D. I. Tsiulyanu and S. I. Marian, *Phys. Stat. Sol.(b)* 195:283 (1996).
16. A. Feltz, in: *Amorphe und Glasartige Anorganische Festkorper*, Akademie-Verlag, Berlin (1983).
17. E. Skordeva, K. Christova, M. Tzolov and Z. Dimitrova, *Appl. Phys. A* 66:103 (1998).
18. K. Tanaka, *Phil. Mag. B* 54:L3 (1986).
19. J. C. Phillips, *Phys. Rev. B* 54:R6807 (1996).
20. R. Kerner and M. Micoulaut, *J. Non-Cryst. Sol.* 210:298 (1997); M. Micoulaut, *Eur.Phys. J. B.* 1:277(1998).
21. S. P. Love and A. J. Sievers, *Chem. Phys. Lett.* 153:379 (1988).
22. S. P. Love, A. J. Sievers, B. L. Halfpap and S. M. Lindsay, *Phys. Rev. Lett.* 65: 1792 (1990).
23. B. Uebbing and A. J. Sievers, *J. Non-Cryst. Sol.* 203:153 (1996).
24. O. Brand and H. Lohneysen, *Europhys. Lett.* 16:455 (1991).
25. X. Liu and H. Lohneysen, *Phys. Rev. B* 48:13486 (1993).
26. A. Srinivasan and K. N. Madhusoodanan, *Sol. State Comm.* 83:163 (1992).
27. R. Bohmer and C. A. Angell, *Phys. Rev. B* 45:10091 (1992).
28. R. Bohmer and C. A. Angell, *Phys. Rev. B* 48:5857 (1993).
29. J. C. Phillips and J. M. Vandenberg, *J. Phys.: Condens. Matter* 9:L251 (1997).
30. A. N. Sreeram, A. K. Varshneya and D. R. Swiler, *J. Non-Cryst. Sol.* 128:294 (1991).
31. X. Feng, W. J. Bresser and P. J. Boolchand, *Phys. Rev. Lett.* 78:4422 (1997).
32. M. Nakamura, O. Matsuda, and K. Murase, *Phys. Rev. B* 57:10228 (1998).
33. I. Takahashi, T. Shimura, and J. Harada, *J. Phys. Cond. Matter* 5:6525 (1993).
34. A. Munkholm, S. Brennan, F. Comin and L. Ortega, *Phys. Rev. Lett.* 75:4254 (1995).
35. C. H. Chen, P. M. Bridenbaugh, J. C. Phillips, and D. A. Aboav, *J. Non-Cryst. Sol.* 65:1 (1984).
36. U. Senapati and A. K. Varshneya, *J. Non-Cryst. Sol.* 185:289 (1995).
37. U. Senspati, K. Firstenberg and A. K. Varshneya, *J. Non-Cryst. Sol.*222:153 (1997).
38. T. C. Lubensky, *Sol. State Comm.* 102:187 (1997).
39. G. A. Thomas, M. A. Paalanen, and T. F Rosenbaum, *Phys. Rev.B* 27: 3897 (1983).
40. K.M. Itoh, E. E. Haller, J. W. Beeman, W. L. Hansen, J. Emes, L.A. Reichertz, E. Kreysa, T. Shutt, A. Cummings, W. Stockwell, B. Sadoulet, J. Muto, J. W. Farmer, and V. I. Ozhogin, *Phys. Rev. Lett.* 77: 4058 (1996).
41. J. C. Phillips, *Proc. Nat. Acad. Sci. 95:*7264 (1998).
42. B. D. Josephson, *Physics Lett. 21*:608 (1966).
43. J. C. Phillips, *Proc. Nat. Acad. Sci. 94*:12771 (1997).

TOPOLOGICALLY DISORDERED NETWORKS OF RIGID POLYTOPES: APPLICATIONS TO NONCRYSTALLINE SOLIDS AND CONSTRAINED VISCOUS SINTERING

Prabhat K. Gupta

Department of Materials Science and Engineering
The Ohio State University
Columbus, OH 43210

I. INTRODUCTION

In 1932, Zachariasen [1] proposed that the structure of a glass consists of an "extended three dimensional network lacking periodicity with an energy content comparable with that of the corresponding crystal". He also argued that for the energy of the glass to be comparable with that of the crystal the coordination polyhedra and the manner in which they are connected must be the same in a glass and in the corresponding crystal. However, unlike crystals, the relative orientations of adjacent polyhedra "vary within rather wide limits" leading to a lack of long range order in glasses. Warren [2] coined the term "random networks" for such disordered structures.

The topology of a network refers to the manner in which the structural units (i.e., the coordination polyhedra) are connected in the network. Figure 1 shows schematically topologically ordered and topologically disordered networks of rigid congruent regular triangular units in 2-dimensions. Crystalline networks, being periodic, are topologically ordered. The structure of a glass is topologically disordered. Topological disorder is necessary for the metastability of glasses as it requires breakage and reformation of bonds for a glass to crystallize. Gupta and Cooper [3] have suggested the term 'topologically disordered' (TD) to describe the structure of glasses as it conveys the essence of the Zachariasen's view of a glass structure more precisely than the term 'random' which leaves the nature of randomness unspecified.

The structural units of a TD network can be polyhedra, polygons, or rods. We use the term 'polytope' to denote a general structural unit of arbitrary dimension δ ($\delta = 1$ for rods, $= 2$ for polygons, and $= 3$ for polyhedra).

While finite size TD networks can always be constructed using any polytope as the

Rigidity Theory and Applications
Edited by Thorpe and Duxbury, Kluwer Academic/Plenum Publishers, 1999

structural unit, the conditions under which a polytope (or a mixture of polytopes) can form an extended (i.e., infinitely large) TD network are not clear. Attempts to construct extended TD networks of regular, congruent, rigid vertex-sharing tetrahedra have demonstrated that the constraints of polytope rigidity and the requirements of connectivity at the vertices make it difficult to construct large TD networks [4,5]. The existence of infinitely large TD networks (at least, for certain combinations of polytopes and patterns of connectivity), therefore, remains questionable.

The conditions for the existence of an extended TD network from regular, congruent, and rigid polytopes with specified connectivity, have been difficult to establish. Zachariasen [1] formulated a set of qualitative rules about the polyhedra and ways of linking them which will allow the formation of extended TD networks. In 1978, Cooper [6] examined quantitatively the conditions under two dimensional infinitely large TD networks can exist. His rationale can be summarized as follows. For a two-dimensional TD network of rigid polygons (each with V vertices) with C polygons sharing every vertex, there are 3/R constraints per angle due to closure of a ring of size R and 1/C constraints per angle due to the condition that all angles at a vertex (C intra-polygonal and C inter-polygonal) must sum to 2 π. Thus the degrees of freedom per angle are $1 - (3/R) - (1/C)$. Since the average ring size is $[VC/(VC-V-C)]$ and since there are C inter-polygonal angles at a vertex, the degree of structural freedom per vertex, f, is given by

$$f = 2 - (C/V)[2V - 3].$$

It follows that a necessary condition for the existence of an infinitely large TD network is that f is non-negative. The analysis was extended to infinitely large TD networks of arbitrary dimension, d, by Gupta and Cooper in 1989 [3,7-9]. Their result is given by the following equation:

$$f = d - (C/V)[\delta V - \delta(\delta + 1)/2] \tag{1}$$

Here V is the number of vertices of a polytope, δ is the polytope dimension, and C (called connectivity) is the number of polytopes connected to a vertex. It should be emphasized that $V \geq (\delta + 1)$ and that $d \geq \delta$ in equation (1).

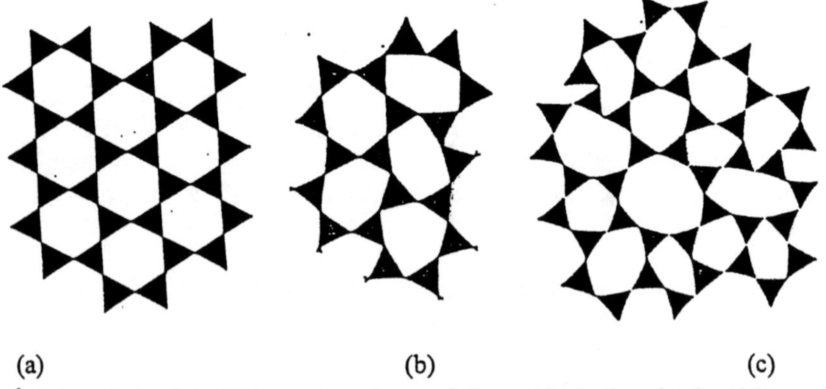

(a) (b) (c)

Figure 1: Schematic drawings of (a) crystalline, (b) generic but topologically ordered, and (c) topologically disordered 2-dimensional networks made of rigid, congruent, regular triangles with two triangles sharing every vertex.

In 1979, using a bond-constraint counting approach in covalent systems, Phillips [10] proposed that the optimum condition for an extended TD network to exist is when the average atomic coordination number, r, is equal to a critical value, r_c. The exact value of r_c depends on the nature of bond constraints. It is equal to 2.4 when all length and angle constraints are intact. It is more than 2.4 when some of the constraints are broken. In 1983, Thorpe [11] reformulated Phillips' approach in terms of rigidity percolation and showed that the critical coordination number, r_c, corresponds to the rigidity percolation threshold. For $r > r_c$, the network is overconstrained (rigid) and for $r < r_c$ the network is underconstrained (floppy). Boolchand and Thorpe [12] have provided corrections for 1-fold coordinated atoms in a network. Table I lists some important milestones in the development of the current understanding of the structure of noncrystalline solids.

Table I: Milestones (<1990) in the development of the current understanding of structure of glasses.

Year	Name (Reference)	Milestone
1932	W. H. Zachariasen (1)	Postulated structure of a glass as an "extended three dimensional network lacking periodicity"
1934	B. E. Warren (2)	Coined the term "random network" and showed agree ment between X-ray diffraction results on simple glasses and the random network model.
1966	D. J. Evans and S. V. King, R. J. Bell and P. Dean (4)	Hand constructed random networks of rigid vertex-sharing tetrahedra as models of silica glass
1971	D. E. Polk (5)	Hand constructed random network model of a-Si.
1978	A. R. Cooper (6)	Derived the condition for the existence of 2-dimensional random networks of vertex-sharing rigid polygons.
1979	J. C. Phillips (10)	Introduced constraint counting/average coordination number approach to rationalize glass forming abilities in covalent systems.
1983	M. F. Thorpe (11)	Reformulated constraint counting approach in terms of rigidity percolation concepts.
1989	P. K. Gupta and A. R. Cooper (3,7)	Derived the condition for the existence in arbitrary dimension of extended topologically disordered networks made of vertex-sharing rigid polytopes.

The two approaches - the *'rigid polytope'* approach of Gupta and Cooper and the *'bond constraint'* approach of Phillips and Thorpe [13] - give identical results under identical set of constraints. This is because bond constraints and polytope rigidity are not independent concepts. When all bond length and angle constraints inside a polytope are intact, the polytope is rigid. Thus when the bond angle constraint at the vertices of the polytopes is broken but all constraints within the polytopes are intact the two approaches become equivalent. The two approaches are also similar in another way. Both require a priori assumptions before they can be applied to a real system. In the 'bond constraint' approach, one needs to decide which, if any, constraints are broken. In the 'rigid unit approach' one has to decide the polytopic shapes of the structural units. In spite of these similarities, the two approaches are conceptually different. The rigid unit approach is not atomistic and is strictly geometrical. Therefore, unlike the bond constraint approach, it is not limited to covalent glasses. It can be applied to any topologically disordered network as long as well-defined structural units can be identified. Indeed, Gupta and Cooper have used it to rationalize glass formation in a large variety of systems. The concept of rigid polytopes has also been used by Dove and coworkers [14] to calculate the number of floppy (zero frequency) modes - called rigid unit modes - in silicate networks.

In the next section, we outline our arguments [15] leading to equation (1) and discuss its extensions to irregular networks (having inhomogeneous vertex connectivity) of regular congruent rigid polytopes, to networks containing multiple (more than one type of) polytopes, to networks involving edge- sharing polytopes, and to networks of distorted polytopes. The applications of this approach to noncrystalline solids are discussed in section III.

Kuo and Gupta [16] have used the rigid polytope approach to examine the rigidity percolation thresholds in composites containing rigid inclusions in a non-rigid matrix. further, they have applied the rigidity percolation concepts to rationalize the kinetics of constrained viscous sintering (densification of porous viscous liquids in presence of rigid inclusions). These results are reviewed in section IV. This is followed by the concluding remarks.

II. TOPOLOGICALLY DISORDERED (TD) NETWORKS OF RIGID POLYTOPES

Extended TD Networks of Vertex-Sharing Regular, Congruent, and Rigid Polytopes:

Consider a d-dimensional finite TD network composed of M regular, congruent, and rigid δ-dimensional polytopes. Suppose a single polytope has V vertices. For convenience, we assume that the connections at the vertices are free as in universal ball joints. Under these conditions, constraints in the network arise only from two sources: the rigidity of the polytopes and network connectivity. We consider the general case of an irregular network where not all vertices have the same connectivity, C. The value of C varies from 1 to a maximum of C_{max}. Let N_j ($j = 1$ to C_{max}) be the number of vertices connected to j polytopes. It follows that

$$M\,V = \Sigma\, j\, N_j \qquad (2)$$

and the total number of vertices, N, is given by

$$N = \Sigma\, N_j \qquad (3)$$

It is easy to see that each rigid polytope contributes a total of $[\delta\, V - \{\delta\,(\delta + 1)/2\}]$ independent internal constraints. This follows from the fact that if the vertices were unconstrained, then there will be $\delta\, V$ 'coordinate' degrees of freedom in the δ dimensional space for the V polytope vertices. On the other hand when the polytope is rigid, it has $\delta(\delta+1)/2$ 'rigid body' degrees of freedom (δ translational and $\delta(\delta-1)/2$ rotational) .

We now make an important assumption that, because the network lacks topological order, the constraints from different polytopes are independent. This assumption is certainly not valid for crystalline networks where, because of symmetry, some of the constraints are not independent. It may also not be valid for generic but topologically ordered networks such as the one shown in figure 1 (b). Thus the total number of constraints, N_c, in a finite TD network are given by

$$N_c = M\,[\delta V - \{\delta(\delta + 1)/2\}]. \qquad (4)$$

When some of the constraints are dependent (i.e., redundant), equation (4) provides an upper bound to the total number of constraints. The structural degrees of freedom F in the finite network are given by

$$F = d\,N - N_c - d(d+1)/2 \qquad (5)$$

The last term on the right hand side accounts for the degrees of freedom of a rigid network. From equations (2) and (3), it follows that:

$$M / N = C_{av} / V,$$

where the average connectivity, C_{av}, is defined by

$$C_{av} = \Sigma\, j\, N_j \;/ \Sigma\, N_j$$

Dividing equation (5) by N, substituting C_{av}/V for M/N, and using equation (4) gives the following result for f(N), the structural degrees of freedom per vertex in a finite TD network:

$$f(N) \equiv (F/N) = d - (C_{av}/V)\,[\delta V - \delta(\delta + 1)/2\,] - [d\,(d+1)/(2N)] \qquad (6)$$

Taking the limit as N becomes large, gives the desired result for the structural degrees of freedom per vertex, f, in an extended TD network:

$$f = d - (C_{av}/V)\,[\delta V - \delta(\delta + 1)/2\,] \qquad (7)$$

Thus, an extended TD network of rigid polytopes exists only if $f \geq 0$.

Equation (7) contains four parameters: polytope parameters δ and V are known once the polytope is specified and network parameters d and C_{av} are known once the network dimension and connectivity are specified. It shows that f decreases a) with increase in C_{av}, b) with decrease in d, c) with increase in V, and d) with increase in δ. The critical connectivity, C^*, (corresponding to $f = 0$) is given by

$$C^* = V\,d / \,[\delta V - \delta(\delta + 1)/2\,] \qquad (8)$$

Table II lists the values of C^* for various polytopes (δ, V) and the network dimension, d.

Extended TD Networks of Multiple Types of Rigid Polytopes

Equation (7) gives the value of f for congruent polytopes. Consider, now, the case when the network contains p different types of polytopes. The i-type of polytopes (i ranging from 1 to p) are specified by δ_i and V_i. Let the number fraction of i-type of polytopes in the network be x_i. Then using arguments similar to equations 2 through 7, it can be shown that the structural degrees of freedom per vertex, f, are given by:

$$f = d - C_{av} \Sigma \{x_i [\delta_i V_i - \delta_i (\delta_i + 1)/2]\} / \{ \Sigma(x_i V_i)\} \tag{9}$$

Equation (9) reduces to equation (7) when there is only one type of polytope.

Table II . Critical connectivity, C* (corresponding to f = 0 in equation (7)) for d-dimensional infinite TD networks composed of δ-dimensional regular rigid polytopes of V vertices each. The value of f is negative when C > C*.

Polytope	δ	V	d	C*
Rod	1	2	1	2
			2	4
			3	6
Triangle	2	3	2	2
			3	3
Square	2	4	2	1.6
			3	2.4
Hexagon	2	6	2	4/3
			3	2
Tetrahedron	3	4	3	2
Octahedron	3	6	3	1.5
Cube	3	8	3	4/3

Extended TD Networks containing Edge- and Face-Sharing Rigid Polytopes

We now discuss networks where the structural units are connected not only at vertices but some of them also share edges. For simplicity, we do not discuss face-sharing here. The analysis is similar to edge-sharing and has been discussed in references [9] and [15]. It is obvious that edge-sharing can only occur in polytopes with $\delta > 1$. To analyze the effect of edge-sharing, we start with the corresponding network without edge-sharing and then calculate corrections as edge-sharing is introduced. When an edge is shared between j polytopes ($j \geq 2$), we refer to it as a j-shared edge. For each j-shared edge, the network has (j-1) fewer edges and (j-1) fewer vertices than the corresponding network without edge-sharing. The reduction of (j-1) edges decreases the total number of constraints by (j-1). On the other hand, reduction of (j-1) vertices causes an increase in the number of constraints by d (j-1). Therefore the net increase in the total number of constraints per j-shared edge is (j-1) (d -1).

Let the fraction of vertices associated with j-shared edges be Y_j. Then the structural degrees of freedom, f, can be expressed as follows:

$$f = d - (C_{av}/V)\,[\delta V - \delta(\delta + 1)/2] - [(d - 1)/2]\,\Sigma\,(j - 1)\,Y_j \qquad (10)$$

Most cases of edge-sharing involve only binary edges (j = 2). In this case equation (10) simplifies to

$$f = d - (C_{av}/V)\,[\delta V - \delta(\delta + 1)/2] - [(d - 1)/2]\,Y_2$$

As expected, equations (9) and (10) show that the effect of edge sharing vanishes when d = 1. These equations also show that edge-sharing always decreases the structural degrees of freedom in a TD network making it more difficult for such a network to exist with edge-sharing than without edge-sharing.

Extended TD Networks of Distorted Polytopes

Excessive strain energy in extended TD networks (due to excess constraints or excess connectivity) can change the network in a variety of ways: crystallization (whereby some constraints are made redundant) may occur, the connectivity may breakdown, or the polytopes may become distorted. We discuss the latter in this section assuming that the first two are absent. One simple and approximate way to examine the existence of an extended TD network of distorted polytopes is to treat the polytopes as if some internal constraint within the polytope is broken. This is an oversimplification of constraint-softening but one that allows simple analysis. A polytope can be distorted if either the center-vertex length-constraint or the vertex-center-vertex angle-constraint is broken. We illustrate the analysis assuming that the angle-constraint at the polytope center is broken but the center to vertex length-constraint is intact. In this situation, the network is best considered as made up of rigid rods ($\delta = 1, V = 2$) with V rods connected at the polytope centers and C_{av} rods connected at the polytope vertices. The network when viewed as made up of rods will have an average connectivity of $C_{av,\,rod}$ which is related to the connectivity C_{av} of distorted polytopes, as follows:

$$C_{av,\,rod} = 2\,V\,C_{av}/\,[V + C_{av}] \qquad (11)$$

Substituting the values of δ (=1) and V(=2) for rods in equation (7) and using equation (11) one obtains for the structural degrees of freedom:

$$f = d - (C_{av,rod}/2). \qquad (12)$$

Equation (12) shows that a 3-dimensional network made of rigid rods with average connectivity of 6 or more will be rigid.

Equation (12) is derived for the case when the polytopes share only vertices We now consider edge-sharing among distorted polytopes. Because the angle constraint at the polytope center is broken, the edges of distorted polytopes do not provide any constraint. The expression for f, by assuming that edges are shared by only two polytopes (j = 2), becomes:

$$f = d - [V C_{av} / (V + C_{av})] - (d/2) Y_2 \quad (13)$$

Using equation (11), equation (13) can also be expressed in terms of the average connectivity , $C_{av,rod}$, of the rods. As in equation (10), equation (13) shows that increase in edge-sharing (i.e., increase in the value of Y_2) decreases structural freedom in the network.

III. APPLICATIONS TO NON-CRYSTALLINE SOLIDS (NCS)

Before discussing applications of the concepts developed in the previous section, it is important to state the meaning of the terms glasses and amorphous (a-) solids as they are used here. Noncrystalline solids (NCS) can be formed either by the conventional method of cooling a liquid or by non-conventional methods such as vapor deposition, sol-gel technique, and a variety of solid state amorphization techniques (e.g., radiation induced amorphization of crystals [17]). Some compositions (such as silica) can be formed by more than one technique. Gupta [18] has proposed the following classification of NCS. When the short range order in an NCS (independent of how it is made) is the same as in the corresponding supercooled liquid (as for example in SiO_2), the NCS is called a 'glass'. When the short range order in an NCS is different from that in the corresponding supercooled liquid (e.g., a-Si), the NCS is called 'a-solid'. This is a useful classification since according to this scheme a-solids do not exhibit glass transition but undergo a first order melting transition to the supercooled liquid state. This has also been suggested by Turnbull and coworkers for a-Si [19]. It is clear from this classification that an a-solid must be formed by a method other than liquid-quenching. We use this terminology here.

Structural Freedom and Glass Transition:

Since the structure of a glass formed by cooling a liquid is the same as the structure of the liquid at the glass transition (or fictive) temperature, Tg, it follows that if the glass structure is an extended TD network, then such a network must exist in the super-cooled liquid state at Tg. This concept has important consequences:

1. At $T = Tg$, f is sufficiently small and consequently the network is rigid enough to freeze within the specified experimental time determined by the cooling rate, q. The exact value of fg ($\equiv f(Tg)$) depends on q; lower q leads to a lower value of fg.
2. The value of f for a supercooled liquid vanishes at the Kauzmann temperature, T_K. The Kauzmann temperature corresponds to the temperature where the configurational entropy of the supercooled liquid vanishes [20]. According to the Adam-Gibbs theory [21], T_K also represents the lower limit of Tg corresponding to $q \rightarrow 0$.
3. For $T > Tg$, f(T) is an increasing function of T for a fixed composition because the fluidity of the super-cooled liquid increases with increase in T. An increase in f can only arise from a decrease in the number of constraints. This implies that either some of the bond constraints must soften or the network connectivity must decrease with increase in T.
4. Generally, there exists a hierarchy in the strengths of atomic bonds [22]. The hierarchy is most apparent in covalent systems where the intra-polytope center-to-vertex length-stretching bonds are the strongest. They are followed by the angular bonds at the polytope centers which in turn are followed by the angular bonds at the polytope vertices. At high enough temperatures, all bonds are broken. As temperature is reduced, different types of

bonds gradually stiffen at different rates. Constraints due to strong intra-polytope bonds become hard (making the polytopes rigid) at temperatures greater than Tg. Constraints due to weak inter-polytope bonds harden only at temperatures near or below Tg. At sufficiently low temperatures, all bond-constraints become hard. In short, which constraints are broken and which are intact, depends on the temperature.

5. For $T < T_K$, weak bonds continue to stiffen and f becomes negative (i.e., the network becomes overconstrained). A negative f indicates the presence of redundant constraints in the network. These constraints, while redundant as far as rigidity is concerned, do contribute to the elastic constants of the network. Therefore, the rigidity transition as measured by the elastic moduli as a function of composition is diluted at sufficiently low temperatures. Indeed, no unambiguous signature of rigidity transition has been observed at T << Tg as a function of composition [23a]. However, there is considerable debate on this issue. Boolchand et. al. [23b] claim that a jump in their Raman data between x = 0.225 and 0.230 in $Ge_xX_{(1-x)}$ (X= S or Se) constitutes direct evidence of stiffness threshold at r = 2.46.

6. Rigidity transition is also washed out at high enough temperatures in the liquid state where the structures are floppy for all compositions of interest. It follows, therefore, that to detect a signature of rigidity transition as a function of composition, the variation of properties with composition should be examined in the vicinity of the glass transition temperatures. Such data have been reported by Tatsumisago et. al., [24] and by Senapati and Varshneya [25] for chalcogenide systems. The results of these studies show a strong positive correlation of Tg with the average coordination number r for r < 2.4, a minimum in the configurational heat capacity, ΔC_p, at r of about 2.4, and an apparent minimum in fragility at about r = 2.67 [25]. These results show clear signs of some sort of transition in the range r = 2.4 to 2.7 in the supercooled liquid state and thus provide support for the notion that the TD network forms in the liquid state.

7. According to the bond-constraint model, f > 0 at T = 0 K for underconstrained systems where r < 2.4. An example is Se for which r = 2. However, we have argued that f < 0 for $T < T_K$ and T_K for Se is known to be 240 K [24]. These two apparently contradictory statements are reconciled by taking into account the secondary interactions which are not included the bond-constraint approach. These secondary interactions are sufficiently strong at T near Tg to initiate freezing of the supercooled liquid and play a critical role in determining the glass transition temperatures of underconstrained systems.

Structural Freedom, Configurational Entropy, and Fragility of Supercooled Liquids

We have stated that in the supercooled liquid state f is an increasing function of T with $f(T_K) = 0$. Gupta and Patton [26] have attempted to relate f(T) to the configurational entropy, ΔS (T), of the supercooled liquid and also to the shear viscosity, η(T) using the Adam-Gibbs Theory [21].

If it is assumed that the energy increase of the second lowest level of a-minima over the lowest a-minima in the potential energy landscape of a supercooled liquid is much greater than RT_K, then the configurational entropy of the supercooled liquid, $\Delta S(T_K) = 0$ and the system stays primarily in the lowest minima at T_K. As T is increased, the system begins to occupy the higher minima. The configurations corresponding to the higher minima give rise to an increase in the configurational entropy for $T > T_K$. These additional configurations are the ones allowed by the increased structural freedom due to constraint softening. Let the number of these additional configurations be Ω (T). Gupta and Patton [26] have argued that $\Omega(T)$ can be related to f(T) by assuming that each floppy mode (corresponding to $f > 0$) allows the system to exist in ω distinct configurations such that:

$$\Omega(T) = \omega^{Nf(T)}$$

Here ω is assumed to be a constant. N is the total number of vertices in the network. The configurational entropy can be expressed as:

$$\Delta S(T) = 1/N\,[\,k\,\mathrm{Ln}\;\Omega(T)] = k\,f(T)\;\mathrm{Ln}\,\omega \qquad (14)$$

Here k is the Boltzmann's constant. Using the Adam-Gibbs theory [21],

$$\mathrm{Ln}\,\eta(T) = \mathrm{Ln}\,\eta(\infty) + A\,/\,[\,T\,\Delta S(T)] \qquad (15)$$

where A is a constant, f(T) can be related to the T-dependence of the shear viscosity, η, of a supercooled liquid. The fragility, m, of a supercooled liquid is defined as [27]

$$m = (1/Tg)\,[\partial\,\mathrm{Ln}\,\eta\,/\,\partial\,(1/T)]\,\big|_{Tg} \qquad (16).$$

It can be shown from equations 14 -16 that

$$m = 40\,[\,1 + ([\partial\,\mathrm{Ln}\,f\,/\,\partial\,\mathrm{Ln}\,T)\,\big|_{Tg}\,] \qquad (17)$$

where we have made use of the fact that Ln $[\eta_g / \eta(\infty)\,] \cong 40$ [27].

Structural Freedom in Real Glasses

Experimental results about the coordination polyhedra and the connectivity at the vertices are available for several simple glass forming compositions from diffraction and NMR experiments [28]. Therefore it is possible to check the prediction whether f - calculated using equation (7) - is non-negative for glass forming compositions. Table III lists the values of the four structural parameters (δ, V, d, and C) obtained from experiments and the calculated values of f.

The tetrahedral compound glass formers such as SiO_2, GeO_2, BeF_2 are extremely good glass formers. Their structure consists of a 3-dimensional TD network composed of rigid tetrahedra with $C = 2$ [28]. According to equation (7), the value of f is zero for these glasses.

B_2O_3 is also a good glass former. Its structure is composed of B_2O_3 triangles and boroxol (B_3O_6) rings. The boroxol rings are large planar triangular units made up of three B_2O_3 triangles. Thus B_2O_3 is essentially a TD network of two sizes of rigid regular triangular units with two triangles sharing a vertex. There is some uncertainty whether the structure is a truly 3-dimensional connected network or consists of twisted intertwined ribbons in which case $d = 2$ [29]. In both cases, f is non-negative (see Table III).

Addition of alkali oxides to silica glass converts some of the bridging oxygens into non-bridging oxygens [30]. Thus, there exist two types of vertices in alkali silicate glasses: bridging with $C = 2$ and non-bridging with $C = 1$. The fraction of non-bridging oxygens in a composition $xNa_2O(1\text{-}x)SiO_2$ is 2x / (2-x). As shown in Table III, the value of C_{av} decreases and of f increases as the amount (x) of alkali is increased. These glasses are easy to form except when x is large (and f is large) when they tend to crystallize readily.

Table III. Structural parameters and the calculated value of f for real glasses.

System	δ	V	d	C_{av}	f
SiO_2	3	4	3	2	0
GeO_2	3	4	3	2	0
BeF_2	3	4	3	2	0
B_2O_3(1)	2	3	3	2	1
B_2O_3(2)	2	3	2	2	0
$xNa_2O.(1\text{-}x)SiO_2$ $(0 \leq x \leq 2/3)$	3	4	3	4(1-x)/(2-x)	3x/(2-x)
P_2O_5	3	4	3	8/5	3/5
a-metals	1	2	3	6	0
$SiO_{(2-x)}N_{(2x/3)}$ $(0 \leq x \leq 2)$	3	4	3	12/(6-x)	-3x/(6-x)

(1) Assuming a 3-dimensional network.
(2) Assuming a twisted intertwined ribbon structure.

The structure of P_2O_5 glass is a TD network of PO_4 tetrahedra with three bridging vertices (C = 2) and one non-bridging vertex with C = 1[31]. The non-bridging vertices correspond to the doubly bonded oxygens. In this case f is positive and P_2O_5 is known to be good glass former.

In silicon oxynitride glasses, nitrogen substitutes for oxygens forming two kinds of vertices: oxygen vertices with C = 2 and nitrogen vertices with C = 3 [32]. Adding nitrogen to silica makes f negative suggesting that nitridation of pure silica glass is difficult. However, nitrogen can be added to alkali-silicate glasses for which f > 0. In fact, one can calculate the maximum amount of nitrogen which can be incorporated in alkali silicate glasses as a function of the alkali content. Consider glass formation in an alkali silicon oxynitride system of the general composition yNa_2O - (1 -y)[$SiO_{(2-x)}$ $N_{(2x/3)}$]. This system has three types of vertices: non-bridging oxygens with C = 1, bridging vertices at the oxygens with C = 2, and bridging vertices at the nitrogens with C = 3. The average connectivity in this system is given by 12 (1-y)/ [(1-y)(6-x)+ y]. The condition f = 0 gives the limiting solubility of nitrogen. It can be solved for the maximum value of x as a function of y: x(max) = y/ (1 - y). The value of x(max) increases with increase in y. While detailed systematic nitridation data on alkali-silicate glasses are not available, it is generally known that nitridation is easier upon incorporation of alkali in silica [33].

The structure of a-metals is approximately described as a dense random packing of hard spheres [28]. The average hard contact coordination number in such packings of monosized hard spheres is 6 [34]. This structure can be transformed into a TD network of rigid rods if the centers of contacting spheres are connected by rigid rods (of lengths equal to the sphere diameters) so that C = 6 at all vertices (i.e. the centers of the spheres) of the network. Table III shows that f = 0 for such a TD network of rigid rods. This is an interesting example as metals are known to be poor glass formers. But it supports our notion that f goes to zero at T_K even in poor glass formers.

TD networks made of regular rigid tetrahedra with C = 2 are optimally constrained (f = 0) and cannot accommodate edge-sharing (which will make f negative). In such systems edge-sharing is generally accompanied with distortion of tetrahedra. According to equation (13), f remains positive in this system even if a large fraction of the vertices are edge-shared provided distortion of tetrahedra is allowed. This is consistent with the observations in $GeSe_2$ [35] and $SiSe_2$ [36] systems where edge-sharing has been reported along with significant distortion of the tetrahedra. When polytopes are not distorted, edge-sharing can occur in an otherwise optimally constrained TD network only if C_{av} is reduced by appropriate modification of the composition.

Structural Freedom in Real Amorphous Solids

The 'Rigid Polytope approach' can be extended to a-solids [18]. This is because the existence conditions derived in section II apply to any extended TD network independent of its method of formation. It is therefore useful to inquire whether the observed structures of real a-solids are consistent with these conditions. We discuss here the important example of a-Si.

The structure of a-Si (or a-Ge) consists of an extended TD network where each silicon is connected to four other silicons [37]. In the radial distribution function (RDF), the first peak gives the Si-Si bond length which is equal to that in c-Si and the width of this peak is fairly narrow indicating that the bond length constraint is intact. The second peak in RDF arises from second neighbor atoms and gives the tetrahedral angle of 109° for the average Si-Si-Si bond angle. However the width of this peak is significantly broad ($\pm 10^o$ [38]). If one assumes that the angle constraint is intact so that the structure of a-Si can be regarded as a TD network of regular tetrahedra with C = 4, then equation (7) gives f = -3. Clearly, such a network can not exist without large strain energy. This strain energy causes either a large spread in the bond angle distribution or the formation of dangling bonds. Using the rigid polytope approach, one can analyze the two extremes : a) by assuming that all strain energy is used in the formation of dangling bonds keeping the angle constraint at the non-dangling bonds intact, and b) by assuming that the bond angle constraint is broken without forming any dangling bonds.

If all strain energy has gone in creating dangling bonds then one can calculate the maximum number of dangling bonds per silicon, x, by equating the length and angle constraint from the non-dangling bonds at a silicon atom to 3:

$$(4 - x)/2 + [2\{4 - x\} - 3\}] = 3$$

This gives x = 8/5. In other words a maximum of 40 % of the bonds are dangling bonds. This result can also be applied to the case of a-$Si_{(1-y)}H_y$, giving y = 61.5 % for complete saturation of all the dangling bonds. However, the best a-Si films have only about 10% hydrogen [38].

If we assume that the angular constraint at the Si is broken and no dangling bonds are formed, we can treat the TD network as composed of rigid rods, as discussed in section II, with C = 4. Equation (12) then gives f = 1. While this result shows that such a network can exist, a value of f = 1 probably implies too much structural freedom in the a-Si network. The real situation lies somewhere in between these two extremes: the angular constraint is softened to some extent and some dangling bonds are formed.

Optimum Composition Range for the Formation of Noncrystalline Solids (NCS)

Formation of noncrystalline solids is controlled by two conditions: a) the existence of an extended TD network which requires a non-negative f and b) stability with respect to crystallization which requires that f be small in the range of external variables (such as temperature) used in forming such a network. These two conditions can be associated with two boundaries in the composition space which bracket the range of easy NCS formation. The 'existence' boundary corresponds to $f = 0$ for networks of rigid polytopes. More generally we label it as f(-) which may take slightly negative value for real systems because the coordination polyhedra are not perfectly rigid. The 'stability' boundary corresponds to some finite positive value of f. It is labeled f(+).

In the case of glass formation from liquids, poor glass formation is associated either with liquids which have negative f so they violate the existence requirement or with liquids for which f increases rapidly above Tg with increase in T so that the stability requirement is violated at T > Tg. (We emphasize that there are exceptions to these statements since f does not determine crystallization kinetics completely. For example, the molecular liquid Cl-S-S-Cl has a large f above its Tg but is a good glass former [39]). The value of f remains close to zero over a large temperature range above T_K for good glass formers such as silica. Figure 2 shows schematically the variation of f with T for good and poor glass forming liquids. As suggested by this figure, both boundaries can be approximately characterized in terms of f_m, the value of f at the melting point. Compositions such as Al_2O_3, MgO, Si_3N_4 are poor glass-formers because they violate the existence boundary - labeled fm (-). For example, f for alumina is -7 [17]. The stability boundary, labeled $f_m(+)$, is dependent on the cooling rate, q. Systems such as metals and molecular liquids do not form glasses easily because they violate the stability boundary. Compositions which form glasses easily are confined to the range: $f_m(-) \leq f \leq f_m(+)$.

It is interesting to compare the NCS-forming abilities of compositions made by the conventional method of cooling from the liquid state and by non-conventional methods such as CVD, sol-gel, or solid state amorphization processes. Experimentally, the cooling rate provides a measure of glass forming ability in the conventional method. However, there is no such simple empirical measure for characterizing NCS-forming ability in non-conventional processes. An exception is radiation induced amorphization where the total radiation dose for amorphization is a reasonable measure of amorphizability [17]. Hobbs [40] has shown that a strong correlation exists between radiation induced amorphizability and the value of f.

For the sol-gel technique, the following results appear to be reasonably well established: several compositions (such as SiO_2 - TiO_2 and Si_3N_4) which are difficult to make by melt-cooling can be made relatively easily by the sol-gel method. On the other hand, according to Zarzycki [41], alkali and alkaline earth modified silicates "are formed easily from the melt, but are very difficult to form by the sol-gel technique". Gupta [42] has rationalized both of these observations by suggesting that in the non-conventional processes, there exist extra degrees of freedom introduced by either anion impurities (such as OH, H, Cl), or microporosity which increases the fraction of non-bridging vertices, or by excessive distortion of polytopes. The net consequence of these extra degrees of freedom is to shift the glass forming region to lower values of f. This explains both observations as to why compositions with slightly negative f can be made easily and compositions with positive f near the stability boundary cannot be made easily by non-conventional processes.

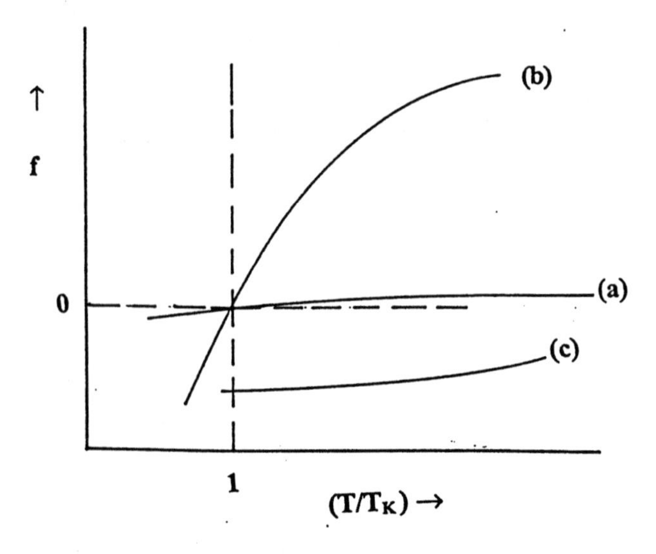

Figure 2: Schematic variation of the structural degrees of freedom, f, in three types of supercooled liquids with respect to temperature, T. The T-axis is normalized with respect to the Kauzmann temperature T_K where f is zero for all glass forming liquids ($f \geq 0$). For good glass formers such as silica (curve a), f is positive but remains close to zero up to fairly high temperatures. For poor glass formers, which violate the stability condition - such as metals and molecular liquids (curve b) - f is zero at T_K but increases rapidly with temperature. These systems tend to crystallize readily at $T > Tg$. For poor glass formers which violate the existence condition - such as alumina (see curve c) - f remains negative, indicating that such systems are poor glass formers because they violate the existence condition.

IV. RIGIDITY PERCOLATION AND CONSTRAINED VISCOUS SINTERING

The concepts of rigidity percolation can be applied to examine the sintering behavior of systems containing rigid inclusions in non-rigid porous viscous melts - a process known as constrained viscous sintering. Results of several investigations [43] have established that the sintering rates drop sharply when the concentration of rigid inclusions reaches a critical value in the range of about 10 - 20 volume % (the exact value depends on the geometric characteristics of the inclusion phase). It is believed that at the critical concentration the inclusions form an extended rigid TD network preventing any further sintering. Most studies have used the conductivity percolation threshold values to rationalize the sinter kinetics in such systems. However, it is the rigidity of the TD inclusion network which inhibits densification.

Rigidity and conductivity (or connectivity) percolation thresholds are equal only under restricted conditions. This can be easily seen by considering a chain of contacting conducting rigid spheres. The chain is conducting but is not rigid. The chain becomes rigid when the contacts between spheres become bonded. This concept of rigidity percolation being dependent on the nature of bonding at the contacts between inclusion particles has been used by Kuo and Gupta [16] to derive rigidity percolation thresholds in powder systems and in continuum matrix composites containing rigid inclusions for both bonded and non-bonded contacts. The analysis uses the concept of extended TD network of rigid rods (of length equal to the diameter of the inclusions) which connect the centers of contacting inclusions. The matrix phase being non-rigid plays no role in determining the percolation threshold. Bonded contacts between inclusions are modeled by angular constraints at the vertices while non-bonded contacts by broken angular constraints. The rigidity percolation threshold for non-bonded contacts (identical to the case of dense random packing of hard spheres) corresponds to a coordination number of 6 in the percolating cluster. The rigidity percolation threshold in the case of bonded contacts corresponds to a coordination number of 2.4. These coordination numbers values can be converted in to critical volume fractions, ρ_i *, of inclusions [16] and correspond to a value of 0.637 (the packing fraction of a dense random packing of monosized hard spheres) for non-bonded contacts and 0.187 for bonded contacts.

In real systems (unless the inclusions have been coated), the contacts are always bonded because necks develop rapidly between contacting particles due to a sharp curvature at the initial points of contact. In this case, the residual porosity at the rigidity percolation threshold where sintering stops is given by [44]:

$$\text{Residual Porosity (volume \%) at the threshold} = 100 \left[1 - (\rho_i^* / \phi) \right]$$

where ϕ is the fraction of the total volume of the condensed phases (excluding pores) occupied by inclusions and ρ_i is the fraction of the total volume (including pores) of the inclusions at the threshold. The value of ϕ is fixed by the composition of the powder mixture chosen at the beginning of sintering.

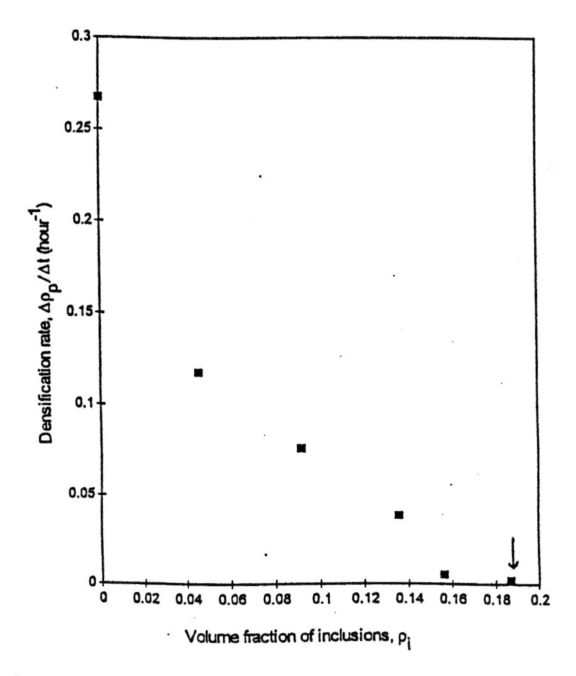

Figure 3: Plot of measured densification rate as a function of the volume fraction of inclusions (ρ_i). The arrow indicates the theoretical value of the rigidity percolation threshold (ρ_i * = 0.187). The densification rate is calculated from the density difference between samples sintered for 1 and 24 hours [44]. The densification rate vanishes at the rigidity percolation threshold.

To verify these results, Kuo [44] studied the sintering kinetics of a supercooled liquid (a lead aluminoborate composition - Schott KZFS-1- above its Tg of 472 C) containing rigid (copper) inclusions as a function of time (up to 24 hours) at a temperature of 700 C in a He atmosphere. Samples were prepared for various values of ϕ (0, 0.05, 0.10, 0.15, 0.20, 0.25) using well-mixed powders containing spherical particles of about the same size (about 40 microns in diameter). After the sintering treatment, the electrical conductivity and the porosity were measured in the samples. Also, their microstructure was observed by Scanning Electron Microscopy. After 24 hours, complete densification was observed only in the case of the sample containing no inclusions. The rate of densification as a function of ρ_i (the volume fraction of the inclusions) is shown in figure 3. Also shown by the arrow is the theoretical value of the rigidity percolation threshold. It is clear that the sintering rate decreases continuously with increase in ρ_i and vanishes at the predicted threshold limit. In addition, the 24 hour samples with $\phi = 0.20$ and $\phi = 0.25$ were observed to be electrically conducting with the $\phi = 0.25$ sample having the higher conductivity. SEM microstructures showed a nonzero dihedral angle between contacting copper inclusions indicative of neck formation at the contacts. Finally, the measured residual porosity after 24 hours for the ϕ = 0.25 sample was 25.2 %, in agreement with the calculated value using equation (18). These results strongly support the notion that the rate of constrained viscous sintering is governed by the formation of a rigid percolating cluster of inclusions during sintering.

V. CONCLUDING REMARKS

Zachariasen's intuitive concept of glass structure as an extended topologically disordered (TD) network continue to guide our present understanding of the structure of non-crystalline solids in general. There are three important questions pertaining to the extended TD networks. These are:

1) The question of existence: Is it possible for a system to exist as a TD network?

2) The question of formation: How can a system be brought into the state of a TD network?

3) The question of stability: How stable (with respect to crystallization) is the TD network?

The theory of extended TD networks of rigid polytopes provides quantitative answers to the question of existence of TD networks. It has been extended to irregular networks, to networks containing multiple polytopes, to networks with edge and face sharing, and to networks with distorted polytopes. This has provided a common basis for rationalization of the structure of a variety of systems (covalent, ionic, metallic, and molecular). The theory is made simple by two facts: a) TD networks are intrinsically generic so that constraint counting is simple and b) for an extended network, surface effects can be neglected. The theory calculates the value of the structural degrees of freedom, f, in the network by counting constraints due to the rigidity of the structural units (polytopes) and connectivity among the polytopes. For a network to exist, $f \geq 0$. Structural units with increased number of vertices and too much connectivity make it difficult for TD networks to exist. Much of the role of composition can be rationalized simply by its influence on either the choice of the polytope or the connectivity of the structural units in the network. Distortion of polytopes can be treated by choosing a simpler structural unit with less number of vertices and of lower dimensionality. The theory applies to any TD network independent of its method of formation.

It is argued that f must be temperature dependent and that the temperature dependence of f plays a key role in the freezing of a liquid structure upon cooling of liquids. The value of f can also be related to the configurational entropy of the supercooled liquid and vanishes at the Kauzmann temperature. Using the Adam-Gibbs theory of the viscosity of liquids, f can also be related to the viscosity. The theory provides qualitative answers to the other two questions. A network cannot be formed if it cannot exist. A large value of f implies low viscosity leading to easy crystallization. Thus the glass forming ability is maximized when f is non-negative but not too large.

We have also used the ideas of the rigidity percolation to examine the technologically important process of viscous sintering in presence of rigid inclusions. The predictions of the theory are borne out by the experiments.

ACKNOWLEDGMENTS

I dedicate this article to my colleague, friend, and advisor Al Cooper who passed away in late 1996. His deep interest in glass science, keen insights, thought provoking questions, and constant encouragement are greatly missed. I also wish to acknowledge my colleague Bruce Patton for many fruitful discussions.

REFERENCES

1. W. H. Zachariasen, J. Am. Chem. Soc., 54:3841 (1932).
2. B. E. Warren, J. Am. Ceram. Soc., 17:249 (1934).
3. P. K. Gupta and A. R. Cooper, J. Noncrystalline Solids, 123:14 (1990).
4. R. J. Bell and P. Dean, Nature, 212:1354 (1966). Also see D. L. Evans and S. V. King, Nature, 212:1353 (1966).
5. D. E. Polk, J. Noncrystalline Solids, 5:365 (1971).
6. A. R. Cooper, Phys. Chem. of Glasses, 19:60 (1978).
7. P. K. Gupta and A. R. Cooper, in: *Proc. Xvth Int. Congress on Glass*, Lenningrad, Vol 1a:13 (1989).
8. P. K. Gupta and A. R. Cooper, in: *The Physics of Noncrystalline Solids*, L. D. Pye, ed., Taylor and Francis, Bristol (1992).
9. P. K. Gupta and A. R. Cooper, in: *Proc. XVIth Int. Congress on Glass*, Madrid, Vol 3:15 (1992).
10. J. C. Phillips, J. Noncrystalline Solids, 34:153 (1979).
11. M. F. Thorpe, J. Noncrystalline Solids, 57:355 (1983).
12. P. Boolchand and M. F. Thorpe, Phys. Rev., B50:10366 (1994).
13. J. C. Phillips and M. F. Thorpe, Solid State Comm., 53:699 (1985).
14. M. T. Dove, M. J. Harris, A. C. Hannon, J. M. Parker, I. P. Swainson, and M. Gambhir, Phys. Rev. Letters, 78:1070 (1997).
15. P. K. Gupta, J. Am. Ceram. Soc., 76:1088 (1993).
16. C. H. Kuo and P. K. Gupta, Acta metall. mater, 43:397 (1995).
17. L. W. Hobbs, A. N. Sreeram, C. E. Jesurum, and B. A. Berger, Nucl. Inst. Methods in Phys. Res., B 116:18 (1996).
18. P. K. Gupta, J. Noncrystalline Solids, 195: 158 (1996).

19. E. P. Donovan, F. Spaepen, D. Turnbull, J. M. Poate, and D. C. Jacobsen, J. Appl Phys., 57:1795 (1985).

20. C. A. Angell in: *Complex Behavior of Glassy Systems*, M. Rubi and C. Perez-Vicente, Ed. Springer, Berlin (1996).

21. G. Adam and J. H. Gibbs, J. Chem. Phys., 43:139 (1965).

22. J. C. Phillips, Phys. Rev., B54: R6807 (1996).

23 a. M. F. Thorpe, J. Noncrystalline Solids, 182:135 (1995).

23 b. X. Feng, W. J. Bresser, and P. Boolchand, Phys. Rev. letters, 78: 4422 (1998).

24. M. Tatsumisago, B. L. Halfpap, J. L. Green, S. M. Lindsay, and C. A. Angell, Phys. Rev. Letters, 64:1549 (1990).

25. U. Senapati and A. K. Varshneya, J. Noncrystalline Solids, 197:210 (1996).

26. P. K. Gupta and B. R. Patton, to be published.

27. R. Bohmer, K. L. Ngai, C. A. Angell, and D. J. Plazek, J. Chem Phys., 99:4201 (1993).

28. S. R. Elliott, *Physics of Amorphous Materials*, Longman, London (1990).

29. P. V. Johnson, A. C. Wright, and R. N. Sinclair, J. Noncrystalline Solids, 50:281 (1982).

30. A. K. Varshneya, *Fundamentals of Inorganic Glasses*, Academic Press (1994).

31. A. C. Wright, R. A. Hulme, D. I. Grimely, R. N. Sinclair, S. W. Martin, D. L. Price, and F. Galeener, J. Noncrystalline Solids, 129:213 (1991).

32. S. H. Risbud, Phys. Chem. Glasses, 22:68 (1981).

33. R. E. Loehman, in: *Treatise Mater. Sci. Technol.* 26:119 (1985).

34. C. H. Bennett, J. Appl. Phys., 43:2727 (1972).

35. I. T. Penfold and P. S. Salmon, Phys. Rev. Letters, 67:97 (1991).

36. L. F. Gladden and S. R. Elliott, J. Noncrystalline Solids, 109:223 (1989).

37. R. Zallen, The Physics of Amorphous Solids, Wiley, New York (1983).

38. R. A. Street, Hydrogenated Amorphous Silicon, Cambridge (1991). See also ref. 26.

39. C. A. Angell, Personal Communication.

40. L. W. Hobbs, J. Noncrystalline Solids, 192&193:79 (1995).

41. J. Zarzycki, Cer. Trans, 29:579 (1993).

42. P. K. Gupta, Glasstech. Ber. Glass Sci., 67C:197 (1994).

43. R. E. Dutton and M. N. Rahaman, J. Am. Ceram. Soc., 75:2146 (1992).

44. C. H. Kuo, *Densification of Non-Reactive Viscous Liquids containing Solid Inclusions*, Ph. D. Thesis, The Ohio State University, (1995).

RIGIDITY CONSTRAINTS IN AMORPHIZATION OF SINGLY- AND MULTIPLY-POLYTOPIC STRUCTURES

Linn W. Hobbs,† C. Esther Jesurum‡ and Bonnie Berger‡

†Department of Materials Science & Engineering
‡Department of Mathematics
Massachusetts Institute of Technology
Cambridge, MA 02139, U.S.A.

Dedicated to the memory of Professor Alfred R. Cooper

INTRODUCTION

The irradiation-induced loss of crystallinity--or, more loosely, amorphization--is a common response of many insulating or semi-insulating solids,[1] as well as some ordered alloys,[2] to atomic displacements accumulated to the order of 1 displacement per atom (dpa). Readily amorphizable non-metallic solids are typically simple covalently-bonded solids or more complex ionic compounds in which lattice energy is less critically influenced by atomic correlations more remote than near-neighbor, and in which the structural alternatives for atomic arrangement, which may thus differ little in energy, become an important factor in the ease of amorphization. The basic problem is to understand how and why atoms can rearrange themselves, within the confines of unaffected surrounding crystal, after a local disordering event. Given few enough constraints and sufficient structural options, the rearrangement can be arbitrary and stochastic, so that the transformation is essentially irreversible and the crystal can be recovered only by full-scale recrystallization--for example, epitaxially at the interface with unaffected crystal--given sufficient atomic mobility at higher temperature. There is therefore a strong temperature dependence for amorphization, so that some amorphizable solids will amorphize only at low or cryogenic temperatures, and there is usually a threshold temperature above which amorphization does not occur. There is likewise a critical radiation fluence, representing a critical atom displacement density or bond rupture density, above which amorphization occurs below the critical temperature.[3]

Rigidity Theory and Applications
Edited by Thorpe and Duxbury, Kluwer Academic/Plenum Publishers, 1999

Amorphization, like glass formation, represents fundamentally a failure to crystallize. Glasses are *topologically-disordered*[4] arrangements which possess neither translational regularity (periodicity) nor rotational invariance. Most amorphized structures are probably topologically disordered, though they may not exhibit a glass transition[5] in the conventional sense, and an even more fundamentally intriguing question amounts to asking what unique property or properties a (damaged) structure must possess to permit a topologically disordered arrangement. The atomic arrangements in both glasses and amorphized solids are not *random* and remain constrained by topological construction rules and steric considerations which derive from the ways structural elements can be connected together. *Connectivity* is therefore a fundamental determinant governing amorphizability, just as it appears to be in glass formation.[6] Radiation-induced amorphization therefore represents a convenient means for exploring the propensity for glass formation which is more readily accessible for most compounds than traditional melt quenching.

Amorphization occurs in a few structures (Si, SiO_2, SiC, silicates) by accumulation of single atom displacements arising from ballistic collisions with light incident particles (like electrons) or a radiolytic mechanism,[7] but in most others from collisions with heavy energetic particles (like fast ions) generating localized displacement cascades involving hundreds to thousands of atoms. Cascade disorder has been simulated using molecular dynamics[8] or topologically-based self-assembly reconstruction algorithms.[9,10] Amorphization susceptibilities range from extremely easy (accumulated displacement density < 0.01 dpa) to extremely difficult (> 50 dpa). Amorphizability has been assessed most commonly by the fast *ion* fluence required to amorphize a material, expressed as the critical deposited energy density (eV/target atom).[11] Table 1 lists the values from ion irradiations carried out by various investigators, for the most part well below the critical amorphization temperature. Monte Carlo computer codes (TRIM-91 *et seq.*[12]), together with accepted values for atom displacement energies,[13] were used to compute displacement energy deposition as a function of depth where only an ion energy and critical ion fluence were supplied.

These critical energy densities vary over *four orders of magnitude*, providing an extremely sensitive measure of amorphizability of solids. Some materials--like alkali halides, or oxides with the rocksalt or fluorite structures--appear impossible or nearly impossible to amorphize, forming localized aggregate defects (dislocations loops, voids, colloids[14]) and even decomposing[15] as a response to radiation disorder, rather than topologically disordering. Others--such as silicas and many silicates--readily amorphize efficiently from 4 K to well above room temperature, rather than forming aggregate defects at all.[16] SiC amorphizes with ions at room temperature at about 0.5 dpa[17] and at 77 K with 1 MeV electrons,[18] but not at 1000 K, where it instead remains crystalline to displacement densities of > 100 dpa and undergoes dislocation climb-driven void swelling.[14]

Attempts have been made[19,20] to correlate such behavioral extremes with a variety of physical parameters--such as structure type, bonding character, melting point, homologous crystallization temperature, structural complexity (as measured by the number of inequivalent cation site environments) and chemical complexity (measured by the number of atom types)--across a wide range of solids. For none of these criteria, however, has the correlation enabled uniformly consistent predictions to be made within a more limited suite of similar structure types. A recent semi-empirical approach[21] has invoked the dependence of crystallization on melt viscosity,[22] within a suite of compounds in the MgO-Al_2O_3-SiO_2 ternary system to arrive at better correlation with an amorphization parameter.

A more fundamental explanation for this surprising variability is the topology of the connectivity, and an assessment of what might be called *structural* or *topological freedom*.[4] The assessment is based essentially on graph theory and dates from more than a century ago to the method of constraint counting introduced by James Clerk Maxwell[23] to investigate the rigidity of mechanical structures. The approach has been applied more recently to rigidity percolation in triangular networks by Thorpe and co-workers,[24-26] and to oxide and chalcogenide glass formation independently by Cooper[6] and Phillips[27] respectively. The principal parameter to be evaluated is available freedom for achieving topologically alternative *extendable* structural arrangements.[28] (This is not the same as existence of polymorphism, though there can be a relationship; some *un*amorphizable alkali halides, after all, exhibit polymorphism.)

In considering the topological foundation of Zachariasen's rules for glass formation, Cooper[29] considered the connectivity of short range-ordered regular, rigid, congruent structuring polytopes; for three-dimensional compounds, these are usually the familiar cation-coordination polyhedra of Pauling's rules which are connected together by combinations of corner, edge or face sharing. Cooper showed that, for some combinations of polytope and connectivity, structures are constrained (usually *over*constrained) to be crystalline, while in other cases they are marginally or *under*constrained and free to adopt aperiodic configurations. Enumeration of these constraints was undertaken by Gupta and Cooper[4] and more comprehensively by Gupta.[30] Hobbs and co-workers[11,31] have firmly established a strong correlation between assessed topological freedom and radiation-induced amorphizability. In this contribution, we extend and correct earlier assessments by considering more carefully the assignments of topological freedom to situations where ambiguities exist, where the chosen assignment warrants a more elaborate discussion, or where a better method of constraint counting has been devised.

TOPOLOGICAL MODELING OF AMORPHIZED NETWORKS

As indicated earlier, the irradiation-induced loss of crystallinity can occur either from accumulation of single atom displacements to a critical density or in the wake of energetic collision cascades. Because the rigidity constraints explored in this paper have their origins in connectivity, it is particularly instructive to model both amorphization mechanisms topologically. The former has been explored to some extent by Thorpe *et al.*[24-26] in two-dimensional triangular networks, and by Thorpe and others[32,33] in three-dimensional depleted diamond-structure networks, in the form of calculation of the critical bond-breakage density required to impart global network floppiness. The latter has been explored by Jesurum *et al.*[9,10] in three-dimensional tetrahedral polytope networks using disordering and reconstruction algorithms which mimic ion irradiation-induced cascade processes.

The first step in the latter is to establish self-assembly algorithms which can generate the initial crystalline assembly, a non-trivial exercise once symmetry is abandoned. Suitable algorithms are based on a set of *local assembly rules*[34-36] which embody the local connectivity of a repeated structuring polytope. The local rules for assembly of α-quartz (SiO_2) are shown in Fig. 1. Each of the four vertices of a constituent [SiO_4] tetrahedron is connected to one other tetrahedron at a common vertex ("4 1" in the rules file). Operationally, the initial tetrahedron is oriented with respect to a canonical reference orientation (24° rotation about the *y*-axis in this case), then replicated, rotated according to a rotation rule (±60° in this case) and connected to the appropriate vertex of the initial tetrahedron. The sequence is repeated with each of the added tetrahedra in turn serving as the initial tetrahedron. With appropriate rules sets, each of the crystalline polymorphs of SiO_2 stable under different conditions of pressure and temperature can be assembled.[36] Analogous rules govern assembly of Si_3N_4 and SiC polymorphs, in which $[SiN_4]$ or $[SiC_4]$ tetrahedra share vertices with respectively two and three other tetrahedra.[37] Non-crystalline structures can be assembled by modifying the rotation rules.[34-37]

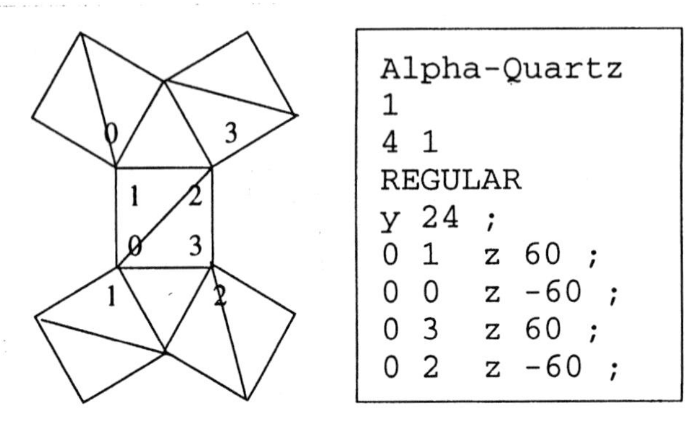

Fig. 1. Local rules for the self-assembly of α-quartz,[36] indicating the 24° rotation of the initial tetrahedron, the ±60° rotations of each added tetrahedron, and a depiction of the assembly scheme for the first five tetrahedra.

Cascade amorphization is modeled[9,10] by breaking connections between all polytopes within a specified volume, imposing a limited random rotation on these polytopes, then reconnecting them according to the same or a different set of assembly rules appropriate to whatever ambient conditions are assumed within the cascade. Springs are used to connect vertices of regular polytopes to provide a vehicle for accommodating (and simulating) necessary polytope distortions (so-called "split-atom" method[68]); the elastic energy of these springs (both extension and torsion) is globally minimized at each stage of the reassembly. Fig. 2 shows the reassembled cascade volumes of α-quartz, β-Si_3N_4 and α-SiC. The SiO_2 cascade is substantially connected with minimal tetrahedral distortion (represented by residual spring segments), but the randomized volumes of Si_3N_4 and especially SiC are difficult to fully connect, and the reassembly is in each case accompanied by large tetrahedral distortions (long residual springs). These difficulties have their origins in restricted topological freedom, as elaborated below.

TOPOLOGICAL FREEDOM AND STRUCTURAL OPTIONS

Fig. 3 depicts six simple two-dimensional examples for which assessment of topological freedom is instructive. Representing by δ the dimensionality of the polytope, by V the number of vertices per polytope, by C the number of polytopes sharing a vertex, and by $\{V,C\}$ the connectivity of the structure, the examples illustrated are $\{2,3\}$, $\{2,4\}$, $\{2,6\}$ $\{3,2\}$, $\{4,2\}$ and $\{3,3\}$ for the rod ($\delta = 1$, $V = 2$), triangle ($\delta = 2$, $V = 3$) and square (δ

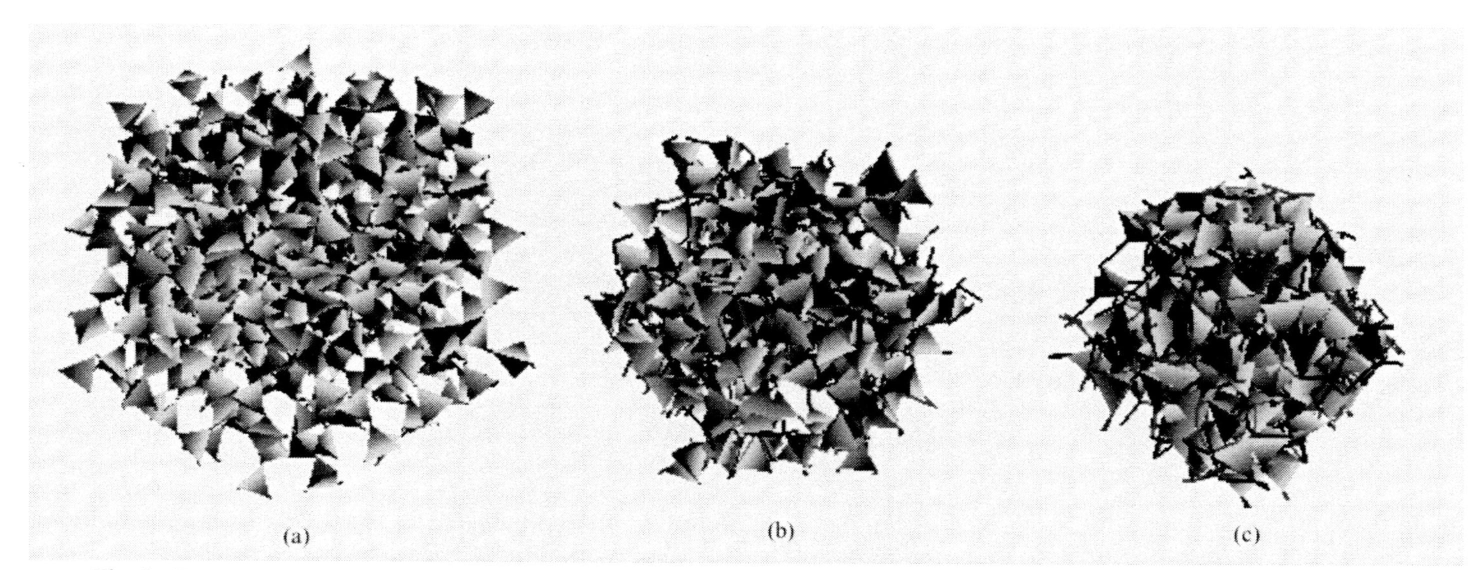

Fig. 2. Reconnected cascade regions for disordered models of a) α-cristobalite SiO_2 reassembled with β-cristobalite rules, b) β-Si_3N_4 reassembled with β-Si_3N_4 rules, and c) α-SiC reassembled with α-SiC rules.

= 2, V = 4) polytopes used. The topological freedom f is represented by the number of degrees of freedom at each vertex--equal to the dimensionality d of the structure--less the number of constraints h imposed by connections to neighboring elements. As is easily verified by a mechanical model, the {2,3} arrangement in two dimensions is underconstrained and free to adopt an arbitrary configuration; the average number of constraints on each vertex (1.5) is a little less than its degrees of freedom (2), as reflected by the value of $f = +0.5$. By contrast, the {2,6} arrangement of rods, for which $f = -1$, is overconstrained, rigid, and clearly required to be crystalline, as verified for the analogous bridge truss considered by Maxwell.[23] Values of $f > 0$ therefore imply freedom to arbitrarily rearrange, while values of $f < 0$ imply rigid crystalline options. The two-dimensional {3,2} arrangement of triangles (the Zachariasen model for a glass[38]), for which $f = 0$, is marginally constrained; there is no freedom once the boundary is set, but altering the boundary allows other possibilities. Finally, the two-dimensional {3,3} arrangement of triangles is again overconstrained and crystalline, with $f = -1$ (and equivalent to the {2,6} arrangement of rods).

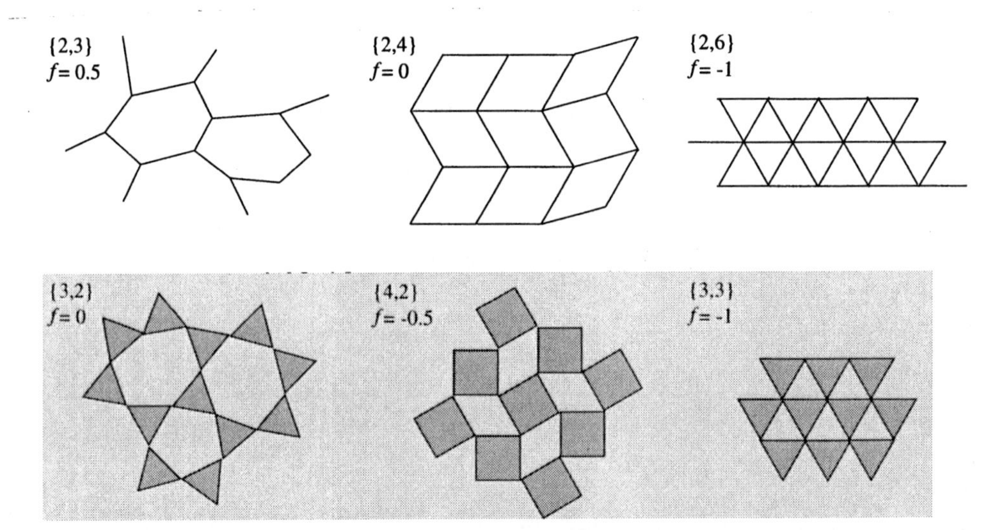

Fig. 3. Two-dimensional networks constructed using one- and two-dimensional structuring polytopes, with respective connectivities {V,C} and topological freedoms f indicated (reproduced from Hobbs *et al.*[31]).

The {4,2} arrangement of square polytopes is a special and instructive example, because it represents an *atypical*, rather than a generic, network. The arrangement shown in Fig. 3d is crystalline but floppy (it can be readily sheared into a checker-board pattern), despite the assigned value of $f = -0.5$. Two-dimensional {4,2} structures can also be rendered in the form of quasicrystals[39] based on Penrose tilings,[40] a non-crystalline possibility probably anticipated by Cooper[6,41] before the experimental evidence for quasicrystals had appeared. (The {4,2} arrangement in Fig. 3 can be a special crystalline arrangement of quasicrystal rhomboid elements.) The failure of the constraint counting approach to evaluate rigidity in this case reveals that the method of evaluation overcounts constraints in the case of crystalline (and perhaps quasicrystalline) arrangements where at least rotational invariance is present, in which case the constraints are not wholly independent. Hence, the application of Maxwellian constraint counting to initially crystalline atom arrangements to assess amorphizability, as proposed in the present paper, must proceed cautiously, with the recognition that overcounting of constraints may occur, whose extent must remain uncertain until a more general treatment of the influence of rotational and translational invariances on rigidity constraints has been formulated.

In three dimensions, the polytopes have dimensionality $\delta = 1, 2$ or 3; the common structuring polytopes are triangles ($\delta = 2, V = 3$, as in B_2O_3), tetrahedra ($\delta = 3, V = 4$, as in SiO_2, P_2O_5, SiC or Si_3N_4), octahedra ($\delta = 3, V = 6$, as in AlF_3, ReO_3, $CaTiO_3$, TiO_2 and MgO) and cubes ($\delta = 3, V = 8$, as in CaF_2, ZrO_2 and UO_2). Their assumed internal rigidity, guaranteed by hybridized covalent bonding orbitals with distinct angular preferences in some cases and less certainly by anion packing about cations in others, subsumes some of the angular constraints. For strictly corner-sharing of rigid congruent regular polytopes, Gupta and Cooper[4] have shown that

$$f = d - h = d - \sum_i C_{avg,i}\{\delta - [\delta(\delta+1)/2V_i]\} \qquad (1)$$

where $C_{avg,i}$ is the average number of polytopes with V_i vertices connected at each vertex. Some of the ways in which triangles, tetrahedra and octahedra can connect together by simple corner sharing are depicted in Fig. 4, together with their connectivities $\{V,C\}$ and consequential topological freedoms f. For the additional possibility of shared polytope edges and faces, Gupta[30] calculates

$$f = d - C\{\delta - [\delta(\delta+1)/2V]\} - (d-1)(Y/2) - [(p-1)d - (2p-3)](Z/p) \qquad (2)$$

where Y is the fraction of vertices with connectivity which participate in edge sharing and Z the fraction of vertices which participate in sharing of p-sided faces. When face sharing is present, the edges involved in the face sharing are not counted towards the edge-sharing complement Y. The formulation (2) strictly applies only to structures comprising a single polytope type; multi-polytope cases can be treated on the basis of average connectivites, as in eqn (1), but the calculation of Y and Z is less straightforward, and some simplifying approaches are discussed below.

Table 1 ranks in order of amorphizability a number of simple structures for which the polytope is obvious and which share only *one* connectivity element (vertices, edges or faces) with other polytopes. The structure of B_2O_3, a facile glass former, is based on a $\{3,2\}$ arrangement of $[BO_3]$ triangles (or of $[B_3O_6]$ boroxyl rings which are just larger triangular polytopes and have identical connectivity); this arrangement is even more unconstrained ($f = +1$) in three dimensions ($d = 3$) than in two. All silicas (except stishovite) and the silica framework of framework silicates are based exclusively on full corner sharing of $[SiO_4]$ tetrahedra (or substituted tetrahedra); such tetrahedral arrangements are marginally constrained ($f = 0$). The corner-sharing octahedra in the ReO_3 structure, a structure which the glass-forming compound AlF_3 adopts and of which pervoskites (*e.g.* $CaTiO_3$) are a stuffed derivative, are only somewhat more constrained ($f = -1$) than the tetrahedra in silica.

Analogous $[PO_4]$ tetrahedra in P_2O_5 and many phosphates share only three of the four available vertices. While the topology is thus identical to B_2O_3 corner-sharing

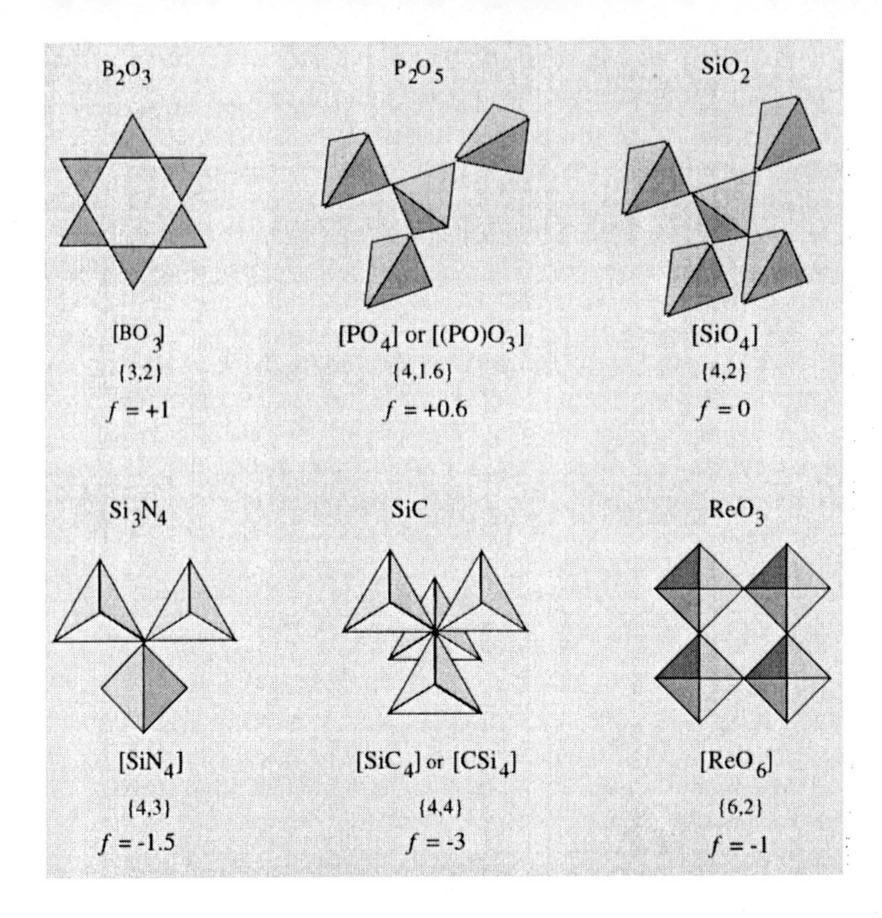

Fig. 4. Connectivities and topological freedom for 2- and 3-dimensional corner-sharing networks of tetrahedral and octahedral polytopes.

triangles, the unshared oxygen can nevertheless participate in reconstruction possibilities and so must be accounted for. (Also, expending energy in breaking the P=O bond to the unshared oxygen during radiation-induced atom displacement does not contribute to a change in topology, so on this basis alone radiation-induced rearrangements in P_2O_5 should be less-efficient than in B_2O_3.) Gupta[30] computes the connectivity of the unshared vertex at $C = 1$ and of the shared vertices as $C = 2$. For each P atom there is exactly one non-bridging (double-bond) oxygen atom, so P_2O_5 may be represented as $(PO)_2O_3$ by analogy to B_2O_3. For each non-bridging oxygen, there are 1.5 bridging oxygens, so that

$$C_{avg} = (1\times1 + 1.5\times2)/(1 + 1.5) = 1.6$$

and $f = +0.6$.

SiO_2, Si_3N_4 and SiC constitute a series illustrating the effect of successively increasing connectivity in strictly corner-sharing tetrahedral structures. In these compounds, $[SiO_4]$, $[SiN_4]$ and $[SiC_4]$ tetrahedra respectively share vertices with one (C = 2), two (C = 3) and three (C = 4) other tetrahedra.[37] Silica crystallizes in six compact tetrahedral polymorphs,[36] all with $f = 0$. Si_3N_4 crystallizes in two polymorphs (α and β),

Table 1. Coordination, connectivity, topological freedom and amorphizability for single polytope compounds with a single polytope-sharing mode

Structure	Polyhedra : sharing	$\{V, C\}$	f	Amorphization dose (eV/atom)[a]	
MgO	Octahedra : edges	{6,6}	−10	5000	[1]
UO_2	Cubes : edges	{8,4}	−7	> 3000	[51]
SiC	Tetrahedra : corners	{4,4}	−3	13	[47]
Si_3N_4	Tetrahedra : corners	{4,3}	−1.5	57	[45]
ReO_3	Octahedra : corners	{6,2}	−1	35[b]	[58]
Be_2SiO_4	Tetrahedra : corners	{4,2}	0	11	[43]
$AlPO_4$	Tetrahedra : corners	{4,2}	0	10	[62]
SiO_2	Tetrahedra : corners	{4,2}	0	7	[43]
P_2O_5	Tetrahedra : corners	{4, 1.6}	+0.6	0.5[c]	[59]
B_2O_3	Triangles : corners	{3,2}	+1	d	

[a]Low-temperature values of critical energy density required for amorphization well below critical amorphization temperature. [b]Value for $KNbO_3$. [c]Value for $Pb_2P_2O_7$. [d]Not measured.

for both of which $f = -1.5$. The many polytypic variants of SiC are all based on stacking of sublayers of the two related polymorphic arrangements (α and β), for both of which $f = -3$. Attempts[10] to amorphize the self-assembled models of SiO_2, β-Si_3N_4 and α-SiC, using the disordering and reassembly algorithms indicated in Fig. 2, parallel the available structural freedoms: SiO_2 can be reassembled in non-periodic form with relative ease; Si_3N_4 reassembles with great difficulty, accompanied by underconnection and significant tetrahedron distortion; and SiC cannot be convincingly reassembled.

Experimentally, at least four silica polymorphs (tridymite, cristobalite and quartz[42,43] and coesite[44]) have been shown to amorphize easily (~7 eV/atom, 0.11 dpa), while Si_3N_4 is considerably more difficult to amorphize (57 eV/atom, 0.9 dpa[45]). Surprisingly, both α and β forms of SiC amorphize with comparative ease (as low as ~13 eV/atom, 0.25 dpa for 77 K ion irradiations[46,47]), both in cascades from ion irradiation[17,46,47] and under fast electron irradiation[18]), though with a critical amorphization temperature (~340 K for electron irradiation, ~500 K for ion irradiation) not far above room temperature. An explanation for this apparent failure of constraint theory to predict amorphizability is advanced in the next section.

Addition of edge- and face-sharing introduces connectivity redundancies and reduction of topological freedom, as seen in eqn (2). A vertex involved in a single shared edge has one degree of freedom less; if involved in two *independent* edges, then some freedom would remain; if involved in two *dependent* edges (belonging to the same polytope), then there may or may not be freedom remaining; if involved in a shared face, then the vertex has no degrees of freedom left. However, confusion can arise in the calculation of Y in eqn (2) if a *global* enumeration of vertices involved in shared edges is performed. If every vertex were involved in at least one shared edge, then $Y = 1$ from the definition in eqn (2); but vertices involved in *two or more* shared edges may be more constrained, and there needs to be some way to differentiate these degrees of constraint. From the more local point of view of a *single polytope*, what is important is whether one of its vertices is involved in an edge share with *that polytope* and another polytope. A similar consideration arises for face sharing and calculation of Z. This point is discussed later for the cases of rutile and alumina.

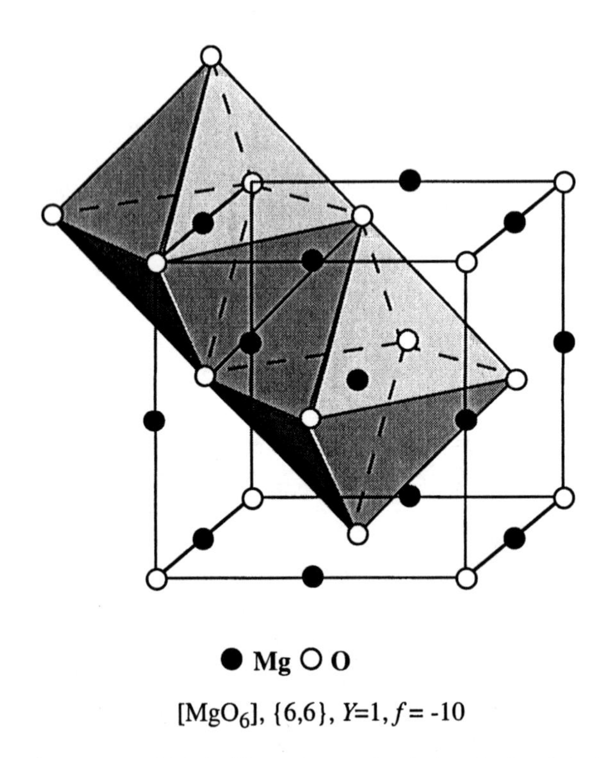

$[MgO_6]$, {6,6}, Y=1, f = -10

Fig. 5. Rocksalt structure of MgO, illustrating mode of edge sharing $[MgO_6]$ octahedra.

$[MX_6]$ octahedra in the {6,6} MX rocksalt structure (MgO, for example, Fig. 5) are fully edge-shared and so seriously overconstrained ($f = -10$) as to be least likely to be found in topologically disordered arrangements in pure compounds (*e.g.* in the absence of chemical effects from implanted ions). The $[MX_8]$ cubes of the {8,4} MX_2 fluorite structure (Fig. 6) are likewise fully edge-shared with a similar result ($f = -7$), a result of some significance to a recent proposal[48] to accommodate Pu nuclear wastes in ZrO_2-based wasteforms which appear to be very unamorphizable.[49] (The usual tetragonal and monoclinic forms of zirconia are lower-temperature displacive modifications of the high-temperature cubic fluorite structure which can be stabilized to room temperature by aliovalent cation additions.[50]) The fluorite structure is also the form adopted by $(U,Pu)O_{2+x}$-based oxide fuels for fission reactors which--like the carbonitride U(C,N) fuels with the rocksalt structure--survive long irradiation periods at high temperatures and are stored as spent fuel nearer room temperature without any evidence of amorphization after hundreds of dpa displacement doses. Ion irradiation of UO_2 has confirmed its strong resistance to amorphization at low temperature (> 3000 eV/atom, > 40 dpa).[51]

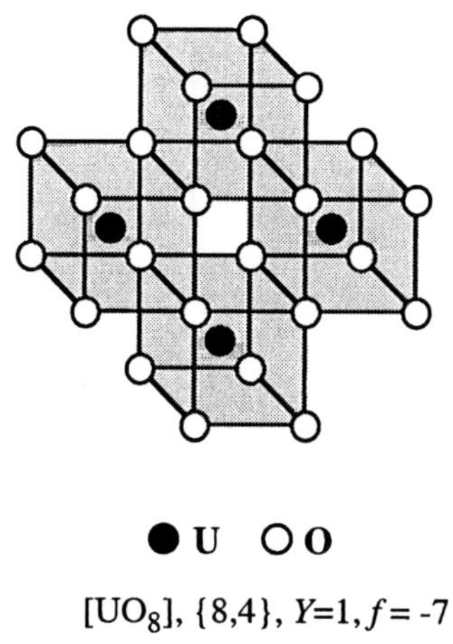

● U ○ O

[UO_8], {8,4}, Y=1, f = -7

Fig. 6. Fluorite structure of UO_2, illustrating mode of edge sharing [UO_8] cube polytopes.

CHOICE OF POLYTOPE

For predictably coordinated compounds, the coordination unit is likely to persist intact and undistorted in any reorganization of the structure, and the choice of polytope is obvious. For monatomic solids, like C or elemental metals (M) where alternative coordinations are possible, the choice is less obvious. For example, for close-packed metals, the coordination is 12, for body-centered metals 8, and for metallic glasses somewhere in between. The minimal polytope choice is an M-M rod ($\delta = 1$) representing an interatomic bond; for this choice, close-packed metals are represented by {2,12}, yielding f = -3, body-centered metals by {2,8} with f = -1 (Table 2). These are slight under-representations of the constraints, because steric exclusions do not permit the rods to adopt arbitrary M-M-M angles. Pure metals and random solid-solution alloys do not appear to amorphize under irradiation, presumably because it is too easy for single atoms without bonding directionality to rearrange into lowest energy configurations; but ordered intermetallic compounds amorphize with surprising ease.[2] The ordering preferences definitely play a topological role, just as do polytopic coordinations in more strongly bonded ceramic compounds, though irradiation may eventually effect a more average site occupation. The intermetallic alloy Zr_3Al, for example, has the $L1_2$-structure (ordered occupation of face-centered cubic lattice sites) and amorphizes at about 63 eV/atom (1 dpa),[52] more or less in keeping with the $f \approx -3$ estimate for a close-packed arrangement.

The underconstraint represented by rod polytopes becomes more evident for monatomic solids of lower coordination with highly directed covalent bonds. In graphite, C atoms are 3-connected through sp^2 hybridized bonds in planar hexagonal nets (Fig. 7) only loosely bound to neighboring sheets. Carbon can also assemble in the 4-coordinated sp^3-bonded tetrahedral diamond structure (Fig. 8). For the choice of [CC] rod polytope, graphite can be represented by {2,3} (δ = 1, d = 2), in which representation f = +1, and diamond by {2,4} (δ = 3, d = 3) with f = +1. Given the experimental amorphizabilities (11-15 eV/atom,[53,54] Table 2) for graphite and silicon (with the diamond structure)--comparable to SiO_2 for which f = 0--both of these representations clearly underestimate the structural constraints. Underrepresentation occurs because the trigonal (or tetrahedral) directionality

Table 2. Coordination, connectivity, topological freedom and amorphizability for some elemental structures and tetrahedral networks

Structure	Polyhedra : sharing	$\{V, C\}$	f	Amorphization dose (eV/atom)[a]
c.p. metal	Rods : ends	{2,12}	−3	63[b] [52]
SiC	Tetrahedra : corners	{4,4}	−3	13 [47]
b.c. metal	Rods : ends	{2,8}	−1	11[c] [70]
SiC	***Tetrahedra : corners***	***{4,2}***	***< 0***	***13*** [47]
C (graphite)	***Triangles : corners***	***{3,2}***	***< 0***	***15*** [53]
Si	***Tetrahedra : corners***	***{4,2}***	***< 0***	***11*** [54]
C (graphite)	Rods : ends	{2,3}	+1	15 [53]
Si	Rods : ends	{2,4}	+1	11 [54]

[a]Low-temperature values of critical energy density required for amorphization. [b]Value for Zr_3Al ($L1_2$ structure). [c]Value for NiTi (B2 structure). ***Italicized entries based on atom-centered polytopes which are not coordination polyhedra.***

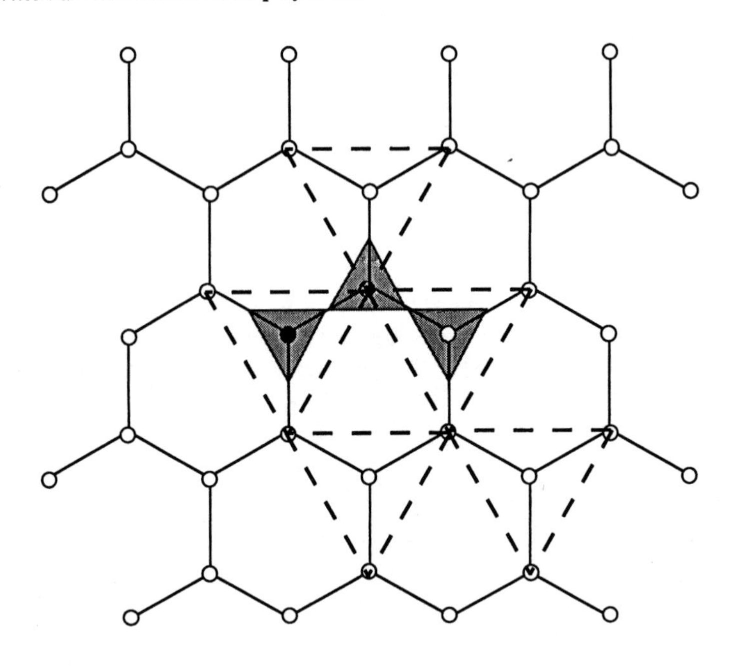

Fig. 7. Hexagonal two-dimensional network of graphite (0001) sheet, with three connectivity representations using [CC] rods, [CC_3] coordination triangles, and [C_{-3}] triangles with vertices at the C-C bond midpoints.

of the C-C (or Si-Si) bonds is ignored: the bonding geometry introduces angular constraints which must be accounted for; these are largely steric and weaker for metallic bonding, as indicated, but are strong constraints for highly directional covalently bonded carbon (or silicon).

More appropriate choices of polytope are an equilateral triangle for graphite and a tetrahedron for diamond. In Fig. 7, a [CC_3] triangular polytope (δ = 2) is seen to be {3,3}-connected, resulting in $f = -1$. The difficulty with this representation, which now overestimates the experimentally-deduced constraints, is that it requires two distinguishable species of carbon atom, one at the polytope center, the other at the vertices, which is unphysical. A still better representation is the [$C-_3$] shaded triangles centered on each C atom with vertices at the bond midpoints; these are {3,2}-connected, with $f = 0$. In this representation, the C-vertex-C angle is required to be 180° (corresponding to the straight [CC] rod), whereas in the {3,2} arrangement of triangles depicted in Fig. 3 this angle is free; hence, there is less freedom in the graphite structure than in the {3,2} representation, and f is somewhat < 0, as suggested by the 15 eV/atom[53] amorphization dose. Flexibility in this C-vertex-C angle corresponds to a distortion of the 120° sp^2 bonds, which means that amorphization of graphite must be accompanied by bond-angle changes.

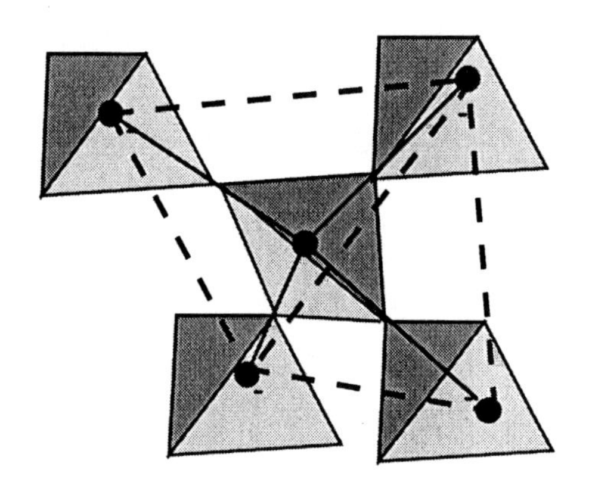

Rods: {2,4}, $\delta = 1, f = +1$

Tetrahedra: {4,4}, $\delta = 3, f = -3$

Tetrahedra (to bond midpoint): {4,2}, $\delta = 3, f = 0$

Fig. 8. Diamond structure of silicon, with three connectivity representations using [SiSi] rods, [$SiSi_4$] coordination tetrahedra, and [$Si-_4$] tetrahedra with vertices at the Si-Si bond midpoints.

In Fig. 8, the tetrahedral structure of silicon is represented by rods ($f = +1$, an underrepresentation of the tetrahedral bond constraints) and then alternatively by two tetrahedron arrangements. The first of the latter, [$SiSi_4$] tetrahedra {4,4}-connected for which $f = -3$, clearly overstates the rigidity of the network for the same reason that [CC_3] triangles fail to represent graphite: the indistinguishability of the Si atoms. The alternative [$Si-_4$] tetrahedra with vertices at the bond midpoints are {4,2}-connected. As in the analogous representation of graphite, the Si-vertex-Si angle is constrained to be 180° if the 109.5° tetrahedral bond angles are to be maintained; to the extent that the tetrahedral angle can be altered, the structure approaches the topological freedom ($f = 0$) of the {4,2} network of SiO_2. The experimental amorphizability of Si (11 eV/atom,[54] Table 2) suggests a value close to $f = 0$. Hence, amorphization of silicon must also be accompanied by bond-angle changes.

By comparison, SiC is highly overconstrained, and viable amorphization of SiC appears topologically difficult or impossible, both from the reassembly exercise of Fig. 2c and from the $f = -3$ value (Table 2) of topological freedom based on {4,3} connectivity of $[SiC_4]$ (or $[CSi_4]$) tetrahedra. Its comparatively facile amorphizability can be explained by the ability of SiC to sustain chemical disorder in the form of C_{Si} or Si_C antisite defects. These yield local regions with silicon- or diamond-like topology; the C_{Si} defect is favored energetically, locally creating a $[CC_4]$ tetrahedron of diamond. Collisional disorder may thus locally generate an *average diamond* structure, and random occupation of sites would lead to topological freedom similar to that of silicon ($f \approx 0$, as argued above). Hence, it is not unreasonable to expect that SiC will amorphize from regions of high anti-site density. The replacement-to-displacement ratio is higher for cascade displacements than for single atom displacements, so it can be expected that ion irradiation will be more efficient (per unit deposited displacement energy) than electron irradiation in inducing amorphization, which appears to be the case.[55] Recent MD simulations of collision cascades in SiC[8] have resulted in anti-site defects, and indeed very recent EXAFS and Raman experiments[46,56] in ion-amorphized SiC have detected substantial $[CC_4]$ coordination. Anti-site defects are likely to anneal thermally at modest temperatures in SiC, accounting for (or at least contributing to) the lower (~290-340 K) observed critical temperature for amorphization by electon irradiation[18] compared to that (~420-498 K) for ion-irradiation-induced amorphization.[47]

MULTIPLE SHARING MODES AND MULTIPLE POLYTOPES

More complex solids usually exhibit multiple polytope-sharing modes (combinations of corner, edge or face sharing) and multiple polytope types (Table 3). What is at issue is how to evaluate the *combined* contributions of each from both a global and local point of view. For the simpler of the two cases, the single polytope structure with combined sharing modes, eqn (2) provides a fair assessment of topological freedom, provided the sharing modes are rather evenly distributed throughout the structure.

A good example is rutile (TiO_2, and also the isomorphous structures of the high pressure silica polymorph stishovite and a polymorph of GeO_2), which features strings of edge-sharing octahedra linked by corner sharing in {6,3} connectivity (Fig. 9). The oxygen atoms at every vertex are common to three octahedra ($C = 3$), and *every* oxygen is involved in a shared edge linking two octahedra; however, a *given* octahedron has only two shared edges, with only 4 of its 6 oxygens involved in *those* edge shares, so $Y = 4/6$, not 1 as defined by Gupta,[30] is the fraction of vertices involved in shared edges for the purposes of eqn (2). The result of this edge sharing is a large reduction (to $f_{avg} = -3.7$) in topological freedom, which is borne out by the relative difficulty (75 eV/atom, 0.75 dpa[57]) of its amorphization.

In the corundum structure of α-Al_2O_3, $[AlO_6]$ octahedra share edges and faces, but the face sharing is very inhomogeneously distributed, confined to isolated pairs of

Table 3. Coordination, connectivity, topological freedom and amorphizability for structures with multiple polytopes and multiple sharing modes

Structure	Polyhedra : sharing	$\{V, C\}$	f_{avg}	f_{sp}	Amorphization dose (eV/atom)[a]	
α-Al_2O_3	Octahedra : faces, edges	{6,4}	−6	**−5**	380	[66]
$MgAl_2O_4$	Octahedra : edges; tetrahedra : corners	{4,4} {6,4}	**−5**		400[b]	[61]
$CaTiO_3$	Octahedra : corners truncated cubes : faces	{6,2}[c] {12,4}	−11.5		89	[58]
TiO_2 (rutile)	Octahedra : edges, corners	{6,3}	**−3.7**		75	[57]
Mg_2SiO_4	Octahedra : edges; tetrahedra : edges, corners	{6,4} {4,4}	−5.2	**−3.4**	79	[22]
Al_2SiO_5 (sillimanite)	Octahedra : edges; tetrahedra : corners[c]	{6,3} {4,3}[c]	**−2.2**		40	[22]
$CaSiO_3$	Octahedra : edges; tetrahedra : corners[c]	{4,2}[c]	**< 0**[c]		11	[43]
$Pb_2P_2O_7$	Tetrahedra : corners[c]; $[PbO_8]$, $[PbO_9]$	{4, 1.6}[c]	**< +0.6**[c]		0.5	[59]

[a]Low-temperature values of critical energy density required for amorphization. [b]Cation sublattice only. [c]Weak-link connectivity. **Preferred values in bold.**

octahedra which are only edge-shared with other octahedra; the arrangement of six such pairs is defined in the depiction of successive basal layers of octahedra in Fig. 10. Every octahedron is paired with another in such a way, face-sharing one of its 8 triangular faces with a second tetrahedron in the adjacent basal layer; of its twelve edges, 3 are consumed in the shared face, 3 are shared with surrounding octahedra, and 6 remain unshared. Eqn (2) thus becomes $f = -5 - Y - Z$. All six oxygen vertices are involved in the shared edges, but half of these are in the shared face, so $Y = 1/2$ and $Z = 1/2$. In this way, eqn (2) calculates $f_{avg} = -6$, but this is in fact an overestimate of the constraint. Because of their isolation, the shared faces contribute nothing to the network rigidity; instead they define a "super-polytope" (illustrated in Fig. 10) which haply offers a neater and less ambiguous way to enumerate the rigidity. The super-polytope comprises two face-sharing octahedra with 9 vertices, an $[Al_2O_9]$ unit, which is connected to 3 other such polytopes by a single edge-sharing mode in which all vertices participate ($Y = 1$). This arrangement yields $f_{sp} = -5$, which we take to correctly represent the topological freedom of the corundum structure.

Structures with two or more polytope types, which are often accompanied by two or more shared modes, present particular difficulties in deciding how to weight the separate

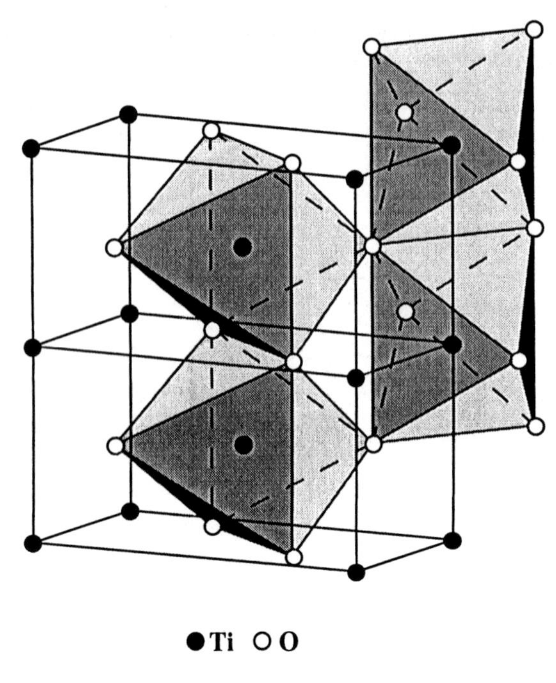

Fig. 9. Rutile structure of TiO_2, showing edge and corner sharing of $[TiO_6]$ octahedra.

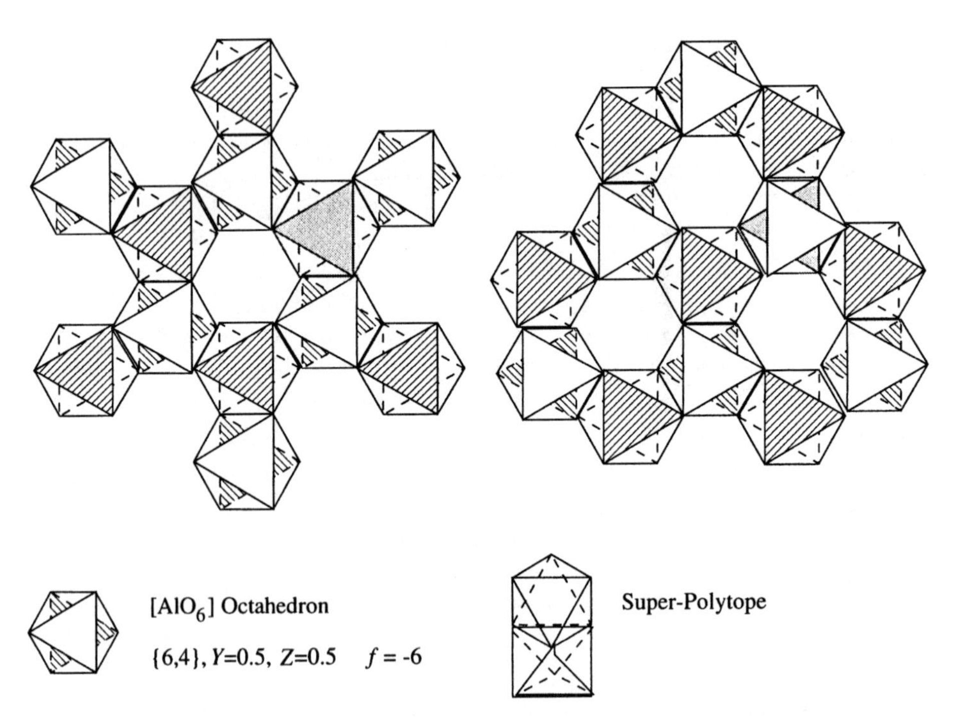

Fig. 10. Successive (0001) layers of corundum structure of α-Al_2O_3, illustrating edge sharing of $[AlO_6]$ octahedra and face sharing (shaded) of octahedra in an $[Al_2O_9]$ super-polytope.

contributions to rigidity. For some of these structures, it is possible to identify a local "weak link" in the connectivity, whose severance causes the structure to unravel. Fig. 11 presents two projections of the structure of sillimanite (Al_2SiO_5, or $Al_2O_3{\cdot}SiO_2$) which depict how columns of edge-sharing [AlO_6] octahedra are linked together by corner-sharing [SiO_4] and [AlO_4] tetrahedra. Evaluating the separate contributions to rigidity made by each of the three polytopes, the edge-sharing ($Y = 4/6$) octahedra are {6,3}-connected and contribute $f_o = -3.7$, while the two kinds of corner-sharing tetrahedra each contribute $f_t =$ -1.5. One could simply *globally* average these contributions ($f_{avg} = -2.2$), but *local* vulnerability is more important to assess. The corner share between [AlO_4] and [SiO_4] tetrahedra is particularly critical from the standpoint of structural integrity and comprises a weak link: breaking just these links leaves the columns of octahedra linked by only rods in a floppy configuration. The question remains how to quantify the vulnerability; compared to rutile, which also has columns of edge-sharing octahedra linked by corner sharing, sillimanite links them with two and sometimes three shared corners and should be much less constrained. One alternative is to consider a network comprising only the O-O polytope edges in the structure, which yields a minimum structural freedom $f_{min} = -2$, which is in any event close to the global average in this case. Similar arguments can be made for $CaSiO_3$, comprising edge-sharing [CaO_8] octahedra linked by corner-sharing [SiO_4] tetrahedra.

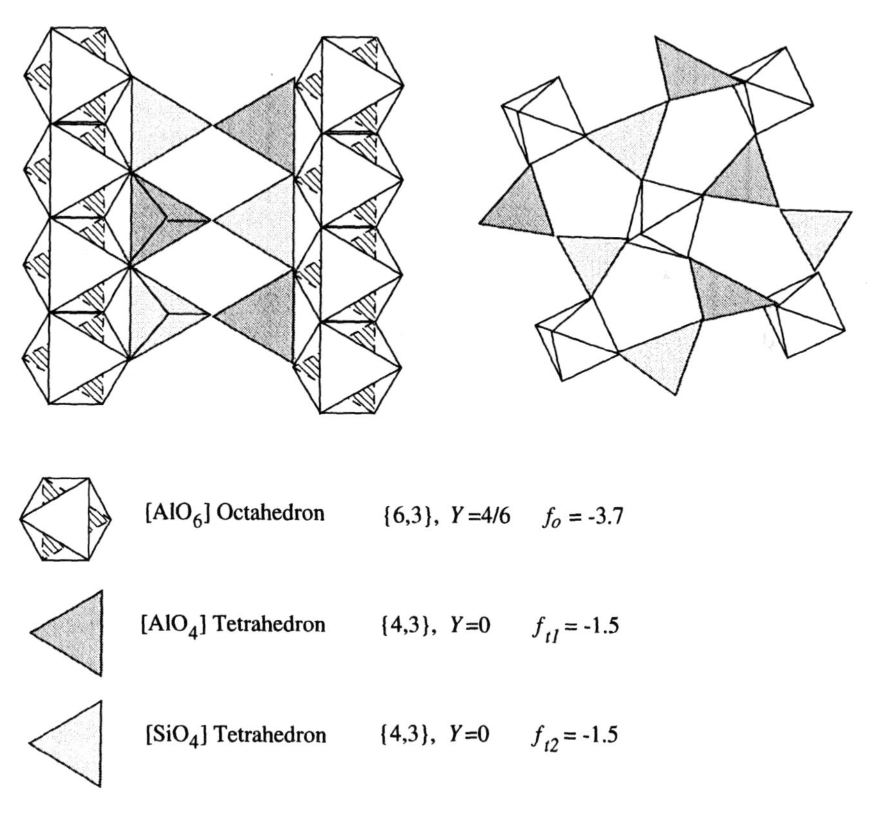

Fig. 11. Two views of the sillimanite form of Al_2SiO_5 ($Al_2O_3{\cdot}SiO_2$), illustrating the edge-shared chains of [AlO_6] octahedra and the corner sharing of [AlO_4] and [SiO_4] tetrahedra.

In many structures, the linkage of smaller lower-coordination polytopes defines larger interstitial holes occupied by cations with higher coordination. The perovskite structure (*e.g.* $CaTiO_3$, Fig. 12) is an example: a stuffed version of the ReO_3 assembly of corner-sharing octahedra (Fig. 4). The large Ca cation is 12-coordinated by oxygen, forming cuboctahedra which share square faces with each other and triangular faces with the [TiO_6] octahedra. The corner-sharing [TiO_6] octahedra, taken alone as in ReO_3 ($V = 6$, $C = 2$, $Y = 0$), yield a contribution $f_o = -1$, while the face-sharing [CaO_{12}] cuboctahedron polytopes ($V = 12$, $C = 4$, $Z = 1$) taken alone yield the contribution $f_{co} = -8$. Considering the impact of both cation polytopes in the overall structure, the parameters for the octahedra ($V = 6$, $C = 6, Z = 1$) yield a contribution $f_o = -10$, while those for the truncated cube ($V =$ 12, $C = 6$, Z - = 1) yield a contribution f_{tc} = -13, for a number-weighted average of $f_{avg} =$ -11.5. Perovskites are moderately amorphizable: $CaTiO_3$ amorphizes after 89 eV/atom, 0.89 dpa[58] and other pervoskites, like $KNbO_3$ (0.35 dpa, assuming similar displacement energies) are even more amorphizable. Given their relative ease of amorphization, neither the -11.5 average, nor even the even the average of the two polytopes taken independently ($f = -4.5$) accurately represents the freedom available to perovskites, which from the order in Tables 1-3 appears to be much closer to the $f = -1$ for the corner-sharing octahedra alone. Given that the Ca^{2+} cation contributes an ionic bond strength of +2/12 = 0.17 to each oxygen, compared to the +4/6= 0.67 contributed from each Ti^{4+}, it seems not unreasonable to assume that the large cuboctahedral cation coordination polytope of the highly-coordinated Ca^{2+} cation does not contribute substantially to structural rigidity. A similar circumstance may hold for $Pb_2P_2O_7$ ($2PbO{\cdot}P_2O_5$), where the amorphizability appears to be largely controlled by the corner-sharing [PO_4] tetrahedra[59] and the contribution of [PbO_8] and [PbO_9] coordination polytopes to the structural stability is minimal.

Oxide spinels (*e.g.* $MgAl_2O_4$, Fig. 13) represent an example of particularly well distributed connectivity. In this structure, cations occupy tetrahedral and octahedral sites in a cubic close packed (ccp) arrangement of oxygen anions. There are twice as many octahedra (which are in {6,4} connectivity) as tetrahedra (in {4,4} connectivity): in the *normal* spinel illustrated, the Mg^{2+} cations occupy the designated tetrahedral sites and Al^{3+} cations the designated octahedral sites, though redistribution is possible, for example during irradiation. All O^{2-} anions are equivalent, each being part of 3 octahedra and 1 tetrahedron and also participating in 3 shared edges. Half the octahedron edges are shared, but *every* vertex in an octahedron participates in that edge sharing (so$Y_o = 1$); the tetrahedra share only corners, and only with the octahedra, and each vertex *with respect to a tetrahedron* is involved only in corner sharing of that tetahedron ($Y_t = 0$). Hence, for the octahedra $f_o =$ -6 and for tetrahedra $f_t = -3$, yielding the globally weighted average $f_{avg} = -5$ Alternatively, applying eqn (2) to vertices in the octahedral and tetrahedral polytopes, $f_{avg} =$ $3 - 6 - 1.5 = -4.5$ less the constraint (+1 if $Y_o = 1$) provided by the octahedral edge sharing, yielding $f_{avg} = -5.5$. The globally weighted value of Y in this case is certainly not, however, as large as unity. We believe that averaging the separate polytopic freedoms is the better method of enumeration and that for spinels $f = -5$ is the correct structural freedom.

Spinels are resilient structures because their connectivity is so well distributed. The edge-sharing octahedra form a very rigid backbone which survives even if the corner-sharing tetrahedra are eliminated. Under ion irradiation, spinels appear to disorder in two stages.[60,61] In the first, the cation sublattice appears to disorder after about 400 eV/atom (4

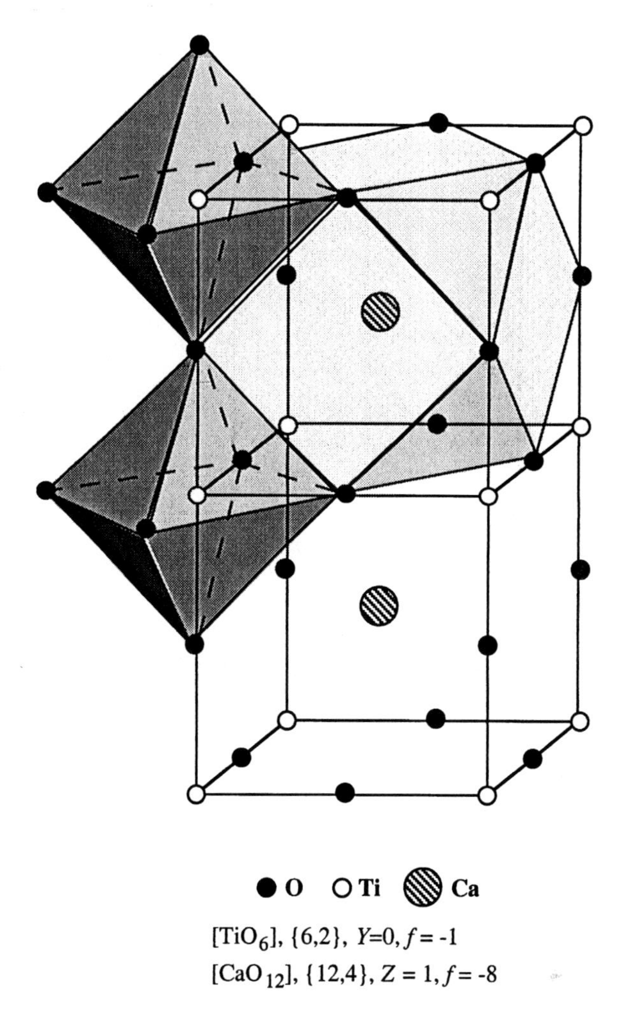

Fig. 12. Perovskite structure of $CaTiO_3$, showing corner sharing of [TiO_6] octahedra, which are in turn face-shared with face-sharing [CaO_{12}] cuboctahedra.

dpa), leaving the oxygen sublattice intact; in the second, complete amorphization is observed at 3500 eV/atom (35 dpa) at cryogenic temperatures. The former figure is in accord with the calculated topological freedom (Table 3). The latter represents the dose to amorphize a structure which is closer to that (Table 1) for the defect rocksalt structure that would obtain if the cations were distributed uniformly over 3/4 of the octahedral interstices of the ccp oxygen sublattice by irradiation-induced disordering.

Forsterite (Mg_2SiO_4), which has the olivine structure (a hexagonal analogue of the cubic spinels), is a compound of identical stoichiometry which is far less resistant to irradiation-induced amorphization[62] (Table 3). The Mg cations form [MgO_6] octahedra arranged in edge-sharing chains, while the Si atoms form isolated [SiO_4] tetrahedra which share 3 of their 6 edges and 1 corner with the chains of [MgO_6] octahedra (Fig. 14). As in spinel, the octahedra share half their edges, and their oxygen vertices are connected to 3 octahedra and 1 tetrahedron. Unlike spinel, there are two different kinds of vertices. The first (75%) are part of 3 octahedra and 1 tetrahedron and all participate in edge sharing (Y_1

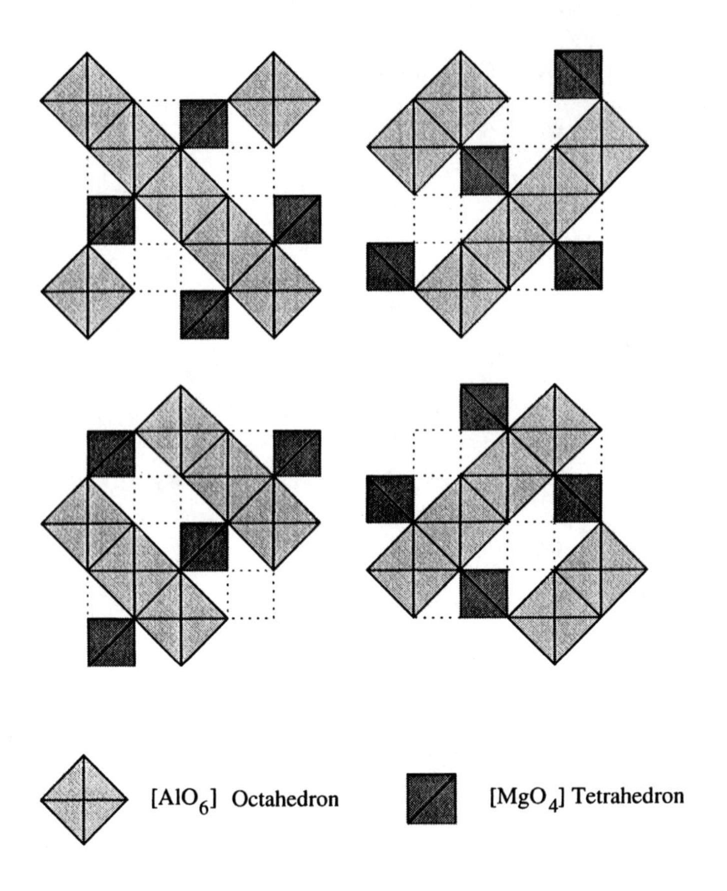

Fig. 13. Four successive (001) layers of the normal oxide spinel $MgAl_2O_4$, showing egde sharing of $[AlO_6]$ octahedra and corner sharing of $[MgO_4]$ tetrahedra. The edge-sharing octahedra form a rigid backbone by themselves.

= 1), yielding a contribution $f_1 = -5.5$, as in spinel; the other (25%) do not participate in edge sharing ($Y_2 = 0$), yielding the contribution $f_2 = -4.5$. The species-weighted average is $f_{avg} = -5.25$.

Alternatively, forsterite can be considered to be comprised of two sorts of $[MgO_6]$ octahedra, distinguishable by their environments ($C = 4$, $Y_{o1} = 1$, $f_{o1} = -5.75$ and $C = 4$, $Y_{o2} = 0.5$, $f_{o2} = -5.5$), and $[SiO_4]$ tetrahedra ($C = 4$, $V = 4$, $Y_t = 0.75$, $f_t = -3.75$), for a weighted average of $f_{avg} = -5.2$, very close to the vertex-based estimate. Both these global approaches, while adequate for spinel, do not correctly represent the susceptibility of the rigidity to removal of oxygens at the second kind of vertex. The super-polytope approach, discussed for alumina, can be advantageously applied to the olivine case in this case to obviate the difficulty. A rigid $[Mg_2SiO_{11}]$ super-polytope can be defined (illustrated in Fig. 14) which comprises two edge-sharing octahedra and 1 tetrahedron and has 11 vertices. Eight of these vertices are shared between three super-polytopes, 2 between two super-polytopes and 1 is non-bridging. Therefore $C_{avg} = 11\ (8/3 + 2/2 + 1) = 2.36$. Also, there are 6 vertices involved in face sharing of the super-polytopes, so that Z = 6/11. The resulting freedom, f_{sp} = −3.4, is much smaller than the global average and more accurately reflects the weak link provided by the vertices which do not participate in edge sharing.

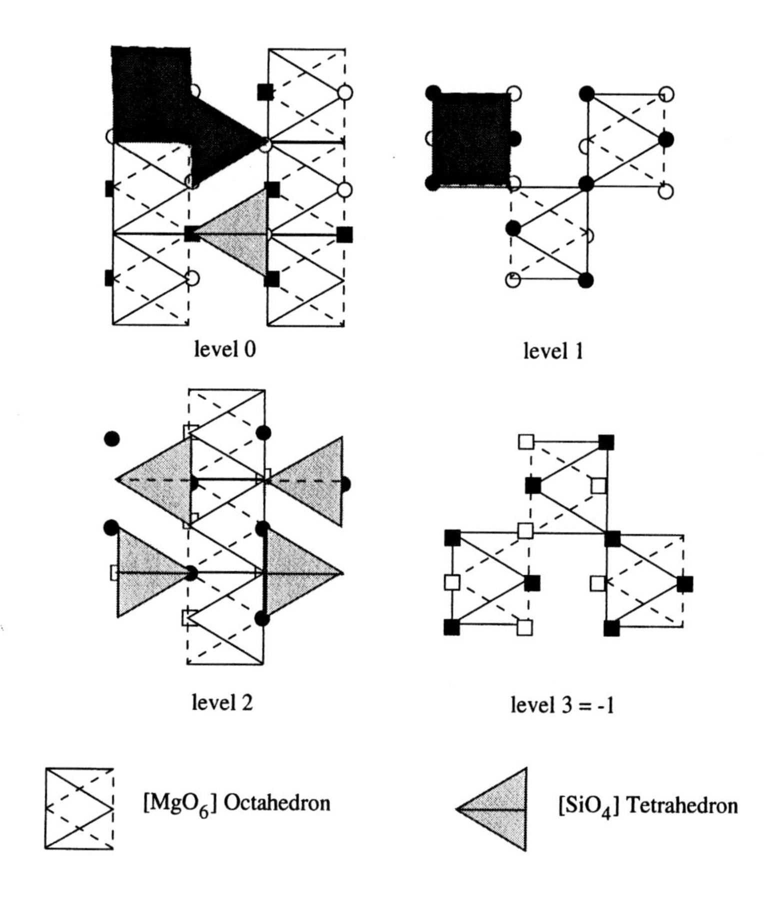

Fig. 14. Four successive layers in the olivine structure of Mg_2SiO_4, showing chains of edge-sharing $[MgO_6]$ octahedra and linking $[SiO_4]$ tetrahedra. The constituents of a $[Mg_2SiO_{11}]$ rigid super-polytope are shaded.

CHEMICAL EFFECTS

The topological freedom approach is purely geometrical, short-range (on the order of a single polytope and its immediate neighbors), and indifferent to the different chemical species in isostructural solids. Chemistry will impose an effect on the critical deposited energy (eV/atom) firstly because bond energies differ with chemical species; reporting irradiation doses for amorphization in displacements per atom (dpa) normalizes the effect of correspondingly different displacement energies, though the latter are not known with any precision in most non-metallic solids.[13] But, chemistry can secondarily effect more subtle changes as well on the longer-range topology which can affect amorphizability. Two examples are $AlPO_4$ (berlinite[62]) and Be_2SiO_4 (phenakite[63]), both of which comprise {4,2] networks of [MO_4] tetrahedra for which $f = 0$. Berlinite is isostructural with the quartz SiO_2 polymorph (Al and P sit on either side of Si in the Periodic Table) and is dominated by 8-rings of alternating [$AlPO_4$] and [PO_4] corner-sharing tetrahedra. It is 50% less amorphizable than quartz for ion irradiation, probably because of the ordering of Al and P ions. Phenakite ($2BeO{\cdot}SiO_2$) comprises 4- and 6-ring network arrangements of [SiO_4] and [BeO_4] tetrahedra and is also somewhat less amorphizable than quartz or other {4,2} silica polymorphs composed of 6- and 8-rings.

As an example of another chemical effect, significant differences in amorphizability have been noted[64] between isostructural fayalite (Fe_2SiO_4) and forsterite (Mg_2SiO_4) which have been attributed to the variable valence of the Fe ion and the possibility of its oxidation from Fe^{2+} to Fe^{3+}. The factor-of-two difference in amorphizabilities of isostructural $CaTiO_3$ and $KNbO_3$ pervoskites has also already been noted. A semi-empirical approach[21] has recently been attempted to fold in some chemistry with the topology to arrive at an empirical amorphizability parameter (incorporating local connectivity, ionic bond strength and polymorph stability) which can reflect these chemical subtleties. Nevertheless, the topological approach alone appears broadly correct across four orders of magnitude of susceptibility to amorphization, as discussed below.

The chemical effect of the implanted ion for ion irradiation-induced amorphization is worthy of passing comment. For an amorphization dose of 1 dpa, the concentration of implanted ions in the implantation zone reaches of order 1%, above the solubility limit for many implanted species. Even for irradiation with self ions, the alteration in stoichiometry can alter relevant topology, unless all constituent ions are implanted to identical ranges in stoichiometric proportions--and even then the stress resulting from accommodation of the added atoms can influence amorphizability. Two approaches to solutions for this problem have been pursued. In the first, specimens much thinner than the ion range are irradiated, so that no incident ions lodge in the sample; the implantation can then be coupled with simulaneous or *post situ* transmission electron microscopy to assess the amorphization threshold.[65] This approach is particularly useful for structures highly resistant to amorphization, like alumina, which requires several dpa (3.8 dpa, 380 eV/atom[66]). In the second, very high energy (~4 MeV) ions are used to measurably separate the peak damage and implantation zones; cross-sectional TEM is then used to examine the damage zone in the relative absence of implanted ion chemical effects.[67] Zinkle[67] has in fact claimed that Si_3N_4 is unamorphizable up to 700 eV/atom (7 dpa) at 300 K using the latter technique with 3.6 MeV Fe^+ ions, though this irradiation temperature could be near or above the critical amorphization temperature; Si_3N_4 has been shown to amorphize at 77 K below 1 dpa using 1 MeV Si^+ implantations.[45]

PREDICTING AMORPHIZABILITY

The rationalized values calculated for topological freedom in Tables 1-3 indicate a strong correlation with measured amorphizabilities. The correlation can be rendered into more quantitative form by plotting the logarithm of $|f - 3|$ against the logarithm of the amorphization dose D (Fig. 15). The result is surprisingly linear across the whole four orders of magnitude in critical amorphization dose and corresponds to the approximate functional relationship

$$f = 3 - 2.51\, D^{0.187}. \quad (3)$$

The topological justification for choice of this form for eqn (3) is that the irradiation successively destroys the constraints on each poltyope vertex (which without constraints would ultimately end up with three degrees of freedom in three-dimensional structures). From a comparison of eqns (1) or (2) with (3), it is evident that the term $2.51\, D^{0.187}$ in eqn (3) represents the destruction of the vertex constraints by irradiation sufficient to render a structure topologically floppy. The reason why a structure like those of silica, with $f = 0$ already, or $Pb_2P_2O_7$ with $f > 0$, should require additional input of bond-rupture energy is that introduction of topological disorder requires rebonding; mere floppiness (*e.g.* presence of soft rigid-unit modes[68]) is not sufficient for amorphization. The particular power law in eqn (3) will depend on the kinetic model[69] adopted for the amorphization process and its back reactions, which must in any case be similar for the range of structures represented.

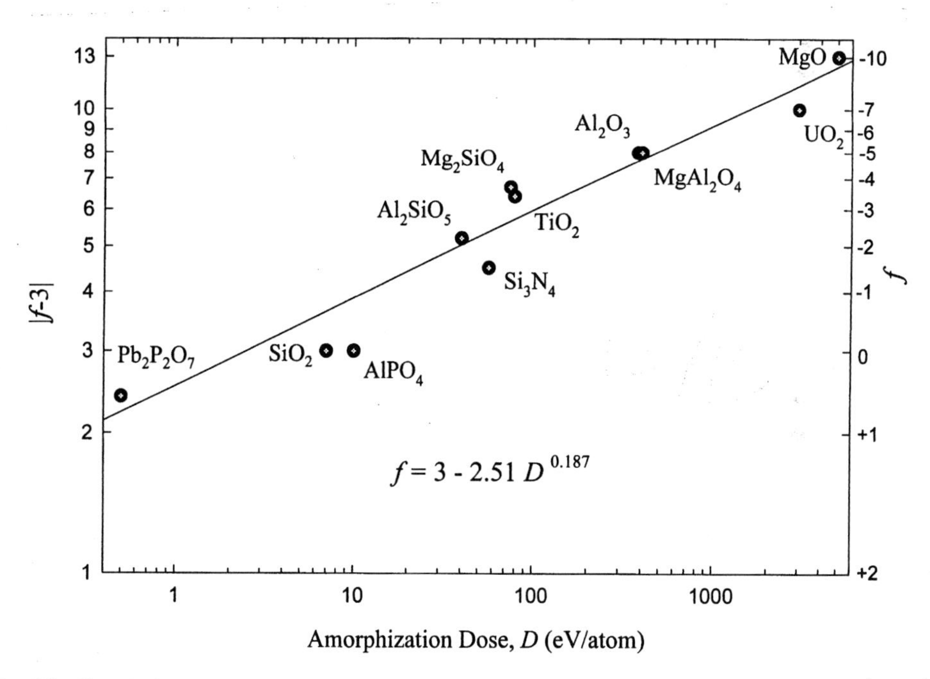

Fig. 15. Correlation of topological freedom f with amorphization dose D in log-log plot of $|f - 3|$ against D. The line is a non-linear least squares fit to the experimental amorphization doses given in eV/atom.

Fig. 15 can be used to predict the effective topological freedom in structures (*e.g.* perovskites) for which the contribution of the several structural elements is unclear and can increase our appreciation of what features confer structural stability. Eqn (3) predicts $f =$ -2.8 for $CaTiO_3$, which suggests that the effect of the highly-coordinated Ca^{2+} ion and its face-sharing [CaO_{12}] cuboctahedral polytopes on structural rigidity of a corner-sharing network of octahedra (for which f = -1) is certainly not negligible, but by no means overriding either.

ACKNOWLEDGMENTS

The authors thank Professor Prabhat Gupta of the Ohio State University for stimulating discussions. Both Professor Gupta and one of us (LWH) are indebted to our late colleague, Professor Alfred R. Cooper, for instilling in us his profound curiosity about the topology of structures. His contribution to the science of rigidity through his interest in glass formation was both seminal and remains underacknowledged. It is an honor to dedicate this paper to him. The authors are grateful to the Office of Basic Energy Sciences, U.S. Department of Energy, for support of this work under grant DE-FG02-89ER45396.

REFERENCES

1. C. J. McHargue, P. S. Sklad and C. W. White, *Nucl. Instrum. Meth.* **B46**:79 (1990).
2. D. E. Luzzi and M. Meshii, *Res Mechanica* **21**:207 (1987); A. T. Motta, *J. Nucl. Mater.* **244**:227 (1997).
3. L. W. Hobbs, F. W. Clinard, S. J. Zinkle and R. C. Ewing, *J. Nucl. Mater.* **216**:291 (1994).
4. P. K. Gupta and A. R. Cooper, *J. Non-Cryst. Solids* **123**:14 (1990).
5. J. Zarzyki, *Glasses and the Vitreous State*, Cambridge Univ. Press, Cambridge (1991), p. 8.
6. A. R. Cooper, *Phys. Chem. Glasses* **19**:60 (1978).
7. L. W. Hobbs and M. R. Pascucci, *J. Physique* (Paris) **41**:C6-273 (1980).
8. T. Diaz de la Rubia, M.-J. Caturla and M. Tobin, *Mat. Res. Soc. Symp. Proc.* **373**:555 (1995); R. Devanathan, W. J. Weber and T. Diaz de la Rubia, *Nucl. Instrum. Meth.* **B141**:118 (1998).
9. C. E. Jesurum, V. Pulim and L. W. Hobbs, *Nucl. Instrum. Meth.* **B114**:25 (1998).
10. C. E. Jesurum, V. Pulim, B. Berger and L. W. Hobbs, "Topological modeling of cascade amorphization in network structures using local rules," *Mater. Sci. Eng.* **A** (in press, 1998).
11. L. W. Hobbs, *Nucl. Instrum. Meth.* **B91**:30 (1994).
12. J. F. Ziegler, J. P. Biersack and U. Littmark, *The Stopping Power of Ions in Solids*, Pergamon, New York (1985).
13. S. J. Zinkle and C. K. Kinoshita, *J. Nucl. Mater.* **251**:200 (1997).
14. F. W. Clinard, Jr. and L. W. Hobbs, in: *Physics of Irradiation Effects in Crystals,* R. A. Johnson and A. N. Orlov, eds., Elsevier, Amsterdam (1986) p. 387.
15. L. W. Hobbs, *Scanning Microscopy,* **Supplement 4**:171 (1990).
16. M. R. Pascucci, J. L. Hutchison and L. W. Hobbs, *Radiat. Effects* **74**:219 (1983).
17. W. J. Weber and L. M. Wang, *Nucl. Instrum. Meth* **B106**:298 (1995); W. J. Weber, L. M. Wang and N. Yu, *Nucl. Instrum. Meth.* **B116**:322 (1996).
18. H. Inui, H. Mori and T. Sakata, *Philos. Mag.* **B65**:1 (1992); *ibid.* **B66**:737 (1992); A. Matsunaga, C. Kinoshita, K. Nakai and Y. Tomokiyo, J. *Nucl. Mater.* **179-181**:457 (1991).

19. H. M. Naguib and R. Kelly, *Radiat. Effects* **25**:1 (1975).
20. Hj. Matzke, *Radiat. Effects* **64**:3 (1982).
21. S. X. Wang, L. M. Wang, R. C. Ewing and R. H. Doremus, "Ion-beam induced amorphization in the MgO-Al_2O_3-SiO_2 system: II--Empirical model," *J. Non.-Cryst. Solids* (in press, 1998).
22. S. X. Wang, L. M. Wang, R. C. Ewing and R. H. Doremus, "Ion-beam induced amorphization in the MgO-Al_2O_3-SiO_2 system: I--Experimental and theoretical basis," *J. Non.-Cryst. Solids* (in press, 1998).
23. J. Clerk Maxwell, *Philos. Mag.* **27**:294 (1864).
24. D. J. Jacobs and M. F. Thorpe, *Phys. Rev. Lett.* **75**:4651 (1995); *Phys. Rev.* **E53**:3682 (1996).
25. M. F. Thorpe, B. D. Djordjevic and D. J. Jacobs, in: *Amorphous Insulators and Semiconductors,* M. F. Thorpe and M. I. Mitkova, eds., Kluwer, Dordrecht (1997), p. 289.
26. M. F. Thorpe, "Rigidity in networks and glasses," *this volume*.
27. J. C. Phillips, *J. Non-Cryst. Solids* **34**:153 (1979); *Physics Today* **35**:27 (1981).
28. L. W. Hobbs, *J. Non-Cryst. Solids* **192**:79 (1995).
29. A. R. Cooper, *J. Non-Cryst. Solids* **49**:1 (1982).
30. P. K. Gupta, *J. Amer. Ceram. Soc.* **76**:1088 (1993).
31. L. W. Hobbs, A. N. Sreeram, C. E. Jesurum and B. Berger, *Nucl. Instrum. Meth.* **B166**:18 (1996).
32. H. He and M. F. Thorpe, *Phys. Rev. Lett.* **54**:2107 (1984).
33. D. S. Franzblau and J. Tersoff, *Phys. Rev. Lett.* **68**:2172 (1992).
34. C. E. Jesurum, *Local Rules-Based Topological Modeling of Ceramic Structures*, Ph.D. thesis, Massachusetts Institute of Technology, Cambridge, MA (1998).
35. L. W. Hobbs, C. E. Jesurum and B. Berger, "The topology of silica networks," in: *Structure and Imperfections in Amorphous and Crystalline SiO_2,* J.-P. Duraud, R. A. B. Divine and E. Doorhyee, eds., Wiley, London (in press, 1998).
36. L. W. Hobbs, C. E. Jesurum, V. Pulim and B. Berger, *Philos. Mag.* **A78**:679 (1998).
37. C. E. Jesurum, V. Pulim and L. W. Hobbs, *J. Nucl. Mater.* **253**:87 (1998).
38. W. H. Zachariasen, *J. Amer. Chem. Soc.* **54**:3841 (1932).
39. C. Janot, *Quasicrystals: A Primer*, Clarendon Press, Oxford (1992).
40. R. Penrose, in: *Introduction to the Mathematics of Quasicrystals*, M. V. Jaric, ed., Academic Press, Boston (1989).
41. A. R. Cooper, *Phys. Chem. Glasses* **17**:38 (1976).
42. L. C. Qin, *Proc. 51st Ann. Mtg. Microscopy Society of America*, eds. G. W. Bailey and C. L. Rieder, San Francisco Press, San Francisco, CA (1993), p. 1102.
43. R. K. Eby, R. C. Ewing and R. C. Birtcher, *J. Mater. Res.* **7**:3080 (1992).
44. W. L. Gong, L. M. Wang, R. C. Ewing and J. Zhang, *Phys. Rev.* **54**: 3800 (1996).
45. W. Bolse, S. D. Peteves and F. W. Saris, *Appl. Phys.* **A58**:493 (1994).
46. W. Bolse, *Nucl. Instrum. Meth.* **B141**:133 (1998).
47. W. J. Weber, N. Yu and Wang, *J. Nucl, Mater.* **253**:53 (1998).
48. C. Degueldre, P. Heimgartner, G. Ledergerber, N. Sasajima, K Hojou, T. Muromura, L. Wang, W. Gong and R. Ewing, *Mater. Res. Soc. Symp. Proc.* **439**:625 (1997); N. Sasajima, T. Matsui, K. Hojou, S. Furuno, H. Otsu, K. Izui and T. Muromura, "Radiation damage in yttria-stabilized zirconia under He or Xe ion irradiation," *Nucl. Instrum. Meth.* **B** (in press, 1998).

49. K. E. Sickafus, Hj. Matzke, K. Yasua, P. Chodak III, R. A. Verrall, P. G. Lucuta, H. R. Andrews, A. Turos, R. Fromknecht and N. P, Baker, *Nucl. Instum. Meth.* **B141**:358 (1998).
50. E. C. Subbarao, in: *Science and Technology of Zirconia*, A. H. Heuer and L. W. Hobbs, eds., *Advances in Ceramics* **3** (American Ceramic Society, Westerville, OH, 1981) p. 1.
51. Hj. Matzke and L. M. Wang, *J. Nucl. Mater.* **231**:155 (1996).
52. L. M. Howe and M. Rainville, *J. Nucl. Mater.* **68**:215 (1977).
53. H. Abe, H. Naramoto and C. Kinoshita, *Mater. Res. Soc. Symp. Proc.* **373**:383 (1995).
54. J. K. N. Lindner, R. Zuschlag and E. H. te Kaat, *Nucl. Instrum. Meth.* **B62**:314 (1992).
55. S. J. Zinkle and L. L. Snead, *Nucl. Instrum. Meth.* **B116**:92 (1996); L. L. Snead and S. J. Zinkle, *Mater. Res. Soc. Symp. Proc.* **439**:595 (1997).
56. W. Bolse, J. Conrad, F. Harbsmeier, M. Borowski and T. Rödle, *Mater. Sci. Forum* **248/249**:319 (1997).
57. T. Hartmann, W. J. Weber, N. Yu, K. E. Sickafus, J. N. Mitchell, C. J. Wetteland, M. G. Hollander and M. Nastasi, *Nucl. Instrum. Meth.* **B141**:398 (1998).
58. A. Meldrum, L. A. Boatner and R. C. Ewing, Nucl. Instrum. Meth. B141:353 (1998).
59. A. N. Sreeram and L. W. Hobbs, *Mater. Res. Soc. Symp. Proc.* **279**:559 (1993); **321**:26 (1994).
60. N. Yu, K. E. Sickafus and M. Nastasi, *Philos. Mag. Lett.* **70**:235 (1994); K. E. Sicakfus, N. Yu and M. Nastasi, *Nucl. Instrum. Meth.* **B116**:85 (1996).
61. K. E. Sickafus, N. Yu and M. Nastasi, "Amorphization of $MgAl_2O_4$ spinel using 1.5 MeV Xe^+ ions under cryogenic irradiation conditions," *J. Nucl. Mater.* (in press, 1998).
62. A. N. Sreeram, L. W. Hobbs, N. Bordes and R. C. Ewing, *Nucl. Instrum. Meth.* **B116**:126 (1996).
63. R. C. Ewing, L. M. Wang and W. J. Weber, *Mater. Res. Soc. Symp. Proc.* **373**:346 (1995).
64. L. M. Wang, W. L. Gong, N. Bordes, R. C. Ewing and Y. Fei, *Mater. Res. Soc. Symp. Proc.* **373**:407 (1995).
65. L.-M. Wang, Nucl. Instrum. Meth. **B141**:312 (1998).
66. R. Devanathan, W. J. Weber, K. E. Sickafus, M. Nastasi, L. M. Wang and S. X. Wang, *Nucl. Instrum. Meth.* **B141**:366 (1998).
67. S. J. Zinkle, *Nucl. Instrum. Meth.* **B91**:234 (1994).
68. M. T. Dove, A. P. Giddy and V. Heine, *Trans. Amer. Cryst. Assn.* **27**:65 (1993); M. T. Dove, V. Heine and K. D. Hammonds, *Mineralogical Magazine* **59**:629 (1995); M. T. Dove, in: *Amorphous Insulators and Semiconductors,* M. F. Thorpe and M. I. Mitkova, eds., Kluwer, Dordrecht (1997), p. 349.
69. W. J. Weber, R. C. Ewing and A. Meldrum, *J. Nucl. Mater.* **250:** 147 (1997).
70. J. L. Brimhall, H. E. Kissinger and A. R. Pelton, *Radiat. Effects* **90**:241 (1985).

FLOPPY MODES IN CRYSTALLINE AND AMORPHOUS SILICATES

Martin T. Dove, Kenton D Hammonds and Kostya Trachenko

Department of Earth Sciences, University of Cambridge, Downing Street, Cambridge CB2 3EQ, UK

INTRODUCTION

The rigidity of network structures has traditionally been an issue for understanding the properties of glasses than crystals,[1–4] but recently the basic concepts have been re-visited in the study of crystalline silicates. The essential idea, as outlined in several other chapters, is that the rigidity of a network is associated with the balance between the numbers of constraints and degrees of freedom, an approach that dates back to Maxwell in 1864.[5] In a network structure, the number of degrees of freedom, N_f, is simply 3 × the number of atoms, and the number of constraints, N_c, may be equal to the number of bonds. For networks containing atoms with different coordination numbers, the number of bonds per atom can be tailored depending on the exact composition, and one of the early results was the demonstration of a phase transition involving the loss of rigidity as a function of the mean coordination number. For a network where the only forces are pair interactions involving bonded atoms, when $N_f > N_c$ there are $N_f - N_c$ zero frequency vibrational modes. These are the so-called *floppy modes*. This idea, recently slightly refined by Jacobs and Thorpe,[3,6] has underpinned a large body of work on the rigidity of network glasses.

When applied to silicate networks containing corner-linked SiO_4 (and AlO_4) tetrahedra, whether crystalline or amorphous, the general Maxwell approach gives the result that $N_f = N_c$ for an infinite system (or a finite system with periodic boundary conditions). This result is obtained in Appendix 1. The point from this result is that the silica system is balanced between the two limits of either being rigid (overconstrained) and floppy (underconstrained). We have found, however, that the general way of counting the constraints in the Maxwell analysis is usually an oversimplification, in that for crystalline silicates we also need to take symmetry into account—this point is developed in Appendix 2. This has the effect of making some of the constraints redundant, thereby decreasing N_c relative to N_f and allowing some of the vibrations to propagate without any distortions of the tetrahedra. We call these vibrations the *Rigid Unit modes* (RUMs), and in most respects they are equivalent to the floppy modes defined for amorphous systems (we consider RUMs to represent a particular type of floppy mode, in that the flexibility of the structure involves movements of rigid units rather than individual atoms). If the only forces in the crystal are those which preserve the rigidity of the tetrahedra, and pairs of linked tetrahedra are able to swing about their common linkage with no energy cost, the RUMs will have zero frequency. Of course, this situation is not realised in practice, since there are interactions between pairs of tetrahedra in addition to other long-range forces, but the stiffness of SiO_4 and AlO_4 tetrahedra is much larger than all other forces, so the RUMs will have low frequencies in relation to the frequencies of other vibrations. The concept of rigid unit modes has been very powerful initially in helping us to understand the origin and characteristics of displacive phase transitions in silicates, and more recently the ideas have been used to explain phenomena such as negative thermal expansion, and to interpret localised structural flexibility of zeolites.

Our purpose in this chapter is to introduce the basic ideas of the RUM model as applied to crystalline silicates, and to then show how the ideas can be applied to silica glass. We first describe the way the idea has been applied to crystalline materials, and then describe a number of applications of the RUM approach (displacive phase transitions, disordered crystalline phases, and negative thermal expansion). Zeolites are then treated as a special topic, with emphasis on how RUMs give a mechanism to form localised distortions of their framework structures. Finally we outline the application of the RUM idea to silica glass. Some of the technical points are given in Appendices 1–3. In Appendix 4 we list some Internet resources, and highlight key references for further reading in Appendix 5.

RIGID UNIT MODES IN FRAMEWORK CRYSTALS

Methodology

In the study of floppy modes in glasses, the approach is to count the total number of zero frequency modes for the system taken as a whole. Crystalline phases are different from glasses in that the translational periodicity imparts a new significance to the wave vector, and so our approach for determining the flexibility of crystals should usefully take account of dependence on wave vector. The theory of lattice dynamics[7] provides the formalism for this. For a crystal containing n atoms in each unit cell, there are $3n$ vibrational modes (travelling waves) for each wave vector, and the frequency of each mode will vary continuously with wave vector. The RUMs will form a subset of these vibrational modes, and as noted above they will have lower frequencies than most other vibrations. In a crystalline silicate, the vibrational frequencies typically range in value up to around 30 THz (those with frequencies from 20–30 THz involve stretching of Si–O and Al–O bonds), whereas a small number of experimental studies indicate that RUMs have frequencies in the range 0–1 THz.[8–10]

In studies of floppy modes in glasses it is conventional to determine the flexibility of a structure in terms the total number of zero-frequency solutions to the dynamical equations that impart the bond constraints. For crystalline silicates it is possible to count the number of zero-frequency modes for particular wave vectors, and the number of RUM wave vectors can be compared between different structures. Our approach has been to cast the problem into the formalism of molecular lattice dynamics,[7] in which the polyhedra are treated as rigid molecular units. In order to develop dynamical equations we have developed the *split-atom method*, which is described in Appendix 3.[11] At first sight this appears to be a rather artificial model, but it turns out to give a good general way of taking the amount of tetrahedral distortion into account. The split-atom method gives us a method to determine directly the number of RUMs for a given wave vector, since the RUMs are the modes with zero frequency within the lattice dynamics implementation of the split-atom method. The split-atom method has been developed into a computer program called CRUSH, which is

Table 1. Rigid unit modes in example aluminosilicates for wave vectors of special symmetry, excluding the trivial acoustic modes at $k = 0$ (taken from reference 13). The "—" indicates that the wave vector is not of special symmetry in the particular structure. The numbers in brackets denote the numbers of RUMs that remain in any lower-symmetry low-temperature phases. The groupings represent points, lines and planes in reciprocal space.

k	Quartz $P6_222$	Cristobalite $Fd3m$	Tridymite $P6_3/mmc$	Sanidine $C2/m$	Leucite $Ia3d$	Cordierite $Cccm$
$0,0,0$	1 (0)	3 (1)	6	0	5 (0)	6
$0,0,\frac{1}{2}$	3 (1)	—	6	1	—	6
$\frac{1}{2},0,0$	2 (1)	—	3	—	—	6
$\frac{1}{3},\frac{1}{3},0$	1 (1)	—	1	—	—	6
$\frac{1}{3},\frac{1}{3},\frac{1}{2}$	1 (1)	—	2	—	—	0
$\frac{1}{2},0,\frac{1}{2}$	1 (1)	—	2	—	4 (0)	2
$0,1,0$	—	2	—	1	—	—
$\frac{1}{2},\frac{1}{2},\frac{1}{2}$	—	3 (0)	—	0	0	—
$0,1,\frac{1}{2}$	—	—	—	1	—	—
$0,0,\xi$	3 (0)	2 (0)	6	—	0	6
$0,\xi,0$	2 (0)	2 (2)	3	1	0	6
$\xi,\xi,0$	1 (1)	1 (0)	1	—	4 (0)	6
ξ,ξ,ξ	—	3 (0)	—	—	0	—
$\frac{1}{2},0,\xi$	1 (0)	—	2	—	0	2
$\xi,\xi,\frac{1}{2}$	1 (1)	—	0	—	0	0
$\frac{1}{2}-\xi,2\xi,0$	1 (1)	—	1	—	—	6
$\frac{1}{2}-\xi,2\xi,\frac{1}{2}$	1 (1)	—	0	—	—	0
$0,\xi,\frac{1}{2}$	0 (0)	—	1	1	—	—
$\xi,1,\xi$	—	1 (0)	—	—	0	—
$\xi,\zeta,0$	1 (0)	0	1	—	0	6
$\xi,0,\zeta$	0 (0)	0	2	1	0	0
$\xi,1,\zeta$	—	0	—	1	0	—
ξ,ξ,ζ	0 (0)	1 (0)	0	—	0	0

freely available together with a set of ancillary programs (see Appendix 4 for details of access to the programs).[11,12]

In many structures the number of modes that are technically RUMs (i.e. with zero frequency) is actually a vanishingly small fraction of the total number of modes (often given by the ratio of the volume of an infinitesimally-thin plane to that of the whole Brillouin zone). In many cases there are low-frequency vibrational modes that are not RUMs by the strict technical definition since they do involve small distortions of the tetrahedra, but these distortions are so small that the mode frequencies are small. In effect, these modes can be considered to be quasi-RUMs (QRUMs), and their importance should not be understated. For example, in a material with a plane of RUMs, the modes of vibration on the same phonon branch as the RUMs and with wave vectors close to the RUM plane will have low frequencies because of the continuity of the vibrational frequencies with wave vector. In this sense the strict number of RUMs only indicates part of the flexibility of the structure, and an alternative approach, which takes account of the contribution of the QRUMs, is to calculate the vibrational density of states, $g(\omega)$, using the split-atom method. Since the frequencies of all vibrations in this approach gives a direct measure of the extent of polyhedral distortion, the calculation of $g(\omega)$ with the split-atom method will give a quantitative measure of the distribution of vibrations across the range of polyhedral distortion. In particular, $g(\omega)$ will give the proportion of modes which involve minimal polyhedral distortions. We will see below that this approach can give valuable insights.

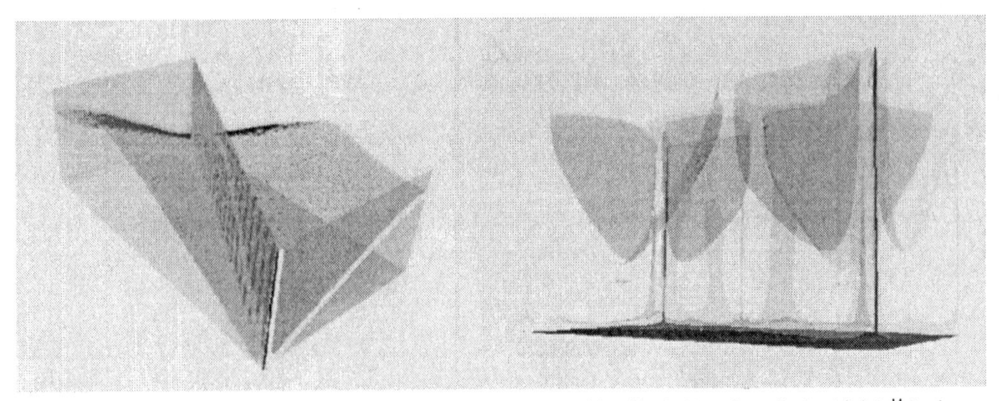

Figure 1. Curved surfaces of wave vectors for RUMs in HP-tridymite (left) and cordierite (right).[14] In the HP-tridymite there are flat planes of RUMs which are indicated in Table 1, but there is also a curved surface whose intersection with one of the planes is clearly indicated. This curved surface has been measured by electron diffraction. In cordierite there are flat planes of RUMs, but there are surfaces in the shape of cups which actually enclose regions where there are one RUM per wave vector.

The existence of RUMs in crystalline silicates

We have calculate the number of RUMs in a wide range of aluminosilicate minerals with framework crystal structures built from corner-linked SiO_4 and AlO_4 tetrahedra,[13] and some of the results are summarised in Table 1. We find that there are always RUMs; in some cases there are RUMs with wave vectors on lines of symmetry in reciprocal space (e.g. leucite, $KAlSi_2O_6$), and in other cases (e.g. cordierite, $MgAl_4Si_5O_{18}$, and the silica phases) there are RUMs on planes of wave vectors.

One of the results of our study concerns the role of symmetry. Many aluminosilicates undergo displacive phase transitions, in which the symmetry of the crystal structure is lowered by small RUM rotations of the tetrahedra (see below). There is always a smaller number of RUMs in the lower symmetry phase. For example, in the high-symmetry cubic β-phase of cristobalite (SiO_2) there are 6 planes of RUMs in reciprocal space, but in the low-symmetry tetragonal α-phase there are only 2 lines of RUMs and some other RUMs of more general wave vector.[8,11,13] In Table 1 we indicate the way that the distributions of RUMs in some of the materials change as a result of displacive phase transitions. Since the symmetry is lowered by the action of one specific RUM distortion being frozen into the structure, we have the implication that certain (but not all) pairs of RUMs may not be able to distort the structure with finite amplitudes together without the tetrahedra having to distort.[13]

One might expect that since the existence of RUMs depends on the specific details of the crystal symmetry, the RUMs may be restricted to wave vectors of special symmetry. Indeed, the data given in Table 1 are for such wave vectors only. However this turns out not to be the case, and it is surprisingly common to find RUMs with wave vectors lying on curved surfaces in reciprocal space.[14] Two examples are shown in Figure 1. The curved surface of HP-tridymite (the high-temperature phase of tridymite, SiO_2) was first noted by measurements of curved surfaces of diffuse scattering in electron diffraction,[15] which were subsequently interpreted as arising from the curved surfaces of RUMs.[16] We have searched for curved surfaces of RUMs in all the examples given in Table 1 and have found that there are surfaces in almost all the examples and their low-symmetry phases.[14] Colour versions of these surface representations are available from one of the Web sites listed in Appendix 4.

One surprising results is that in some silicates there are RUMs for every wave vector. The first example is the highest-symmetry phase of sodalite,[13] which has one RUM for

Figure 2. Crystal structures of high-symmetry sodalite, space group *Im*3*m* (left) and zeolite-LTA, space group *Pm*3*m* (right), showing the outline of the unit cell in both cases.

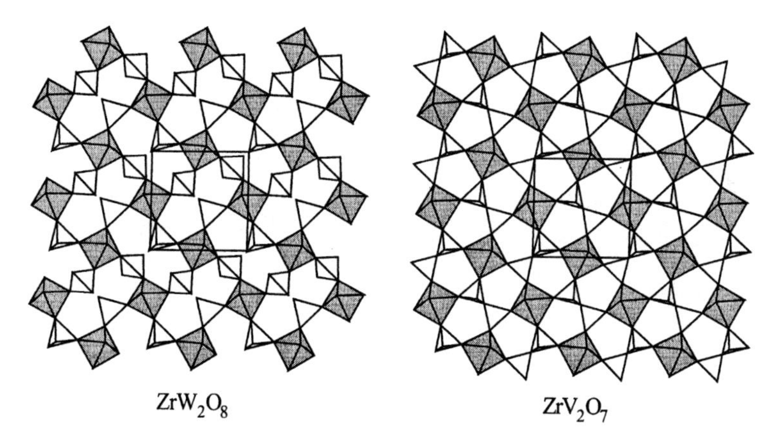

Figure 3. Crystal structures of ZrW_2O_8 and ZrV_2O_7, showing ZrO_6 octahedra and WO_4 and VO_4 tetrahedra as shaded objects. Note that the two tetrahedra in the centres of the unit cells are linked to each other in ZrV_2O_7, but are separated in ZrW_2O_8.

every wave vector. Many zeolites have crystals structures built from the basic sodalite structure, and some of these have several RUMs for each wave vector.[17–19] One example is Zeolite-LTA, which has 4 RUMs for each wave vector. The existence of one RUM for each wave vector gives the possibility to form localised RUM distortions, and this is described below.[17–19]

The existence of RUMs in other crystalline materials

The classic example of a RUM is the octahedral rotational distortion in the perovskite structure, Figure 3. In the cubic phase there is a single RUM for the lines of wave vectors along the edges of the cubic Brillouin zone, i.e. for wave vectors of the form $\{^1/_2, ^1/_2, \xi\}$. The general Maxwell counting would suggest that a structure containing linked octahedra would be overconstrained, but the problem with the general approach is outlined in Appendix 1.

Another example of interest is the cubic phase of ZrW_2O_8.[20,21] The crystal structure contains both ZrO_6 octahedra and WO_4 tetrahedra within a corner-linked framework (Figure 3). One of the vertices of each the WO_4 tetrahedra is not linked to other polyhedra, and as a result the general Maxwell counting gives the exact balance $N_c = N_f$. Rather surprisingly

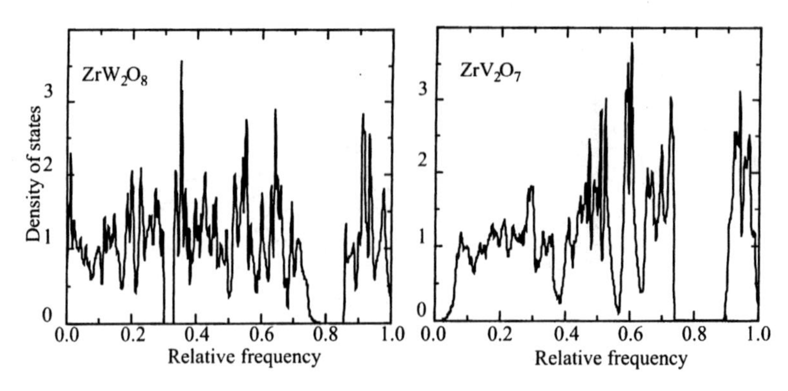

Figure 4. Vibrational density of states for ZrW_2O_8 and ZrV_2O_7 calculated using the split-atom method.[20]

there are no RUMs on any special wave vector, but there is a complex curved surface of RUMs.[20,21] The importance of this result lay in an attempt to explain the negative thermal expansion in ZrW_2O_8.[20]

ZrV_2O_7 has a similar structure to ZrW_2O_8, Figure 3. The difference between the two materials is in the number of oxygen atoms, and this is accommodated in the structure by the linking together of the non-bridging vertices of two close tetrahedra. This additional linkage acts to give a cross-bracing to the structure, as reflected in the new condition that the general Maxwell counting gives $N_c/N_f = 7/6$. The difference in the flexibility of the two structures is highlighted in the calculated density of states from the split atom method, as shown in Figure 4.[20] The less-flexible ZrV_2O_7 has the normal Debye density of states, $g(\omega) \propto \omega^2$ as $\omega \rightarrow 0$, whereas for ZrW_2O_8 the RUMs give the result that $g(\omega) \rightarrow$ constant as $\omega \rightarrow 0$. This difference in the limiting behaviour of $g(\omega)$ within the split-atom method can be used to define the difference between systems with and without RUMs.[22]

Although we have found that most systems containing only corner-linked tetrahedra have RUMs,[13] we find that many systems containing corner-linked octahedra have no RUM flexibility, even if they also contain tetrahedra linked within the framework structure.[22] The one clear exception is the perovskite structure, but in this case it is the simplicity of the structure that allows some of the general constraints to be redundant, as discussed in Appendix 2. Another exception is As_2O_5, which contains corner-linked AsO_4 tetrahedra and AsO_6 octahedra. This system has a few RUMs, including a line of acoustic mode RUMs which give rise to a ferroelastic phase transition.[23]

APPLICATIONS

Displacive phase transitions

The first application of the RUM model was to help understand the origin of displacive phase transitions in silicates.[11,13,24,25] The example of the displacive phase transition in quartz is shown in Figure 5.

In quartz the phase transition occurs as a result of a soft optic mode at $\mathbf{k} = 0$. This mode is a RUM, and it is because of this that it can have the low frequency necessary to drive a phase transition.[26,27] In fact this soft mode lies along a line of RUMs in the [1,0,0] direction in reciprocal space. This line of RUMs softens more-or-less uniformly, and the

Figure 5. The structures of the low-temperature low-symmetry α-quartz (left) and high-temperature high-symmetry β-quartz (right), showing the SiO_4 tetrahedra as shaded objects. The difference between the two structures is the presence of the 6-fold symmetry in β-quartz, which is lost through the phase transition as a result of the RUM rotations of the tetrahedra.

Figure 6. The structures of the low-temperature low-symmetry α-cristobalite (left) and high-temperature high-symmetry β-cristobalite (right), showing the SiO_4 tetrahedra as shaded objects. The difference between the two structures arises from large RUM rotations of the tetrahedra.

interaction of this line of RUMs with one of the transverse acoustic modes drives an incommensurate phase transition slightly ahead of the α–β phase transition.[26–28]

Another example is cristobalite, which is shown in Figure 5. The high-temperature β-phase has a high-symmetry cubic structure which is based on the diamond structure (the Si atoms have the normal positions of the C atoms, and the O atoms lie midway between all nearest-neighbour Si atoms). This structure has linear Si–O–Si bonds, which are energetically unfavourable. This gives rise to considerable short-range disorder (see below). There are RUMs in β-cristobalite for all wave vectors in the $\langle 1\bar{1}0 \rangle$ zones, and it is the double-degenerate RUM at $\mathbf{k} = (1,1,0)$ that acts as a soft-mode for the phase transition to the α-phase.[8,13]

A third phase of silica, tridymite, has the hexagonal analogue of the β-cristobalite structure, which again has a disordered high-temperature phase (HP) with linear Si–O–Si bonds in the average structure. Tridymite has a relatively complicated sequence of phase transitions, and the structures of some of the phases are shown in Figure 7.

The transition sequences are detailed in Figure 8. It is found that all the low-temperature phases can be derived from higher-symmetry phase by a RUM distortion.[29] However, the interesting point is that there are several separate chains of phase transitions rather than one continuous sequence in changing temperature. The sequence HP → OC flows quite easily on cooling (although some details, including the exact phase transition temperatures, remain uncertain). The transition to the OS phase involves an

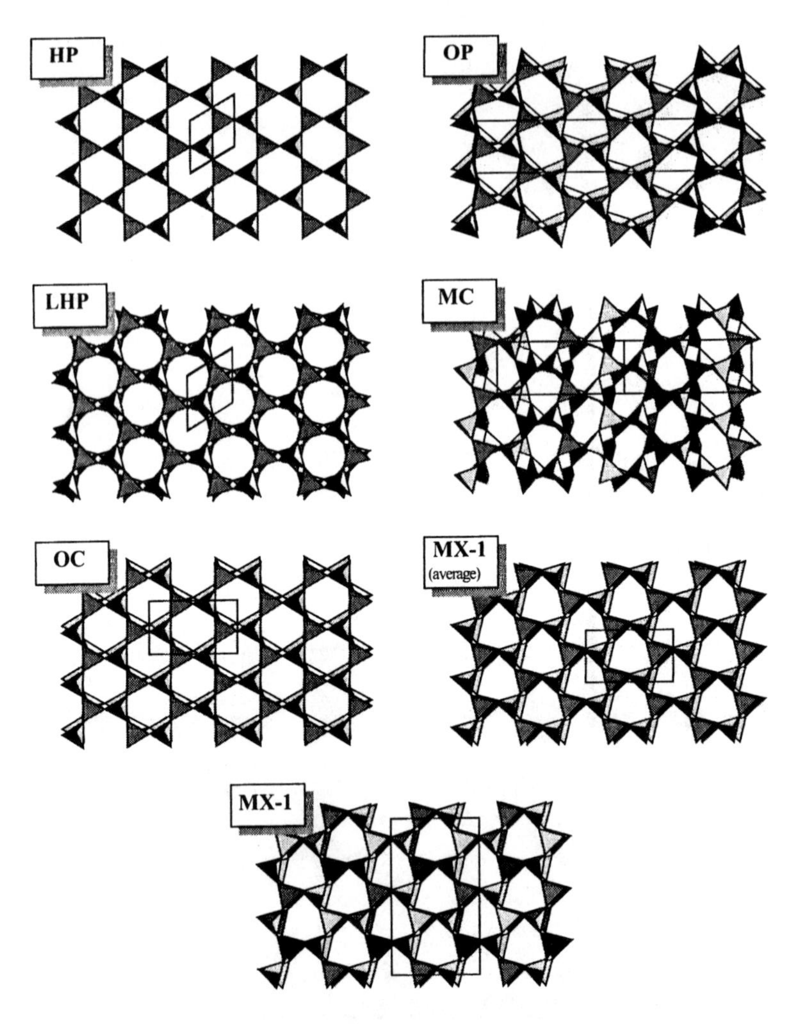

Figure 7. Crystal structures of the various phases of tridymite, showing the SiO_4 tetrahedra as shaded units, and giving the outlines of the conventional unit cells in each case. HP and LHP phases have hexagonal lattices; OC and OP phases have orthorhombic structures; MC and MX-1 have monoclinic structures.[29]

incommensurate distortion, which is then followed by a lock-in phase transition to the orthorhombic OP phase.[30] At lower temperatures still there is a first-order phase transition to the monoclinic MC phase. The RUM analysis and a general symmetry analysis suggest that this phase arises from a distortion of the parent HP phase rather than following from the sequence that leads to the OP phase. The monoclinic MX-1 phases, which include an incommensurate phase, have been observed in samples with stacking faults (e.g. from grinding), and these are also derived as RUM modulations of the parent HP phase.

The discussion of the displacive phase transitions in these polymorphs of silica show that the RUM model can rationalise the distortions that accompany the phase transitions, and can complement general symmetry analysis in determining transition sequences. The fact that the phase transitions arise from RUM distortions means that the distortions have little energy cost, a fact that is particularly apparent in tridymite. Two other results follow from the RUM model, which we briefly mention here.[10,24,31] The first is that the transition temperatures have been shown to arise directly from the stiffness of the polyhedra. Whereas one tends to think of the polyhedra in RUM systems as being perfectly rigid, in fact there is a finite (albeit large) energy associated with the distortions of the polyhedra,

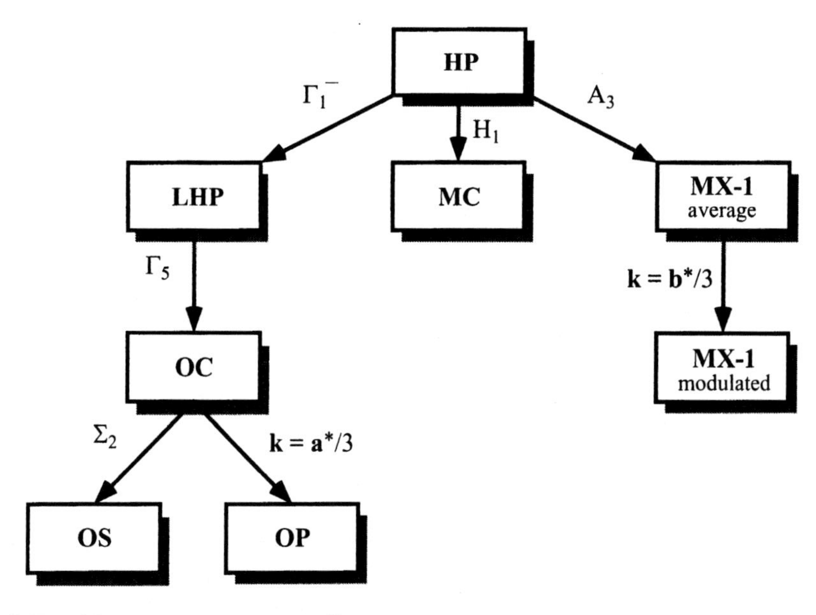

Figure 8. Transition sequences in tridymite.[29] The symbols give the symmetry of the RUM in the parent phase.

and this energy determines the overall range of vibrational frequencies. It turns out that this energy directly determines the values of the phase transition temperatures. The other factor in a simple equation for the transition temperature[10,24,31] is the maximum distortion that is possible, which in turn is controlled by the local steric effects as neighbouring polyhedra re pushed towards each other as the framework structure is buckled by the phase transition. The second result from the RUM model concerns the length scales over which the fluctuations associated with the displacive phase transitions in these systems are correlated. In practice there is considerable anisotropy: in some directions the fluctuations are long-range correlations, and in others the correlations are short-range. Without going into details here, we note that for any particular system the RUM spectrum determines this anisotropy, which in turn determines whether critical fluctuations may be expected to be important.[25,31]

High-temperature disordered phases

We noted above that the high-temperature phases of cristobalite and tridymite appear to have linear Si–O–Si bonds. This configuration has a high energy, and it is probably that on a short length scale the structure distorts to allow local bond angles to be lowered to more reasonable values (140–150°). The linear bonds then arise as a result of averaging over the short-range disorder, and evidence that this is so comes from the fact that the average Si–O bond lengths in these phases are anomalously short.[33–35]

The RUM model provides a mechanism for short-range disorder to exist.[8,13,25,25,34] Since there are planes of RUMs in reciprocal space, it will be possible for the superposition of all RUM distortions to create localised dynamic disorder. Since the RUMs are lie on planes of wave vectors rather than having wave vectors everywhere in the Brillouin zone, there will be longer-range correlations in directions normal to these planes, but on the other hand there are several orientations of the planes of RUMs so the effects of these correlations will be minimised. An example of the possibility to create dynamic disorder is given in Figure 9, which shows a configuration of SiO_4 tetrahedra in HP-tridymite generated by RMC modelling based on neutron total scattering measurements.[34,35] Each of six tetrahedra should form a perfect hexagon (compare with the average structure of HP-

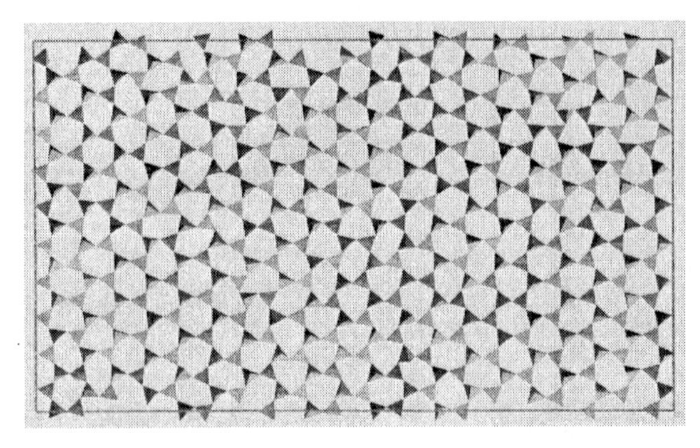

Figure 9. Configuration of HP-tridymite obtained by RMC modelling with constraints that preserve the shape and size of the SiO_4 tetrahedra.[34,35]

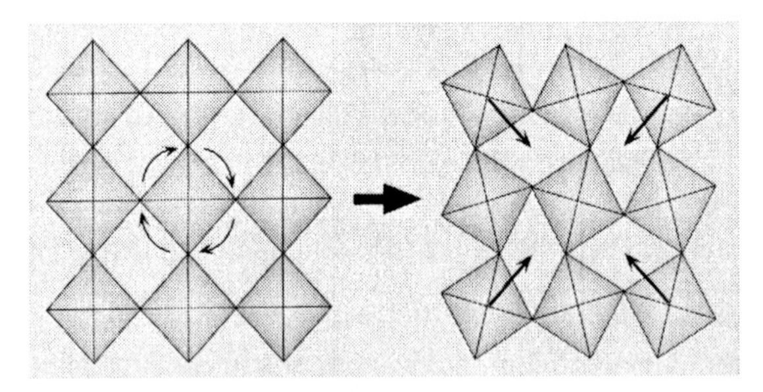

Figure 10. Rotation of the squares in a cubic two-dimensional perovskite structure pulls the structure in on itself.

tridymite shown in Figure 7), but these hexagons are distorted as a result of RUM rotations and displacements of tetrahedra. This plane actually corresponds to a plane of RUMs, so any correlations between the distortions of individual rings is allowed to be of short range. The same picture emerges for β-cristobalite by both RMC modelling[34,35] and molecular dynamics simulations.[36–38] There is now a large body of theoretical and experimental evidence to show that the RUMs create the dynamic disorder required to remove the linear Si–O–Si atomic alignments, and the evidence is against earlier interpretations based on the formation of domains. This point is discussed in much more detail elsewhere.

We remind the reader of a point made earlier, namely that when one RUM distortion is imposed on a structure, some (or most) of the other vibrational modes that were RUMs can no longer propagate without the polyhedra distorting. Thus one can ask whether a complete set of RUMs vibrating with finite amplitude can exist together, or whether their mutual interaction raises their frequencies to the point whereby the RUM viewpoint is no longer appropriate.[39] This issue has been addressed in detail for β-cristobalite through molecular dynamics simulations,[37,38] and it is clear that the RUM picture remains a good model even when each RUM has a fairly large amplitude.

Thermal expansion

A number of materials with crystal structures having corner linked polyhedra have negative thermal expansion, or else anomalously low thermal expansion. β-quartz and cordierite are silicates with negative thermal expansion along one direction, β-cristobalite

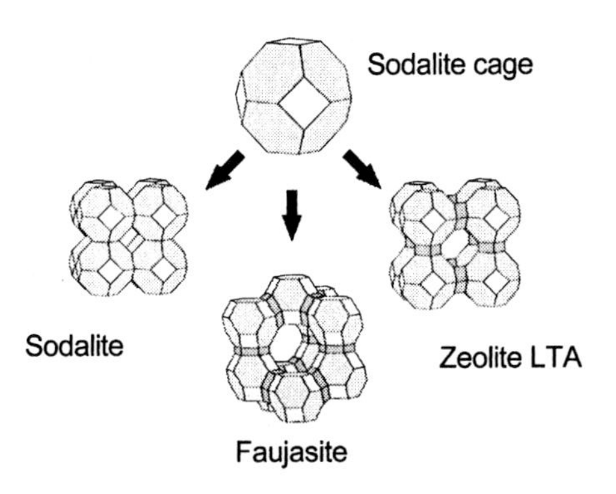

Figure 11. Structures of two zeolites built from the basic sodalite cage. The cages are connected by additional rings of tetrahedra. The corners of the faces represent the centres of the tetrahedra.

has nearly zero thermal expansion at high temperatures, and ZrW_2O_8 is one example of a ceramic with negative thermal expansion. The RUM model provides an intuitive explanation of the origin of negative thermal expansion. Consider the simple example shown in Figure 10. As one rigid unit rotates, it causes its neighbours to rotate, but also drags its neighbours inwards, giving an overall reduction in the volume of the crystal.

This volume reduction is well-known to follow displacive phase transitions where the RUM generates a static distortion, but the RUM vibrations will generate a similar volume reduction. Consider a RUM vibration with a certain amplitude that generates a mean-square rotation of the polyhedra, $\langle\theta^2\rangle$. This will lead to a change in volume, $\Delta V \propto -\langle\theta^2\rangle$. Since $\langle\theta^2\rangle$ arises from thermal vibrations, we have

$$\begin{aligned} \langle\theta^2\rangle &\propto \frac{k_B T}{\omega^2} \\ \Rightarrow \Delta V \propto -\langle\theta^2\rangle &\propto -\frac{k_B T}{\omega^2} \end{aligned} \tag{1}$$

There will be many modes contributing to $\langle\theta^2\rangle$, each with its own frequency and constant of proportionality linking $\langle\theta^2\rangle$ to ΔV. This theoretical outline has been worked out in more detail elsewhere.[40,41] In practice we find that there are many more modes that give a negative ΔV than only the RUMs, because even modes that are not RUMs may have a significant component of $\langle\theta^2\rangle$, but the low value of ω^2 that RUMs give to the denominators in the above equations means that the RUMs or quasi-RUMs give the largest contributions to ΔV. The ideas have been applied to β-quartz,[41] ZrW_2O_8,[20,21] β-eucryptite[42] and cordierite[42] through detailed calculations based on empirical model interatomic potentials.

ZEOLITES AND LOCALISED RUM DISTORTIONS

Zeolites are open structures built from corner linked SiO_4 and AlO_4 tetrahedra. The basic building blocks are cages of tetrahedra, which are then linked together by additional rings of tetrahedra. The way that the cages are linked together then creates channels between the cages. One series of examples is shown in Figure 11. It should be noted that all tetrahedra are fully linked to four other tetrahedra, as in the denser silicates, so that the general Maxwell counting will give $N_c = N_f$.

We mentioned above the remarkable fact (at least, remarkable in the light of the general Maxwell counting for framework silicates) that some zeolites have one or more RUMs for each wave vector.[17–19] Calculations on a range of zeolites have shown that even when there

Table 2: Percentage of all normal modes in some selected zeolites that are either RUMs or QRUMs

Zeolite	% RUMs	% QRUMs
LTA	2.8	6
RHO (*Im*3*m*)	1.4	7
RHO (*I*43*m*)	0.7	5
Sodalite	2.8	6
Faujasite	1.4	5
UTD-1	0	8
Paulingite	0.6	6
Chabasite	0	6
Natrolite	0	5
ZSM-5	0	4

is not a RUM for each wave vector, there may be bands of vibrations throughout reciprocal space with very low, nearly zero, frequencies; we classify these modes as quasi-RUMs, QRUMs. Examples showing how the RUMs and QRUMs form a significant fraction of the total number of vibrations are given in Table 2.

The interesting point about having one or more RUMs or QRUMs for each wave vector is that it is possible to add up waves of all wave vectors to create a distortion that is localised in real space. For simplicity we consider the case where there is one RUM for each wave vector **k**, giving a deformation pattern $\mathbf{R}_\mathbf{k}$ (which defines the displacements and rotations of each tetrahedron in the unit cell due to the RUM) with amplitude $A_\mathbf{k}$. When all RUM deformations are added together, each with phase factor $\gamma_\mathbf{k}$, a deformation in real space, **L**, is obtained:

$$\mathbf{L} = \sum_\mathbf{k} A_\mathbf{k} \mathbf{R}_\mathbf{k} \exp(i\gamma_\mathbf{k}) \tag{2}$$

There is complete freedom in the choice of $A_\mathbf{k}$ and $\gamma_\mathbf{k}$. However, this does not mean that there is a similar freedom to create any localised deformation pattern—the localised deformation **L** is still restricted by the types of deformation given by the set of $\mathbf{R}_\mathbf{k}$. It should also be stressed that the localised deformations we have in mind are quite different from a deformation that is repeated in each unit cell—this type of deformation would be equivalent to a combination of $\mathbf{k} = 0$ RUM deformations, but here we are considering combining RUMs with all wave vectors.

We have developed a FORTRAN program called LOCALRUM to allow the calculation of the local RUMs for a given zeolite structure using the eigenvectors $\mathbf{R}_\mathbf{k}$ calculated by CRUSH.[19] The approach taken within LOCALRUM is to create a supercell of the zeolite structure and to work with the set of wave vectors consistent with the size of this supercell (for example, if a supercell is chosen to be 4 times larger than the conventional unit cell in each direction, the set of wave vectors will contain multiples of $\mathbf{a}^*/4$ etc.). The most useful way to proceed is to select a specific part of the zeolite framework on which to attempt to localise a deformation of the structure. This could be a single tetrahedron, a ring of tetrahedra, or a part of an internal surface. The program will then adjust the sets of values of $A_\mathbf{k}$ and $\gamma_\mathbf{k}$ in order to obtain the greatest possible degree of localisation. The criterion used for this is to maximise a pseudo-intensity I,[19] defined as

$$I = \sum_{ij} \left| L_{ij} \right|^2 \tag{3}$$

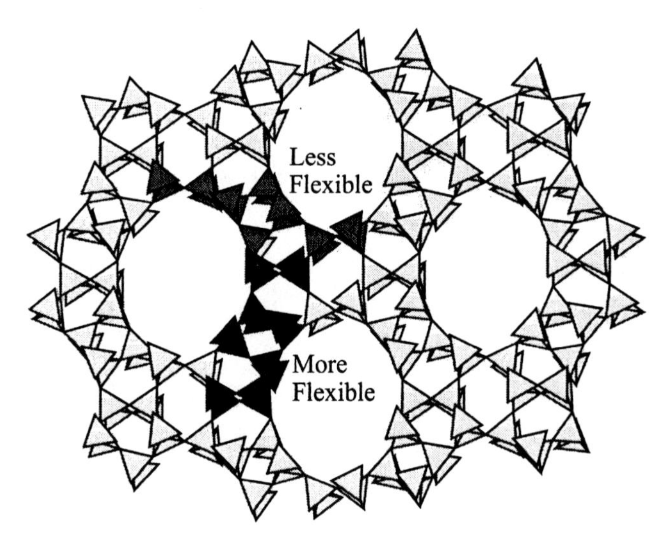

Figure 12. The UTD-1 zeolite, showing a part of the structure that is reasonably flexible to allow localised RUM deformations, and another part that is rather less flexible.

where i represents the tetrahedron in the region selected as the site for localised deformation, and j represents the component of the motion. If **L** has been normalised appropriately, the sum over all tetrahedra in the supercell will be equal to unity. If the sum is limited to a small region, closer the value of the pseudo-intensity is to unity the greater the degree of localisation. It is possible to form starting estimates of the sets of values of A_k and γ_k based on a trial deformation, but the LOCALRUM program can work equally as well with initial random values for the amplitudes and phases of the different RUMs.

As noted above, if there is more than one RUM per wave vector, all RUMs can be incorporated into the LOCALRUM procedure. The procedure can also use low-frequency QRUMs, and the number of bands of QRUMs to be used can be chosen in advance.

The degree of localisation can be quantified using the participation number[19]

$$\mathcal{N} = \frac{1}{\sum_i \left(\sum_j \left| L_{ij}^2 \right| \right)^2} \tag{4}$$

If **L** is normalised such that $|\mathbf{L}|^2 = 1$, $\mathcal{N}$ gives the number of tetrahedra involved in a localised deformation.

An example of the LOCALRUM approach is a set of calculations on the zeolite UTD-1,[18,19] which contains 14-membered rings of tetrahedra. In particular, this example illustrates the way in which the constraints of the RUM eigenvectors impose limitations on the degree of overall flexibility. A section of the structure is shown in Figure 12. Calculations were performed using different parts of the structure as trial centres for localised deformations; the two parts shown in Figure 12 contain 16 tetrahedra each. Figure 12 summarises the main result that some parts of the structure are rather more flexible than others. A deformation localised on the more flexible part of the structure involves 120 tetrahedra, whereas a deformation localised on the less flexible part involves 220 tetrahedra. Although these two regions appear to be very similar, the eigenvectors of the RUMs and QRUMs used in the calculation clearly differentiate between the flexibility of the two regions, and the localised deformations are tightly bound by the actual eigenvectors of the RUMs and QRUMs.

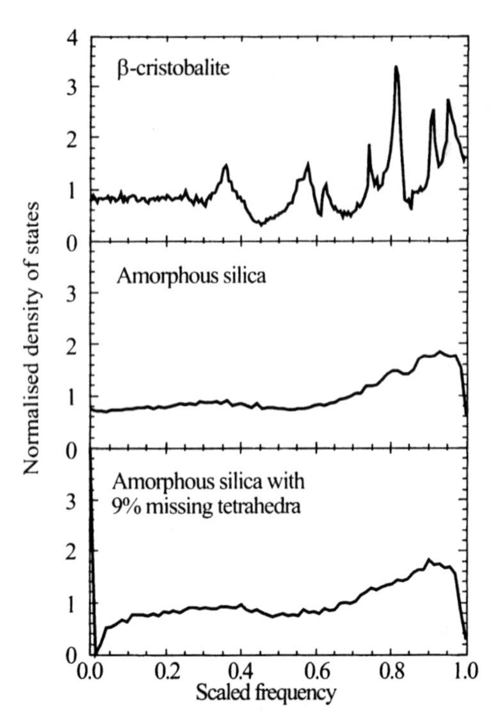

Figure 13. Vibrational density of states within the split-atom method for our model of silica glass compared with that for β-cristobalite, and for silica glass with a number of missing tetrahedra.

SILICA GLASS

Low-energy vibrations in silica glass

Let us recall some earlier comments about RUMs in crystalline silicates. For frameworks of linked tetrahedra the general Maxwell counting gives $N_c = N_f$, so there needs to be something special about the framework structure to allow some of the constraints to become redundant. This has a relationship to symmetry, because it is found that for two structures with the same topology but different symmetry (perhaps related through a displacive phase transition), the structure with the higher symmetry has greater RUM flexibility. Silica glass has the same result from the general Maxwell counting as the framework crystalline silicates, but it does not have any of the symmetry. Therefore it might be imagined that there would be no RUMs in silica glass.

This prediction turns out to be surprisingly wide of the mark. In order to investigate the flexibility of amorphous silica we constructed a configuration of silica starting from a set of configurations for amorphous silica produced using the Wooten–Weaire algorithm[43] and kindly provided by one of the editors of this volume (Prof. M F Thorpe). We placed an oxygen atom midway between each Si–Si bond. This gave linear Si–O–Si bonds, so we allowed the structure to relax to give more realistic bond angles using molecular dynamics simulations with the silica interatomic potential of Tsuneyuki *et al.*[44] The final configuration gave an overall radial distribution function $T(r)$ that was in good agreement with neutron diffraction, and individual pair distribution functions $g(r)$ in agreement with those obtained from RMC modelling. Each configuration had periodic boundary conditions, and every tetrahedron was linked to 4 others. Using this method we obtained configurations with different numbers of SiO_4 tetrahedra between 216–4096, although the results presented here are relatively insensitive to the size of the configuration.

The flexibility of the amorphous silica network is illustrated by a calculation of the density of states, $g(\omega)$, using the split-atom method, Figure 13.[45] Recall the point made earlier that the form of $g(\omega)$ in the limit $\omega \rightarrow 0$ gives a good indication of the RUM flexibility of a structure. In Figure 13 it can be seen that by this criterion amorphous silica has a RUM flexibility that is comparable to that of β-cristobalite. In fact $g(\omega)$ for amorphous silica resembles a lower-resolution version of that for β-cristobalite.

The vibrations of amorphous silica were characterised by calculations of the participation ratio, $\mathcal{P}$. If $\mathbf{u}_j$ represents the displacement of atom j in any given vibration, the participation ratio is defined as

$$\mathcal{P} = \frac{\left(\sum |\mathbf{u}_j|^2\right)^2}{N\sum |\mathbf{u}_j|^4} \tag{5}$$

For a vibration in which all atoms participate equally, $\mathcal{P} \sim 1$, whereas for a vibration involving only a single atom $\mathcal{P} \sim 1/N$. The formal similarities between $N\mathcal{P}$ and the real-space version $\mathcal{N}$ defined in equation (4) is easy to appreciate. For the $g(\omega)$ of silica glass shown in Figure 13 $\mathcal{P} \sim 0.7$–0.8 for all ω, including the low-frequency modes. This means that the low-frequency RUM-like vibrations involve all tetrahedra in the glass, and is not localised to particularly flexible segments of the glass structure.

For silicates with non-bridging Si–O bonds, i.e. when there are tetrahedra that are linked to less than four other tetrahedra, the general Maxwell counting gives $N_f > N_c$. This implies that there should be a finite number of RUMs when some of the tetrahedra are removed from the configurations of amorphous silica, and in Figure 13 it can be seen that this is indeed the case. A similar picture emerged for a range of concentrations of missing tetrahedra. The presence of defects gives a peak in $g(\omega)$ at $\omega = 0$. What is interesting is that the modes within this peak come from the low-frequency distribution of modes in the fully linked network, and as the number of missing tetrahedra increases the number of zero frequency modes increases and the gap in $g(\omega)$ at low frequencies opens up. The new RUMs have participation ratios that are uniformly distributed between 0 and 0.8, implying that some of these are highly localised, whereas others involve the whole configuration.

Possible tunnelling modes

There has been considerable interest in the possibility of the existence of large scale atomic rearrangements in silica glass that may act as two-level tunnelling states at low temperatures. We have searched for the existence of these states by investigating the molecular dynamics simulations of our configurations.[45] Using specific search algorithms of the atomic trajectories we detected events such as those shown in Figure 14, where we plot coordinates of some of the oxygen atoms as functions of time for simulations runs performed using different initial velocities. It is clear from these plots that some atoms undergo sudden changes in their positions. The displacements shown in these plots are of about 1.4 Å, which is a long way for an oxygen atom to move (it may correspond to a rotation of about 50°).

The changes in the local configuration highlighted by one of the events in Figure 14 can be seen in Figure 15, which shows snapshots of the associated group of tetrahedra before and after the jump event superimposed on each other. This highlights the size of the rotations of the tetrahedra. Calculations of the participation ratios using the displacements of the atoms $\mathbf{u}_j$ through the jumps suggest that a typical jump event involves around 30

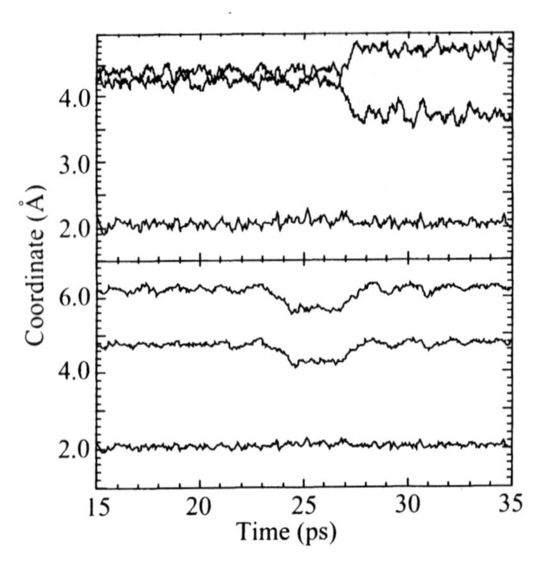

Figure 14. Coordinates of two oxygen atoms in different simulation runs as functions of time, showing sudden jumps at around 25 ps.[45]

Figure 15. A group of tetrahedra with positions of the atoms before and after a large-amplitude jump event superimposed over each other.[45]

tetrahedra. The simulations of a couple of jump events have been animated, and can be viewed from the web page given in Appendix 4.

The flexibility against large-scale changes in the structure was investigated further by comparing different structures equilibrated from the initial configurations with linear Si–O–Si bonds using different initial atomic velocities.[45] The interesting result was that parts of the structure always went into the same relaxed structure, but other parts of the structure were able to relax to give different orientations of the tetrahedra. In fact these more-flexible parts of the structure were those that took part in the jump events highlighted above. This suggests that a network of amorphous silica has regions that are more or less flexible.

CONCLUSIONS

In this chapter we have outlined a number of feature of the RUM flexibility of crystalline networks of linked tetrahedra and octahedra. As we have stressed, the initial value of the RUM approach was to give an understanding of the origin and characteristics of displacive phase transitions. However, since the RUM approach also gives use the means to understand the whole area of low-frequency dynamics of network structures containing rigid units, and the insights from this have been useful to understand dynamically disordered crystalline phases, zeolites, and now silica glass.

APPENDIX 1. MAXWELL COUNTING IN FRAMEWORK SILICATES

Our approach is to count the degrees of freedom and constraints treating the SiO_4 tetrahedra as rigid bodies. For three dimensional motion each tetrahedron will have $N_f = 6$ degrees of freedom, three translations and three rotations. The constraints in the system are those associated with holding the vertex of one tetrahedron at the same place in three-dimensional space as the common vertex of the linked tetrahedron, giving three constraints for each shared vertex. In counting the total number of constraints in the system we assign half of these constraints to each of the two linked tetrahedra. Since there are four vertices on each tetrahedron, the number of constraints per tetrahedron is $N_c = 4 \times 3/2 = 6$. This argument makes no supposition about the symmetry of the network, and is therefore applicable both to crystalline and amorphous forms.

It is worth noting three generalisations. First, for a system of linked octahedra, as in the perovskite structure, the number of vertices of each octahedron is 6 rather than 4, so that $N_c = 6 \times 3/2 = 9$. In this sense we predict that the perovskite structure is overconstrained, but when we take symmetry into account we find that the perovskite structure can have some degree of flexibility. Second, for two polyhedra with a shared edge, the number of constraints from the edge is 5 rather than 6, 3 to place one vertex from each polyhedra at a common place, and 2 to make the two edges parallel (the sixth constraint is redundant since we already have the condition that the two edges are of equal length). Similarly, for two polyhedra with a shared face there are 6 constraints, 3 which put the faces at the same place and 3 to make them parallel. Third, in two dimensions, $N_f = 3$ per rigid body, and there are 2 constraints associated with each shared vertex.

In the literature on floppy modes in glasses, the common way to perform the Maxwell counting is from the perspective of the structure containing atoms and bonds. Each atom has 3 degrees of freedom, and each bond gives 1 constraint. For an SiO_4 tetrahedron, we need to ensure that we do not count redundant constraints. If we take one constraint from each Si–O bond, the tetrahedron requires only 5 of the O–O bonds to give independent constraints to be completely defined. Therefore for each SiO_2 formula unit we have $N_f = N_c = 9$, which, as expected, gives the same balance as our counting scheme based on rigid units.

In both approaches, we have assumed that there are no non-bridging bonds. If the structure does not implicitly have non-bridging bonds, this assumption is equivalent to the condition that we have no free surfaces.

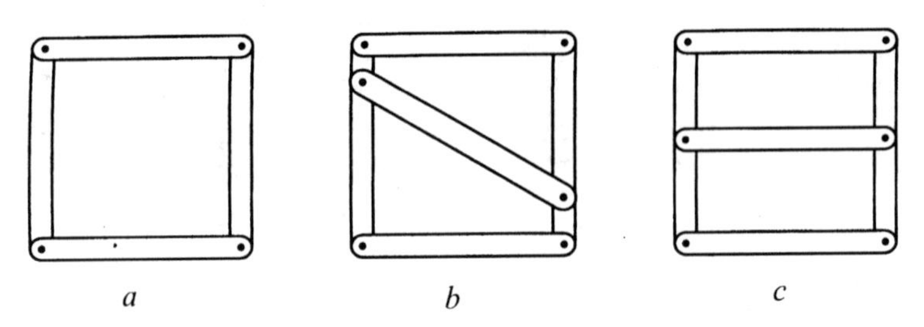

Figure 16. Simple two-dimensional objects build from rods linked at corners.

APPENDIX 2. AN EXAMPLE OF THE ROLE OF SYMMETRY

The role of symmetry in reducing the number of independent constraints is illustrated in the following simple example of linked rods.[10,11,24] We compare three two-dimensional objects shown in Figure 16. Each bar has three degrees of freedom, two translations and one rotation, and each linkage gives two constraints. For case *a* in Figure 16 we have $N_f = 4 \times 3 = 12$, and $N_c = 4 \times 2 = 8$, giving $N_f - N_c = 4$. These 4 zero-frequency modes are the two uniform translations of the whole object, one uniform rotation, and the shear mode which represents the floppy mode for the system. By inspection it is clear that case *b* in Figure 16 has been made rigid by the addition of a cross-bracing fifth rod. The counting scheme gives have $N_f = 5 \times 3 = 15$, and $N_c = 6 \times 2 = 12$, so that $N_f - N_c = 3$. Thus the only zero-frequency modes are the three rigid-body motions. The interesting example is case *c* in Figure 16. This has the same topology as case *b*, and the same simple counting would yield the result that this structure is rigid, but we know that if the fifth bar is parallel to the bottom bars the structure will have the shear floppy mode. So we need to count the constraints more carefully. We start with the four outside rods as counted for case *a*. Addition of the fifth rod simply increases the number of degrees of freedom by 5, as in case b, giving $N_f = 15$. To count the additional constraints given by the fifth rod, we first add the two constraints that tie the end of the rod to one of the outer rods. However, we do not need to count both the constraints at the other end. Instead it is sufficient to note that if the fifth rod is has the same length as the top and bottom rods, we only need the single constraint that it is parallel to the top rod. Therefore symmetry has lowered the number of independent constraints by one, giving $N_c = 12 + 1 = 13$, and $N_f - N_c = 4$. Thus we have recovered the shear mode by ensuring that we have not counted redundant constraints.

A similar picture emerges when we think about a two-dimensional array of squares, Figure 17. This represents the cubic perovskite structure, for which there is a well-known floppy mode in which neighbouring squares rotate by equal amount but with opposite signs: this mode provides the common octahedral-tilting displacive phase transitions in many materials with the perovskite structure.

A similar picture emerges when we think about a two-dimensional array of squares, Figure 17. This represents the cubic perovskite structure, for which there is a well-known floppy mode in which neighbouring squares rotate by equal amount but with opposite signs: this mode provides the common octahedral-tilting displacive phase transitions in many materials with the perovskite structure.

The simple way of counting constraints is to assign two constraints to each corner. This gives $N_c = 4$ for each square, compared to $N_f = 3$. This suggests that the structure is overconstrained, and does not admit the possibility of having any zero-frequency modes. Having free surfaces does not help either. For a general case with n_y rows of n_x squares,

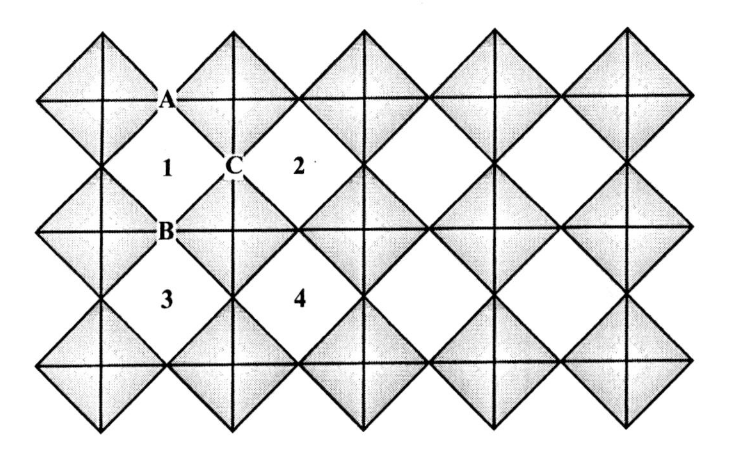

Figure 17. Two-dimensional crystal of squares representing the cubic perovskite structure. The numbers are used in the text.

giving $n_x \times n_y$ squares, we have $N_f = 3 \times n_x \times n_y$. The total number of constraints can be counted by going along each row counting the constraints between squares along the row and between the row below, giving

$$N_c = 2n_x(n_y - 1) + 2n_y(n_x - 1) = 4n_xn_y - 2(n_x + n_y) \tag{6}$$

This is larger that N_f, whereas we require $N_f - N_c = 4$ to allow for the floppy rotational mode.

To count the constraints taking account of redundancies, we follow the approach we used for the rods. We take the first group of 4 squares in Figure 17, labelled **1**. The two squares on the left of this group share 2 constraints providing the common linkage. When we add the remaining two squares, we first add two constraints for each of linkages marked **A** and **B**. To get these two tetrahedra to join at linkage **C** we have one constraint on the orientations of each the two tetrahedra. Adding all these constraints together gives 7 for this group, 1 less than the number of degrees of freedom. When we extend the row by adding the group marked **2**, we simply duplicate the constraints involved in forming linkages **A** – **C**, so that each additional group adds the same number of constraints as degrees of freedom. Thus the first row of groups of 4 squares has $2 + 6(n_x - 1)$ constraints. Now we consider the next row below. To construct group **3** we again add 6 degrees of freedom and 6 constraints. To complete group **4** we only need to add one square, for which we need 2 constraints to tie one corner in place, and one constraint to give it the correct orientation to link up its other corner. In order to complete this new row, we have to add squares one at a time to complete groups of 4 squares, so that each new square brings in the same number of constraints as degrees of freedom. And so the analysis repeats row by row, giving the actual number of independent constraints. Each new row adds $3n_x$ constraints. Adding the constraints from n_y rows gives

$$N_c = 2 + 6(n_x - 1) + 3n_x(n_y - 2) = 3n_xn_y - 4 \tag{7}$$

This is the correct number to give the three rigid-body motions and the rotational floppy mode.

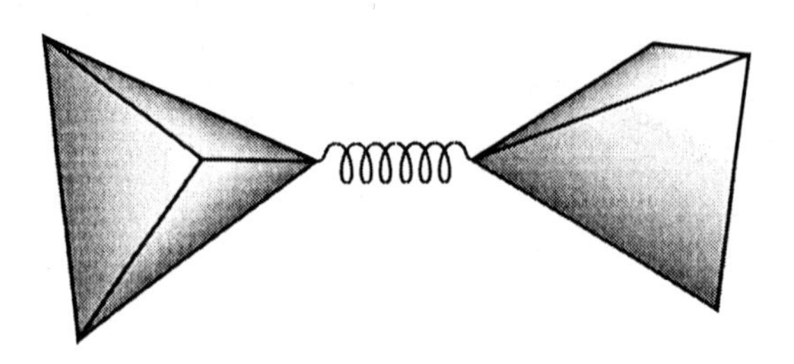

Figure 18. The split-atom representation of a pair of linked tetrahedra. An oxygen atom that is shared by two tetrahedra is represented as a pair of atoms with an equilibrium separation of zero. The spring represents the force required to separate the split atoms, and is analogous to the forces acting when two linked tetrahedra move in a manner that requires the distortions of the tetrahedra.

APPENDIX 3. THE SPLIT-ATOM METHOD

We introduced the split-atom method in order to be able to solve the dynamical equations for a general example of a crystal containing linked rigid polyhedra.[11] There are two types of constraints that need to be treated explicitly, namely the polyhedra have to be rigid, and the positions of two linked vertices need to be kept at the same position. Our approach has been to treat the constraints on the rigidity of the polyhedra as *strict* constraints, and to treat the constraints on the positions of the vertices as *slack* constraints. This involves inventing a potential energy term that is harmonic in any separation of two vertices that are to be constrained to have the same place: this is illustrated in Figure 18. We can imagine that by allowing two vertices to separate we have split the atom into two halves. Formally the separating of the atoms is equivalent to distorting the two tetrahedra as they move with respect to each other, if one considers the mid-point of the two tetrahedra to be the position of the shared vertex of the two distorted tetrahedra, and the stiffness of the spring in the split-atom model is equivalent to the stiffness of the tetrahedra. Indeed, we tune the value of the spring to reproduce the range of vibrational frequencies in a typical silicate.

The split-atom method gives us a set of dynamical equations for the vibrations of a periodic array of tetrahedra that are equivalent to the formalism of molecular lattice dynamics, and we have developed a program for their solution. This involves calculating the dynamical matrix for a given wave vector, and diagonalising to give the set of vibrational frequencies for all modes of the given wave vector and the associated mode eigenvectors (which correspond to the displacements and rotations of the polyhedra associated with each vibration). The zero-frequency solutions to the dynamical equations correspond to the actual RUMs for the given wave vector.

APPENDIX 4. INTERNET RESOURCES

We have established a WWW site in association with our RUMs project, http://www.esc.cam.ac.uk/rums. This contains links to a number of pages, including the following:

http://www.esc.cam.ac.uk/mineral_sciences/crush, which describes the CRUSH suite of programs[11,12] (including the LOCALRUM code for zeolite studies[19]) and allows the programs to be downloaded;

http://www.esc.cam.ac.uk/~martin/movies.html, which shows animations of simulations of disordered crystalline and amorphous phases undergoing RUM motions,[24] and of animations of possible tunnelling states[45] (using Quicktime® formats);

http://www.esc.cam.ac.uk/~martin/surfaces.html, which shows many examples of the colour representations of the three-dimensional surfaces of RUM distributions in the Brillouin zone described in this chapter.[14]

The main rums page also gives abstracts to all pertinent references, some of which are cited in this chapter.

APPENDIX 5. FURTHER READING

In this chapter there are a number of ideas which can only be touched on briefly. Further details on specific aspects of the topics in this chapter can be obtained in the following references:

11 The split-atom method and its implementation
13 Displacive phase transitions in crystalline aluminosilicates (also 24,25)
34 RUMs and high-temperature disordered phases
40 Negative thermal expansion
19 RUMs in zeolites
45 RUMs in silica glass

ACKNOWLEDGEMENTS

We are pleased to be able to acknowledge the considerable contribution of our present and former colleagues in Cambridge: Volker Heine, Andrew Giddy, Ian Swainson, Patrick Welche, Alix Pryde, and Manoj Gambhir. We are also pleased to acknowledge the collaboration with Mark Harris and David Keen (ISIS) for parallel experimental work. Parts of this work have been supported by the EPSRC and NERC (UK), and the Cambridge Overseas Trust. Some of the calculations have been performed using the Hitachi computers of the Cambridge High Performance Computing Facility.

REFERENCES

1. H. He and M.F. Thorpe. *Phys. Rev. Lett.* 54:2107 (1985)
2. Y. Cai and M.F. Thorpe. *Phys. Rev.* B40:10535 (1989)
3. M.F. Thorpe. in: *Amorphous Insulators And Semiconductors*, Thorpe, M.F. and Mitkova, M.I ed. Kluwer, Amsterdam pp 289–328 (1997)
4. J.C. Phillips. *J. Non-Cryst. Sol.* 34:153 (1979)
5. J.C. Maxwell. *Phil. Mag.* 27:294 (1864)
6. D.J. Jacobs and M.F. Thorpe. *Phys. Rev.* 53:3682 (1996)
7. M.T. Dove. *Introduction to Lattice Dynamics*. Cambridge University Press, Cambridge (1993)
8. I.P. Swainson and M.T. Dove. *Phys. Rev. Lett.* 71:193 (1993)
9. H. Boysen. in: *Ferroelastic and Co-elastic Crystals*; Salje, E.K.H., Cambridge University Press, Cambridge pp 334–349 (1990)
10. M.T. Dove, V. Heine, and K.D. Hammonds. *Min. Mag.* 59:629 (1995)
11. A.P. Giddy, M.T. Dove, G.S. Pawley, and V. Heine. *Acta Cryst.* A49:697 (1993)
12. K.D. Hammonds, M.T. Dove, A.P. Giddy and V Heine. *Am. Min.* 79:1207 (1994)
13. K.D. Hammonds, M.T. Dove, A.P. Giddy, V. Heine, and B. Winkler. *Am. Min.* 81:1057 (1996)
14. A.K.A. Pryde, M.T. Dove, V. Heine and K.D. Hammonds. *Min. Mag.* (in press)
15. R.L. Withers, J.G. Thompson, Y. Xiao, and R.J. Kirkpatrick. *Phys. Chem. Min.* 21:421 (1995)
16. M.T. Dove, K.D. Hammonds, V. Heine, R.L. Withers, Y. Xiao, and R.J. Kirkpatrick. *Phys. Chem. Min.* 23:55 (1996)

17. K.D. Hammonds, H. Deng, V. Heine, and M.T. Dove. *Phys. Rev. Lett.* 78:3701 (1997)
18. K.D. Hammonds, V. Heine, and M.T. Dove. *Phase Trans.* 61:155 (1997)
19. K.D. Hammonds, V. Heine, and M.T. Dove. *J. Phys. Chem.* B 102:1759 (1998)
20. A.K.A. Pryde, K.D. Hammonds, M.T. Dove, V. Heine, J.D. Gale, and M.C. Warren. *J. Phys.: Cond. Matt.* 8:10973 (1996)
21. A.K.A. Pryde, K.D. Hammonds, M.T. Dove, V. Heine, J.D. Gale, and M.C. Warren. *Phase Trans.* 61:141 (1997)
22. K.D. Hammonds, A. Bosenick, M.T. Dove, and V. Heine. *Am. Min.* 83:476 (1998)
23. J.M. Perez-Mato (personal communication)
24. M.T. Dove. in: *Amorphous Insulators And Semiconductors*, Thorpe, M.F. and Mitkova, M.I ed. Kluwer, Amsterdam pp 349–383 (1997)
25. M.T. Dove. *Am. Min.* 82:213 (1997)
26. B. Berge, J.P. Bachheimer, G. Dolino, M. Vallade, and C.M.E. Zeyen. *Ferroelectrics.* 66:73 (1986)
27. M. Vallade, B. Berge, and G. Dolino. *Journal de Physique, I.* 2:1481 (1992)
28. F.S. Tautz, V. Heine, M.T. Dove, and X. Chen. *Phys. Chem. Min.* 18:326 (1991)
29. A.K.A. Pryde and M.T. Dove. *Phys. Chem. Min.* (submitted)
30. A.K.A. Pryde and V. Heine. *Phys. Chem. Min.* (in press)
31. M.T. Dove, M. Gambhir, and V. Heine. *Phys. Chem. Min.* (in press)
32. P. Sollich, V. Heine, and M.T. Dove. *J. Phys.: Cond. Matt.* 6:3171 (1994)
33. M.T. Dove, D.A. Keen, A.C. Hannon, and I.P. Swainson. *Phys. Chem. Min.* 24:311 (1997)
34. M.T. Dove, V. Heine, K.D. Hammonds, M. Gambhir, and A.K.A. Pryde. in: *Local structure from diffraction*; Thorpe, M.F. and Billinge, S. ed. Plenum pp 253–272 (1998)
35. D.A. Keen. in: *Local structure from diffraction*; Thorpe, M.F. and Billinge, S. ed. Plenum pp 101–120 (1998)
36. I.P. Swainson and M.T. Dove. *J. Phys.: Cond. Matt.* 7:1771 (1995)
37. M. Gambhir, V. Heine, and M.T. Dove. *Phase Trans.* 61:125 (1997)
38. M. Gambhir, V. Heine, and M.T. Dove. *Phys. Chem. Min.* (in press)
39. F. Liu, S.H. Garofalini, R.D. Kingsmith, and D. Vanderbilt. *Phys. Rev. Lett.* 71:3611 (1993)
40. V. Heine, P.R.L. Welche, and M.T. Dove. *J. Am. Cer. Soc.* (in press)
41. P.R.L. Welche, V. Heine, and M.T. Dove. *Phys. Chem. Min.* (in press)
42. P.R.L. Welche (personal communication)
43. F. Wooten and D. Weaire. *Solid State Physics* 40:1 (1987)
44. S. Tsuneyuki, M. Tsukada, H. Aoki, and Y. Matsui. *Phys. Rev. Lett.* 64:776 (1990)
45. K. Trachenko, M.T. Dove, K.D. Hammonds, M.J. Harris, and V. Heine. *Phys. Rev. Lett.* (submitted)

GENERIC RIGIDITY OF NETWORK GLASSES

M.F. Thorpe, D.J. Jacobs, N.V. Chubynsky and A.J. Rader

Department of Physics and Astronomy,
and Center for Fundamental Materials Research,
Michigan State University, East Lansing, Michigan 48824

1. INTRODUCTION

In this article, we discuss continuous random network models for glasses and their mechanical properties. We have collected together most of the ideas that have been put forward over the past fifteen years or so, on constraint counting and floppy modes in random networks. We show how model systems can help us better understand glasses *via* rigidity percolation. These ideas involving rigidity percolation have been tested experimentally in bulk glasses. Also included in this article are some key results obtained in the last three years, which go beyond simple mean field theory or *Maxwell constraint counting*. These results, which involve a new graph theoretical approach called the *Pebble game*, show that Maxwell counting is quite reliable in locating the transition from rigid to floppy. Exact constraint counting within the pebble game allows us to study the nature of the rigidity transition in very great detail. There are indications that the rigid transition may be first order if the network is deficient in small rings, while normally it is second order in glasses.

The study of network structures has fascinated scientists in many areas - ranging from engineering and mechanics to the material and biological sciences. Going back more than a century, Maxwell was intrigued with the conditions under which mechanical structures made out of struts, joined together at their ends, would be stable (or unstable) [1]. To determine the stability, without doing any detailed calculations (that would have been impossible then except for the simplest structures), Maxwell used the method of *constraint counting*. This counting is an approximate method that proves to be accurate for structures where the density (of struts or joints) is roughly uniform. Maxwell's constraint counting method is exact for some geometries that we will discuss in this paper. The idea of a constraint in a mechanical system goes back to Lagrange [2] who used the concept of holonomic constraints to reduce the effective dimensionality of the space. The difficult part is to determine which constraints are *linearly independent*. If the linearly independent constraints can be identified, then the problem is solved – however in most large systems

Rigidity Theory and Applications
Edited by Thorpe and Duxbury, Kluwer Academic/Plenum Publishers, 1999

this identification is not possible except using a numerical procedure on an actual realization.

The problem under consideration is a static one – given a mechanical system, how many independent deformations are possible without any cost in energy? These are the zero frequency modes, which we prefer to refer to as *floppy modes* because in any real system there will usually be some weak restoring force associated with the motion.

Sometimes it is convenient to look at the system as a *dynamical* one, and assign potentials or spring constants to deformations involving the various struts (bonds) and angles. It does not matter whether these potentials are harmonic or not, as the displacements are virtual. However it is convenient to use harmonic potentials so that the system is linear. It is then possible to set up a Lagrangian for the system and hence define a dynamical matrix, which is a real symmetric matrix having therefore real eigenvalues. These eigenvalues are either positive or zero. The number of finite (non–zero) eigenvalues defines the rank of the matrix. Thus our counting problem is rigorously reduced to finding the *rank* of the dynamical matrix. The rank of a matrix is also the number of linearly independent rows or columns in the matrix. Neither of these definitions is of much practical help, and a numerical determination of the rank of a large matrix is difficult and of course requires a particular realization of the network to be constructed in the computer. Nevertheless the rank is a useful notion as it defines the mathematical framework within which the problem is well posed.

The genius of Maxwell [1] was to devise the simple constraint counting method that allows us to *estimate* the rank of the dynamical matrix and hence the number of floppy modes. We discuss the application of these ideas to bulk covalent network glasses.

Until recently it has not been possible to improve on the approximate Maxwell constraint counting method, except on small systems using brute force numerical methods. Recently a powerful combinatorial algorithm, called the *Pebble Game* [3] has become available. This allows very large systems to be analyzed in two-dimensional central-force networks and in three-dimensional bond-bending networks. These terms will be described in more detail later.

The layout of this article is as follows. In the next section we give a brief history of the construction of random network models for covalent glasses. In section 3 we describe the powerful yet simple ideas of constraint counting, and in section 4 we discuss the exact algorithm - the pebble game - and the insights that it has provided into rigidity percolation. Up to this point the work is review, but in the next two sections we include some notes, equations and graphs of work in progress that relate to whether the transition is first or second order. We thought it important to include this work in progress as it relates to some significant recent experimental results on glasses. In section 5, we introduce a different but equivalent approach where we treat the bond-bending network as a *body-bar* network. We then discuss an exactly solvable model that can be analyzed using this approach. In section 6, we introduce the random bond model, which is useful as it contains no rings of bonds in the thermodynamic limit. In section 7, we describe the current state of experiments on glasses that are relevant to this topic. Throughout this paper we focus on central force networks in two dimensions and bond-bending networks that have central and non-central forces in three dimensions as these are most relevant to glasses. The reader should be aware that we do flip back and forth between these two model systems, as appropriate, in order to illustrate various points. The central force networks have a transition close to a mean coordination of 4, whereas the bond-bending networks have a transition close to a mean coordination of 2.4.

2. CONTINUOUS RANDOM NETWORKS

A perfect crystal is a structure in which the atoms, or groups of atoms, are arranged in a pattern that repeats itself periodically to form an infinite solid. For instance, crystalline diamond makes a lattice which is produced by the periodic repetition of the 8 atom diamond cubic cell (or the smaller 2 atom primitive face-centered-cubic unit cell) in all three spatial directions.

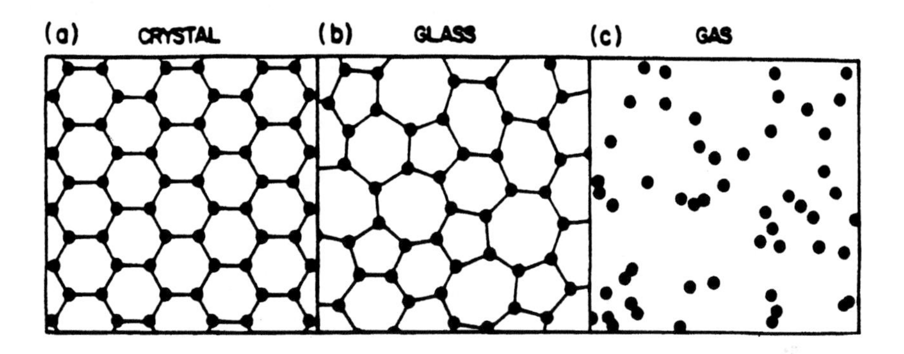

Figure 1. Schematic sketches of the atomic arrangements in (a) a crystalline solid, (b) an amorphous solid, and (c) a gas. (Zallen [4])

Amorphous materials do not have the periodicity (long range order) characteristic of a crystal. Starting from a crystalline structure and slightly displacing every atom in a random manner, the periodicity is certainly destroyed, but the random structure obtained is not yet amorphous. An amorphous structure is *topologically* distinct from a crystalline one. Thus, to obtain an amorphous structure from a crystalline one, it is necessary not only to introduce randomness in the atomic positions, but also to change the *topology* of the original perfect lattice. We can schematically sketch atomic arrangements in three typical cases: a crystal, an amorphous solid, and a gas, as in Figure 1.

In Figure 2 we show a piece of the three dimensional *amorphous* diamond structure which was obtained in a computer simulation [5]. By amorphous diamond, we mean a structure that is 4-fold coordinated everywhere and therefore similar to a-Si and a-Ge, except for the relatively stronger bond bending forces.

Binary compounds like SiO_2, can be found both in the crystalline and in the amorphous state, the later being topologically different from crystalline SiO_2. Schematic two-dimensional representations of the crystalline and amorphous structures for a hypothetical A_2B_3 compound, which is an analogous structure in two dimensions, are shown in Figure 3. Similarly, the crystalline diamond network possess only 6-fold rings of atoms, while the amorphous network also has 5-fold, 7-fold, and higher order rings, and consequently the number of 6-fold rings is reduced from its crystalline value.

The main experimental tool for verifying the structure of glasses is via *the Radial Distribution Function* that can be obtained by Fourier transforming X-ray or neutron diffraction data.

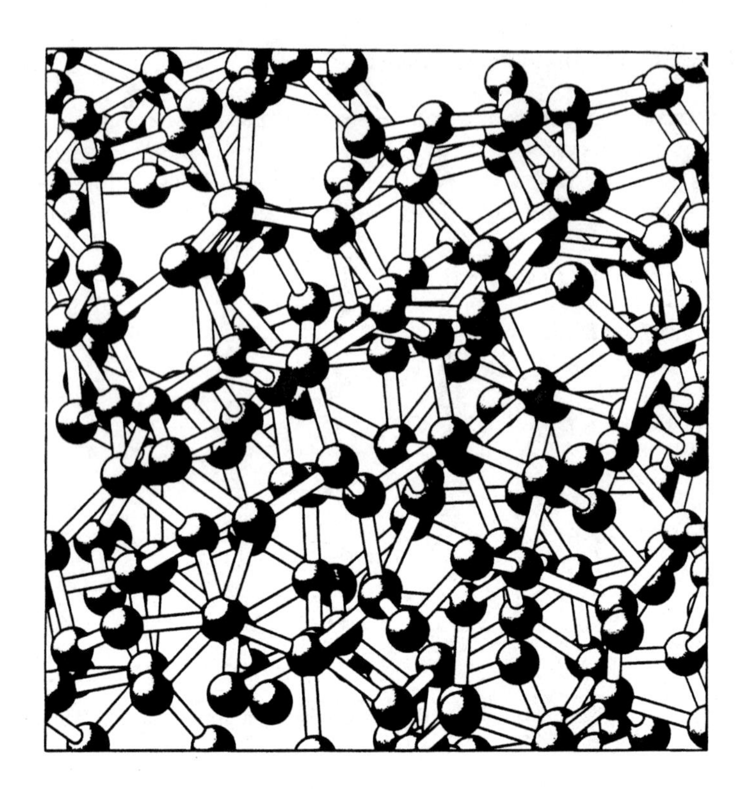

Figure 2. A piece of an amorphous diamond structure.

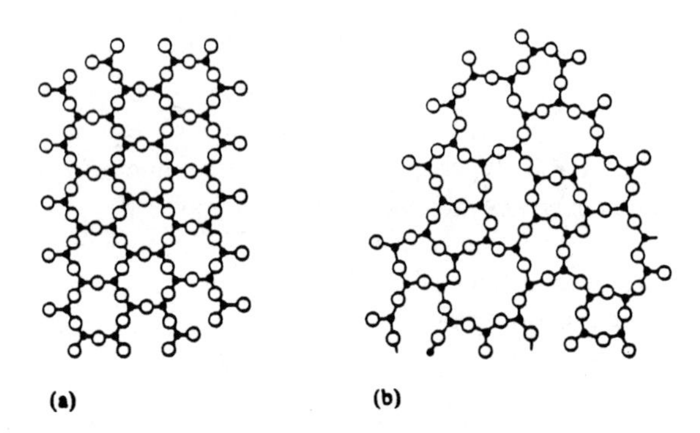

Figure 3. Crystalline and amorphous form of a binary compound. (a) Hypothetical crystalline compound A_2B_3. (b) Amorphous form of the same compound. (Elliott 1984 [6])

In this article, we shall focus our attention only on the *Continuous Random Network* models, which are suitable for describing the structures of materials with predominantly covalent *directed* bonds, namely, covalent glasses. The prototypical covalent amorphous solids are amorphous silicon, (a-Si), and amorphous germanium, (a-Ge) which belong to the semiconductors from Group IV in the periodic table. Every atom in these structures has four nearest neighbors, i.e. it is tetrahedrally coordinated.

We note here that the use of the term *continuous* in connection with random networks is somewhat unfortunate, because random networks are, in fact *discrete* in their structure. By *continuous* it is meant that one can build an amorphous network continuously, starting from the short-range order unit, up to an indefinite size, without including unsatisfied bonds or breaking the structure. In other words, *continuous* means that there are no identifiable boundaries in the structure, separating regions of distinctly different type of structure.

3. CONSTRAINT COUNTING

We start by examining a large covalent network that contains no dangling bonds. We can describe such a network by the chemical formula $Ge_xAs_ySe_{1-x-y}$ where the chemical element, Ge, stands for *any* fourfold bonded atom, As for *any* threefold bonded atom and Se for *any* twofold bonded atom. Every atom has its full complement of nearest neighbors and we consider the system in the thermodynamic limit where the number of atoms $N \to \infty$. There are no surfaces or voids and the chemical distribution of the elements is not relevant (although remarks in section 5) except that we assume there are no isolated pieces, like a ring of Se atoms. The total number of atoms is N and there are n_r atoms with coordination r (r = 2, 3 or 4), then

$$N = \sum_{r=2}^{4} n_r \tag{1}$$

and we can define the mean coordination

$$\langle r \rangle = \frac{\sum_{r=2}^{4} r n_r}{\sum_{r=2}^{4} n_r} = 2 + 2x + y . \tag{2}$$

We note that $\langle r \rangle$ (where $2<\langle r \rangle<4$) gives a partial but very important description of the network. Indeed, when questions of connectivity are involved the average coordination is the key quantity.

In covalent networks like $Ge_xAs_ySe_{1-x-y}$, the bond lengths and angles are well defined. Small displacements from the equilibrium structure can be described by a Kirkwood [7] or Keating [8] potential, which we can write *schematically* as

$$V = \frac{\alpha}{2}(\Delta l)^2 + \frac{\beta l}{2}(\Delta \theta)^2 \tag{3}$$

The mean the bond length is l and Δl is the change in the bond length and $\Delta \theta$ is the change in the bond angle. The bond-bending force (β) is essential to the constraint counting approach for stability, in addition to the bond stretching term (α). The other terms in the potential are assumed to be much smaller and can be neglected at this stage. If floppy modes are present in the system, then these smaller terms in the potential will give the floppy modes a small finite frequency. For more details see Thorpe [9]. If the modes already have a finite frequency, these extra small terms will produce a small, and rather uninteresting, shift in the frequency. This division into *strong* and *weak* forces is essential if the constraint counting approach is to be useful. It is for this reason that it is of little, if any, use in metals and ionic solids. It is fortunate that this approach provides a very reasonable starting point in many covalent glasses.

We will regard the solution of the eigenmodes of the potential of Eq. (3) as a problem in classical mechanics [1, 9, 10]. The dynamical matrix has a dimensionality of $3N$, which corresponds to the $3N$ degrees of freedom in the system. In a stable rigid network we would expect all the squared eigenfrequencies $\omega^2 > 0$ with six modes being at zero frequency. These six modes are just the three rigid translations and the three rigid rotations. We are assuming that our (large) network has free boundary conditions. Of course these 6 modes have no weight in the thermodynamic limit. The total number of zero frequency modes can be estimated by Maxwell counting [1], as was first done by Thorpe [9] following the work of J. C. Phillips [11, 12] on ideal coordinations for glass formation.

The constraint counting proceeds as follows. There is a single constraint associated with each bond. We assign $r/2$ constraints associated with each $r-$ coordinated atom. In addition there are constraints associated with the angular forces in Eq. (3). For a twofold coordinated atom there is a single angular constraint; for an r–fold coordinated atom there are a total of $2r-3$ angular constraints. The total number of constraints is therefore

$$\sum_{r=2}^{4} n_r\left[r/2+(2r-3)\right]. \tag{4}$$

The *fraction* f of zero–frequency modes is given by

$$f=\left[3N-\sum_{r=2}^{4} n_r\left[r/2+(2r-3)\right]\right]/3N \tag{5}$$

This expression can be conveniently rewritten in the compact form

$$f=2-\frac{5}{6}\langle r\rangle, \tag{6}$$

where $\langle r\rangle$ is defined in Eq. (2). Note that this result only depends upon the combination $2x+y=\langle r\rangle$, which is the only relevant variable. When $\langle r\rangle=2$ (e.g. Se chains), then $f=1/3$; that is one third of all the modes are floppy. As atoms with higher coordination than two are added to the network as cross-links, f drops and goes to zero at $\langle r\rangle_c=2.4$. The network becomes rigid, as it goes through a phase transition from *floppy* to *rigid*. This mean field approach has been quite successful in covalent glasses and helps explain a number of experiments as will be described later. Also in later sections, we discuss the results of computer experiments and show that they are rather well described by the results of this section.

4. GENERIC RIGIDITY PERCOLATION

The elastic properties of random networks of Hooke springs has been studied over the past 15 years [9, 10, 13-16]. One of the most interesting findings has been that effective medium theory describes the behavior of the elastic constants and the number of floppy modes remarkably well [13-16] except very close to the phase transition from a rigid to a floppy structure.

Early attempts to study the critical behavior in central-force networks were not very satisfactory [13-19], and the question of the universality class of the rigidity transition has only recently been resolved. This question is fundamental to understanding the nature of the rigidity transition, and may have important implications as to how the character of the glass transition is affected by the mean coordination, as has been discussed recently via fragile and strong glass formers [20]. We show here how substantial progress can be made in understanding the geometrical nature of generic rigidity percolation [3, 21-24].

There are two important differences between rigidity and connectivity percolation. The first difference is that rigidity percolation is a vector (not a scalar) problem, and secondly, there is an inherent long-range aspect to rigidity percolation. These differences make the rigidity problem become successively more difficult as the dimensionality of the network increases. In two dimensions, Figure 4(a) shows four distinct rigid clusters consisting of two rigid bodies attached together by two rods connecting at pivot joints. Now the placement of one additional rod, as shown in Figure 4(b), locks the previous four clusters into a single rigid cluster. This non-local character allows a single rod (or bond) on one end of the network to affect the rigidity all across the network from one side to the other.
Using concepts from graph theory, we set up generic networks where the connectivity or topology is uniquely defined but the bond lengths and bond angles are arbitrary. A generic network does not contain any geometric singularities [25], which occur when certain geometries lead to null projections of reaction forces. Null projections are caused by special symmetries, such as, the presence of *parallel bonds* or *connected collinear bonds*. Rather than these *atypical* cases, their *generic* counterparts as shown in Figures 4(b) and (c) will be present. This ensures that all infinitesimal floppy motions carry over to finite motions [25-27].

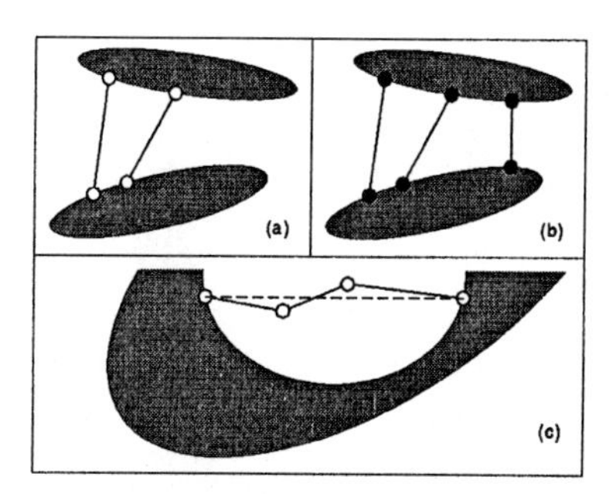

Figure 4. The shaded regions represent 2d rigid bodies. The (closed, open) circles denote pivot-joints that are members of (one, more than one) rigid body. (a) A floppy piece of network with four distinct rigid clusters. (b) Three generic cross links between two rigid bodies make the whole structure rigid. If the bonds were parallel, the structure would not be rigid to shear [64]. (c) Three non-collinear connected rods connecting across a rigid body is generic and contains one internal floppy mode. If they were collinear (along the dotted line), then there would be two infinitesimal (not finite) floppy motions, and under a horizontal compression buckling would occur.

Early studies on rigidity percolation were on regular (non-generic) lattices, which as we now know inadvertently delayed a proper understanding of the rigidity transition. In non-generic (referred to as atypical) networks, many geometrical singularities occur which lead to non-linear effects. For example, a diode-like problem frequently occurs in atypical networks where a string of collinear bonds can only be extended with a cost in energy but can be compressed with no cost in energy due to buckling [e.g. Figure 4(c)]. The diode effect complicates studies because it leads to the breakdown of linear elasticity theory, which must be reversible. A simple way to view a generic network is to take a regular lattice structure and randomly displace each site location by a small amount. This introduces local distortions throughout the lattice and is in itself a good physical model for amorphous and glassy materials. Many early studies involved the non-linear effects arising from geometrical singularities [13-19], and should therefore be regarded as a separate problem.

By considering generic networks, the diode effect and the problematic geometric singularities are completely eliminated. Therefore, the problem of rigidity percolation on generic networks leads to many conceptual advantages because all geometrical properties are robust. Moreover, real glass networks have local distortions, and are modeled better by generic networks. In two dimensions, there exist efficient, exact combinatorial algorithms allowing for the possibility of an in depth study of rigidity percolation. Similar algorithms have been extended to three dimensions, as discussed later, for some important special cases.

The rigidity of a network glass is related to how amenable the glass is to continuous deformations that require very little cost in energy. A small energy cost will always arise from weak forces, which are present in addition to the hard covalent forces that involve bond-lengths and bond-angles. These small energies can be ignored because the degree to which the network deforms is well quantified by just the number of floppy modes [19] within the system. A mental picture of floppy and rigid regions within the network has led to the idea of rigidity percolation [13, 3].

Much understanding of the general phenomena of central-force rigidity percolation can be obtained by studying a random network of Hooke springs. To be specific, we begin by considering a network of Hooke springs characterized by the potential

$$V = \tfrac{1}{2}\sum_{\langle ij\rangle} \alpha_{ij}\eta_{ij}\left(l_{ij} - l_{ij}^{o}\right)^2 \tag{7}$$

where the sum is over all bonds $\langle ij\rangle$ connecting sites i and j in the network. A bond connecting sites i and j is present if $\eta_{ij} = 1$ with probability p and absent if $\eta_{ij} = 0$ with probability 1-p. The spring constants, α_{ij}, and the equilibrium bond lengths, l_{ij}^{0}, are positive real numbers but are left arbitrary. The site locations are also arbitrary, because the network is generic. Note that rigidity is a static concept, involving virtual displacements, so that while it is convenient to use harmonic potentials as done in Eq. (7), any set of pair potentials would give the same results for the geometric aspects of rigidity. A collection of sites form a *rigid cluster* when no relative motion within that cluster can be achieved without a cost in energy. Conversely, *floppy modes* correspond to finite motions within the system, which do not cost energy. Therefore, the geometrical properties and the number of floppy modes can be determined by an equivalent bar and joint structure [19]. Note that a d-dimensional system always has at least $d(d+1)/2$ floppy modes due to d global translations and $d(d-1)/2$ global rotations.

The number of floppy modes in d dimensions is given by the total number of degrees of freedom for N sites minus the number of independent constraints. A redundant bond can only add additional reinforcement and/or cause internal stress in an existing rigid body. A key quantity is the number of floppy modes, F, in the network, or normalized per degree of freedom, $f = F/dN$. By defining the number of *redundant bonds* per degree of freedom as n_r, we can write quite generally,

$$f = \frac{dN - \left(\tfrac{1}{2}Nzp - dNn_r\right)}{dN} = 1 - \frac{p}{p^*} + n_r \tag{8}$$

where $p^* = 2d/z$ and z is the lattice coordination. Neglecting the redundant bonds, as first done by Maxwell [1], we find that f is linear in the bond concentration, p, and goes to zero at the Maxwell approximation, p^*, for the threshold. The Maxwell approximation gives a good account of the location of the phase transition and the number of floppy modes, but it ultimately fails since the number of independent constraints is not just the total number of bonds as some bonds are dependent.

We will focus on the geometrical aspects of rigidity percolation. Previous studies have used numerical methods on networks containing $N \approx 10^4$ sites, but were faced with difficult challenges. For example, relaxation methods are well suited for calculating elastic constants, but not for characterizing the geometric structure. This is because numerically one cannot identify which bond has exactly zero stress or if a bond accidentally has zero stress. However, until the pebble game, this was the only approach available in determining the stress carrying backbone.

The Pebble Game

We have been able to study networks containing more than 10^6 sites, using an integer algorithm, which gives exact and unique answers to the geometric properties of generic rigidity percolation. Because of the non-local characteristic of rigidity percolation, [e.g. Figures 4(a)] burning-type algorithms [28] commonly used in connectivity percolation are useless. This implies that the entire structure needs to be specified (stored in memory) since rigidity of a given region may depend on bonds far away.

A very efficient combinatorial algorithm, as first suggested by Hendrickson [25], has been implemented to (i) calculate the number of floppy modes, (ii) locate over-constrained regions and (iii) identify all rigid clusters for $2d$ generic bar-joint networks. The crux of the algorithm is based on a theorem by Laman [26] from graph theory.

Theorem: *A generic network in two dimensions with N sites and B bonds (defining a graph) does not have a redundant bond iff no subset of the network containing n sites and b bonds (defining a subgraph) violates* $b \leq 2n-3$.

By simple constraint counting it can be seen that there must be redundant bond(s) when Laman's condition is violated. This necessary part generalizes to all dimensions such that if $b > dn - d(d+1)/2$ then there is a redundant bond for $n \geq d$. For $n < d$ it follows that if $b > n(n-1)/2$ then there is a redundant bond. Note that $n = 1$ is an excluded case in Laman's theorem, because two sites are required for a bond to be present. The essence of Laman's theorem is that in two dimensions finding $b > 2n-3$ is the only way redundant bonds can appear. This sufficient part does not generalize to higher dimensions [25].

The basic structure of the algorithm is to apply Laman's theorem recursively by building the network up one bond at a time. Only the topology of the network is specified, not the geometry. Because of the recursion, only the subgraphs that contain the newly added bond need to be checked. If each of these subgraphs satisfy the Laman condition, $b \leq 2n-3$, then the last bond placed is independent, otherwise it is redundant. By counting the number of redundant bonds, the exact number of floppy modes is determined.

Searching over the subgraphs is accomplished by constructing a pebble game [3, 21]. Each site in the network has two pebbles tethered to it. A pebble is either free when it is on a site or *anchored* when it is covering a bond. A free pebble represents a single motion that a site can undertake. Consider a single site having two free pebbles, representing two translations. If two additional free pebbles can be found at a different site, then the distance between this pair of sites is not fixed. Placing a bond between this pair of sites will constrain their distance of separation. To record this constraint, one of the four free pebbles is anchored to the bond. Once the bond is covered, only three free pebbles can be shared between that pair of sites. After a bond is determined to be independent, it will always remain independent and covered.

We begin with a network of N isolated sites each having two free pebbles. The system will always have $2N$ pebbles; initially two free pebbles per site. We place one bond at a time in the network connecting pairs of sites. The topological placement of either the sites or bonds will depend on the model under study such as the site or bond diluted generic triangular lattice described here. Only independent bonds are covered by pebbles. Therefore, before a bond can be covered it must be tested for independence. For each bond placed in the network, four pebbles (two on each site at the ends of the bond) must be free for the bond to be independent. When a bond is determined to be independent, any one of the four pebbles can be anchored to that bond. In general, all four pebbles across an added bond will not be free because they are already anchored to other bonds. These anchored pebbles may possibly become free at the expense of anchoring a neighboring free pebble while keeping a particular independent bond covered. In other words, pebbles may be shuffled around the network provided all independent bonds remain covered.

It is always possible to free up three pebbles across a bond, since they correspond to its rigid body motion. When a fourth pebble across a bond cannot be found, then that bond is redundant and it is not covered. In Figure 5 an example of how pebbles are shuffled is schematically shown on a small generic structure. Two distinct pebbles are associated with each site for which each pebble can either be used to cover a bond or is free to cover a bond. As schematically drawn in Figure 5, the two pebbles closest to a given site are the pebbles that are tethered to that site. Thus a pebble may either be on a site (free pebble) or on a bond (anchored pebble) but it always remains tethered to a given site regardless of how the pebbles are shuffled. Note that a bond may be covered by a pebble from either of its end sites. Therefore, free pebbles can be moved across the network by exchanging the site from which a pebble is used to cover a bond.

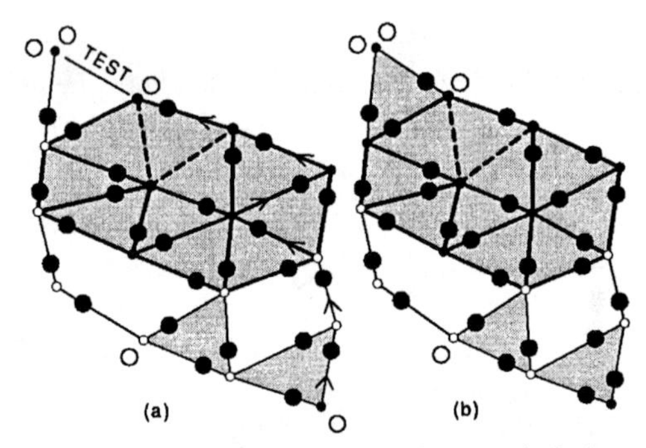

Figure 5. A demonstration of the pebble game on a generic network. Independent (redundant) bonds are shown with solid (dashed) lines which are (are not) covered by a pebble. Large (filled, open) circles denote (anchored, free) pebbles on (bonds, sites). The two closest pebbles to a given site are tethered to that site. Small (filled, open) circles denote sites belonging to (one, more than one) rigid cluster. Over-constrained bonds are shown with heavy dark lines. Shaded regions denote 2d rigid bodies. (a) There are seven rigid clusters and the five free pebbles indicate 5 floppy modes until a new bond is added and tested for independence. A fourth free pebble is found via the path traced by arrows. (b) The added bond is independent and thus covered. There are now six rigid clusters and four floppy modes.

Over-constrained regions are recorded each time a dependent bond is found. These regions correspond to the set of bonds that were searched in trying to free the fourth pebble but failed. These regions, called Laman subgraphs, violate the condition $b \leq 2n-3$. An bond added to a Laman subgraph will be redundant.

We identify all the rigid clusters after the network is completely built. First, we identify isolated sites. Then the rigidity of all other sites is tested with respect to a reference bond. If a test-bond between either one of the pair of sites forming the reference bond and the site in question is found to be (dependent, independent) then that site (is, is not) rigid with respect to the reference bond. The test-bond is actually never added to the network. Since a bond can only belong to one cluster (unlike sites), all the bonds within a rigid cluster are ascribed to a particular reference bond. A systematic search is made to map out all rigid clusters.

We show in Figure 5(b) the end result of the pebble game applied to a simple structure. Many aspects of rigidity are displayed. It can be seen that:

(1) The exact number of floppy modes is determined by the number of free pebbles remaining. A depletion or excess of pebbles to cover a set of bonds distinguishes the over-constrained regions from the floppy regions; unlike the approximate global counting of Maxwell.
(2) This network is uniquely decomposed into a set of six distinct rigid clusters, although the clusters are not disconnected.
(3) The free pebble along the bottom edge cannot be shuffled over to the rigid body at the top, which already has three free pebbles. This free pebble is shared among three bars and two triangles. Generally, free pebbles get trapped in floppy regions consisting of many rigid clusters giving arise to complex collective floppy motion.
(4) The number of redundant bonds is unique, whereas their locations are not unique since this depends on the order of placing the bonds. Nevertheless, each redundant bond belongs to a unique over-constrained region (Laman subgraph). For example, there are 19 over-constrained bonds in the rigid cluster at the top of the structure in Figure 5(b), while having only two redundant bonds.
(5) A rigid cluster will generally have sub-regions that are over-constrained. If any bond that is over-constrained is removed, the rigidity of the network is unchanged.

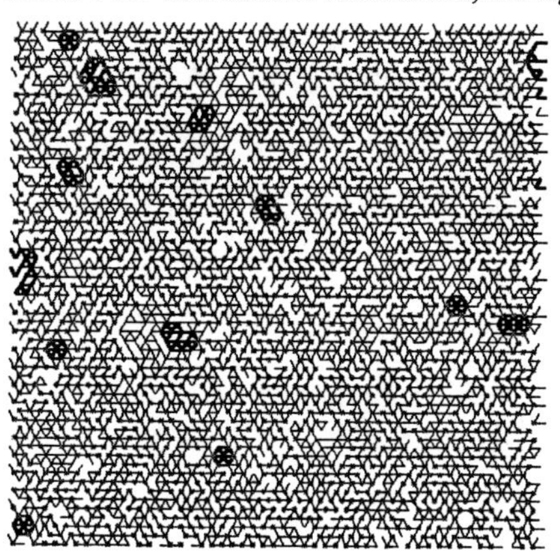

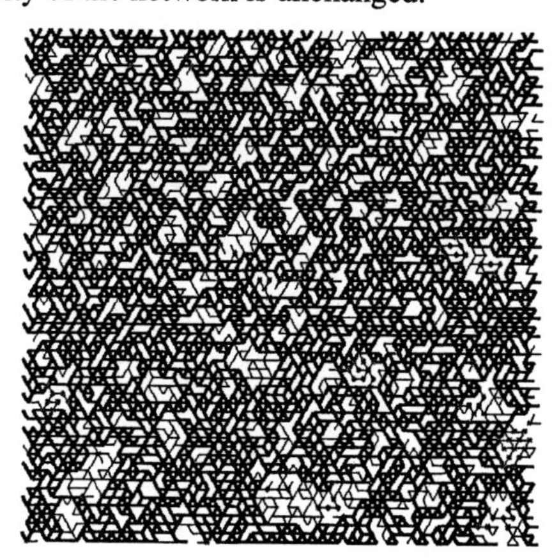

Figure 6. The topology of a typical section from a bond-diluted generic network at p=0.62 (below percolation) and at p=0.70 (above percolation). A particular realization would have local distortions (not shown), thus making the network generic. The heavy dark lines correspond to over-constrained regions, and the light lines to isostatic regions. The open circles correspond to sites that are acting as pivots between two or more rigid bodies.

Sections of a large network on the bond-diluted generic triangular lattice are shown in Figure 6 after the pebble game was applied. Below the transition the network can be macroscopically deformed as the floppy region percolates across the sample. Above the rigidity transition, stress will propagate across the sample. However, below the transition there are clearly pockets of large rigid clusters and over-constrained regions, while above the transition there are clearly pockets of floppy inclusions within the network.

Two Dimensional Central Force Networks

In this section, we review recent results for central-force generic rigidity percolation on both the bond diluted triangular net. We begin by working out a better estimate for the bond rigidity threshold. The Maxwell counting prediction for the rigidity threshold gives $p^* = 2d/z = 2/3$ for bond dilution on the triangular net, which has dimension $d = 2$ and $z = 6$ nearest neighbor bonds. To get a more accurate estimate for the rigidity threshold, the presence of redundant bonds and floppy inclusions must be accounted for.

In a low concentration expansion the first diagram to contribute redundant bonds is shown in Figure 7, where only 11 of the 12 bonds are independent. This diagram leads to a correction for the number of floppy modes as $n_r = \frac{1}{2}p^{12} + O(p^{18})$ and for bond dilution. In a high concentration expansion, the first two dominant contributing diagrams are shown in Figures 7(b) and (c) corresponding to dangling ends and isolated sites.

From these two leading diagrams the fraction of floppy modes, f, at high concentrations are given by

$$f = 3(1-p)^5 - 2(1-p)^6 + O\left((1-p)^8\right). \tag{9}$$

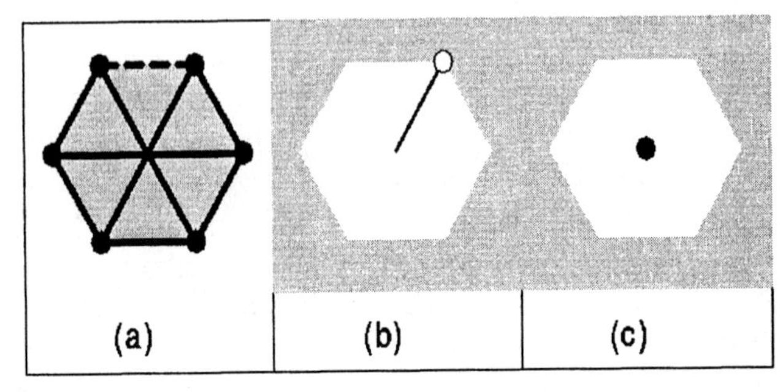

Figure 7. All shaded areas represent a 2d rigid region. The topology is shown for: (a) The lowest order diagram on a triangular network to have a dependent bond (dashed line). All 12 bonds are over-constrained. (b) The lowest order diagram for a floppy inclusion corresponds to a dangling bond. (c) An isolated site is counted as a floppy inclusion and contributes two floppy modes at the next lowest order.

These types of corrections will shift the transition from the Maxwell threshold of 2/3, and they will be responsible for a non mean-field-like critical behavior. We equate the number of floppy modes from the truncated low and high concentration expansions. For bond dilution we estimate the threshold to be 0.6622. Thus we find a small downward shift. The pebble game reveals that the rigidity thresholds *shift* about 50% more than the above estimate to $p_{cen} = 0.6602 \pm 0.0003$. The site diluted version of this problem has been studied by Moukarzel and Duxbury [23] who discuss some of this work in their article in this book.

A sharp peak in the second derivative of the number of floppy modes shown in Figure 8 appears without any signs of a discontinuity. The peak most resembles a simple cusp. As can be seen there is virtually no difference between the data for linear system sizes L=680, 960 and 1150. Only very slight system size dependence has been observed. The trend is for smaller systems to show a cusp-like singularity in the second derivative $f^{(2)}$ as well, but with the peak slightly shifted to the left with a smaller amplitude.

The behavior of the second derivative suggests that the number of floppy modes is analogous for rigidity and connectivity percolation. In the case of connectivity percolation, the number of floppy modes is simply equal to the total number of clusters, which corresponds to the free energy [9, 29-31]. It would be nice if a similar result holds for rigidity percolation. We find that the second derivative of the total number of clusters changes sign across the transition, thus violating convexity requirements. Noting that typically rigid clusters are not disconnected, we suggest that the number of floppy modes generalizes as an appropriate free energy [9, 29-31]. With this assumption, we have estimated the exponent α in the usual context of a *heat capacity* critical exponent.

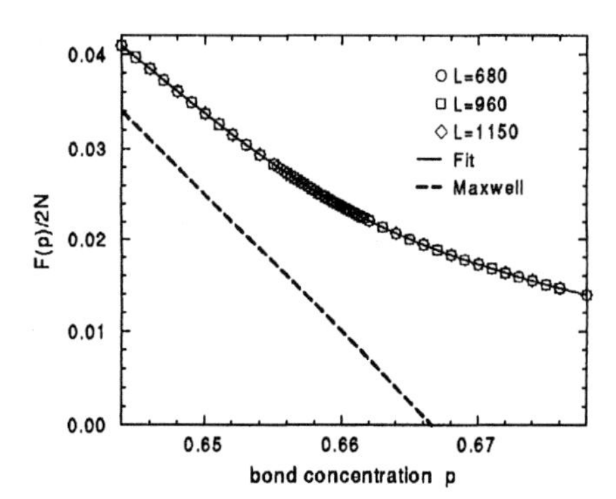

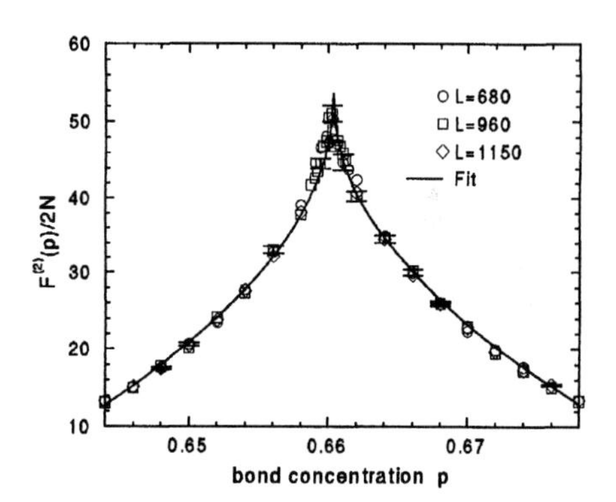

Figure 8. (a) Simulation results for the fraction of floppy modes, *f=F/2N*, for a bond diluted generic network compared to the Maxwell prediction. All error bars are smaller than the symbols. (b) The second derivative of the fraction of floppy modes for a bond diluted generic network as calculated from Monte Carlo sampling. The fitting results for the cusp yields p_{cen} = 0.6603 $\pm$ 0.0003 and an exponent of 0.48 $\pm$ 0.05. Typical error bars are shown which reflect both the statistical errors in the Monte Carlo sampling and the ensemble averaging.

In addition to calculating the number of floppy modes, we analyzed the rigid cluster statistics [21] for bond dilution with free and periodic boundary conditions and site dilution with periodic boundary conditions. The free boundary condition data was generated mainly to check boundary effects. Finite size scaling techniques as used in percolation theory [28] are applied here assuming only a single relevant length scale exists.

We also look at the geometrical properties of the over-constrained regions within the network. These over-constrained regions are not necessary to sustain rigidity. An isostatic framework [21], for example, has just the right positioning of bonds to form a rigid cluster without a redundant constraint. However, when external forces are applied to a rigid isostatic framework, using rigid busbars for example, over-constrained regions will be induced across it, which forms a percolating stressed region. Within a random environment there will be redundant bonds scattered throughout the network. Here the redundant bonds are essentially acting as external forces on an underlying isostatic framework. Physically, the resulting over-constrained regions characterize internal stress caused by bond mismatch. Thus, the over-constrained regions propagate stress (without externally applied forces), and are most closely analogous to the current carrying backbone in connectivity percolation. We find that monitoring the probability for a network of linear size L to contain either a spanning rigid cluster or spanning stressed backbone, leads to the same correlation length exponent, ν, and critical threshold.

From extrapolation using finite size scaling techniques, [21] we find p_{cen} = 0.6602 ± 0.0003 for the bond diluted triangular lattice which is in excellent agreement with that obtained from the cusp singularity in the second derivative of f as shown in Figure 17.. Moreover, the rigidity transition is second order, but in a different universality class than connectivity percolation, with the exponents; α= -0.48±0.05, β= 0.175±0.02 and ν= 1.21±0.06. The fractal dimension of the spanning clusters and the spanning stressed regions at the critical threshold are found to be d_f= 1.86±0.02 and d_{BB} = 1.80±0.03 respectively. (See Refs. [28, 21] for definition of all the above critical exponents)

It has been suggested by Duxbury and co-workers that the rigidity transition might be weakly first order on triangular networks. This observation was guided by results on Bethe lattices which we will discuss later in the context of glasses. While we think this is unlikely, it cannot be completely ruled out at present and is discussed in more detail by Moukarzel and Duxbury [23] and discussed in these proceedings.

Three Dimensional Bond Bending Networks

Unfortunately it is not possible to extend the pebble game to three dimensional central-force networks, as this algorithm is based on Laman's theorem which does not generalize to three dimensions. This is because of the existence of banana graphs as shown in Figure 9.

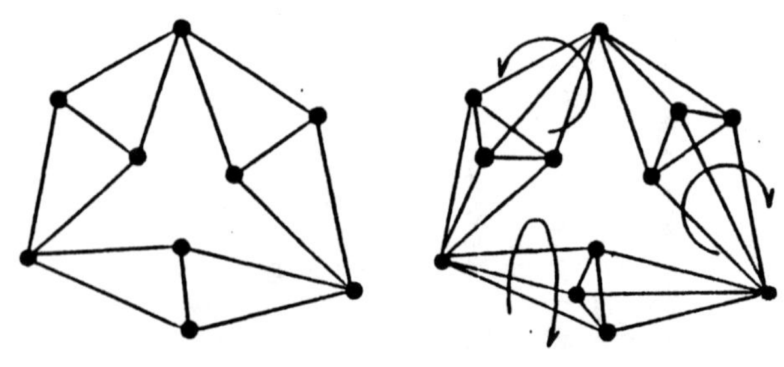

Figure 9. (Left panel) A single rigid cluster in two dimensions. (Right panel) Showing 3 bananas in a three dimensional network. There are four rigid clusters here - three bananas plus an additional rigid cluster consisting of the three sites at the corners which are connected by implied bonds. Such non-contiguous rigid clusters cannot occur in two dimensional networks, and are responsible for the breakdown of Laman's theorem in three dimensions.

While the pebble game described in this paper is only applicable in 2d, we have generalized the rules for a certain class of three-dimensional networks. Although the Laman condition is not generally sufficient [25-27, 32] in three dimensions, we recently have shown [33] that it can be generalized within bond bending networks. While this paper does not give a rigorous mathematical proof, it does explain why bond-bending networks can almost certainly be treated with pebbles, based on a Laman type theorem. The problem of the Laman bananas, as shown in Figure 9, is conveniently eliminated once angular forces are included as in a Kirkwood or Keating potential Eq. (3). As a result, we are able to construct a three dimensional pebble game for the bond-bending model, having nearest and next nearest neighbor forces. Fortunately, the bond-bending model is precisely the class of models that is applicable to the study of many covalent glass-networks.

For the most part, the three dimensional pebble game rules come from naively generalizing the 2d rules (such as using three pebbles per site instead of two). However, there are some subtle differences caused by restricting the three dimensional networks to be in a certain class. The order of bond placement in building a network now becomes important, so that the bond-bending model at each step of the process is preserved and thereby insuring that at no time will banana structures form. A somewhat longer discussion of the three dimensional pebble game is given by Jacobs, Kuhn and Thorpe in these proceedings, where reference should be made in particular to Figure 4, which shows a particular graph and its associated pebbles.

For purposes of testing rigidity in generic three-dimensional bond-bending networks, it is only necessary to specify the network topology or connectivity of the central-force (CF) bonds, since the second nearest neighbors via CF bonds define the associated bond-bending constraints. Here, we have considered two test models. In the first model, a unit cell is defined from our realistic computer generated network of amorphous silicon [5] consisting of 4,096 atoms having periodic boundary conditions. Larger completely four coordinated periodic networks containing 32,768, 262,144 and 884,736 atoms are then constructed from the amorphous 4,096-atom unit cell.

The four-coordinated network is randomly diluted by removing CF bonds one at a time with the constraint that no site can be less than two-coordinated. That is, a CF bond is randomly selected to be removed. If upon removal either of its incident atoms becomes less than two coordinated, then it is not removed and another CF bond is randomly selected from the remaining pool of possibilities. The order of removing CF bonds is recorded. This process is carried out until all remaining CF bonds cannot be removed, leading to as low an average coordination number as possible. All CF bonds that were successfully removed are marked. This method of bond dilution gives a simple prescription for generating a very large model of a continuous random $Ge_xAs_ySe_{1-x-y}$ type of network. For comparison, a second test model, a diamond lattice, was diluted in the same way and contained 32,768, 262,144 and 10^6 atoms.

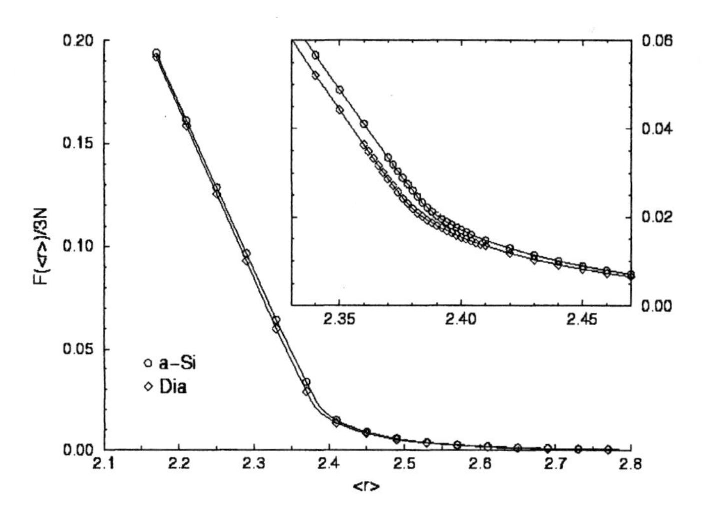

Figure 10. Showing the number of floppy modes F as a function of the mean coordination $\langle r \rangle$ for two bond diluted models; based on the diamond lattice and on amorphous Si. The inset shows a blow up in the critical region.

In the application of the three dimensional pebble game, the network is built up one constraint at a time. The first step is to place all the unmarked CF bonds in the network. Once a CF bond is placed in the network, all its associated bond-bending constraints must also be placed before the next CF bond can be considered. Then the marked CF bonds (including their associated bond-bending constraints) are placed in the network in the reverse order from that when they were removed as the network was randomly diluted. In this way, the rigidity properties of the network can be monitored as a function of the average coordination number which typically ranges from ($\langle r \rangle$= 2.2 to 4.0), and the network type is always like that of $Ge_xAs_ySe_{1-x-y}$.

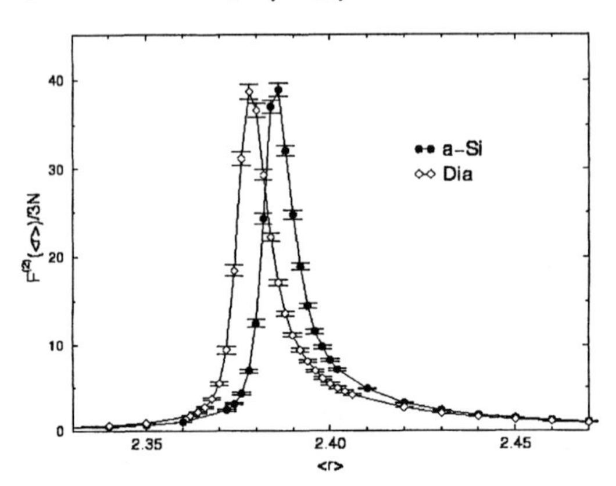

Figure 11. Showing the second derivative of the number floppy modes $F(2)$ as a function of the mean coordination <r> for bond diluted models; based on the diamond lattice and on amorphous Si.

The ensemble average over many realizations of bond dilution for the fraction of floppy modes and its derivatives have been numerically obtained for different size networks. The fraction of floppy modes is calculated exactly for each realization, its first derivative is obtained by Monte Carlo sampling and the second derivative is determined by one numerical differentiation. The fraction of floppy modes and its second derivative for both the bond diluted amorphous silicon and diamond lattices are shown in Figures 10 and 11 respectively. The bond-diluted generic diamond lattice behaves in nearly the same way as the bond-diluted amorphous silicon. They both have a rigidity transition slightly below the simple Maxwell constraint counting estimate of 2.4. The rigidity transition can be accurately found from the sharp peak in the second derivative in the fraction of floppy modes. In particular, for the diamond lattice $\langle r \rangle_c = 2.375 \pm 0.003$ and for a-Si $\langle r \rangle_c = 2.385 \pm 0.003$. Remarkably, the Maxwell constraint counting estimate is accurate to about 1% in locating the threshold in both cases.

In Figure 12 we show the rigid and floppy regions of a typical section of the bond-diluted amorphous silicon model below and above the rigidity transition at an average coordination number of 2.37 and 2.40 respectively. It is useful to study these two figures, and compare them with Figure 6 showing a similar map of rigid and floppy regions in a two dimensional central force network. It can be shown [33] that the only floppy element in a three dimensional bond-bending network is a hinge joint. Hinge joints can only occur through a CF bond and are always shared by two rigid clusters – allowing one degree of freedom of rotation through a dihedral angle. Consequently, it is possible for a given r-coordinated atom to be at most a member of $r + 1$ rigid clusters. Note that in two dimensional bond diluted central force generic networks, sites that belong to more than one cluster act as a pivot joint, and more than two rigid clusters can share a pivot joint.

In Figure 12 it is seen that just below the rigidity transition, a-Si is still very floppy with small overconstrained regions sparsely scattered throughout the network – mostly due to the presence of four and five fold rings. Unlike in two dimensions, large isostatic rigid regions are not seen. Although most of the CF bonds are acting as hinge joints, it is worth noting that the average number of hinge joints per floppy mode is about 15, indicating that these modes are not localized on just one or two bonds. With an increase in average coordination number of about 1.25% a-Si becomes rigid with a spanning overconstrained region, as also shown in figure 12. There are still very few isostatic rigid regions found, although many floppy inclusions remain. The average number of hinge joints per floppy mode is again about 15 after passing through a maximum at the rigidity transition. Note that the rigidity transition discussed in this section on bond diluted networks, is always second order.

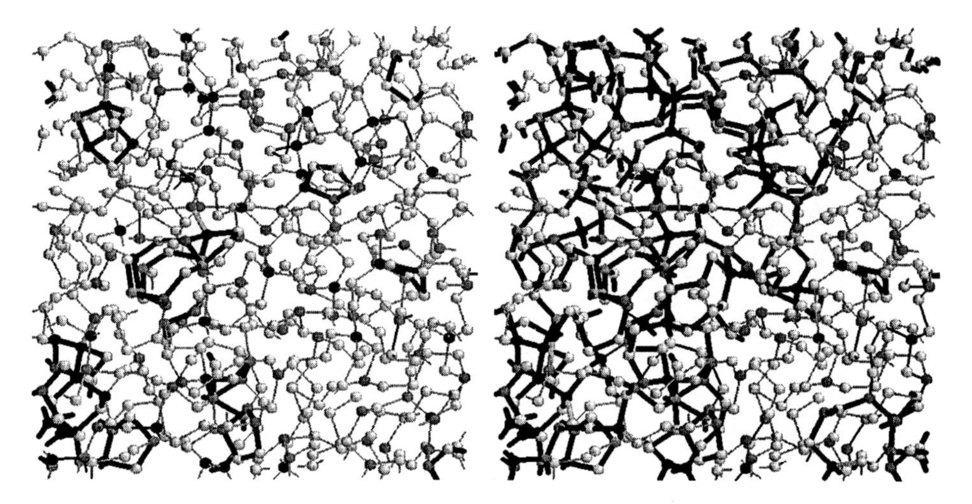

Figure 12. Typical section of a random network with a low mean coordination (2.37), where rigidity has not percolated. (left) and with a high mean coordination (2.40) where rigidity has percolated (right). The wide black bonds are over-constrained, while the thin black bonds are isostatic. The hinge joints are thin gray lines if isostatic and wide gray lines if over-constrained.

5. BODY-BAR NETWORKS

Up to now in this article, we have regarded the elements of the networks as points with d degrees of freedom, connected up with a single *bar* between points. Each bar designates a single distance constraint, representing either a central force between nearest neighbors, or an angular force between second nearest neighbors. There are some advantages to using a different, yet equivalent representation. A *body*, rather than a point, is used with $g = d(d+1)/2$ degrees of freedom. In $d = 3$ dimensions, this describes the 3 translational degrees of freedom and the 3 rotational degrees of freedom. These bodies are then connected with multiple bars, rather than just a single bar, where there are b bars connecting nearest neighbor bodies.

This representation is usually referred to as the body-bar model [34,35]. The formal advantage is that we can describe a covalent random network with only nearest neighbor interactions. A mapping to go between a bond-bending network and a body-bar network in three dimensions has been proposed [33] in three dimensions. For the case of a continuous random network having two or higher coordinated sites only, all sites are simply treated as bodies with $g = 6$ degrees of freedom, and $b = 5$ bars are generically placed between nearest neighbors, as defined via the central-force bonding. Thus we may say that there is a Laman type theorem for this case.

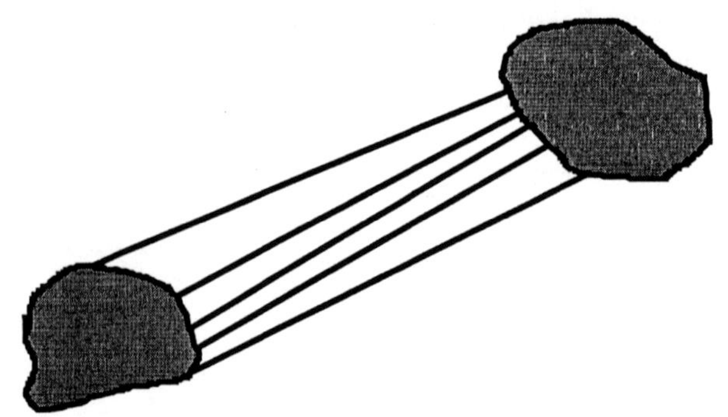

Figure 13. A sketch of two bodies joined by five bars (b =5) generically.

The fraction of floppy modes (for general g and b) is given by

$$f = \left[gN - \sum_{r=2}^{4} n_r br/2 \right] / gN = 1 - \frac{b}{2g}\langle r \rangle \qquad (10)$$

in a similar way to (5). Using (10), the number of floppy modes goes to zero at $\langle r \rangle_c = 2g/b$. If we choose a body with $g = d(d+1)/2$ in a d dimensional space, and have the number of bars between nearest neighbors *one* less than this, then we have the transition at

$$\langle r \rangle_c = [2d(d+1)]/[(d-1)(d+2)]. \qquad (11)$$

This arrangement with $g = b + 1 = d(d+1)/2$ means that a pair of sites and the connecting bars has a single residual degree of freedom, which in three-dimensions represents the dihedral angle twist. For $d = 3$, the expression (11) becomes $\langle r \rangle_c = 2.4$ in agreement with previous results. In passing we note that (11) gives $\langle r \rangle_c = 3$ in two dimensions. This would represent bodies with 3 degrees of freedom each connected by 2 bars. As the dimension d gets higher, the expression (11) tends monotonically to $\langle r \rangle_c = 2$ so that eventually connectivity implies rigidity.

A dangling bond is a rather strange object in the body-bar network as it has a twist associated with it that must be eliminated. We can conveniently eliminate the twist by using the same number of bars as there are degrees of freedom on the dangling atom. This scheme leads to an exact mapping back to the bond-bending network. Using Maxwell constraint counting, the transition occurs at

$$\langle r \rangle_c = 2.4 - 0.4\left(\frac{n_1}{N}\right), \qquad (12)$$

where (n_1 / N) is the fraction of singly coordinated sites entering as a simple correction term. This result has been found previously by Boolchand and Thorpe [36] using a bond-bending approach and it is derived later in this article and shown as Eq. (48).

A body-bar representation can be applied to other types of networks besides bond-bending networks. Another case of interest that can be handled by the body-bar formalism is when angular forces are *not* present at some atoms within the network. This is particularly important as it applies to the proto-typical glass SiO_2, which does not have significant angular forces at the two-coordinated oxygen atoms.

It is convenient to use the designation A for atoms that have 2-, 3- and 4- nearest neighbors and the associated angular forces between atom A and its neighbors. We will assign the symbol A to represent *any* of the following group of elements, C, Si, Ge, As, Se and S. These elements occur in covalent glasses so that the C, Si and Ge atoms are four-fold coordinated, As is three-fold coordinated, and Se and S are two-fold coordinated. For these we have $g_A = 6$ degrees of freedom and $b_{AA} = 5$ bars as already discussed.

Oxygen atoms are two-fold coordinated but lack an angular force and H is a singly coordinated atom, or dangling end. For O and H atoms there are a number of possible schemes involving integer values of g and b. We give the simplest and most physical here. We choose $g_O = g_H = 3$ degrees of freedom and $b_{AO} = b_{AH} = 3$ bars. We also have $b_{OO} = b_{OH} = b_{HH} = 1$ bar. We give the later for completeness as of course an H_2 molecule is not part of the network and has 2(3) - 1 = 5 degrees of freedom as rotation along an axis through the bond is not allowed. Another trivial example would be a methane CH_4 molecule that has 6 + 4(3) - 4(3) = 6 degrees of freedom.

Thus the net effect is to treat the A-atoms as *bodies* with $g = 6$, and the O and H atoms as *points* with $g = 3$. Metal atoms that are incorporated into the network can be treated in a similar way, realizing that they have central forces only. Thus we have $g_M = 3$ degrees of freedom where the metal atom, M, acts as a point object also. In addition there are $b_{MA} = 3$ bars and $b_{MM} = b_{MO} = b_{MH} = 1$ bar.

In general such body-bar problems are hard to solve because some bodies are in fact points, which potentially brings back the problems of banana structures. In the special case where we restrict ourselves to only A-type atoms, as described previously, there is a Laman-type theorem, and a pebble game can be devised to do the bookkeeping to perform exact constraint counting.

The body-bar formalism gives us a nice scheme that makes calculations on Bethe lattices much simpler. This, as will be described in the next section, is the direct result of eliminating the need to consider next nearest neighbor interactions. It remains to be seen if Laman type theorems can be established for any of the cases above, to put this kind of mapping on a useful and rigorous basis. In the next section we give an example of how the body-bar formalism can be used on a Bethe lattice. The formalism is sufficiently simple that correlations between the coordinations of nearest neighbors can also be included.

Bethe Lattices

There is an important class of models for which the equations describing quantities associated with rigidity percolation can be obtained analytically. These are Bethe-lattice-type models with applied rigid busbar boundary conditions. This approach has been discussed for connectivity and rigidity in recent publications [24, 31] and is summarized in contributions to these proceedings [see articles by Leath and Zeng, and also by Moukarzel and Duxbury in these proceedings]. A Bethe lattice is shown in Figure 14, where the bodies are connected to the busbar in a tree-like fashion. The busbar is used to *nucleate* rigidity, but we obtain a solution asymptotically far from the busbar, where the busbar has no direct influence.

Previously random bond or site diluted Bethe lattices were considered analytically [24, 31]. For random bond dilution one considers a network in which all sites initially have the same prescribed coordination number z and then a certain fraction of the nearest neighbor bonds are removed randomly. In such a network the coordinations of sites from zero up to z are possible. The whole network is characterized by a single parameter p, the fraction of bonds remaining, and the concentrations of sites of given coordination is expressed in terms of this parameter. Usually, though, we have a glass of certain composition, where there is a *prescribed* number of atoms of given coordinations. Such networks are considered in the rest of this section.

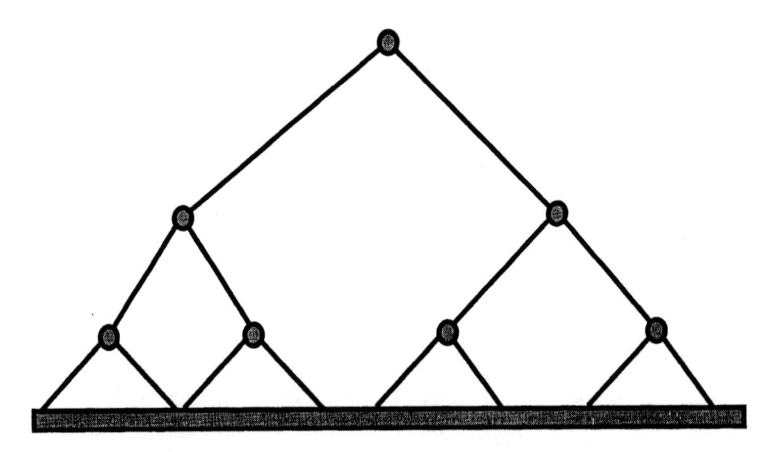

Figure 14. A piece of a generic Bethe lattice with a busbar and bodies with $b = 1$ bars and $z = 3$ neighbors.

With Maxwell constraint counting, it is found that for a three-dimensional network, with angular constraints, the rigidity transition occurs at an average coordination $\langle r \rangle_c = 2.4$. So the simplest approach to get the transition is to assume we have a network with two types of sites only, namely, 2- and 3-coordinated (for brevity, 2- and 3-sites). The concentration of 3-sites is denoted as x, then that of 2-sites is $1-x$ and the average coordination is $\langle r \rangle = 2+x$.

The parameter x (or $\langle r \rangle$) is not the only one needed to characterize a network of this type. Besides the average coordination, another issue of relevance is how the *chemical bonding* occurs. For example, an atom of a given type may tend to be preferentially connected to an atom of the *same* or *different* type. Actual situations may range from complete phase separation to perfect chemical ordering where there is always an atom of the opposite type between two atoms of the same type.

To describe these possible situations, we consider the numbers of bonds N_{ij} connecting atoms of types i and j (i-j bonds for brevity). We have 3 quantities: N_{22}, N_{33} and N_{23} of interest that must obey a conservation law. In the case of two-coordinated atoms, it follows that twice the number of 2-2 bonds plus the number of 2-3 bonds gives the total number of bonds from 2-sites (i.e. twice the number of 2-sites). Similarly it is with the three-coordinated atoms, so that we have

$$\begin{aligned} 2N_{22} + N_{23} &= 2N_2 \\ 2N_{33} + N_{23} &= 3N_3, \end{aligned} \tag{13}$$

where N_i is the number of i- sites. Since N_i can be expressed in terms of the concentration x, and the total number of sites N $[N_3 = xN,\ N_2 = (1-x)N]$, only one of the quantities N_{ij} is independent.

The total number of bonds is $N_B = N_{22} + N_{33} + N_{23}$ [this *sum rule* is not independent of (13)]. It is convenient to use *probabilities* for bonds of various types y_{22}, y_{23} and y_{33} defined as

$$y_{ij} = \frac{N_{ij}}{N_B} \tag{14}$$

instead of N_{ij}. Dividing (13) by N_B and using

$$N_B = \frac{\langle r \rangle N}{2} = \frac{2+x}{2} N, \tag{15}$$

we obtain relations for y_{ij}

$$\begin{aligned} 2y_{22} + y_{23} &= \frac{4(1-x)}{2+x}, \\ 2y_{33} + y_{23} &= \frac{6x}{2+x}, \end{aligned} \tag{16}$$

so that again we have two equations for 3 quantities, and these quantities can be expressed in terms of a single parameter. It is convenient to introduce the correlation parameter θ such that

$$y_{23} = 2\theta, \tag{17}$$

then

$$\begin{aligned} y_{22} &= \frac{2(1-x)}{2+x} - \theta, \\ y_{33} &= \frac{3x}{2+x} - \theta. \end{aligned} \tag{18}$$

We also have the sum rule on the probabilities

$$y_{22} + y_{23} + y_{33} = 1. \tag{19}$$

The parameter θ characterizes the chemical bonding properties of the network. At $\theta = 0$ there are no 2-3 bonds, thereby having complete phase separation (two separate networks, one 2-coordinated and another 3-coordinated). The maximum value of $\theta = 0.5$ is possible only at $x = 0.4$, because this corresponds to having only 2-3 bonds in the network, thereby having perfect chemical ordering, where every 2-site has only 3-sites as neighbors and vice versa.

Random Dilution. We consider how the number of 2-3 bonds is changed when dN_B bonds of the type 3-3 are removed (dN_B is presumed to be small compared to the total number of bonds). When we remove a 3-3 bond, the sites that were connected by this bond are converted to 2-coordinated from 3-coordinated. So if these sites were connected with other 3-sites, the bonds will convert from 3-3 type to 2-3 type and the number of 2-3 bonds will increase. On the other hand, if these sites were connected to 2-sites, the bonds will be converted to 2-2 from 2-3 and the number of 2-3 bonds will decrease. We do not allow any dangling ends. Then

$$dN_{23} = 4(p_3 - p_2)dN_B, \tag{20}$$

where p_i are the portions of i-sites connected to a former 3-site. The factor of 4 is because there are 4 untouched bonds going out of the two 3-sites between which the (fifth) bond is removed. We can find p_i from the following considerations. The total number of bonds going out of 3-sites is $(2y_{33} + y_{23})N_B$, of which $2y_{33}N_B$ are connected to 3-sites and $y_{23}N_B$ are connected to 2-sites. Then

$$\begin{aligned} p_2 &= \frac{y_{23}}{2y_{33} + y_{23}} = \frac{(2+x)\theta}{3x}, \\ p_3 &= \frac{2y_{33}}{2y_{33} + y_{23}} = 1 - \frac{(2+x)\theta}{3x}. \end{aligned} \tag{21}$$

On the other hand, ΔN_{23} can also be obtained using the definitions of y_{23}, θ and expression (15) for N_B:

$$dN_{23} = (2+x)Nd\theta + N\theta dx. \tag{22}$$

Using (15) again, we obtain

$$dN_B = \frac{N}{2}dx \tag{23}$$

and, finally, using (20), (21), (22) and (23), we obtain the differential equation

$$(2+x)\frac{d\theta}{dx} = -2 + \frac{8+x}{3x}\theta. \tag{24}$$

The starting network is fully 3-coordinated (i.e. when $x=1$, $\theta=0$). This is the initial condition for (10) and the solution is

$$\theta = \frac{6x(1-x^{1/3})}{2+x}. \tag{25}$$

This solution is shown in Figure 15 as the trajectory marked 3, which we may think of as *random bond dilution.* Figure 15 presents the phase diagram of the system. We use $\langle r \rangle = 2+x$ and $y_{23} = 2\theta$ as convenient axes. The lines labeled 1 and 2 correspond to equations $y_{33} = 0$ and $y_{22} = 0$ respectively. Outside the area bounded by them and the $\langle r \rangle$ axis, at least one of these probabilities becomes negative, so only points inside have a physical meaning (we refer to this area as the allowable region).

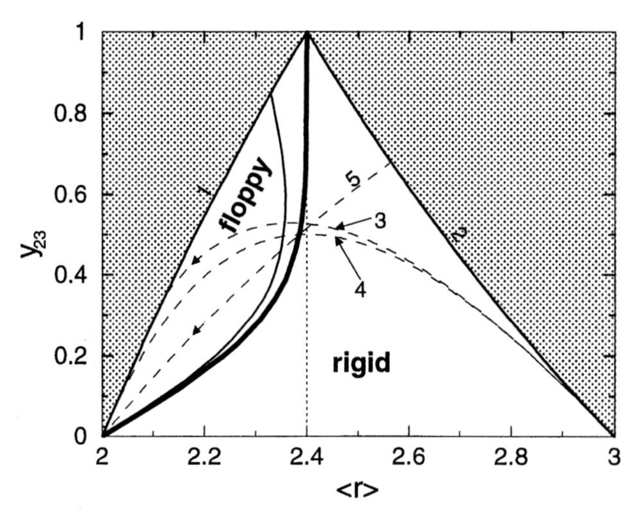

Figure 15. The phase diagram for a 2-3 network. The heavy line divides the rigid and floppy regions. The lines marked 1 and 2, mark the edge of the allowable region, and the trajectories 3 and 4 are described in the text. The heavy line is the first order phase boundary and the lighter line is the spinodal line. The trajectory marked 5 is described in the text.

Using the pebble game on Random Bond Model, the most convenient procedure to obtain a sequence of networks is bond dilution as described above. We can use this point in the phase diagram as a check as described below.

Later in this article, we study the Random Bond Model [RBM] which in the thermodynamic limit becomes a Bethe lattice as there are no loops or rings of bonds. We consider a three-dimensional RBM with angular constraints - the bond-bending model discussed earlier. In practice it is often convenient, especially on the RBM, to start with a fully 3-coordinated network and remove 3-3 bonds only one by one, so that no dangling bonds are created. The network obtained in such a way is *not* completely random, but is the *random bond dilution* situation discussed in this section. We consider this situation here only to make the comparison with pebble game results on the RBM possible.

Random Bonding. We now discuss completely *random bonding,* where bonds are linked up in a random fashion, without regard for their coordinations. Assume for a while that the bonds are directed (i.e. we distinguish between their beginnings and ends). Then random bonding means that the probability that a randomly chosen bond ends at a given site is proportional to the number of bonds going out of this site (i.e. its coordination number). Then $y_{ii} \propto i^2x_i^2$, $y_{ij} \propto 2ijx_ix_j$, where x_i is the concentration of a site of type i, so that $x_2 = 1-x$, $x_3 = x$. The factor of 2 is because 2-3 bonds may end either with a 2- or with a 3-site. The normalization constant can be found from the sum rule (19), and the final expression for θ for random bonding is

$$\theta = \frac{6x(1-x)}{(2+x)^2}. \tag{26}$$

The trajectory 4 represents Eq. (26), which corresponds to random networks as described above. The trajectories 3 and 4 are relatively close to each other, but the notable feature of trajectory 3 is that it does not reach $\langle r \rangle = 2$, but crosses the boundary 1 of the allowable region at $\langle r \rangle = 2.125$. Physically this means that even though at this concentration 3-sites are left in the network, there are no more 3-3 bonds, so that further dilution becomes impossible.

The general idea is to map a RBM onto a Bethe lattice equivalent to the RBM in the sense that the bulk rigidity properties of the tree are the same as the network. This equivalence is based on the fact that both have no rings, so they are equivalent as long as the bonding properties (i.e. the probabilities of certain types of bonds) are the same.

The properties of a Bethe lattice are the following. It has some 2-sites and some 3-sites. All sites always have one bond directed *away* from the busbar shown in Figure 14 and the rest *towards* the busbar. The probabilities for various kinds of site to be connected to the site under consideration are the same as in a RBM, so that for a 3-site they are given by (21)

$$n_2^{(3)} \equiv p_2 = \frac{y_{23}}{2y_{33} + y_{23}} = \frac{(2+x)\theta}{3x},$$
$$n_3^{(3)} \equiv p_3 = \frac{2y_{33}}{2y_{33} + y_{23}} = 1 - \frac{(2+x)\theta}{3x}. \quad (27)$$

and for a 2-site they can be obtained in a similar way:

$$n_2^{(2)} = \frac{2y_{22}}{2y_{22} + y_{23}} = 1 - \frac{\theta(2+x)}{2(1-x)},$$
$$n_3^{(2)} = \frac{y_{23}}{2y_{22} + y_{23}} = \frac{\theta(2+x)}{2(1-x)}. \quad (28)$$

Here $n_i^{(j)}$ is the probability that a randomly chosen bond from a site that is known to be of a type j connects it to the site of a type i. For the *completely random network* $n_i^{(j)}$ does not depend on j, so that $n_2^{(3)} = n_2^{(2)}$ and $n_3^{(3)} = n_3^{(2)}$ when θ is given by (26).

Bethe Lattice Solution

Obtaining equations for this correlated Bethe lattice model is a rather difficult task if one needs to deal with both nearest-neighbor and next-nearest-neighbor constraints (the latter representing angular constraints). However, it is possible to find another equivalent model which has no angular constraints. This is where the body-bar formalism for type A atoms described previously is very useful.

Now the model is similar to the Bethe lattice model considered previously [24, 31], but with pre-assigned correlations. We introduce quantities $T_{i(j)}$ $(i = 0,\ldots,6\ ,\ \ j = 2, 3)$. These are the probabilities that a j-coordinated site has i degrees of freedom with respect to the busbar if this bond going *away* from the busbar is ignored as if it were non-existent when the motion of the site is considered. Such a definition allows us to ignore the levels of the Bethe lattice other than the considered one, and write down the probabilities for the given level in terms of those for the previous level. Then, since the Bethe lattice solution, a long way from the busbar is sought, the probabilities for the given and previous levels are set equal. Then the equations are

$$
\begin{aligned}
T_{6(2)} &= n_2^{(2)} X_{(2)} + n_3^{(2)} X_{(3)} \\
T_{5(2)} &= n_2^{(2)} T_{4(2)} + n_3^{(2)} T_{4(3)} \\
T_{4(2)} &= n_2^{(2)} T_{3(2)} + n_3^{(2)} T_{3(3)} \\
T_{3(2)} &= n_2^{(2)} T_{2(2)} + n_3^{(2)} T_{2(3)} \\
T_{2(2)} &= n_2^{(2)} T_{1(2)} + n_3^{(2)} T_{1(3)} \\
T_{1(2)} &= n_2^{(2)} T_{0(2)} + n_3^{(2)} T_{0(3)} \\
T_{0(2)} &= 1 - T_{1(2)} - T_{2(2)} - T_{3(2)} - T_{4(2)} - X \\
T_{6(3)} &= \left(n_2^{(3)} X_{(2)} + n_3^{(3)} X_{(3)}\right)^2 \\
T_{5(3)} &= 2\left(n_2^{(3)} T_{4(2)} + n_3^{(3)} T_{4(3)}\right)\left(n_2^{(3)} X_{(2)} + n_3^{(3)} X_{(3)}\right) \\
T_{4(3)} &= 2\left(n_2^{(3)} T_{3(2)} + n_3^{(3)} T_{3(3)}\right)\left(n_2^{(3)} X_{(2)} + n_3^{(3)} X_{(3)}\right) + \left(n_2^{(3)} T_{4(2)} + n_3^{(3)} T_{4(3)}\right)^2 \\
T_{3(3)} &= 2\left(n_2^{(3)} T_{2(2)} + n_3^{(3)} T_{2(3)}\right)\left(n_2^{(3)} X_{(2)} + n_3^{(3)} X_{(3)}\right) + 2\left(n_2^{(3)} T_{3(2)} + n_3^{(3)} T_{3(3)}\right)\left(n_2^{(3)} T_{4(2)} + n_3^{(3)} T_{4(3)}\right) \\
T_{2(3)} &= 2\left(n_2^{(3)} T_{1(2)} + n_3^{(3)} T_{1(3)}\right)\left(n_2^{(3)} X_{(2)} + n_3^{(3)} X_{(3)}\right) + 2\left(n_2^{(3)} T_{2(2)} + n_3^{(3)} T_{2(3)}\right)\left(n_2^{(3)} T_{4(2)} + n_3^{(3)} T_{4(3)}\right) \\
&\quad + \left(n_2^{(3)} T_{3(2)} + n_3^{(3)} T_{3(3)}\right)^2 \\
T_{1(3)} &= 2\left(n_2^{(3)} T_{0(2)} + n_3^{(3)} T_{0(3)}\right)\left(n_2^{(3)} X_{(2)} + n_3^{(3)} X_{(3)}\right) + 2\left(n_2^{(3)} T_{1(2)} + n_3^{(3)} T_{1(3)}\right)\left(n_2^{(3)} T_{4(2)} + n_3^{(3)} T_{4(3)}\right) \\
&\quad + 2\left(n_2^{(3)} T_{2(2)} + n_3^{(3)} T_{2(3)}\right)\left(n_2^{(3)} T_{3(2)} + n_3^{(3)} T_{3(3)}\right),
\end{aligned}
\tag{29}
$$

where

$$
\begin{aligned}
X_{(2)} &= T_{5(2)} + T_{6(2)} \\
X_{(3)} &= T_{5(3)} + T_{6(3)}.
\end{aligned}
\tag{30}
$$

It is convenient to introduce the quantities

$$
\begin{aligned}
T_i^{(j)} &= n_2^{(j)} T_{i(2)} + n_3^{(j)} T_{i(3)} \\
X^{(j)} &= n_2^{(j)} X_{(2)} + n_3^{(j)} X_{(3)},
\end{aligned}
\tag{31}
$$

then

$$
\begin{aligned}
T_6^{(j)} &= n_2^{(j)} X^{(2)} + n_3^{(j)} X^{(3)2} \\
T_5^{(j)} &= n_2^{(j)} T_4^{(2)} + 2 n_3^{(j)} T_4^{(3)} X^{(3)} \\
T_4^{(j)} &= n_2^{(j)} T_3^{(2)} + n_3^{(j)} \left(2 T_3^{(3)} X^{(3)} + \left(T_4^{(3)}\right)^2\right) \\
T_3^{(j)} &= n_2^{(j)} T_2^{(2)} + 2 n_3^{(j)} \left(T_2^{(3)} X^{(3)} + T_3^{(3)} T_4^{(3)}\right) \\
T_2^{(j)} &= n_2^{(j)} T_1^{(2)} + n_3^{(j)} \left(2\left(T_1^{(3)} X^{(3)} + T_2^{(3)} T_4^{(3)}\right) + \left(T_3^{(3)}\right)^2\right) \\
T_1^{(j)} &= n_2^{(j)} T_0^{(2)} + 2 n_3^{(j)} \left(T_0^{(3)} X^{(3)} + T_1^{(3)} T_4^{(3)} + T_2^{(3)} T_3^{(3)}\right) \\
T_0^{(j)} &= 1 - T_1^{(j)} - T_2^{(j)} - T_3^{(j)} - T_4^{(j)} - X^{(j)}.
\end{aligned}
\tag{32}
$$

The physical meaning of definitions (32) is that we average $T_{i(j)}$ over the possible values of j with the weights equal to the probabilities that the j-site is connected to a site; the type of which is determined by the upper index in (31). The set of Eqs. (32) is closed and can therefore can be solved. The rigidity properties, such as the number of floppy modes per site in the network, can be determined from $T_i^{(j)}$.

A question regarding the Bethe lattice solution is the following. In Bethe lattices not all the bonds are equivalent; there are the bonds that are directed *towards* the busbar and ones that are directed *away* from the busbar. Each site *always* has one bond of the second kind, while the number of the bonds of the first kind depends on the coordination of that site. Alternatively for a 2-site we could choose 2 bonds randomly out of the maximum of 3 bonds of both kind, so that the bond of the second kind would sometimes exist and sometimes not exist. Which is correct, in that the procedure should agree with the RBM in the thermodynamic limit? The answer is that this is just a matter of choice. Choosing the model as we do, means we define $T_{i(j)}$ as the probabilities for i degrees of freedom when one bond (the one directed away from the busbar) is always ignored, with a different definition we would ignore it sometimes and not ignore it at other times, so $T_{i(j)}$ would be modified, but the final answer would be the same. Our choice turns out to be the most convenient one, and we will understand later why.

For completely random networks $n_i^{(j)}$, as was mentioned above, do not depend on j. Then Eqs. (32) are also independent of the upper indices, so the number of independent equations is halved. Eqs. (32) always have a trivial solution, i.e. $T_i^{(j)} = 0$ for $i = 0,\ldots,5$ and $T_6^{(j)} = 1$. This corresponds to a floppy network. There is also a region where there are non-trivial solution(s). The boundary between these regions is the spinodal line [24].

We can get the non-trivial solution of (32) for two important limiting cases. First of all, when there are no 2-3 bonds ($\theta = 0$), $n_2^{(2)} = n_3^{(3)} = 1$, $n_2^{(3)} = n_3^{(2)} = 0$ and it is easy to check that $T_0^{(3)} = 1$ and all the other $T_i^{(j)} = 0$. In this case we actually have 2 separate networks, one 2-coordinated and one 3-coordinated, and the first one is, of course, floppy, while the second one is rigid with all sites having 0 degrees of freedom.

The other case, which will be important to us later, is the right boundary of the allowable region, where there are no 2-2 bonds. In this case $n_2^{(2)} = 0$, $n_3^{(2)} = 1$, $n_2^{(3)} = 2(1-x)/3x$, $n_3^{(3)} = 1 - 2(1-x)/3x$ and again, it is easy to check that $T_0^{(2)} = 1$, $T_0^{(3)} = n_3^{(3)}$, $T_1^{(3)} = n_2^{(3)}$. It seems from this solution that some 3-sites have 1 degree of freedom, but we should remember that we ignore some bonds, and actually there are no floppy modes in the network at all.

We would like to determine the number of Floppy modes F everywhere in the allowable region. The simplest way to do this we could find is the following. We start with the network corresponding to some point with $x = x_0$ on the right boundary. We know that everywhere at the right boundary F is zero. Now we pick an arbitrary 2-site and insert another site (which will become a 2-site) into any of the two bonds going out of the chosen site. This is a way to increase the fraction of 2-sites. It is clear that when the number of sites goes to infinity, there will be an infinitesimal portion of 3-sites, so $x = 0$.

Now we are going to find the trajectories in the $x - \theta$ plane corresponding to site insertion described above. The first relation we use is that the change in the number of sites equals the change in the number of bonds:

$$\Delta N = \Delta N_B \tag{33}$$

The change in the number of 2-3 bonds is 0, because we always insert a site into a 2-2 bond (in which case two 2-2 bonds replace one 2-2 bond, no change in 2-3) or 2-3 bond (in which case instead of 2-3 bond we get 2-2 and 2-3 bonds, so again no change in the number of 2-3). On the other hand,

$$N_{23} = y_{23}N_B = (2+x)N\theta, \tag{34}$$

so

$$\Delta N_{23} = N\theta\Delta x + \theta(2+x)\Delta N + (2+x)N\Delta\theta = 0 \tag{35}$$

Now we use $N_B = N(2+x)/2$ (which was also used in (34)).

$$\Delta N = \Delta N_B = \frac{2+x}{2}\Delta N + \frac{N}{2}\Delta x,$$
$$x\Delta N + N\Delta x = 0 \Rightarrow Nx = \text{const},$$

$$N = N_0\frac{x_0}{x}, \quad \Delta N = -\frac{N_0 x_0}{x^2}\Delta x. \tag{36}$$

Inserting (36) into (35), we obtain

$$\begin{aligned}&\theta\Delta x - \frac{\theta(2+x)}{x}\Delta x + (2+x)\Delta\theta = 0,\\ &(2+x)\frac{d\theta}{dx} = \frac{2\theta}{x}\end{aligned} \tag{37}$$

$$\theta = \frac{Cx}{2+x} \tag{38}$$

The value of C is determined by x_0. The latter can be found as the crossing point of (38) and the right boundary. When x_0 is changed from 0.4 to 1, C changes from 3 down to zero. The value of θ in Eq. (38) for $C=3$ coincides with the left boundary, while $C=0$ corresponds to the bottom boundary ($\theta=0$), so while C is changed from 0 to 3 we cover the whole allowable region. An example of one of the curves (38) is shown as the trajectory 5 in Figure 15.

It is easy to find the average change in F when one site is inserted. Just by counting floppy modes, it turns out that it is not changed when one of the neighbors of a 2-site near which the insertion occurs has 0 degrees of freedom when the connection with the 2-site is ignored (the corresponding probability is $T_0^{(2)}$, hence the convenience of averaging and ignoring a bond in defining $T_i^{(j)}$) and another has 0, 1, 2 or 3 degrees of freedom in the same way; or one site has 1 degree of freedom and another has 1 or 2. Otherwise the F is increased by 1. So in a single insertion act the change in F is

$$\Delta F = 1 - T_0^{(2)}\left(T_0^{(2)} + 2\left(T_1^{(2)} + T_2^{(2)} + T_3^{(2)}\right)\right) - T_1^{(2)}\left(T_1^{(2)} + 2T_2^{(2)}\right) \tag{39}$$

Of course, F itself depends on the network size, so we should use some size-independent quantity instead, we use $f = F/3N$, which may vary from 0 to 1. Since the number of sites is changed in the process of insertion, F and f are *not* just proportional.

It is convenient to calculate the derivative df/dx. Using the definition of f and (40), we find that

$$\frac{df}{dx} = \frac{-(\Delta F - 3f)}{3x}. \tag{41}$$

Then f can be determined by the (numerical) integration of (41) as

$$f(x) = \int_x^1 \frac{df}{dx}\, dx \tag{42}$$

The result will be correct as long as we choose the correct branch of the solution of (32) in calculating (39). We know the solution of (32) on the right boundary. We can continue it smoothly inside the allowable region (in principle, until we reach the spinodal line). We know, however, that it will be valid only before we reach the transition, at the transition the $T_i^{(j)}$ jump to their trivial values. We do not know a priori where the transition occurs, but we can determine this knowing, firstly, f in the floppy region (where the trivial solution is realized) – Maxwell counting gives the correct result

$$f = 2 - \tfrac{5}{6}\langle r\rangle \tag{43}$$

in this region, and secondly, that f does not jump at the transition (the derivative f' does). So we can just plot (42) and (43) so that the point where they intersect is the transition point. This procedure is shown in Figure 16, where f is plotted for the trajectory shown as trajectory 5 in Figure 15.

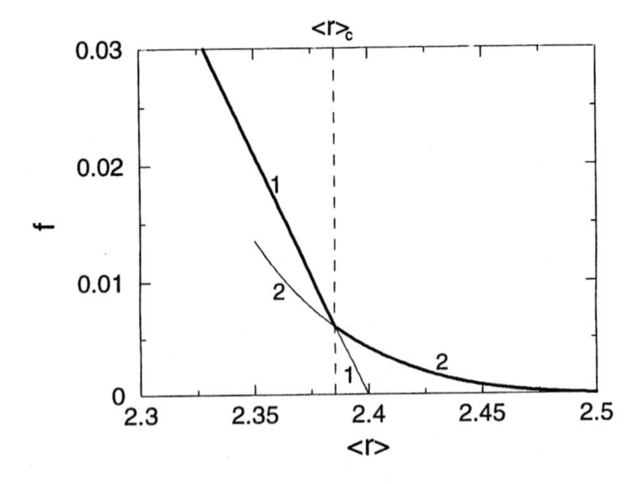

Figure 16. Showing the crossing of the two solutions for f at the first order rigidity transition.

Changing C in (38), we can get the whole transition line in the $x-\theta$ plane. The spinodal line is shown as the thinner line just left of the phase transition line. An interesting property of the phase boundary is that it goes from one corner of the allowable region to another. Starting at the lower left corner is quite obvious, because all the networks on the $y_{23} = 0$ line consist of the floppy 2-coordinated part and rigid 3-coordinated part, the latter is infinite and propagates through the whole system, so these networks are rigid. The less obvious thing is that the line ends at the upper corner, where $y_{23} = 1$ and $\langle r \rangle_c = 2.4$, so that Maxwell's constraint counting is exact at this point. This follows from the fact that the non-trivial solution gives $f = 0$ on the whole right boundary, the trivial solution has $f = 0$ at $\langle r \rangle = 2.4$, and the right boundary crosses $\langle r \rangle = 2.4$ line right in the upper corner. This point is particularly interesting as it corresponds to chemically alternating networks like As_2Se_3 and As_2S_3.

To derive the expression for ΔF for bond dilution, we first define the probabilities T_i for a 3-site to have i degrees of freedom with respect to the busbar if one of the bonds going out of it is ignored. These are given by

$$
\begin{aligned}
T_6 &= \left(X^{(3)}\right)^2 \\
T_5 &= 2X^{(3)}T_4^{(3)} \\
T_4 &= 2X^{(3)}T_3^{(3)} + \left(T_4^{(3)}\right)^2 \\
T_3 &= 2\left(X^{(3)}T_2^{(3)} + T_3^{(3)}T_4^{(3)}\right) \\
T_2 &= 2\left(X^{(3)}T_1^{(3)} + T_2^{(3)}T_4^{(3)}\right) + \left(T_3^{(3)}\right)^2 \\
T_1 &= 2\left(X^{(3)}T_0^{(3)} + T_1^{(3)}T_4^{(3)} + T_2^{(3)}T_3^{(3)}\right) \\
T_0 &= 1 - T_1 - T_2 - T_3 - T_4 - T_5 - T_6
\end{aligned}
\tag{43}
$$

When we remove a bond between two sites that both have 0 degrees of freedom in the above-mentioned sense (which corresponds to T_0), the number of floppy modes does not change. If one site has 0 degrees of freedom and another site has one, the change is by 1. In general, for sites having i and j degrees of freedom, the change is by minimum of the two numbers: $i + j$ and 5 (the number of bonds in a bar). Thus

$$\Delta F = 5 - 5T_0^2 - 8T_0T_1 - 3T_1^2 - 6T_0T_2 - 4T_0T_3 - 4T_1T_2 - 2T_0T_4 - 2T_1T_3 - T_2^2 \tag{44}$$

and using the definition of f and the fact that N is constant in this procedure, we have

$$\frac{df}{dx} = -\frac{\Delta F}{6}. \tag{45}$$

Eq. (45), unlike (40), does not contain f, so we do not need to know f (obtained by integration), for which we would need to know df / dr everywhere on the curve. All we need to calculate df / dr at a given point are the values of T_i at the same point. Of course, this is not true if we need f or the transition point.

This section has been long as we wanted to give the details of this Bethe solution of rigidity for a body-bar network with chemical ordering between neighbors. This is the first calculation of its kind and shows that the Maxwell estimate of $\langle r \rangle_c = 2.4$ is exact for chemically alternating 2-3 networks and only slightly high for networks which have completely random bonding where $\langle r \rangle_c = 2.3855$. For networks where there is a tendency for like-atoms to cluster, which would be very unusual in covalent glassy materials, the critical point may lie anywhere between 2 and 2.3855. It is clear that any body-bar type network of interest can be solved using the techniques in this section, which include can include appropriate chemical clustering parameters. The drawback is that the transition appears to be always first order, but as we will see in the next section, the *location* of the transition seems to be very robust and practically independent of the order of the transition.

6. RANDOM BOND MODELS

We have constructed networks for glasses like $Ge_xAs_ySe_{1-x-y}$ by randomly positioning points in the plane as shown in Figure 17.

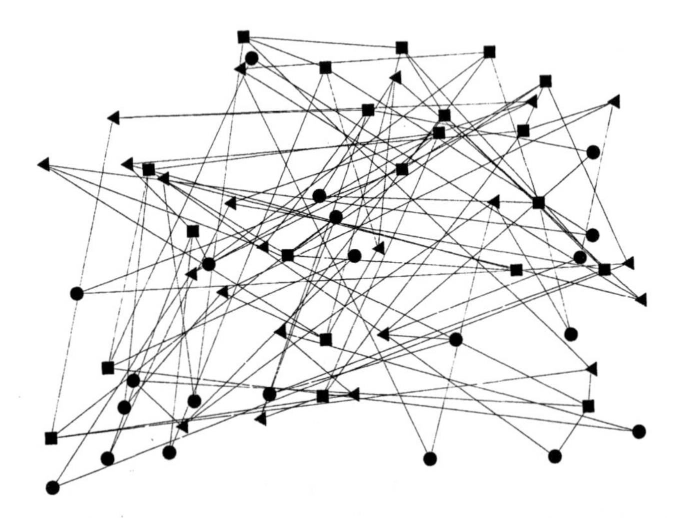

Figure 17. The squares are four-fold coordinated sites, the triangles are three-coordinated sites and the circles are two-fold coordinated sites.

The main feature of these Random Bond Models (RBM) is that there are no loops or rings of bonds in the thermodynamic limit [37] and so they become equivalent to the Bethe lattice solution which was described in the previous section. The way rigidity is nucleated is rather interesting and a little different in the two cases. The transition is usually first order on these networks without loops, and in the Bethe lattice, rigidity nucleates from the busbar. In the RBM shown in Figure 17, there is no busbar and so rigidity must nucleate using the (few) large rings that are present on any finite RBM.

In Figure 18, we compare results for the 2-3 network obtained in the previous section for random bond dilution (along trajectory 3) with the results obtained for the Bethe lattice in the previous section. It can be seen that the agreement is precise, as anticipated. The small area around the first order transition is blown up and shown in Figure 19, where it can be seen that the transition sharpens up as the number of sites in the RBM is increased as would be expected.

The transition from rigid to floppy occurs at $\langle r \rangle_c = 2.3893$. This result is useful as it serves as a useful check of the complicated equations of the previous section, and further establishes the equivalence of the Bethe lattice solution and the RBM.

We note that the point where trajectory 3 crosses the phase boundary in Figure 15, is above the completely random case shown as trajectory 4, where the transition is slightly lower at $\langle r \rangle_c = 2.3855$

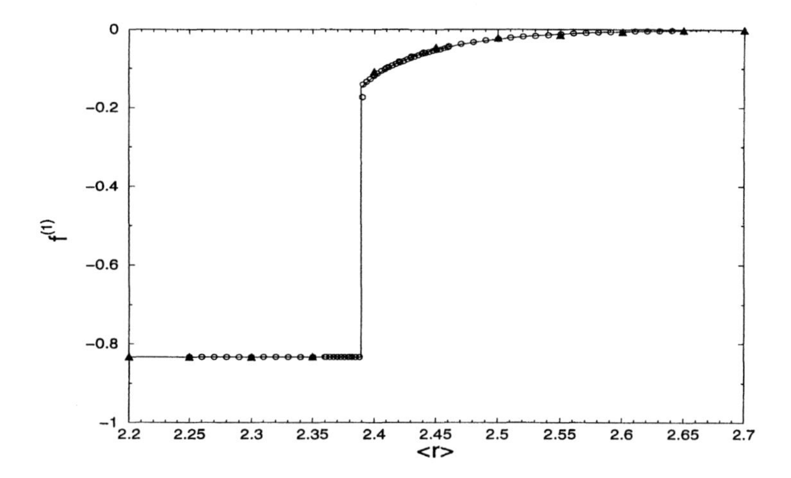

Figure 18. A plot of the derivative of the fraction of floppy modes, $f^{(1)}$, plotted against the mean coordination $\langle r \rangle$. This is a comparison of the Bethe lattice solution for the 2-3 network along the bond dilution trajectory marked 3 in Figure 15, with the RBM. The solid triangles are for a sample with 32,768 sites and the open circles for a sample with 8,000 sites.

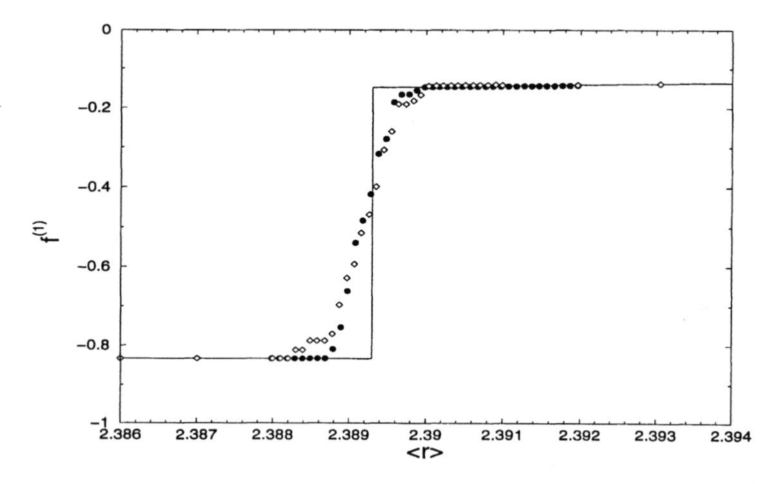

Figure 19. A blow up of the transition region in Figure 18. This is a comparison of the Bethe lattice solution for the 2-3 network along the bond dilution trajectory marked 3 in Figure 15, with the random bond model. The open diamonds are for a sample with 103,823 sites (averaged over 30 samples) and the solid circles are for a sample with 262,144 sites (averaged over 37 samples).

Nucleating Rings. It is interesting to speculate how the first order transition on networks without rings would evolve into the second order transition on *real* lattices with rings. We have attempted to answer this from both ends. We have introduced lots of triangles into a RBM with nearest neighbor central forces, by connecting up the sites which were *initially twelve-fold* coordinated so that there were lots of triangles. We could vary the number of triangles up to some maximum. The network was then randomly diluted and the jump at the first order transition was monitored, as shown in Figure 20. It can be seen that indeed the discontinuity does decrease continuously until the number of triangles per site

$N_3 \approx 0.2$, where there appears to be a tricritical point and the transition becomes second order. These are preliminary results obtained by F. Babalievsky, but are very suggestive that the number of small nucleating rings is the controlling factor in whether the transition is first or second order.

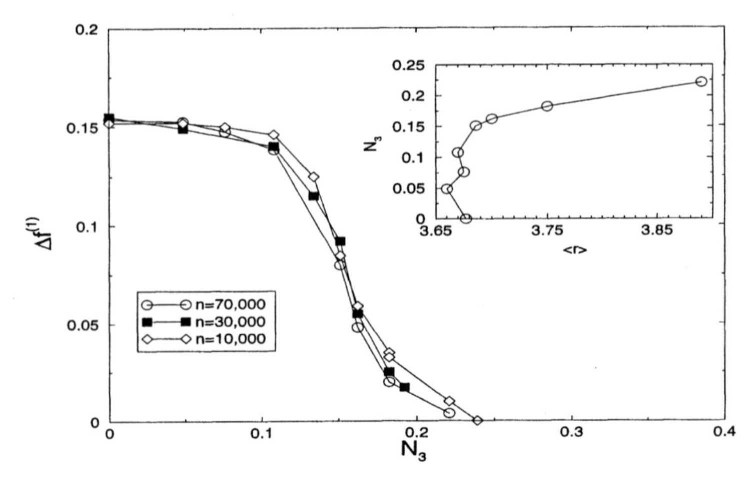

Figure 20. Showing the jump in $f^{(1)}$ as the number of triangles in a nearest neighbor central force model is increased. A tricritical point appears when N_3, the number of triangles per site, is around 0.2. The insert shows the first order part of the phase boundary up to the tricritical point.

We therefore introduce the notion of a *nucleating ring*, which is a ring that would be rigid in isolation, and can act as a nucleation center for rigidity. For nearest neighbor central force networks, this restricts us to triangles, which are isostatic, as all larger polygons are floppy. It was for this reason that we introduced *triangles* into the RBM. For bond-bending networks, all rings of size 6 or less are rigid, while rings of size seven and larger are floppy. A six-fold ring is isostatic, a five-fold ring has one overconstrained bond etc. Not all nucleating rings are *relevant*. This is illustrated in Figure 21.

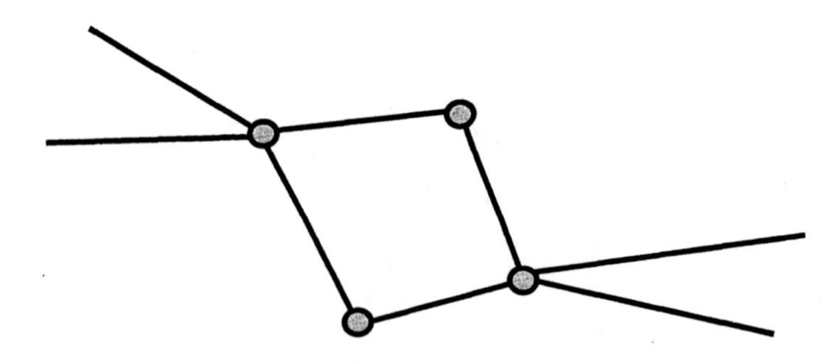

Figure 21. Showing a four-fold ring, which is rigid in a bond-bending network, but with only two pairs of bonds connecting to the rest of the network. This ring can be renormalized to a point and is therefore irrelevant.

In an attempt to induce a first order transition in a three dimensional bond-bending network, we have taken a diamond lattice and removed bonds in a random manner until we have a network containing *no* six and eight membered rings. There are therefore *no* nucleating rings as defined in this section. In order to have a parameter to tune, we could also leave some rings in the network.

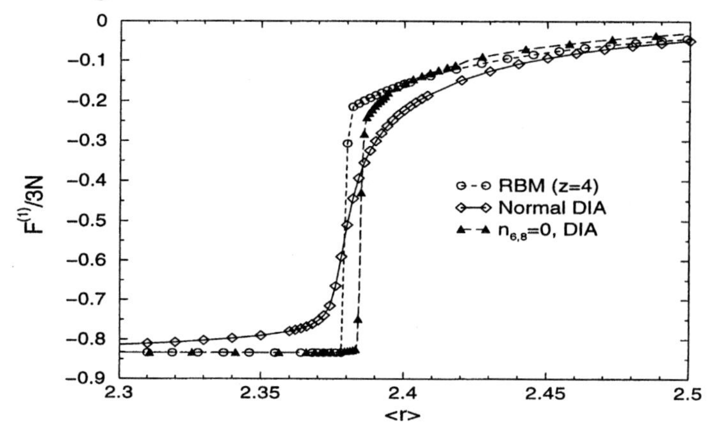

Figure 22 Showing the first derivative of the number of floppy modes $F^{(1)}$ as a function of the mean coordination $\langle r \rangle$. The open circles are the RBM, and the solid triangles are from the diluted diamond lattice that contains *no* six and eight fold rings. The open diamonds are from a diluted diamond lattice.

It can be seen in Figure 22 that in the absence of six and eight membered rings we have a first order transition that is very close to that obtained in a RBM with initial coordination of four. This substantiates our assertion that it is the small nucleating rings that determine whether the transition is first or second order. We monitored the transition as a function of the number of nucleating rings at the transition. This is shown in Figure 23.

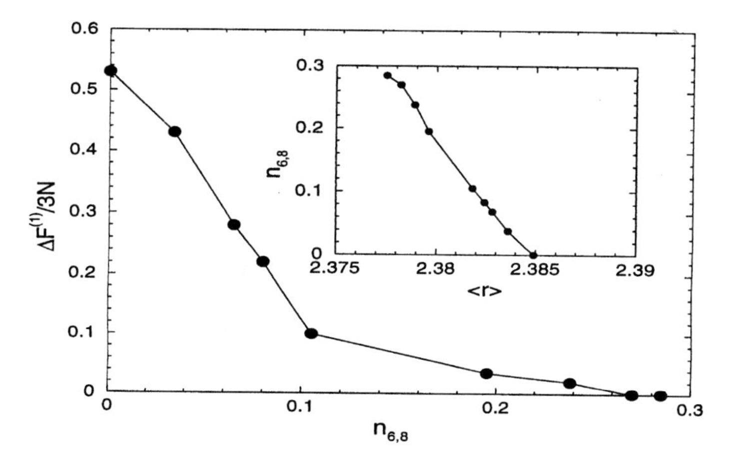

Figure 23. Showing the transition on a bond-bending network with a variable number of nucleating rings $n_{6,8}$ at the transition. These preliminary results suggest a tricritical point at $n_{6,8} \approx 0.28$, when the location of the transition has moved down from $\langle r \rangle = 2.385$ to $\langle r \rangle = 2.376$, when the transition becomes second order. The ordinate is the estimated jump at the first order transition in the first derivative of the number of floppy modes $F^{(1)}$. The insert shows the first order part of the phase boundary up to the tricritical point.

We see that the concept of nucleating rings is perhaps a bit too simplistic, and that larger nucleating regions probably may need to be considered. An example of an isostatically rigid region in a central force network is shown in Figure 24. Note that this piece of network contains no triangles (nucleating rings).

Nevertheless it seems that the evidence is fairly strong that it is the presence of small nucleating rings that determines whether the transition is first or second order, with the nucleating rings required to prevent rigidity spreading in a catastrophic way as the coordination is increased. It is interesting to note that there is some recent evidence for a first order transition found in very accurate measurements of the Raman scattering in chalcogenide glasses and reported on at this meeting by Boolchand [38]. We note that edge sharing tetrahedra have rings as shown in Figure 21, which are *not* nucleating and so would favor a first order rigidity transition. Experiments are very difficult and time consuming in this area as a new sample has to be made for each desired mean coordination, and it is necessary to have very many samples to go through the transition in small steps.

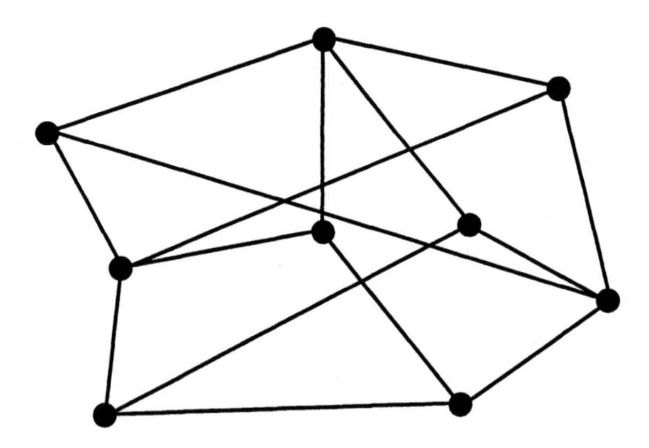

Figure 24. An isostatically rigid piece of central force network that contains no triangles.

7. EXPERIMENTS

Bulk Materials

The above findings have been confirmed using computations on model networks. Some of these results are shown in Figure 25. The computed number of floppy modes (shown in the insert) closely follows Eq. (6) and the elastic constants approach zero from the high coordination side, also at $\langle r \rangle_c = 2.4$, which is referred to as the point at which *rigidity percolation* occurs. This is a phase transition from rigid to floppy, driven not by temperature, but rather by the mean coordination.

Measurements of the elastic constants [39] appear to be influenced considerably by the weak forces, and the phase transition is washed out. The best experimental confirmation of these ideas to date comes from inelastic neutron scattering measurements of the density of states [39], shown in Figure 26. The agreement between theory and experiment is good, even when the weak forces are included in a very simple way and adjusted to bring the zero–frequency modes to the correct (low) frequency. Note that the weight in the floppy modes, given by constraint counting, is unaffected by the weak forces, even though there is some background response to contend with.

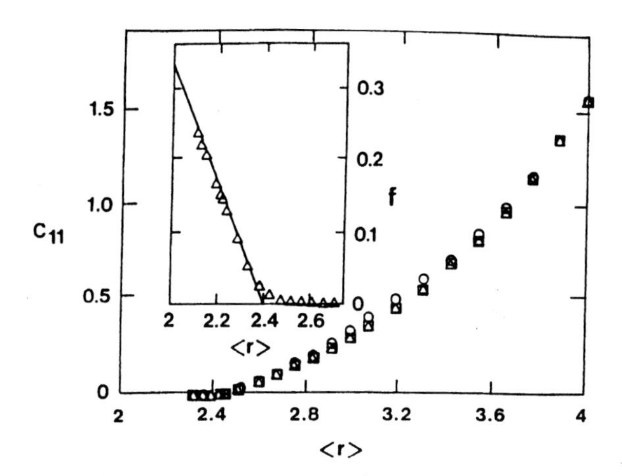

Figure 25. Showing the elastic constant c_{11} for a model network as a function of the mean coordination $\langle r \rangle$ for three different series of random networks. The insert shows the fraction of floppy modes, f. The points are from computer simulation [16], and the solid line in the insert is the straight line given by Eq. (6)

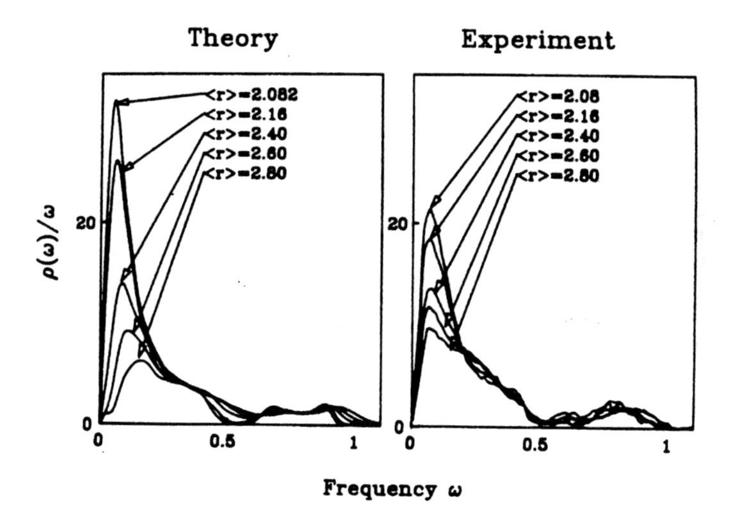

Figure 26. The density of states divided by the frequency, $\rho(\omega)/\omega$, for various values of the mean coordination, $\langle r \rangle$. The left hand set of graphs are theoretical [10] and the right set of graphs are from inelastic neutron scattering experiments [39] on Ge_xSe_{1-x}. The frequency is in units where the maximum frequency is unity.

Correction For Dangling Bonds

The previous section describes the situation when there are *no* dangling bonds present. When dangling bonds are present Eq. (6) needs correction because the expression for the number of angular forces $2r-3$ gives -1 when $r=1$, instead of the correct answer of zero. This has recently been done in a compact way by a number of authors [40-42] following earlier efforts [43, 9].

It is rather straightforward to extend Eq. (6) to include the summation over the dangling bonds ($r=1$) and correct for the miscounting of the angular constraints to give

$$f=\left[3N-\sum_{r=1}^{4} n_r[r/2+(2r-3)]-n_1\right]/3N \tag{46}$$

which now leads to the form

$$f=2-\frac{5}{6}\langle r\rangle-\frac{n_1}{3N} \tag{47}$$

where the definition of $\langle r\rangle$ is extended from (2) to include the dangling bonds. The transition now takes place at a lower mean coordination $\langle r\rangle$ which is given by

$$\langle r\rangle_c=2.4-0.4\frac{n_1}{N}. \tag{48}$$

It is not surprising that the transition takes place at a lower mean coordination $\langle r\rangle$, because the dangling ends play no role in the network connectivity. Indeed another conceptual approach is to strip the dangling ends away and define a *skeleton network* that has only 2, 3 and 4 coordinated atoms. The theory described in this section can then be applied to the skeleton network. Equation (48) has recently been applied to networks containing iodine, which forms a dangling end [41, 42].

These ideas have also been applied to amorphous carbon networks by Tamor and are described in [39]. Amorphous carbon networks can be thought of as consisting of three kinds of atoms: fourfold (diamond–like) carbon, threefold (graphitic) carbon, and often considerable amounts of atomic hydrogen that ties off dangling ends and so is singly coordinated. Suppose that there are N atoms in the network, with a fraction x_4 of fourfold coordinated carbon, a fraction x_3 of threefold coordinated carbon and x_1 of singly boned atomic hydrogen. Then by definition, we have

$$x_4+x_3+x_1=1. \tag{49}$$

To illustrate the use of the skeleton network, we apply it to this case. The mean coordination of the skeleton network, with the hydrogen removed is

$$\langle r\rangle=\frac{2(\textit{Number of bonds})}{\textit{Number of sites}}=\frac{(4Nx_4+3Nx_3+Nx_1)-2Nx_1}{Nx_4+Nx_3} \tag{50}$$

which gives

$$\langle r\rangle=\frac{4x_4+3x_3}{1-x_1}-\frac{x_1}{1-x_1} \tag{51}$$

The important new term is the $2Nx_1$ in the numerator of Eq. (51), which removes the bonds between hydrogen and the rest of the network that are not present in the skeleton network. In deriving Eq. (52), we have assumed that there are no molecular fragments, like for example methane CH_4, that get detached from the network. If this were the case, then these fragments should also be eliminated for purposes of counting. It appears that the elasticity and especially the hardness properties of carbon networks [41] follows Eq. (52) closely. Indeed the hardness goes all the way from close to that of crystalline diamond at the one extreme, to that of mush when too much hydrogen is present.

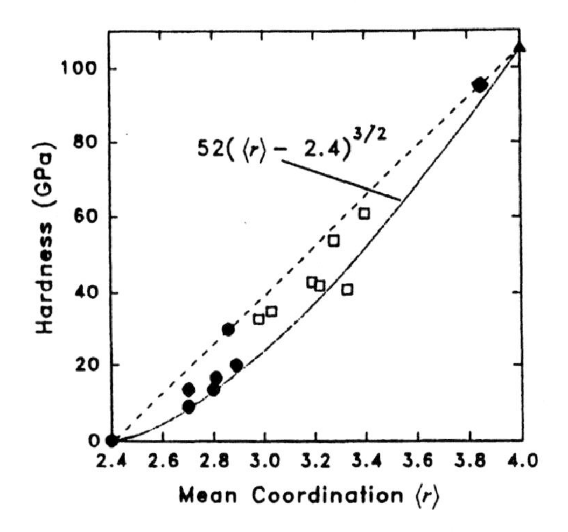

Figure 27. The hardness (measured by nano-indentation) of various diamond films, some containing some hydrogen, as a function of the mean coordination as defined by Eq. (52). The data were compiled and presented in this form by Tamor [41] based on previous work by Tamor and others [41] using various preparation techniques. The solid triangle is the result for crystalline diamond.

The transition from rigid to floppy occurs when the number of floppy modes, F, goes to zero. The most interesting experiment to date measures the hardness of networks containing carbon as a function of the mean coordination. The results seem to be in quite dramatic agreement with the ideas here as shown in Figure 27.

It has recently been suggested that it is better to plot physical properties, like hardness, not against the mean coordination of the skeleton network, but rather to use the original network [42] which contains dangling ends. This may be connected with the fact that it takes energy to move the dangling bonds apart in an experiment that measures hardness. It would be interesting to have measurements of other physical quantities to see if this observation still persists.

Silicate Networks

Although the floppy modes that we have been discussing emanate from the surface, they are in no sense surface modes. The amplitude is not damped away from the surface. Instead, they are bulk in extent and involve the whole solid. It is only the total number that scales like the surface area. This concept is important in other marginal structures. Another 2d example is the kagome lattice which consists of triangles joined at the corners. The most important three dimensional example is provided by silicates [SiO_2] where the SiO_4 tetrahedra are corner–sharing. This case has been examined extensively, both theoretically and experimentally [44] and is reviewed in these proceedings by Dove, Hammonds and Trachenko. These authors refer to these modes as rigid unit modes [RUM's] in which the SiO_4 tetrahedra are not distorted. These modes can be seen using inelastic neutron scattering. This experiment is not so straightforward as in the chalcogenide glasses, because the number of RUM's is proportional to the surface area [$N^{\frac{2}{3}}$], rather than the bulk [N], where N is the number of atoms.

It is not yet clear if the exact results given for the number of surface floppy modes in this section, are exactly or only approximately generalisable to the kagome and silicate structures.

8. SUMMARY

In this paper, we have tried to collect together some of the more important results regarding floppy modes and constraint counting in glasses that have emerged in the past decade. This work has provided a useful conceptual framework within which to discuss some of the physical properties of glasses. Some of the arguments in this area are subtle and still controversial.

Now that a three dimensional pebble game for bond-bending networks is available, we are in a good position to test the level of accuracy of Maxwell counting, and accurately pinpoint the location as well as study the nature of the rigidity transition. Our preliminary results suggest that the rigidity transition for generic three dimensional bond-bending networks is second order if many small nucleating rings are present, but first order if there is an absence of nucleating sites. It appears that Maxwell counting predicts remarkably well the location of this transition in either case. There is much more experimental and theoretical work that needs to be done to understand the nature of the rigidity transition better.

Work on rigidity is also finding applications in biological systems, see the article by Jacobs, Kuhn and Thorpe in these proceedings, where the ability of molecules to be flexible at little cost in energy probably has important implications for enzyme activity and other biologically important interactions.

9. ACKNOWLEDGEMENTS

This research was supported by the NSF under grants No. DMR–9632182 and CHE–9224102. We should like to thank P. Boolchand and P.M. Duxbury for continuing conversations on this topic. We would also like to thank F. Babalievsky for doing the programming to introduce triangles into the RBM and for Figure 20. A copy of the FORTRAN program *the Pebble game* for analyzing two dimensional generic networks is available upon request from M. F. Thorpe.

REFERENCES

1. J.C. Maxwell, *Philos. Mag.* **27**, 294- 299 (1864).
2. J.L. Lagrange, *Mécanique Analytique*, Paris (1788).
3. D.J. Jacobs and M.F. Thorpe, *Phys. Rev. Lett.* **75**, 4051 (1995).
4. R. Zallen, The Physics of Amorphous Solids, John Wiley & Sons, New York (1983).
5. B.R. Djordjević, M.F. Thorpe, and F. Wooten, , Phys. Rev. B 52, 5685 (1995).

6. S.R. Elliott, *Physics of Amorphous Materials*, Longman, London and New York (1984).
7. J.G. Kirkwood, *J. Chem. Phys*. **7**, 505 (1939).
8. P.N. Keating, *Phys. Rev*. **145**, 637(1966).
9. M. F. Thorpe, *J. Non–Cryst. Solids*, **57**, 355 (1983).
10. Y. Cai and M. F. Thorpe, *Phys. Rev. B* **40**, 10535 (1989).
11. J.C. Phillips, *J. Non-Cryst. Solids* **34**, 153 (1979).
12. J.C. Phillips, *J. Non-Cryst. Solids* **43**, 37(1981).
13. S. Feng, and P. Sen, *Phys. Rev. Letts.* **52**, 216 (1984).
14. S. Feng, M. F. Thorpe and E. J. Garboczi, *Phys. Rev. B* **31**, 276 (1985).
15. A. R. Day, R. R. Tremblay and , A–M. S. Tremblay, *Phys. Rev. Lett*. **56**, 2501 (1986).
16. H. He, and M. F. Thorpe, *Phys. Rev. Lett.*, **54**, 2107(1985).
17. A. Hansen, and S. Roux, *Phys. Rev. B* **40**, 749 (1989). see especially Figs. 1 and 3.
18. M. A. Knackstedt, and M. Sahimi, *J. Stat. Phys.* **69**, 887 (1992); S. Arbabi, and M. Sahimi, *Phys. Rev. B* **47**, 695 (1993).
19. E., Guyon, S. Roux, A. Hansen, D. Bideau, J.–P. Trodec, and H. Crapo, *Rep. Prog. Phys.* **53**, 373 (1990).
20. M. Tatsumisago, B. L. Halfpap, J. L. Green, S. M. Lindsay and C. A. Angell, *Phys. Rev. Lett* **64**, 1549 (1990); R. Böhmer and C. A. Angell, *Phys. Rev. B* **45**, 10091 (1992).
21. D. Jacobs, and M. F. Thorpe, *Phys. Rev.* **E 53**, 3682 (1996).
22. C. Moukarzel, P.M. Duxbury and P.L. Leath, *Phys. Rev. Lett.* **78**, 1480 (1997).
23. C. Moukarzel and P.M. Duxbury, *Phys. Rev. Lett.* **75**, 4055 (1995).
24. C. Moukarzel, P.M. Duxbury and P.L. Leath, *Phys. Rev.* **E55**, 5800 (1997).
25. B. Hendrickson, *SIAM J. Comput*. **21**, 65-84 (1992) and private communications.
26. G. Laman, *J. Engrg. Math.* **4**, 331 (1970).
27. L. Lovasz and Y. Yemini, *SIAM J. Alg. Disc. Meth.* **3**, 91 (1982).
28. D. Stauffer, *Introduction to Percolation Theory*, (Taylor and Francis, London) (1985).
29. C.M.Fortuin, and P.W. Kasteleyn, *Physica* **57**, 536 (1972); P. W. Kasteleyn and C. M. Fortuin, *J. Phys Soc. Japan,* **26**, 11 (1969).
30. J.W. Essam, *Rep. Prog. Phys.* **43**, 833 (1980).
31. P.M. Duxbury, D.J. Jacobs, M.F. Thorpe and C. Moukarzel, *to be published.*
32. D. S. Franzblau, *Siam J. on Discrete Math* **8**, 388 (1995); D. S. Franzblau and J. Tersoff, *Phys. Rev. Lett.* **68**, 2172 (1992).
33. D J Jacobs *J. Phys. A: Math. Gen*. 31 6653 (1998).
34. W. Whiteley, *Structural Topology*, **1**, 46 (1979).
35. C. Moukarzel, *J. Phys. A.: Math. Gen*. **29**, 8079 (1996).
36. P. Boolchand and M. F. Thorpe, *Phys. Rev.* **B 50**, 10366 (1994).
37. D.J. Jacobs and M.F. Thorpe, *Phys. Rev. Lett*. **80**, 5451 (1998).
38. X. Feng, W.J. Bresser and P. Boolchand, *Phys. Rev. Lett.* **78**, 4422 (1997).
39. M. F. Thorpe, *J. Non–Cryst. Solids*, **182**, 355 (1995). This mini-review contains references to many experimental results.
40. J. C. Angus, and F. Jansen, *J. Vac. Sci. Technol. A* **6**, 1778 (1988).
41. P. Boolchand, M. Zhang, and B. Goodman, *Phys Rev B* **53**, 11488 (1996).
42. G. H. Döhler, R. Dandaloff, and H. Bilz, *J. Non–Cryst. Solids*. **42**, 87 (1981).
43. M. Tamor, private communication.
44. M. T. Dove, A. P. Giddy and V. Heine, *Amer Crystal Assoc.* **27**, 65 (1993); A.P. Giddy, M.T. Dove, G.S. Pawley and V. Heine, *Acta Crystallographica A* **49**, 697 (1993).

RIGIDITY TRANSITION IN CHALCOGENIDE GLASSES

P. Boolchand, Xingwei Feng, D. Selvanathan, and W.J. Bresser

Dept. of Electrical, Computer Engineering and Computer Science
University of Cincinnati
Cincinnati, Ohio 45221-0030

I. INTRODUCTION

The prediction of a connectivity induced rigidity transition in network glasses was made nearly two decades ago.[1,2] Over the years these simple but elegant ideas have stimulated much theoretical and experimental interest in glass science. The theoretical aspects are captured in two seminal papers[1,2] and are reviewed by the authors in chapters elsewhere in this publication. The experimental aspects of the subject are discussed in the present chapter. The connection between *glass forming tendency* and the *rigidity transition* appears to be quite intimate. A basic understanding of the manner in which a glass network becomes rigid upon crosslinking may have an important bearing on the nature of molecular rearrangements and relaxation processes that follow upon cooling (freezing) a liquid past T_g to form a glass when the system becomes non-ergodic. In the past five years, important experimental developments have occurred, including availability of completely *digital data acquisition systems* for Raman scattering, and evolution of *temperature modulated differential scanning calorimetry*, methods that have made feasible to detect *details of the rigidity transition*[3,4] for the first time. These results correlate well with earlier Mössbauer spectroscopy work as illustrated in this review. We shall provide some of this evidence and discuss their implications. We begin by providing a brief overview of the basic ideas and proceed to discuss possible glass systems and experimental methods that have proved to be useful to examine the stiffness transition.

Mean-Field-Prediction of Rigidity Transition

Historically, ideas on the rigidity transition in covalently bonded networks evolved in 1979 from first attempts to develop a microscopic theory[1] of glass formation and application of percolation theory[2] to describe onset of rigidity. A covalently bonded network constrained by bond-stretching (α) and bond-bending (β) forces will, in a mean-field sense, spontaneously become rigid when the number of cyclical or floppy modes/atom, $f \rightarrow 0$. Enumeration of α- and β-constraints for a 3d network shows that the number of floppy modes/atom, f, is related to the mean coordination number $\bar{r}$ as follows,

$$f = 6 - \frac{5}{2}\bar{r} \tag{1}$$

The condition $f \rightarrow 0$ is realized when the mean-coordination number $\bar{r}$ increases to 2.40.

In a mean-field theory, the variation of f with $\bar{r}$ is thus linear (eq. 1). More realistic variations of $f(\bar{r})$ have emerged from numerical simulations on generic random network in 2d using a powerful integer algorithm.[2] Simulations in 3d on atypical (triangular, diamond) random networks have shown $f(\bar{r})$ to be more or less linear with $\bar{r}$ at $\bar{r} < 2.40$, and to have an exponential tail at $r > 2.40$. The second derivative of f with $\bar{r}$ localizes[2] the rigidity transition in such random networks at $\bar{r} = 2.385$, close to the mean-field value (2.40). The absence of restoring forces in the floppy regime leads elastic constants to vanish. Elastic constants in the rigid regime, however, display a power-law[5,6] with a value p=1.4 or 1.5. Thus numerical simulations in random networks predict a solitary stiffness transition near $\bar{r}$ = 2.40 with the elastic response vanishing below the transition ($\bar{r} \leq 2.40$) and displaying a power-law behavior above the transition ($\bar{r} \geq 2.40$). In viewing the experimental landscape, these robust signatures of the stiffness transition for *random networks* would be useful to recall.

Glass Systems

Chalcogenide and chalcohalide glasses are attractive systems to probe the rigidity transition. Unlike the oxide glasses where the backbone can only be modified by additives, the chalcogenide glasses form over wide range of compositions and the connectivity of the backbone can be changed in a reproducible manner by changing glass compositions. Perhaps simplest and most appealing systems amongst the chalcogenides[4, 7-11] include the *IV-VI binary glasses* such as Ge-Se, Si-Se, Si-Te. The simplicity derives from the binary nature of the glasses and the connectivity of the backbone which can be changed over the range $2 < \bar{r} < 2.7$ by compositional tunning. These systems are appealing because crystalline compound formation occurs at $\bar{r}$ = 2, 2.67, 3 far removed from the anticipated rigidity transition near $\bar{r}$ = 2.40. These considerations permit separating the subtle long-range mechanical from the gross short-range chemical ordering effects.

Chalcohalide glasses are ternaries formed by alloying a halogen (Cl, Br, I) with a chalcogen (S, Se or Te) or multicomponent chalcogenide glass system. The glass forming compositions in several dozens of chalcohalide glasses have been documented[12] and found to largely conform to constraint counting algorithms extended to include the role of the *one-fold coordinated halogen atoms* explicitly.[13] Many of these glasses which form over wide compositions are attractive systems to probe the rigidity transition.[14,15] Recent work along these lines has focused on the Ge-S-I ternary where Raman scattering measurements[14] have provided direct evidence of an anomaly in the corner-sharing (CS) $Ge(S_{1/2})_4$ mode frequency as the iodine concentration of ternary glasses is varied. The observed threshold composition is close to the predicted mean field value, although the underlying transition possesses a finite width. Such measurements also reveal that replacement of S by the oversized iodine leads to considerable internal pressure on the network backbone. The internal pressure leads to a sizeable blue shift of the Raman active A_1 mode frequency of CS $Ge(S_{1/2})_4$ units. Temperature moduled differential scanning calorimetry measurements have also been performed on such ternary glasses and reveal the non-reversing heat flow to display a minimum at the stiffness threshold.

Experimental Probes

The rigidity transition can be expected to manifest directly in the elastic response of a network. Thus *elastic constants* probed by *ultrasonic echoes,*[16-19] *Brillouin scattering* and *Raman scattering*[3,4,9] appear to be some of the most direct methods to probe the transition. This indeed is the case, particularly in Raman scattering measurements, as we shall demonstrate with some examples in this work.

Low-frequency vibrational excitations in glasses can reveal *floppy modes*[20] that can be observed directly in inelastic neutron scattering measurements. If a glass network is undercoordinated ($\bar{r} < 2.40$), such as a network of entangled chains like in the case of Se-glass, the *neutron vibrational density of states* (VDOS)[20] reveals a low frequency mode centered at 5 meV. This mode possesses approximately 1/3rd the scattering strength of the integral VDOS. In a Se chain every atom has two nearest neighbors, and there are two constraints per atom, $n_c = 2$. The number of floppy modes per atom, $f = n_d - n_c = 1$, where n_d represents the degrees of freedom per atom revealing that 1/3rd of the vibrational modes are floppy.[20] Thus the 5 meV mode is regarded to be a floppy mode in Se glass.

The low-T ($T \rightarrow 0$) limit of the *Lamb-Mössbauer factor* (f_o) provides the first-inverse moment[21] of the VDOS, $\langle \omega^{-1} \rangle$ and will, in general, reveal a reduction of the *nuclear-resonant absorption signal* due to the presence of floppy modes. Thus, for example, systematic measurements[21] of the Lamb-Mössbauer factor, $f_o(x)$ in ^{119}Sn doped Ge_xSe_{1-x} glasses has revealed a qualitative softening as $x \rightarrow 0$, and a threshold behavior (a sharp kink) as x increases to $x_c \sim 0.20$ identified with the rigidity transition. Thus Lamb-Mössbauer factors can also serve as powerful probe of the stiffness transition, particularly in those cases where a suitable Mössbauer probe nucleus is available. Recently, we have also examined the ^{125}Te Lamb-Mössbauer factor in Ge_xTe_{1-x} glasses and find evidence of an anomaly in the first inverse moment of the VDOS near $x = 0.20$.

It has become popular[22] to discuss the freezing of a glass forming liquid in terms of an energy landscape. There are good reasons to believe that an optimally coordinated supercooled melt is trapped largely in the lowest energy state of the landscape. One can access this information directly from the *non-reversing heat flow* ΔH_{nr} upon heating a glass to T_g in a T-modulated Differential Scanning Calorimetry measurement.[3] Such measurements in several glass systems[3,4] have now been performed and reveal a pattern; $\Delta H_{nr}(\bar{r})$ is a minimum when a glass is optimally coordinated, and it increases by almost an order of magnitude in glass networks that are either over- or under-coordinated. The minimum in ΔH_{nr} ($\simeq 0$) for optimally coordinated ($\bar{r} \simeq 2.4$) networks suggests that such glass networks are trapped in the lowest energy minimum of the energy landscape and that changes in structure as a function of temperature are minimal in the transition region. Temperature modulated DSC, a relatively new tool in glass science, provides thus a direct means to *thermally* probe the stiffness transition in glass $\rightarrow$ liquid transformation.

Molar volumes measured as a function of network connectivity $V_M(\bar{r})$ display a pattern that mimics the observed variation in $\Delta H_{nr}(\bar{r})$. It appears that when a glass network becomes optimally coordinated, not only is network strain minimized in a global sense, but network packing becomes most compact as well.

In Table 1, we summarize the principal experimental probes that have displayed anomalies near the rigidity transition in chalcogenide glasses. Some of these anomalies are elastic in origin, some vibrational in character, others structure related[7,28]. While still others such as the activation energy for stress relaxation[23], non-radiactive decay rates[29] of H_2O guest molecules and the semiconductor to metal transition pressures[10], collectively point towards an unusual morphological state of matter in the *transition region*.

Table 1. Experimental Probes of Stiffness Transition in Glasses

Physical Property	Method	Result	Author (Yr) Ref
Elasticity acoustic	Ultrasonic Moduli	Weak evidence for anomaly in C_{11} and C_{44} near $\bar{r} = 2.40$ in Ge-Se-Sb	Sreeram et al. (1991) ref. 19
acoustic	Brillouin scattering	No results	--
optical	Raman scattering	Vibrational thresholds at $\bar{r} = 2.40, 2.46$	Feng et. al (1996) ref. 3 and present work
Kinetic Heat Flow	T-modulated DSC	Vibrational thresholds at $\bar{r} = 2.40, 2.46$	Feng et al. (1996) ref. 3 and present work
Activation energy for Stress Relaxation	DMA	Shows min. near $\bar{r} = 2.40$	Bohmer & Angell (1993) ref. 23
Thermal Expansion	DMA	Shows min. near $\bar{r} = 2.40$	Senapati & Varshneya (1995) ref. 24
Network Packing or Molar Volumes	Density-Archimedis method	$V_M(\bar{r})$ show min. near $\bar{r}=2.40$ in Ge_xSe_{1-x}	Feltz et. al. (1983) ref. 25 Feng et al. (1986) ref. 3 and present work
Floppy Modes	Neutron VDOS	8 meV mode in g-Se	Kamitakahara et al. (1991) ref. 20
First- and second-inverse moments of VDOS	Lamb-Mössbauer factors	^{119}Sn in Ge_xSe_{1-x} show a threshold at $\bar{r}=2.40$	Boolchand et al. (1990) ref. 21
		^{125}Te in $(Na_2O)_x(TeO_2)_{1-x}$ show a threshold at x=0.18	Zhang & Boolchand (1994) ref. 25
		^{125}Te in Ge_xTe_{1-x} glasses and amorphous films	Enzweiler et al. ref. 26
Network Dimensionality and morphology	Kohlrausch Fractional Exponent β	β increases with $\bar{r}$ and saturates at $\bar{r} \geq 2.40$	Bohmer & Angell (1992) ref. 26
Insulator-Metal Transition Pressures (P_T)	Resistivity with pressure	P_T shows kink at $\bar{r} = 2.40$	Asokan et al. (1988) ref. 10
Network Morphology and local strain	^{129}I Mössbauer Spectroscopy	Site intensities display a local max. at $\bar{r}=2.46$	Bresser et al. (1986) ref. 7, 28

II. RIGIDITY TRANSITION IN Ge_xSe_{1-x} GLASSES

Raman scattering

Figure 1 reproduces representative spectra[3] of the glasses in the Ge composition range $0.15 < x < 0.34$. In these measurements the scattering was excited using the 647.1nm line from a Kr ion laser, and the back-scattered radiation analyzed with a triple monochrometer system (model T64000) from Instruments S.A., Inc. with a microscope attachment and a CCD detector. The system was calibrated using a Ne discharge lamp which provided atomic transitions at $84.83cm^{-1}$, $146.38cm^{-1}$, and $299.63cm^{-1}$ in the CCD spectral window. The same window was used to store spectra of glasses. The stability of the spectrometer was tracked periodically during the measurements. Spectra of the glasses were taken with 750 μWatts of exciting laser power leading to sample temperature of 90 ± 8°C as established by Raman Stokes/anti-Stokes intensity ratios. In subsequent measurements, more samples were synthesized in the transition region, $0.21 < x < 0.24$ at smaller Ge concentration steps of $\Delta x = 0.0025$. Raman spectra of these samples were studied at a lower exciting laser power of 82 μW, for which sample temperatures were close to room temperature (22±5°C). At an intermediate (250 μW) laser power, sample temperatures increased slightly above room temperature to 57±5°C.

In the studied composition range, Raman lineshape reveal largely three modes, at $250cm^{-1}$, $215cm^{-1}$ and at $200cm^{-1}$. These modes have been respectively identified[9] as an A_1 mode of distorted Se_n-chains, of edge-sharing (ES) $Ge(Se_{1/2})_4$ tetrahedra, and of CS $Ge(Se_{1/2})_4$ tetrahedra. We carried forward deconvolution of observed lineshapes in terms of requisite number of Gaussians, with the three lineshape parameters, viz, line-centroid or mode frequency, linewidth and scattering strength of each Gaussian kept unrestricted. In our analysis of the lineshape, we have paid particular attention to the mode frequency shift of the CS mode.

Figure 2 provides a summary of mode frequency variation for the CS mode in Ge_xSe_{1-x} glasses from a deconvolution of Raman lineshapes. One finds that the CS-mode frequency variation, $\upsilon_{cs}(x)$, is linear at low-x ($x < 0.22$), displays[3] a step-like increase in going from $x = 0.22$ to 0.23 and thereafter acquires a superlinear variation[3] when the exciting laser power is 750μWatts. The underlying elasticity (mode frequency squared) variation

$$\nu^2 - \nu_c^2 = A(\bar{r} - \bar{r}_c)^p \qquad (2)$$

can be characterized by a power-law p=1.33. In equation (2) ν_c is the Raman mode frequency corresponding to a value of $x = x_c = 0.23$, corresponding to a mean coordination number $r_c=2(1+x_c)$ of 2.46, since Ge(Se) possess a coordination of 4(2), respectively.

The Raman results of Fig. 2 also reveal that in the transition region $0.21 < x < 0.24$, at low- and intermediate-exciting laser power (82μW, 250μW), the CS-mode frequencies blue shift, and more importantly, the $\upsilon_{cs}(x)$ variation is now continuous and displays two-kinks[30] one at $x = x_c(1) \simeq 0.20$ and a second at $x = x_c(2) = 0.23$.

The frequency variation $\nu_{cs}(x)$ with laser-power has both a *thermal* and a *photo-structural* contribution. We know from our measurements of the Stokes/Antistokes intensity ratios that sample temperatures decrease (90°C → 22°C) as the exciting laser-power is reduced (750μW → 82μW). The blue shift of ν_{cs} upon reducing laser power, suggests presence of a thermal component. The sign of ν_{cs} shift with temperature is certainly consistent with presence of anharmonic effects in glasses. Such effects are

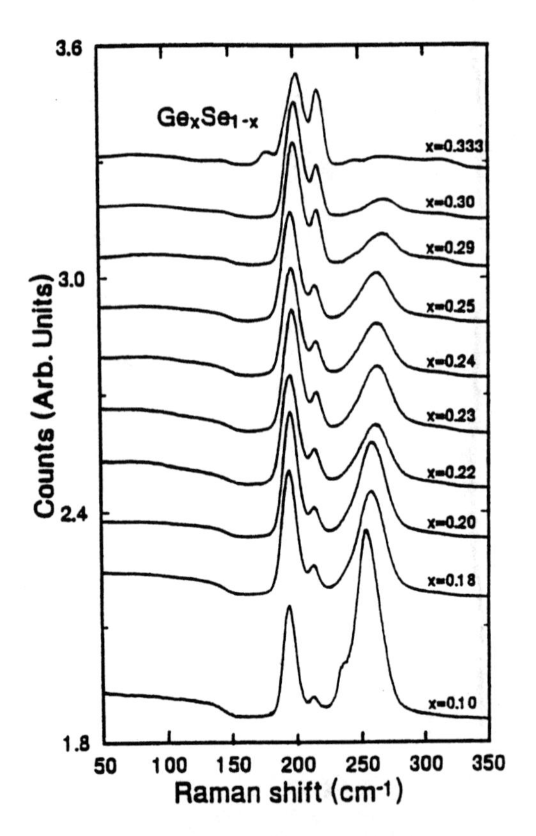

Fig. 1. Representative Raman spectra of Ge_xSe_{1-x} glasses taken as a function of Ge concentration x. See text for details.

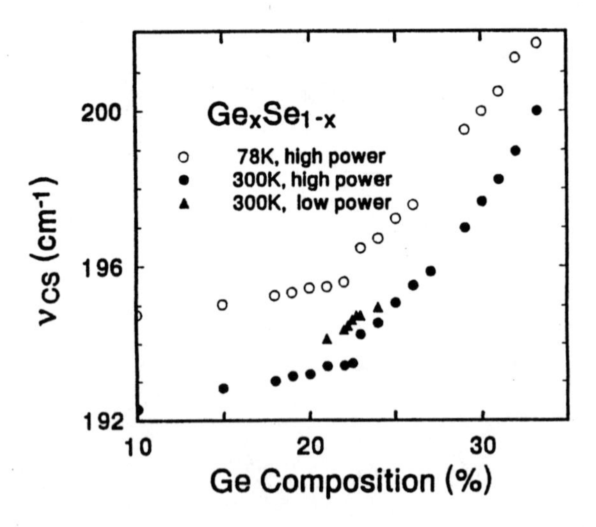

Fig. 2. Corner-sharing $Ge(Se_{1/2})_4$ mode frequency υ_{cs} (x) deduced from Raman scattering lineshape analysis plotted as a function of x. Filled data points represent results obtained at room temperature while open circles represent results obtained at 78K. The filled triangle (circle) represent results obtained at 82μWatts (750μWatts) exciting power on the 647.1nm line from a Kr ion laser.

known[31] to increase optic-mode frequencies as lattice temperatures are decreased in Raman scattering measurements on crystalline semiconductors. However, there must in addition be photoinduced structural effects in these glasses that accompany an increase in laser-power, since the functional variation of $\upsilon_{cs}(x)$ undergoes a change in character from being a continuous one (at 82μW) to becoming a step-like change at higher-powers (750μW). The photostructural effects[32] in these chalcogenides involve re-arrangement of chemical bonds of the type recently proposed to understand the athermal photo-induced fluidity[33] effects in glasses.

Temperature Modulated Differential Scanning Calorimetry

In a conventional DSC measurement the signature of softening of a glass is an endothermic heat flow with respect to an inert reference sample, as the temperature of the glass and reference sample is swept linearly in time at a controlled rate. Temperature modulated DSC is a more recent variant of the method. By programming a sinusoidal temperature variation[34] over the linear T-ramp, it is possible to deconvolute the *heat flow* into two components, one that tracks the sinusoidal T-variation and is therefore called the *reversing heat flow* and the remainder that does not track the periodic T-variation and is thus called the *non-reversing* heat flow[34]. Figure 3 provides MDSC scans of a $GeSe_2$ glass and a $Ge_{0.23}\ Se_{0.77}$ glass illustrating the deconvolution of *heat flow* into the *non-reversing* and *reversing* components. These are two noteworthy features that emerge from these scans. First, the apparent glass transition temperature (T_g^{app}) deduced from the *heat flow* is, in general, lower than the glass transition deduced from the *reversing heat-flow.* The shift T_g-T_g^{app} is only about 3°C for the x = 0.23 glass sample, but it is 12°C for the $GeSe_2$ glass sample. Second, the non-reversing heat flow is miniscule for the optimally coordinated glass sample (x = 0.23) but it is an order of magnitude larger for the stoichiometric glass ($GeSe_2$). The first observation is actually a consequence of the second. Figure 4 provides a summary of the Ge-concentration dependence of $\Delta C_p(x)$ and $\Delta H_{nr}(x)$ deduced from the *reversing* and *non-reversing* heat flows, respectively. The experimental results reveal that $\Delta H_{nr}(x)$ has a deep minimum[3] centered near x = 0.23. We also find $\Delta C_p(x)$ deduced from the reversing heat flow to be more or less flat displaying only a shallow minimum near x = 0.22. The shallow ΔC_p minimum accessed from present MDSC measurements is in contrast to previous results[35] on these glasses obtained from DSC. In DSC measurements, the deep minimum[35] in ΔC_p accessed from heat flow is due primarily to the non-reversing heat component[3], and is a point that is generally not known. In fact, the minimum in ΔC_p is *not* a *generic feature* of the stiffness transition in chalcogenide glasses. Results on the Si_xSe_{1-x} binary, for example, show[4] a monotonic decrease in $\Delta C_p(x)$ with x in the $0.15 < x < 0.33$ range with no evidence of a minimum near the stiffness threshold.

The glass transition temperatures (T_g) established from the reversing heat flow in such MDSC measurements are unaffected by non-reversing or kinetic heat flow effects. These $T_g(x)$ trends provide a convenient means to calibrate glass compositions. T_gs accessed from total heat flow endotherms such as in a DSC measurement, in under- and over-coordinated glass, will in general be lower than the actual T_gs due to the presence of a sizeable non-reversing heat flow as discussed elsewhere[4]. Figure 5 provides $T_g(x)$ variation in Ge_xSe_{1-x} glasses deduced from MDSC scans. Recently a stochastic model has been developed[36] to account for T_g variation with cross-linking and it would be of interest to establish up to what Ge concentrations can such a random network model account for the observed $T_g(x)$ variation in the present glasses.

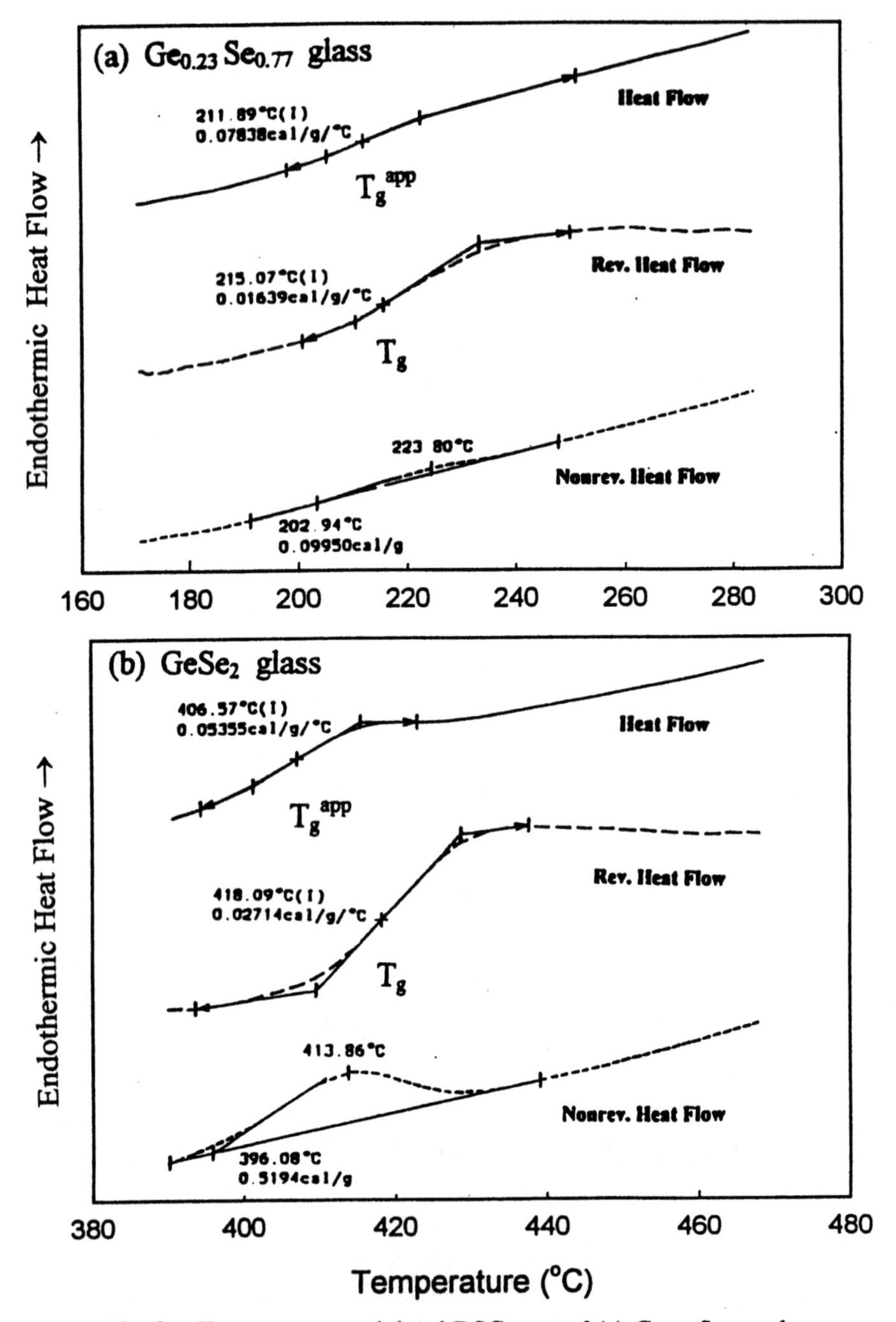

Fig. 3. Temperature modulated DSC scan of (a) $Ge_{0.23}Se_{0.77}$ glass and (b) $GeSe_2$ glass taken with a TA Instrument MDSC model 2920. The scan rate was kept at 3°C/minute and the temperature modulation at 1°C per 100sec. The apparent Tg^{app} from the heat flow and the Tg from the reversing heat flow are defined as the points of highest slope and indicated by a + sign on respective scans. ΔC_p from the reversing heat flow and ΔH_{nr} from the non-reversing heat flow are also labelled on respective scans. Note the drastic reduction in ΔH_{nr} for the $Ge_{0.23}Se_{0.77}$ glass sample.

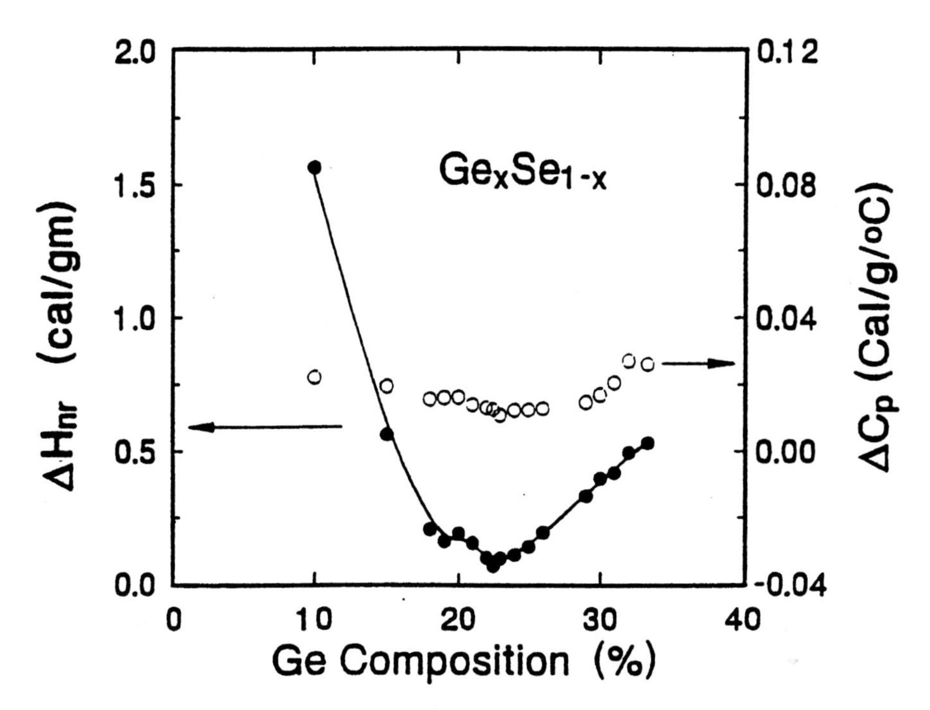

Fig. 4. Trends in $\Delta C_p(x)$ – open circles and $\Delta H_{nr}(x)$ – filled circles, plotted as a function of x in indicated glasses. ΔH_{nr} displays a global minimum at $x_c(2)$ and a local one near xc(1).

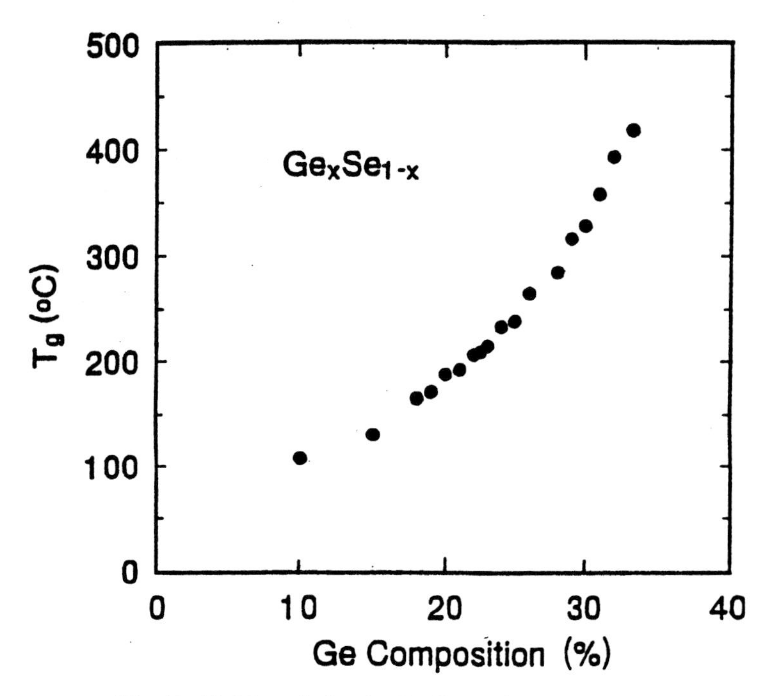

Fig. 5. $T_g(x)$ variation in Ge_xSe_{1-x} glasses plotted as a function of x. $T_g(x)$ were deduced from the reversing heat flow. The observed variation is smooth and monotonic as a function of x and displays no obvious kinks either at $x_c(1)$ or at $x_c(2)$.

Molar Volumes

Molar volumes ($V_M = M/\rho$, where ρ = mass density in gms/cm^3 and M the molecular weight in gms) can be established by measuring ρ, the mass density using Archimedes principle. Such measurements provide a simple and direct way to establish packing of matter in a bulk solid. In bulk glasses, such measurements provide a means to establish the compactness of the network backbone or the available free volume. Feltz and coworkers[37] were one of the first to recognize that V_M of binary Ge_xSe_{1-x} reveal a local minimum centered at x = 0.22 or $\bar{r}$ = 2.44. Since the early work of Feltz et al., other groups[19] have reproduced this trend. In our laboratory, we have measured mass densities of Ge_xSe_{1-x} bulk glasses using a microbalance (resolution 0.1 mg) to an accuracy of about a part in a 1000, and Fig. 6b provides $V_M(x)$ trends for our bulk Ge_xSe_{1-x} glasses obtained from water quenched melts. The results show a systematic decrease in $V_M(x)$ as x increases to 0.23, and thereafter $V_M(x)$ increase with x up to x = 1/3. The minimum in $V_M(x)$ coincides with the anomalies in optical elasticities (Fig. 6a) and non-reversing heat flow $\Delta H_{nr}(x)$ (Fig. 6b). These anomalies localized near x = 0.23 or $\bar{r}$ = 2.44, also coincide with Lamb-Mossbauer factors in these glasses as we discuss next.

Mössbauer Spectroscopy

Two complementary Mössbauer spectroscopic probes have been used to examine[38] Ge_xSe_{1-x} glasses, the results of which bear on the *stiffness transition*. These include ^{119}Sn absorption spectroscopy as a vibrational probe[39] of Ge (cation sites) and ^{129}I emission spectroscopy as a chemical probe[7,28] of Se (anion sites) of the network backbone. We comment on the results of these experiments separately.

^{119}Sn absorption spectroscopy

Sn as a dilute dopant in Ge_xSe_{1-x} is tetrahedrally coordinated and replaces Ge in the network backbone as revealed by ^{119}Sn isomer shifts[32,40] (that are found to be characteristic of tetrahedral covalent Sn) and thermal cycling studies[40]. In the latter investigations melt-quenched $(Ge_{0.99}Sn_{0.01})_xSe_{1-x}$ glassy alloys at x = 1/3, when cycled through T_g, reveal, as expected, no change in ^{119}Sn Mössbauer lineshape, suggesting that the dopant is incorporated homogeneously in the network. On the other hand, when such glasses are heated past T_x, understandably irreversible changes in the lineshape occur as crystallization occurs. About half of the alloyed Sn precipitates as crystalline $SnSe_2$ inclusions while the remainder becomes substitutional in the 2d-form of crystalline $GeSe_2$. In c-$SnSe_2$, Sn is octahedrally coordinated and reveals a smaller[40] isomer-shift than in the glasses (where Sn is tetrahedral).

The full T-dependence of the 119Sn Lamb-Mössbauer factors f(x, T) in $(Ge_{0.99}Sn_{0.01})_xSe_{1-x}$ glasses, at several compositions x in the range $0 < x < 1/3$ has been established[21]. Figure 7 provides a plot of the log of the Lamb-Mössbauer factor (familiarly known as the mean-squared displacement, MSD) as a function of x. The results reveal that the MSD decreases with increasing x, and displays a kink, i.e. a change in slope near x = 0.20.

At all glass compositions studied in the range $0 < x < 0.30$, Sn is tetrahedrally coordinated in $Sn(Se_{1/2})_4$ units and reveals a narrow line in the Mössbauer spectra. At low x, these tetrahedra are formed at crosslinks of entangled Se_n chains in a rather floppy network, and thus display a large MSD. As x increases, the network becomes progressively crosslinked, and stiffer. The MSD of the tetrahedra then systematically decrease leading to

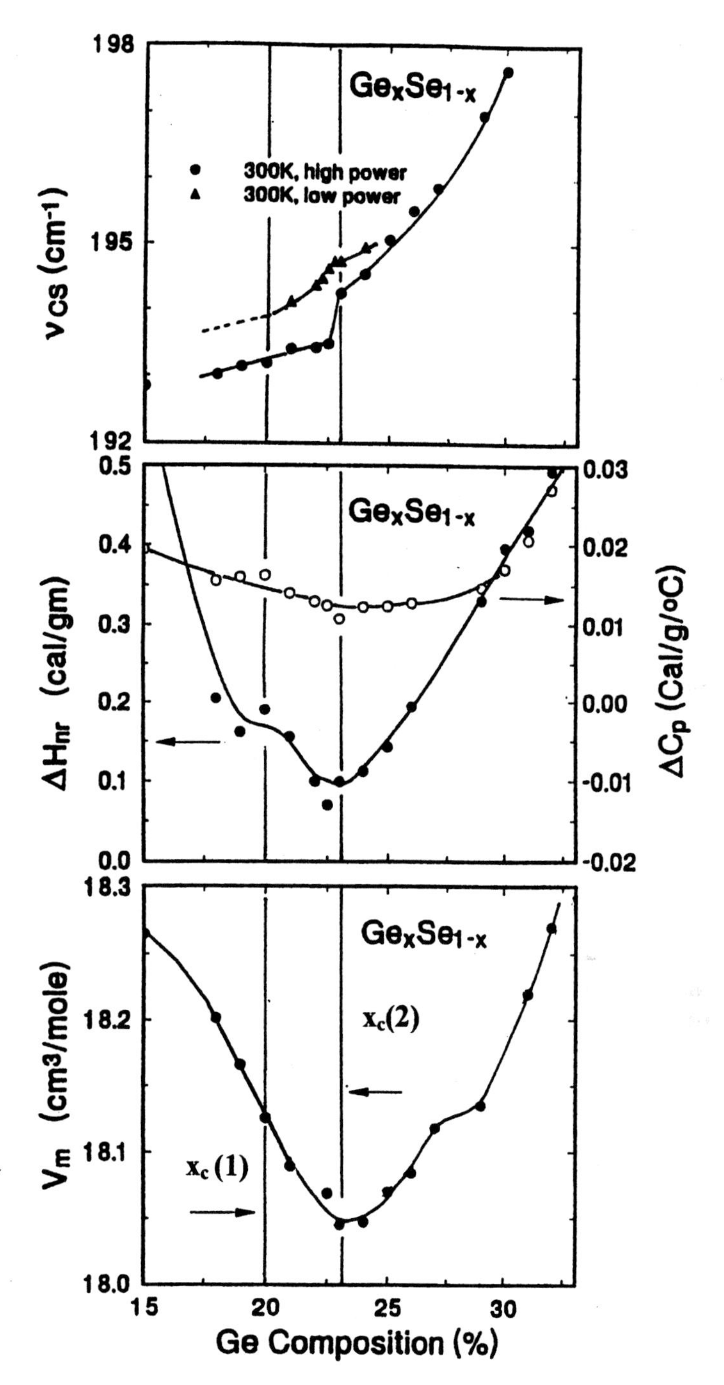

Fig. 6. Ge concentration dependence of (a) CS-mode frequency υ_{CS}, (b) Molar volumes $V_M(x)$ and (c) non-reversing heat flow $\Delta H_{nr}(x)$ in Ge_xSe_{1-x} glasses compared. The compositions $x_c(1)$ and $x_c(2)$ represent the onset and completion of rigidity in the glasses, respectively. In the Raman scattering results of $\nu_{CS}(x)$, the variation observed at low laser power displaying kinks at $x_c(1)$ and $x_c(2)$ is thought to be the intrinsic behavior of the glasses.

a reduced slope once $x > 0.20$, which represents the rigidity onset point. A change in slope of the MSD near $x = 0.20$ is also simulated[41] by vibrational analysis of a bond-depleted diamond lattice.

Smaller incremental steps in x near the observed threshold ($x = 0.20$) will permit elucidating details of the *transition region* and are planned in the near future on several chalcogenide glasses. Given the wealth of Raman results on the Ge_xSe_{1-x} binary now available, the motivation to pursue ^{119}Sn Mössbauer spectroscopy measurements in the transition region becomes persuasive. The covalent radius of Sn (1.41Å) is 16% larger than that of Ge (1.22Å). It is likely that Sn replaces Ge in the more flexible CS units rather than ES units. If this is the case, then changes in MSD with x should also display a dual transition, as seen for the Raman A_1 mode frequency (Fig. 2). Since Mössbauer spectroscopy measurements are immune from photostructural effects due to the exciting gamma-ray beam, it is of interest to carry forward such measurements to complement the Raman measurements and independently ascertain details of the stiffness transitions.

^{129}I emission spectroscopy

The novel physical characteristics of chalcogenides including glass formation derives largely from the unusual bonding chemistry of chalcogen atoms. These atoms usually take on a 2-fold coordination with a pair of electrons forming a non-bonding lone pair. Te as a dopant in selenide and sulfide glasses or crystals can replace the lighter chalcogen atoms in the backbone, and serve as a probe of their local bonding chemistry, as elegantly illustrated[7,28] by ^{129}I Mössbauer emission spectroscopy using ^{129m}Te parent as a dopant through nuclear quadrupole interactions[28].

The covalent radius of Te (1.36Å) is 17% larger than that of Se (1.16Å). The *oversized nature* of the dopant has some interesting *doping consequences.* In network structures that possess a multimodal distribution of chalcogen sites (S or Se), the oversized dopant may select to replace Se sites in local environments where either free volume is readily available or bonds can locally distort with little or no expense of strain energy, to fulfill the larger space requirements of the dopant.

^{129}I spectra of bulk Ge_xSe_{1-x} glasses can be deconvoluted[7,28] into two chemically inequivalent local environments, labelled as sites A and B. The former site is identified with an iodine species σ-bonded to a Ge nearest neighbor (nn), while the latter site to an I σ-bonded to a Se nn. In a Se glass only B-type of a local environment is expected while in a *chemically ordered* $GeSe_2$ network only A-type of a local environment is expected as discussed elsewhere[7,28]. The Mössbauer site intensity ratio $I_B/I_A(x)$ variation deduced from the experiments appears in Fig. 8. There are two outstanding features of these $I_B/I_A(x)$ trends, *first* a monotonic reduction of $I_B/I_A(x)$ with increasing x in the $0.05 < x < 0.33$ range reflecting a baseline variation and *second,* a peak centered at $x = x_c(2) = 0.23(2)$ superposed on this base line.

For a purely stochastic network description[7] of these glasses in which cross-linking of Se_n chains by Ge atoms can proceed in a *random fashion,* the expected $I_B/I_A(x)$ variation is also shown in Fig. 8 as the broken line. A stochastic network description of these glasses is clearly an appropriate description of these glasses at low x (< 0.15) where the observed and calculated trends are qualitatively similar. Beginning at $x > 0.15$, the observed $I_B/I_A(x)$ variation differs *qualitatively* from the projected variation of a random network. Broadly, these results are signature of network inhomogeneities in the form of rigid units nucleating in a floppy network. Specifically, the oversized dopant is expelled from the rigid-regions, where A site occupancy would have occurred in ES units, thus qualitatively altering the random occupancy ratio of Se B to A sites in the network.

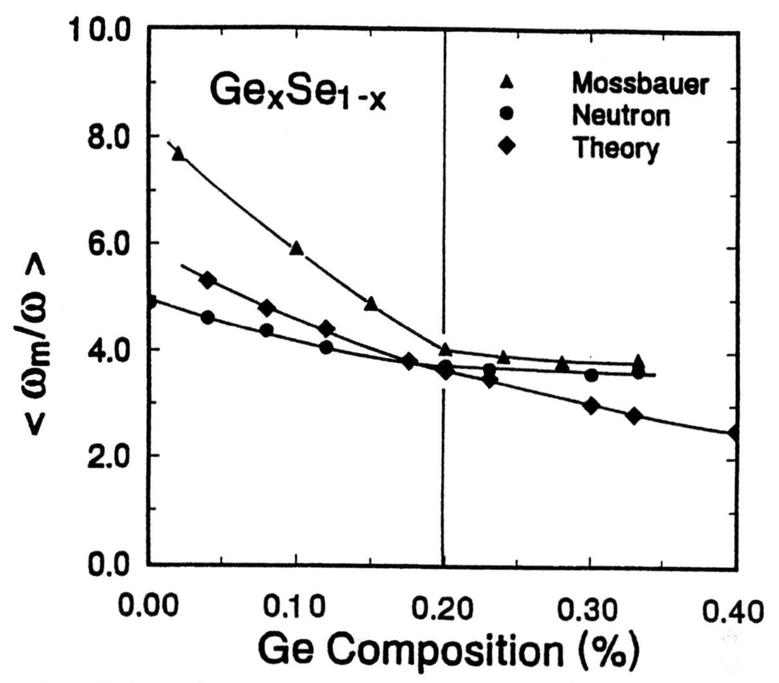

Fig. 7. Log of the T → 0 limit of the Lamb-Mössbauer factors in $(Ge_{0.99}Sn_{0.01})_xSe_{1-x}$ glasses plotted as a function of x, displaying a kink near x = 0.20, ascribed to rigidity onset. Figure is taken from ref. 21.

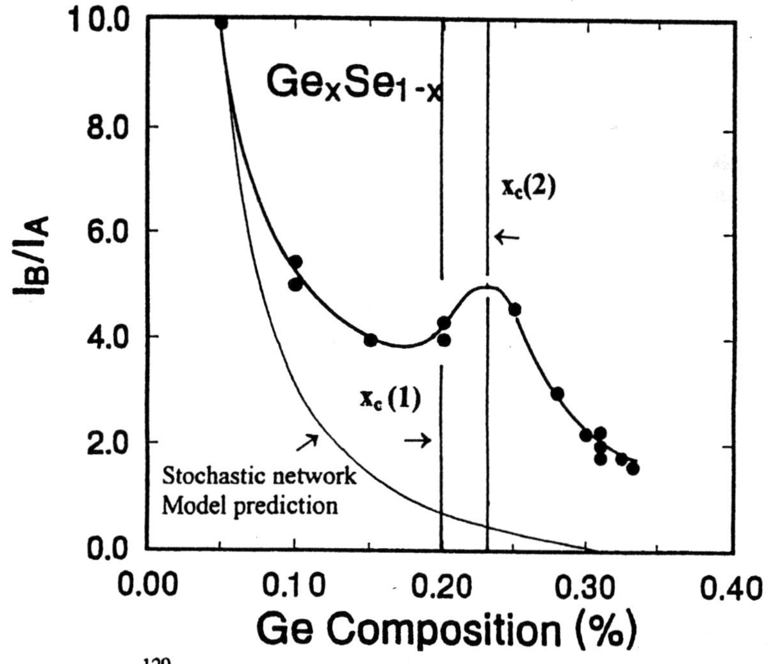

Fig. 8. ^{129}I Mössbauer site-intensity ratio $I_B/I_A(x)$ variation in $Ge_x(Se_{0.99}Te_{0.01})_{1-x}$ glasses plotted as a function of x (bold line). The thin line is the projected $I_B/I_A(x)$ variation for a chemically ordered stochastic Ge_xSe_{1-x} network taken from ref. 7. See text for details.

Globally, although network strain is a minimum at $x_c = 0.23$ as revealed by $\Delta H_{nr}(x)$ trends, locally the dispersion of the floppy- (Se_n chains) and rigid-fragments (ES- and CS-units) of the network backbone are the *finest* at the stiffness transition. This leads to the largest surface to volume ratio of the rigid fragments which is reflected in the *peak* of the Mössbauer site intensity ratio $I_B/I_A(x)$ near $x = 0.23(2)$. At $x > 0.23$, the rigid-fragments grow in extent, reducing the surface to volume ratio of the rigid regions and thus I_B/I_A. But even as x increases to 1/3, I_B/I_A does not extrapolate to zero (as required by a chemically ordered continuous random network) and acquires a finite value (about 2) because rigid fragments do not *completely* fill the network backbone. The stoichiometric glass, $GeSe_2$, is only partially polymerized, and phase-separated into large and small molecular clusters as demonstrated by these Mössbauer spectroscopy results more than a decade ago[38] and as confirmed by Raman scattering measurements[42,43].

Stiffness Transition

The equilibrium phase diagram of the Ge_xSe_{1-x} binary reveals[44] a eutectic at x = 0.08 where the liquidus T_ℓ is about 200°C, and reveals a local minimum. The glass forming tendency in the Ge_xSe_{1-x} binary is apparently optimized in the $0.20 < x < 0.24$ composition range where we have synthesized homogeneous and strain-free bulk glasses by cooling melts at a rather slow rate of 2.5°C/minute (or 0.04 °C/sec) in a T-programmed resistive furnace. These cooling rates are at least three orders of magnitude lower than rates normally used (40°C/sec) in synthesizing glasses by a water quench of melts. At the outset, it would be useful to thus indicate that the glass forming tendency in Ge_xSe_{1-x} binary melts that is *optimized near* $x = 0.22$ has no bearing on eutectic related effects.

The simplest interpretation of these observations is that near x = 0.22, the mean constraints per atom, $\bar{n}_c$ for a binary Ge_xSe_{1-x} glass network, nearly equals 3, the degrees of freedom per atom, n_d, i.e. $\bar{n}_c = n_d$, and the underlying Lagrange mechanical criticality[1] is responsible for the optimized glass forming tendency. The present Raman-, MDSC-, V_M- and Mössbauer measurements refine the manner in which rigidity onsets in the binary Ge_xSe_{1-x} glass network. A comparison of $\nu_{cs}(x)$, $\Delta H_{nr}(x)$ and $V_M(x)$ on the same samples is afforded by the plot of Fig. 6. The plot makes the *two-step nature* of the transition more persuasive. In our Raman scattering results for the CS-mode frequency $\upsilon_{cs}(x)$, taken at low exciting laser power, one observes two kinks; one at $x = x_c(1) \simeq 0.20$ or $\bar{r}_c(1) = 2.40$ and a second one at $x = x_c(2) = 0.23$ or $\bar{r}_c(2) = 2.46$. These Raman results correlate well with the non-reversing heat flow $\Delta H_{nr}(x)$ variation. The latter show a final decrease starting at $x \geq x_c(1)$ and an increase at $x \geq x_c(2)$ yielding a global minimum near $x_c(2) = 0.23$. Molar volumes of these glasses also reveal a minimum centered near x = 0.23. These results suggest that rigidity *onsets* at $\bar{r} = \bar{r}_c(1)$ and the network becomes percolatively rigid at $\bar{r} = \bar{r}_c(2)$, with the composition interval $\bar{r}_c(1) < \bar{r} < \bar{r}_c(2)$ serving as a *transition region*.

CS-units form part of the rigid fragments in the present glasses but only at $x > x_c(2)$ when the underlying elasticity displays a power-law. It is possible that rigidity is nucleated in molecular fragments consisting of dimeric ES-units cross-linking chains of CS-units to form an 8-tetrahedral unit fragment. Numerical simulations of networks possessing medium range structure consisting of such 8-unit fragments would be helpful in understanding the present results.

The Raman scattering results obtained at a high laser power (750μWatts) reveal[3] for the CS-mode frequency only one transition, which occurs at $x = x_c(2)$ with a step-like increase (first order) in the mode frequency. These results are in contrast to those obtained at a low laser power (82μWatts) where two transitions manifest. The two-transitions probably represent the intrinsic behavior of the system, while the solitary first-order transition at a high laser power a *photostructural assisted modification* of the transitions. Thus one finds in the presence of a high photon flux a glass at $x = 0.22$ softens considerably more than one at $x = 0.23$ or 0.24. In Fig. 6a, note the significant reduction in υ_{cs} (x) mode frequency, when laser power is increased from 82 to 750μwatts. On the other hand, photostructural assisted softening is *qualitatively suppressed* at $x = 0.23$ and 0.24 when the glass network becomes rigid. Thus, the presence of photostructural effects accompanying Raman scattering and particularly in the *transition region*, may provide new insights into the manner in which a chalcogenide glass network become rigid.

III. CONCLUDING REMARKS

In the present work, we have focused on the stiffness transition in binary Ge_xSe_{1-x} glasses, where new experimental results are now available from Raman scattering, MDSC and Molar Volumes that correlate well with earlier Mössbauer spectroscopy work. These results taken together provide evidence of a multiplicity of stiffness transition, an *onset* point near $\bar{r}_c(1) = 2.40$ and a *completion* point near $\bar{r}_c(2) = 2.46$. Bulk elastic constants on Ge_xSe_{1-x} glasses from ultrasonic moduli have, to date, shown[16-19] a linear behavior with $\bar{r}$ near $\bar{r} = 2.4$. Such a result could either be due to the intrinsic limitations on the accuracy of performing such bulk measurements on glassy samples or the role of impurities and sample porosity changing the compositional trends. The addition of 5 to 10 at.% Sb in such glasses results in evidence of some structure[19] in bulk moduli and longitudinal elastic constants at $\bar{r}$ = 2.40 and 2.46, respectively, which are discussed elsewhere[45].

Rather complete Raman, MDSC and molar volumes results on binary Si_xSe_{1-x} glasses have now become available, and also reveal[4] evidence of two transitions, an *onset* point near $\bar{r}_c(1) = 2.40$ and a *completion* point near $\bar{r}_c(2) = 2.54$. These stiffness transition results are parallel to those found in the Ge_xSe_{1-x} binary. Taken together, the available results reveal that in both the Ge_xSe_{1-x} and Si_xSe_{1-x} binaries, the first transition at $\bar{r}_c(1)$ occurs[4] near the mean-field predicted value of 2.40. The second transition at $\bar{r}_c(2)$, on the other hand, occurs at 2.46 in the Ge-bearing glasses and at 2.54 in the Si-bearing glasses. We are thus led to believe that the manner in which a glass network becomes rigid appears to be tied to elements of medium range structure. Both families of IV-VI glasses possess closely similar local building blocks which consist of Se_n chains, CS- and ES-tetrahedra. The principal difference between the two glass systems pertains to the manner in which these building blocks couple to each other to define elements of medium range structure[3,4,7,28]. There has been a suggestion[39] that mean-field predictions are likely to be inappropriate near phase transitions, and in fact that there may be two stiffness transitions in network glasses, one *hydrostatic* in nature and a second *shear* in nature. These new ideas are timely and would clearly need to be tested by design of suitable experiments.

Ternary Ge-As-Se glasses containing equal concentrations of Ge and As and up to 30 molar percent of these cations, largely consist of random networks of tetrahedral $Ge(Se_{1/2})_4$ and pyramidal $As(Se_{1/2})_3$ building blocks. These glasses are probably good realizations of a Zachariasen glass as are SiO_2[46] and $Ge_{0.35}Sn_{0.65}Se_2$.[47] It is possible that in such model random networks, one observes a narrow transition region, i.e. $r_c(1) = r_c(2) = \bar{r}_c$ = 2.40 leading to a solitary stiffness transition as predicted by numerical simulations on *random networks*. Certainly such a prediction would be well worth testing.

Acknowledgements

It is a pleasure to acknowledge discussions with Jim Phillips and Mike Thorpe. This work was supported by NSF grants DMR-97-02189 and DMR-94-24556.

REFERENCES

1. J.C. Phillips, *J. Non-Cryst. Solids* 34:153 (1979) and *J. Non-Cryst. Solids*, 43:37 (1981).
2. M.F. Thorpe, *J. Non-Cryst. Solids* 57:355 (1983) and M.F. Thorpe in *Insulating and Semiconducting Glases*, World Scientific Press, Singapore (1999).
3. X. Feng, W.J. Bresser and P. Boolchand, *Phys. Rev. Lett.* 78:4422 (1997), X. Feng, *Masters Thesis, University of Cincinnati*, (1997) unpublished.
4. D. Selvanathan, *Masters Thesis, University of Cincinnati*, (1998) unpublished, D. Selvanathan, W.J. Bresser and P. Boolchand (unpublished).
5. H. He and M.F. Thorpe, *Phys. Rev. Lett.* 54:2107 (1985).
6. D.S. Franzblau and J. Tersoff, *Phys. Rev. Lett.* 68:2172 (1992).
7. W.J. Bresser, P. Boolchand and P. Suranyi, *Phys. Rev. Lett.* 56:2493 (1986).
8. B. Norban, D. Persing, R.N. Enzweiler, P. Boolchand, J.E. Griffiths and J.C. Phillips, *Phys. Rev.* B36:8109 (1987).
9. K. Murase and T. Fukunaga, in *Defects in Glasses*, Mat. Res. Soc. Symp. Proc. 61:101 (1986).
10. S. Asokan, M.V.N. Prasad, G. Parthasarathy and E.S.R. Gopal, *Phys. Rev. Lett.* 62:808 (1989).
11. S. Mahadevan and A. Giridhar, *J. Non-Cryst. Solids* 143:52 (1992), S. Mahadevan, A. Giridhar and A.K. Singh, *J. Non-Cryst. Solids* 57:423 (1983).
12. M. Mitkova and P. Boolchand, *J. Non-Cryst. Solids* 240:1 (1998).
13. P. Boolchand and M.F. Thorpe, *Phys. Rev.* B50:10366 (1994); P. Boolchand, M. Zhang and B. Goodman, *Phys. Rev.* B53:11488 (1996).
14. J. Wells, W.J. Bresser, P. Boolchand, *Bull. Am. Phys. Soc.* 42:249 (1997) and to be published.
15. A.B. Seddon, *J. Non-Cryst Solids* 213/214:22 (1997).
16. S.S. Yun, H. Li, R.L. Cappelletti, R.N. Enzweiler and P. Boolchand, *Phys. Rev.* B39:8720 (1989).
17. J.Y. Duquesne and G. Bellessa, *J. Phys. (Paris) Colloq.* 46:C10-445 (1985).
18. R. Ota, T. Yamate, N. Soga and M. Kunugi, *J. Non-Cryst. Solids* 29:67 (1978).
19. A.N. Sreeram, A.K. Varshneya and D.R. Swiler, *J. Non-Cryst. Solids* 128:294 (1991).
20. W.A. Kamitakahara, R.L. Cappelletti, P. Boolchand, F. Gompf, D.A. Neumann and H.D Mutka, *Phys. Rev. B*44:94 (1991).
21. P. Boolchand, R.N. Enzweiler, R.L. Cappelletti, W.A. Kamitakahara, Y. Cai and M.F. Thorpe, *Solid State Ionics* 39:81 (1990).
22. C.A. Angell, *Science* 267:1924 (1995).
23. R. Bohmer and C.A. Angell, *Phys. Rev. B* 48:5857 (1993).
24. U. Senapati and A.K. Varshneya, *J. Non-Cryst. Solids* 185:289 (1995).
25. M. Zhang and P. Boolchand, *Science* 266:1355 (1994).
26. R. Enzweiler, D. Selvanathan and P. Boolchand (unpublished).
27. R. Bohmer and C.A. Angell, *Phys. Rev.* B45:10091 (1992).
28. W.J. Bresser, P. Boolchand and P.Suranyi and J.P. deNeufville, *Phys. Rev. Lett.* 46:1689 (1981); P. Boolchand, *Z. Naturforsch.* 51a:572 (1995).
29. B. Uebbing and A.J. Sievers, *Phys. Rev. Lett.* 76:932 (1996).
30. X. Feng, W.J. Bresser, D. Selvanathan, M. Zhang and P. Boolchand (unpublished).
31. J. Menendez and M. Cardona, *Phys. Rev.* B29:2051 (1984).
32. H. Fritzsche, *Solid State Commun.* 99;153 (1996).

33. H. Hisakuni and Ke. Tanaka, *Science* 270:975 (1995).
34. B. Wunderlich, Y. Jin and A. Boller, *Thermochim. Acta* 238:277 (1994).
35. M. Tatsumisago, B.L. Halfpap, J.L. Green, S.M. Lindsay and C.A. Angell, *Phys. Rev. Lett.* 64:1549 (1990).
36. R. Kerner and M. Micoulaut, *J. Non-Cryst. Solids* 210:298 (1997).
37. A. Feltz, H. Aust, and A. Bleyer, *J. Non-Cryst. Solids* 55:179 (1983).
38. P. Boolchand in *Physical Properties of Amorphous Materials*, Plenum Press, New York (1995) p. 221.
39. P. Boolchand, J. Grothaus, W.J. Bresser, and P. Suranyi, *Phys Rev.* B25:2975 (1982).
40. P. Boolchand and M. Stevens, *Phys. Rev.* B29:1 (1984).
41. Y. Cai and M.F. Thorpe, *Phys. Rev.* B40:10535 (1989).
42. P.M. Bridenbaugh, G.P. Espinosa, J.E. Griffiths, J.C. Phillips, and J.P. Remeika, *Phys. Rev.* B20:4140 (1979).
43. K. Murase, T. Fukanaga, Y. Tanaka, K. Yakushiji and I. Yunoki, *Physica* 117B and 118B:962 (1983) and also K. Murase in *Insulating and Semiconducting Glasses*, World Scientific (1999).
44. L. Ross and H. Bourgon, *Can J. Chem.* 47:2555 (1969).
45. J.C. Phillips in *Rigidity Theory and Applications*, Plenum, New York (1999).
46. W.H. Zachariasen, *J. Am. Chem. Soc.* 54:3841 (1932), K. Vollmayr, W. Kob and K. Binder, *Phys. Rev.* B54:15808 (1996).
47. M. Stevens, P. Boolchand and J.G. Hernandez, *Phys. Rev.* B31:981 (1985).

RIGIDITY, FRAGILITY, BOND MODELS AND THE "ENERGY LANDSCAPE" FOR COVALENT GLASSFORMERS

C. A. Angell

Department of Chemistry
Arizona State University
Tempe, AZ 85287-1604

INTRODUCTION

One of the earliest applications of the Maxwell concept of constraint counting to the field of materials science, was that of Phillips [1] who was interested in the reasons that certain liquid chalcogenide systems are kinetically very stable against crystallization [2]. Phillips concluded that stability of the glass would be maximized when the **bond density** (number of bonded neighbors per atom in the structure) averaged 2.4.

Subsequent theoretical [3-5] and experimental [6-15] applications of the concept in this area have been more concerned with the description of the physical properties of these glassy materials, in particular with respect to behavior in the vicinity of the composition at which the constraints and degrees of freedom equalize. The constraints are evaluated at the molecular level in terms of the fixed bond lengths and bond angles, and it is predicted that the equalization condition is met when the bond density, usually called the average coordination number and designated $<r>$, is equal to 2.4. (Here coordination number must be understood in the restricted sense of the coordination of each atom with respect to *bonded* neighbors only, which is why we introduce the less ambiguous term). In binary and multicomponent systems, the bond density can be varied continuously by varying the content of elements with different bonding propensities, e.g. 2 for chalcogenide elements S, Se and often Te, 3 for picnides P, As, Sb, and often Bi, and 4 for Si, Ge, and often Sn.

Properties, like the glass transition temperature and the heat capacity jump at T_g, have been studied for a number of different multicomponent systems [2,7,8,9], and breaks of one sort or another have been observed in the vicinity of the "rigidity percolation threshold" [2] , the name assigned to the theoretical transition at which floppy regions become isolated within a continuous rigid matrix. However, the relation of these breaks to the theoretical expectations needs to be looked at carefully.

The theoretical treatments of rigidity percolation in glassy systems have always been concerned with idealized structures in which all possible bonds are intact and all bond angle rules respected. The reason that liquids form from glasses on heating is that a fraction of these constraints, which is a Boltzmann function of temperature, is lifted. The state of excitation is most sensitively monitored by the structural relaxation time which determines the liquid viscosity η through the Maxwell equation connecting it to the shear modulus measured at high frequency G_∞ and the shear relaxation time τ

$$\eta = G_\infty \tau. \qquad (1)$$

G_∞, which measures shear rigidity, depends linearly on temperature via the broken constraint fraction and τ depends on it more strongly, indeed exponentially, as will be discussed further below. Inversely, on cooling, the constraints tend to be reestablished. However, a fraction remain broken, i.e. are frozen in, when the relaxation time reaches the value at which the equilibrium state can no longer be maintained for the cooling rate in question. (We say "ergodicity is broken"). This means that in the real glasses used to test the predictions of the models, the fraction of constraints acting on the system of atoms cannot be the number calculated from the composition. Therefore the effects predicted by the theory cannot be expected to occur at the theoretical percolation threshold. So far this problem has not been mentioned in attempts [4,5] to explain the observations that the observed increases in rigidity with increasing bond density occur at values higher than the simple theory predicts [10,11,14].

The main purpose of this paper is to try to put this "broken constraint" factor into the picture. We will concentrate on the case of the three component chalcogenide glass system Ge-As-Se for which data are available in convenient form for our analysis but comparable information is available for other systems [8-15]. It will be important to realize that the effect of the fraction of constraints frozen-in, on properties like the glass transition temperature, will be largest *just* in the vicinity of the transition that is anticipated by the theory. A subsidiary purpose will be to relate these effects to the "energy landscape" approach to discussion of liquid properties [16,17], which has been repeatedly invoked in the discussion of glasses [18] and more recently has been shown to be intimately related to the details of the relaxation function [19] and also to the fragility of the liquid [20].

The effect of allowing for the effect of frozen-in constraints on the glass transition vs. $\langle r \rangle$ relation will be evaluated from previously published information on liquids and glasses in the system Ge-As-Se [7,12], and data from the liquid states of these systems will then be used to parameterize a simple but relevant treatment of the effect of temperature on the thermodynamic and relaxational behavior of these systems. These results, and the parameters which control them will then be shown to contain a description of the configuration space "energy landscapes" for these systems.

IDEAL GLASS TRANSITIONS FOR Ge-As-Se SYSTEM

The state of the liquid system from which no further entropy can be lost by configurational rearrangements, is known as an *ideal glass* [21]. All other glasses are, in principle, isothermally rearrangeable into this ideal state with a decrease in probability, hence in free energy. It is only a matter of kinetics. The ideal glass is the glass in which the energy has the lowest possible value at 0K, hence is that in which the constraints on the configuration are realized to the maximum extent possible. It is of interest to decide the temperature at which it would be reached during infinitely slow cooling of the equilibrated liquid state. This can be achieved in different ways, of which the most general is by analysis of the temperature dependence of the relaxation or, alternatively and less satisfactorily, of the viscosity.

Supposing that the liquid relaxation times, and the viscosity, can be described by the Adam-Gibbs equation [22], and that the Adam-Gibbs (AG) equation can be equated to the Vogel-Fulcher-Tammann (VFT) equation, then the ideal glass temperature, (now usually designated T_K) can be obtained from data as follows:

1. $\tau = \tau_o \exp(B/[T-T_o])$ VFT equation $= \tau_o \exp(DT_o/[T-T_o])$ (2)

After introduction of the glass transition temperature T_g, defined at a relaxation time 16 orders of magnitude longer than τ_o, Eq. (2) can be converted to

$$T_g/T_o = 1 + D/\ln(10) \log \tau_g/\tau_o = 1 + D/16 \ln(10) \qquad (3a)$$

from relaxation times, and

$$= 1 + D/\ln(10) \log \eta_g/\eta_o = 1 + D/17 \ln(10) \qquad (3b)$$

from viscosity.

2. The Adam-Gibbs equation expresses the relaxation time in terms of the amount of entropy introduced into the liquid by excitation above the ground state temperature T_K, according to the expression

$$\tau = \tau_o \exp(C/TS_c) \qquad (4)$$

3. Evaluation of S_c in Eq. (4) yields Eq. (1) if the heat capacity has a particular simple form, viz., hyperbolic in T, as often seems to be the case for molecular liquids [23,24] and we assign $S_c = 0$ when $T = T_K$. This procedure identifies the ground state temperature of the Adam-Gibbs analysis with the T_o of the Vogel-Fulcher-Tammann equation. Thus we can obtain T_K from transport data through T_g and the fragility.

The fragilities of glassforming liquids have been quantified both by the parameter D, and by the slope m of the relaxation time (or viscosity) near T_g according to

$$m = d\log\tau / d\log\,[T_g/T] \tag{5}$$

Both D and m values have been quoted in the chalcogenide glass literature, and they may be interconverted using the relation

$$m = 16 + 590/D \text{ (for relaxation times)} \tag{5}$$

and $$m = 17 + 590/D \text{ (for viscosities)}$$

The values of T_g which are to be "corrected" according to these relations, are shown in Fig. 1. The values of D or m needed to obtain the ground state temperatures, are available from the study of viscosity in ref. 7., and from the study of mechanical relaxation times in ref. 12. The data are also available in principle [25] from the widths

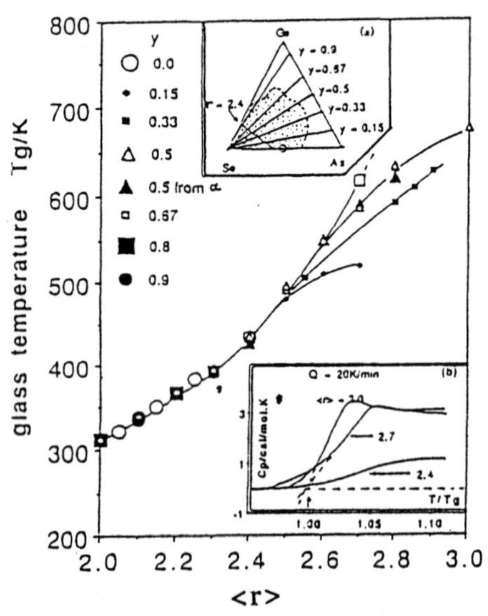

Fig. 1. Glass transition temperatures for Ge-As-Se alloys vs bond density <r>. Inserts show glassforming composition range, and calorimetric forms.

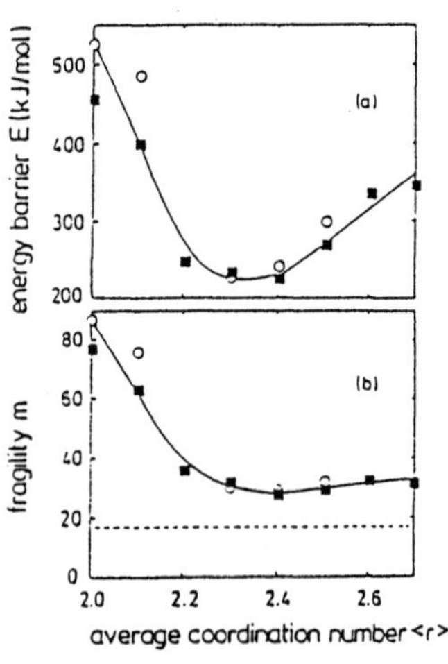

Fig. 2 Activation energies for viscosity and mechanical relaxation, and derived fragilities, as functions of <r>.

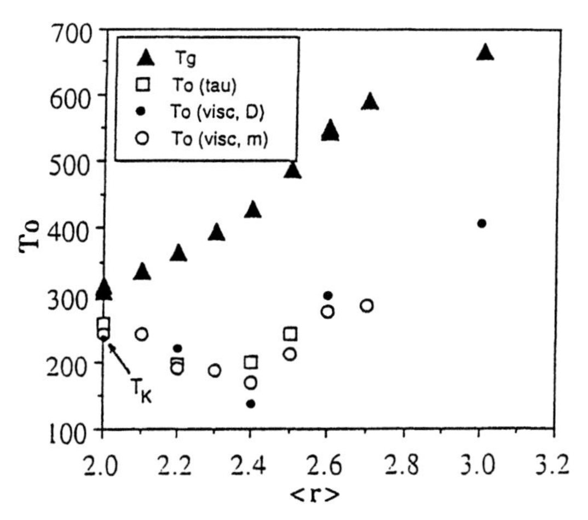

Fig. 3. Comparison of ideal glass transition temperatures $T_{o,i}$ with laboratory glass transition temperatures Tg for Ge-As-Se alloys along the join with [Ge]/[As] = 1 (i.e. Y = 1, in Fig. 1, inset). The spread of values for viscosity To's between D and m-based values indicates the uncertainty in fragility determination by these methods since the raw data are the same. The confirmation by the quite independent mechanical relaxation measurements is reassuring and the special nature of the bond density 2.4 is clearly seen. For pure Se, an ideal temperature, T_K, is also obtainable from purely thermodynamic data.

of the glass transition though these were not systematically measured in ref. 7. The widths of the transitions seen in Fig. 1 have been correlated with the <r> in a subsequent paper [27] and their utility in this respect will be mentioned later in this paper. The glass transition temperatures for a variety of compositions in the Ge-As-Se system are shown in Fig. 1 and the viscosities from which the D values are taken are reproduced in Fig. 2. Derived values of ideal glass transition temperature, based on the identity of T_o and T_K (demonstrated most recently in [20]), are plotted vs. <r> in Fig. 3 together with measured glass transition temperatures. Some additional points based on the more appropriate relaxation time temperature dependences [12] are included. The value of T_K for Se, which is obtained from purely calorimetric data [7], is included in Fig. 3. It will later be seen that a finite T_K (and likewise T_o) is likely an artifact of extrapolation of observables, but it remains a useful characterizing parameter irrespective of behavior below T_g which can only be determined by theory.

DISCUSSION

Percolation at <r> = 2.4

Fig. 3 gives a quite different impression of the effect of constraints, represented by <r>, on the rigidification of the system than does the glass transition temperature displayed in Fig. 1. Insofar as it shows a weak dependence on <r> up to the value 2.4, and thereafter a steep increase, it is more in accord with theoretical expectations. Of course it needs to be substantiated by more detailed relaxation time studies over a wider range of substances. Data needed to extend the present type of analysis seem to exist in the work of Senapati et al [8] on the system Ge-Sb-Se, and Nemilov [9] on the system Ge-Se.

In the meantime we proceed to a theoretical description of the observations at the simplest possible level of statistical thermodynamics in order to show how an account of the observed differences between Figs. 1 and 3 can be reduced to an account of the composition dependence of two parameters of familiar physical significance.

Quasi-lattice model for the observed behavior

To provide a first approximation account of the way the observed behavior obscures the fundamental accord with rigidity theory, we turn to the "bond lattice" description of the thermodynamic excitation of a system of bonded particles. This approach is in fact more appropriate for the covalenty bonded chalcogenides than for almost any other type of glassy system. A previous case in which the equations of this treatment seem appropriate, and indeed provide an adequate account of the observations, is the partly covalent system $ZnCl_2$ [29].

In the bond lattice approach [30], the system of strongly bonded particles is transposed to its "bond lattice" in which the elements of the lattice can justifiably be considered as weakly interacting and can therefore be treated, in the first approximation, as being independently excitable. The number of bonds in the bond lattice depends on the coordination number of the particles.

In the rigidity arguments for chalcogenide systems, bond angles are considered equal constraining as bonds, and we will maintain this simplification in our treatment here. It is, however, a simple matter to introduce states of different excitation energy and to describe the multistate excitation. However, it introduces extra parameters without changing the qualitative behavior in important ways unless the constraints have very different excitation energies or entropies.

A system with intact constraints clearly has a lower enthalpy per mole of constrained particles than one with the constraints broken. Focusing on the constraints themselves we assign an enthalpy of constraint breaking of ΔH. Then the distribution of constraints across the "constraint quasi-lattice" will vary with temperature and pressure according to the usual two state thermodynamic relations described originally by Schottky [31] for low temperature magnetic systems and since applied to various physical systems by many workers [30-35].

In the systems described by Schottky, which give rise to smeared out heat capacity bumps known as Schottky anomalies, the flipping of a spin causes no entropy change other than that associated with the distribution of different spins among the N atoms in the structure. This seems to be almost the same situation in an optimally constrained bonded lattice, as will be seen. However, in under-constrained (and also, it seems, in overconstrained systems) the lifting of a single constraint may give rise to more alternative configurations than are indicated by the standard distribution across the constraint lattice. This can be accommodated by including an entropy of excitation term $S_2 - S_1$ ($= \Delta S$). A small contribution to ΔS will also be made if the excitation is accompanied by a decrease in average vibration frequency for the quasi-lattice region containing the "defect". ($\Delta S_v = R \ln \nu_1/\nu_2$)

The formal two state thermodynamic development follows as:

STATE I <=======> STATE II

i.e., (bond or angle constraint intact) <=======> (broken constraint)

denoted A <=======> B

for which, in first approximation, (ideal mixing, or independent constraint-breaking), the equilibrium constant K_{eq}

$$K_{eq} = [A]/[B] = X_B/(1 - X_B) \qquad (7)$$

and

$$\Delta G = \Delta H - T\Delta S = RT\ln K_{eq} \qquad (8)$$

where ΔH is the enthalpy per mole of constraint-breaking and ΔS is the entropy change in excess of the entropy increase due to distribution of broken constraints across the "constraint lattice". The ΔS parameter proves to be a key parameter in that it determines the fragility of the liquid, as will be seen.

From Eq. (8) the mole fraction of broken constraints is found to be

$$X_B = [1 + \exp(\Delta H - T\Delta S)/RT]^{-1} \qquad (9)$$

and the associated heat capacity is

$$C_p = (\partial H/\partial T)_p = R(\Delta H/RT)^2 \cdot X_B(1 - X_B) \qquad (10)$$

This heat capacity is a dome with onset commencing at a temperature determined by the molar enthalpy increment per constraint break, ΔH. When non-optimally constrained, the extra entropy term ΔS causes a more rapid increase and then the heat capacity has a maximum value which is determined by the magnitude of ΔS. The general behavior for different ΔH, ΔS combinations may be seen in Figs. 4, 5, and 6. The ability to fit the data on liquid $ZnCl_2$, which is an intermediate fragility liquid, is indicated in the figure. The value of ΔS required is quite small. Note in Fig. 5, how high heat capacity comes directly from a large constraint-breaking entropy increment, ΔS of Eq. (8).

In ref. 32 it was argued that the probability of a rearrangement of atoms, such as is needed for a fundamental diffusive event or flow event to occur, must depend on the presence of a critical fluctuation in the local concentration of broken constraints. Invoking the Lagrangian undetermined multipliers treatment of constrained maxima [36], this probability is found to be an exponential function of the fraction of broken constraints at each temperature. This gives rise to a two parameter expression for the temperature dependence of the relaxation probability W(T)

$$W(T) \sim \exp\left(f^*/X_B(T)\right) \qquad (11)$$

where f^* is a critical local broken constraint fraction - which may be assigned the value unity - and X_B is the overall broken constraint fraction, determined by the two parameters ΔH and ΔS. of Eq. 8. X_B has been plotted in Fig. 4. This expression, which becomes a transcendental equation with parameters ΔH and ΔS when X_B is substituted by Eq. (9), is indistinguishable in fitting ability from the two parameter exponent of the VFT equation, Eq. (2). In fact, in the range of T near T_g, X_B is linear in T, hence Eq. (2) becomes the VFT equation. The parameter T_o is the extrapolated intersection with the T axis at $X_B = 0$, and the parameter B is the inverse of the slope, dX_B/dT.

The parameter D of Eq. (2) which describes the "strength" of the liquid, is found to be determined entirely by the magnitude of the ΔS parameter, and to maximize when ΔS is zero as in the Schottky anomaly. This is shown in Fig. 7 where B is seen to be a linear function of T_o, hence $D = B/T_o$ as in Eq. (2). T_o on the other hand is determined almost entirely by the ΔH parameter, as shown by the linear extrapolation to $X_B = 0$ indicated in Fig. 4. This was demonstrated [32] before the "strong/fragile" classification of liquids was expressly formulated and has not been discussed in relation to fragility before.

Two state thermodynamic treatments fall at the lowest level of sophistication in statistical thermodynamics, and we would not introduce them here were it not for two features that we think provide considerable insight into the physics of fragile liquids and their current interpretation in terms of "energy landscapes".

The first is the obvious one that the two state analysis provides a one parameter (ΔS) interpretation of fragility in chalcogenide liquids.

This in turn should encourage experimental and theoretical efforts to understand the origin of the extra degeneracy introduced when local excitations (constraint-breaks) occur in liquids. (The increasing magnitude of the constraint-breaking enthalpy ΔH of Eq. (9) in

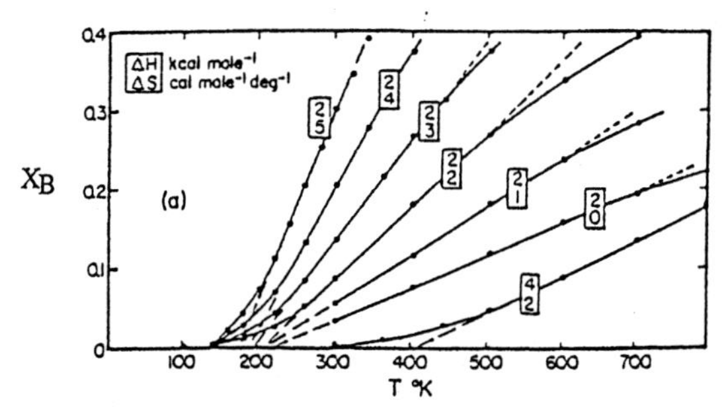

Fig. 4. Fraction of constraints broken at different temperature for the Eq. 9 parameter sets indicated in boxes alongside the plots. Note that extrapolation of the linear portions to low temperatures defines an operational ground state temperature, a Kauzmann temperature, which is seen to be determined almost entirely by the value of ΔH. The rate of excitation above the T_K, which determines the thermodynamic fragility (and also the relaxational fragility - see below), however is determined by the value of ΔS.

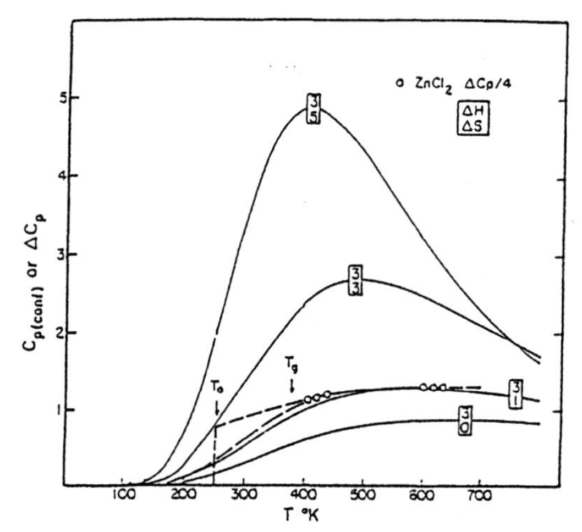

Fig. 5. Variation of the heat capacity with temperature according to Eq. (10) for the parameter sets indicated alongside the curves. Comparison is made with the case of ZnCl2 for data obtained near the glass transition temperature and above the melting point, and the variation of the theoretical curve from that assumed in calculating the Kauzmann temperature for this substance [29], is noted.

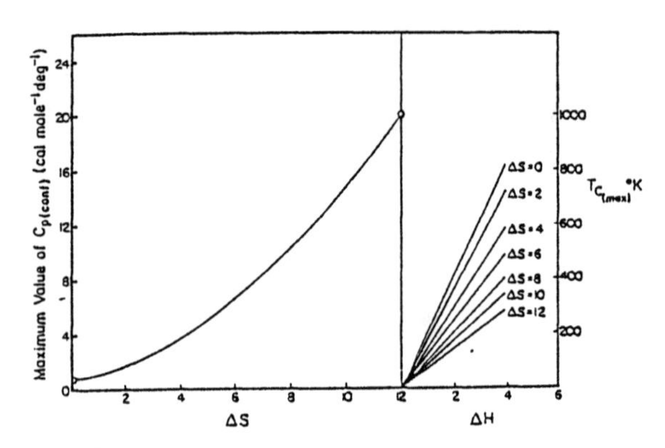

Fig. 6. Dependence of maximum heat capacity of Fig 5 on the value of ΔS, and dependence of the temperature at which it is reached, on the value of ΔH.

overconstrained systems, indicated by the rise in T_0 for <r> > 2.4 seen in Fig. 3, must relate to the rigidity percolation.)

The second is more provocative and relates to the light that this analysis shines on recent attempts to quantify the behavior of liquids in the rather sophisticated collective co-ordinate analysis involved in the "energy landscape approach" to description of complex systems.

THE ENERGY LANDSCAPE VIEW OF LIQUIDS AND GLASSES

The "Landscape Paradigm" is a conceptual framework within which qualitative aspects of a wide range of complex system phenomenologies have been discussed [37]. The name derives from the multitude of temporarily stable states that are permitted and the multitudinal possibilities for transitions via saddle points between the different states. The system is represented by a point which moves on or above the surface according to rules decided by the nature of the problem under consideration. Problems addressed, meaning the types of system which move on the surface, range from economics and evolution, through neural networks, spin glasses and proteins, to the subject of molecular clusters and viscous liquids with which we are here concerned. For each problem there has developed a different language, e.g. "free energy", which should be minimized for a system of molecules, becomes "fitness" which should be maximized for an evolving species.

In a recent letter, Sastry, Debenedetti and Stillinger [19] showed how changes in the dynamic properties of liquids noted in recent experimental [38] and simulation [39] studies can be related to changes in the nature of the energy landscape being explored by the system. The variation of the energy of the landscape minima being sampled at each temperature, as revealed by this computationally very intense study, resembles strikingly the fractional excitation profile of the

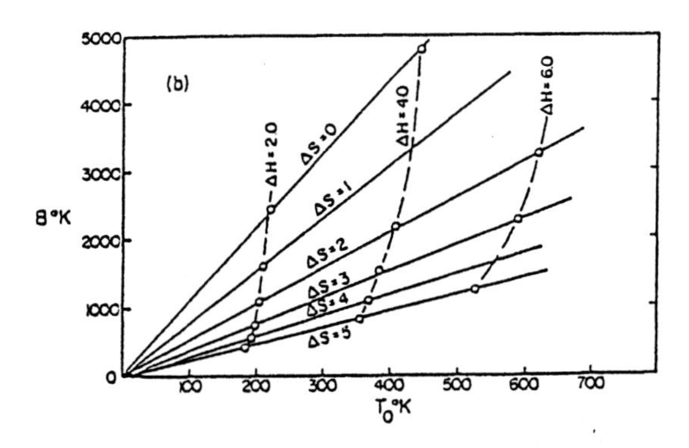

Fig. 7. Relation of T_o , defined by the Fig. 4 plot extrapolations, and the slope of the constraint excitation plots of Fig. 4, (which becomes the parameter B^{-1} of Eq. (2) according to the relaxation probability model, Eq. (11) in the case that $f^* = 1$). Points are obtained for different combinations of ΔH and ΔS, and show how the model requires B to be a linear function of T_o, the slope of which determines the "strength" parameter of the "strong/fragile" liquids classification. This in turn is seen to be determined almost entirely by the value of ΔS.

constraint quasi-lattice described with rudimentary theoretical tools in the foregoing section. We elaborate on the comparison after detailing a little further the landscape interpretation of liquid vitrification.

For the discipline of viscous liquids and glasses, the system point is the potential energy of the collection of N particles in a particular arrangement, the energy of which is determined by the exact positions of all the particles. The energy landscape is a multidimensional (3N+1) surface, hence impossible to conceptualize properly, nevertheless a two dimensional representation of its topology (see Fig. 8, panel (a)) is widely used in discussing the system behavior in response to stresses imposed from outside (i.e. by us).

A glassy state of the system is represented by the system point being stuck in a particular minimum: equilibrium corresponds to the system moving between ("sampling") minima in a region of energy determined by the temperature (imposed by us). When the temperature is increased the system will spend most of its time "higher up" on the energy landscape because that way it gains access to more possible states. This is in accord with the principle that fixed volume systems in equilibrium are systems in their lowest free energy states defined by $A = E - TS$, where A is the Helmholtz free energy, E is the energy and S is the entropy. S is large when the number of states that the system can visit is large. The relation is that inscribed on Boltzmann's tomb, $S = k_B \ln W$ where W is the number of possible states (here energy minima on the landscape). Crystals, which (when perfect) occupy just one (deep) minimum, undergo melting when T exceeds T_m

in order to lower their free energies by gaining access to all the minima of the landscape (in constant pressure systems, in which the volume can change, the $T\Delta S$ just balances the ΔH at T_m). Glassformers are liquids which fail to "find their way back" to the crystal minimum on cooling hence wander "down" among the myriad minima of the landscape as they supercool, ("down" because there then are fewer minima accessible and S is smaller). Eventually the system gets "stuck" in one of the lower minima - not as low as for the crystal - and we have a glass.

The system gets stuck because the vibrational energy (which is always present above zero Kelvin and which adds on to the landscape energy to "float" the system a little above the bottom of any minimum it occupies) becomes insufficient to "activate" (by fluctuations) the system from one minimum to the next. Bringing the temperature up a little allows the system to slowly explore the surrounding minima, hence to restore an equilibrium state (minimum in A). This is called "relaxation" or "annealing" or "aging" according to the sub- field of liquids, glasses or polymers that one works in.

In the study of Sastry et al [18], the part of the landscape that could be fully described was only the upper part in terms of relaxation times (in log units), but in terms of energy changes it is the major part. The system got stuck, for lack of computation speed, at a temperature somewhat above T_c of the mode coupling theory which, according to empirical observations on fragile liquids [20,40], is about 1.6 T_K. This part of the excitation profile is shown in panel (b) of Fig. 8 and continued down to the ground state energy (to be reached at $T_K = T_{c(MCT)}/1.6$) by a dashed line.

Panel (c) provides, for comparison with the panel (b) profile, the excitation profile given by Eq. (9) for a system with ΔH = 2 kcal/mole and ΔS = 2 cal/mol-K. This case was chosen because it has the same "reduced width" as the mixed LJ system when Kelvin T units are used. The reduced width is the ratio of temperatures obtained by extension of the steep linear part of the excitation profile to the "top of the landscape" of panel 2 on the one hand and to the ground state level on the other (see the vertical dashed lines in panels (b) and (c) which delineate the "widths") [41]. The top of the landscape panel (c) is the high temperature limit of X_B, which is 1.0 whenever ΔS differs from 0.

The resemblance of the excitation profiles for the mixed LJ system, (panel (b)) and the simple two state model (panel (c)) is quite impressive and, we think, instructive insofar as one is obtained from a configuration space representation of the system thermodynamics and the other is based on a real space representation. Note that we have had to invoke a rather large excitation entropy in order to match the width of the mixed LJ profile defined as in Fig. 8 panel (b). The ΔC_{pmax} value corresponding to this ΔS value is (Fig. 5) 4.8 cal/mol-K which corresponds closely with the mixed LJ value for ΔC_p at the MD glass transition, namely 4.3 cal/mole.K [44] if there is only one constraint per particle. The corresponding D value, according to Fig. 7

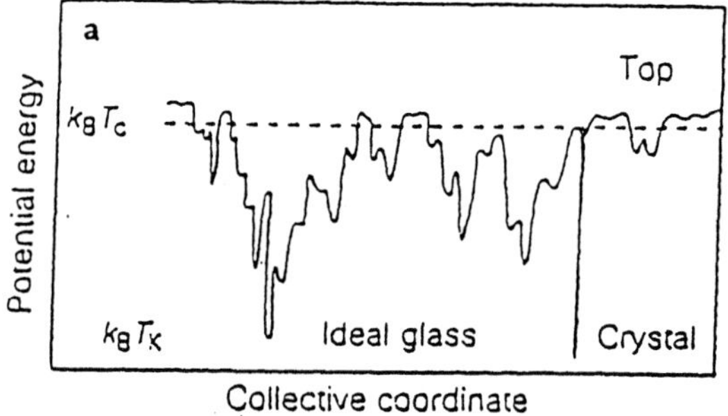

Fig. 8. (a) The 2D representation of the energy "landscape" for a system of interacting particles, indicating the relation between crystal, liquid and "ideal glass" states. The figure suggests that the "top of the landscape" falls near the mode-coupling critical temperature T_c.

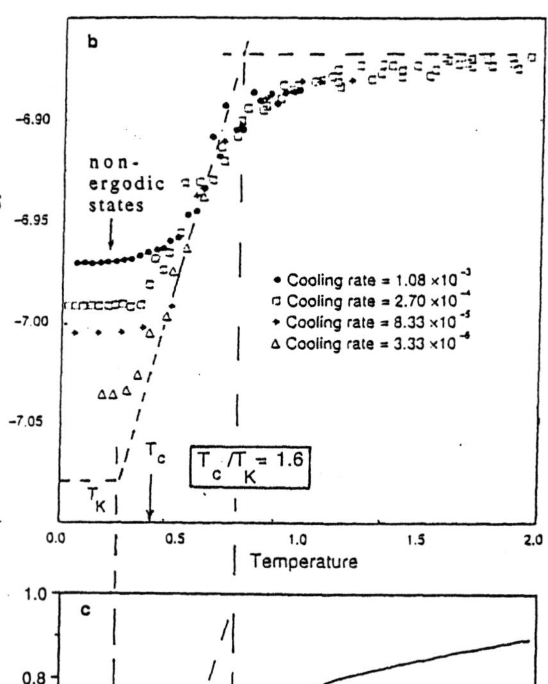

(b) The relation between the portion of the energy landscape visited by the system most frequently, and the temperature of the system, according to recent MD computer simulation inherent structures studies [19]. To obtain the ergodic behavior at lower temperatures, the profile has been extrapolated linearly to the temperature T_K according to the relation $T_c/T_K = 1.6$ observed for a variety of fragile liquids in laboratory studies [20]. T_c, according to the MD studies of ref. 39, is 0.435 in Fig. 8(b) units. The "width" of the profile defined by the linear extrapolation of the steep part of the profile to the ground state and to the "top" respectively, is indicated by vertical dashed lines.

(c) The Eq. (9) excitation profile for parameters (in box) and T units which approximately match the width of the mixed LJ system profile of Fig. 8(b). For Eq. (9), the "top" of the excitation profile is only reached at $T = \infty$. The general similarity of the Fig. 8(b) and Fig. 8(c) excitation profiles can be taken to advantage to simplify the descrip-tion of liquid properties which are necessarily complex in multidimen-sional energy landscape terminology. Note the ergodic behavior near T_K. The gradient of the excitation profile gives a measure of the density of configurational states for the liquid.

($D = B/T_O$), should be approximately 3, c.f. D = 8 for Se, the most fragile liquid in Ge-As-Se system. The value 3 is characteristic of a very fragile liquid [20] as would be expected for mixed LJ. The observed behavior of mixed LJ (unlike the case of $ZnCl_2$ seen in Fig. 5), however, shows no maximum or rapid decrease (as in Fig. 5) in the range of study, and in this respect behaves more like the "terraced landscape" model [45].

Note the position of the mode coupling theory T_c on the profile of the mixed LJ system, Fig. 8(b). It is more than half way down to the ground state energy, and close to the inflection point of Eq. (9) seen in Fig. 8(c). This is consistent with the observations of Fischer [35] who fitted equations based on a two state model to the experimental data for several molecular liquids for which T_c had been determined by other workers. (It is below this energy that most of the many orders of magnitude change of relaxation time occur en route to the glass transition). A number of studies [20,40,46,47)] have identified the temperature T_c with the temperature at which there is some crossover in the temperature dependence of relaxation times and, most recently, some breakdown in the agreement of Vogel-Fulcher and Adam-Gibbs descriptions which fit the low temperature relaxation data so well for most liquids [48]. The departure from near-linearity in temperature of the function X_B which is the denominator of the relaxation probability expression, provides a rather direct explanation for the failure of the Vogel Fulcher equation at higher temperatures.

From the Fig. 8 comparison, we can go on to describe at least qualitatively, the energy landscape excitation profiles for chalcogenide liquids of different <r> values using the fragilities measured in refs. 7 and 12, and the Eq. (9) "excitation profiles" for the associated constraint-breaking parameters ΔH and ΔS. This analysis will be presented in more detail when a study of the simple two component system Ge-Se, currently in progress, has been completed [49].

The bond lattice model gives us a simple tool for understanding the density of configuron states for liquids. The density of states is to be assessed from the gradient of the excitation profile, and it peaks where the second differential vanishes. This density must maximize at the inflection point, at which the entropic drive towards higher energy states with increasing temperature must maximize in the "independent bond" approximation. The inflection point will occur at higher values of X_B for larger values of the ΔS parameter, (and more so in the terraced landscape modification of the model [45]). On this basis it is only in the case of extremely fragile liquids that the inflection point, hence the value of T_c, will approach the "top of the landscape" as implied in a recent paper [20] and now seen in a simulation of a model of the very fragile liquid, o-terphenyl [50].

Further to the ΔS influence on the position of the inflection point on the excitation profile we note that the peak density will be minimized, and the width of the distribution maximized, as constraints are optimized and ΔS of the elementary excitation approaches zero. This will occur near $<r> = 2.4$ for systems with strong bond angle as well as bond length constraints (Fig. 4), but at larger $<r>$ values for systems where the bond angle constraints are weak (e.g. water and silica [1,2]).

It is worthwhile to compare these observations with those of Speedy for hard sphere systems. Speedy [51-54] has argued plausibly, on the basis of simulations on a variety of hard sphere systems, that the density of states (alternative packing densities) is most likely Gaussian (and in this respect a little different from that of our simple model). In one of these systems, the "tetravalent hard sphere" system"[46], constraints were applied. The Gaussian for this case was much broader than for the simple hard sphere system, so the tetravalent system is a "stronger" liquid in our terminology - consistent with what we see in the constraint lattice model, and in the corresponding laboratory chalcogenide glasses, as constraints are optimized. However Speedy [54] examined the effect of the constrained coordination number on the properties of the hard sphere system with out finding any behavior reminiscent of Fig. 3. The origin of such differences between hard sphere fluids and the model chalcogenide systems may be due to the abscence of angular constraints in the former. A systematic study of these factors should do much to enhance our understanding of fragility in the liquid state.

CONCLUDING REMARKS

Of additional interest in the chalcogenide systems is the existence of a maximum T_g value as Ge content increases and the system becomes severely overconstrained. The existence of a general maximum can be deduced from the knowledge that the diffusivity of crystalline Ge exceeds that characteristic of a substance at its glass transition [10^{-22} m^2 s^{-1}] at 550K [55] so that amorphous Ge must presumably have a T_g value no higher than 550K. (This is close to its observed recrystallization temperature.) The maximum has been directly observed in the case of the Ge-Se system [15, 56] and can be associated with an effective reduction of the Ge oxidation state towards +2 (GeSe). A T_g of 550K is reached at 35% Ge in the binary system $GeSe_2$ and should be realized in any cut through the ternary at comparable Ge fractions. Although it has not been reported to date, this would probably be followed by the splitting out of a pure Ge phase with about the same T_g value. In this domain the independent constraint-breaking assumption made in the two-state treatment we have given must break down, and interesting analogies with the landscape interpretations of polyamorphism [43] will present themselves.

ACKNOWLEDGMENTS

This work has been carried out under the auspices of the NSF under Solid State Chemistry grant no. DMR 9614531. The author has benefited from helpful discussions with Burkhardt Geil and Robin Speedy.

REFERENCES

1. J. C. Philipps, *J. Non-Cryst. Sol.* 34:153 (1979).
2. P. J. Weber and J. A. Savage, *J. Non-Cryst. Sol.* 20:271 (1976).
3. M. F. Thorpe, *J. Non-Cryst. Sol.* 57:355 (1983); J. C. Philipps and M. F. Thorpe, *Sol. State Commun.* 53:699 (1985).
4. H. He and M. F. Thorpe, *Phys. Rev. Lett.* 54:2107 (1985).
5. M. F Thorpe in *Amorphous Insulators and Semiconductors*, eds. M. F. Thorpe and M. I. Mitkova, NATO-ASI Series, Plenum Press (1997) pp. 289-328.
6. B. L. Halfpap and S. M. Lindsay, *Phys. Rev. Lett.* 57:847 (1986).
7. M. Tatsumisago, B. L Halfpap, J. L. Green, S. M. Lindsay, and C. A. Angell, *Phys. Rev. Lett.* 64:1549 (1990).
8. V. Senapati and A. K. Varshneya, *J. Non-Cryst. Sol.* 185:289 (1995); A. K. Varshnaya, A. N. Sreeram and D. R. Swiler, *Phys. Chem. Glasses* 34:179 (1993).
9. S. V. Nemilov, *Zhur. Prikl. Khim.* 37:1020 (1964).
10. K. Tamaka, *Solid State Commun.* 60:295 (1986).
11. S. Asokan and E. S. R. Gopal, *Rev. Solid State Sci.* 3:273 (1989).
12. R. Böhmer and C.A. Angell, *Phys. Rev. B.* 45:10091-10094 (1992).
13. (a) P. Boolchand et al, *J. Non-Cryst. Sol.* 182:143-154 (1995).
    (b) P. Boolchand and M. F. Thorpe, *Phys. Rev.* B50:10366-10368 (1994).
    (c) P. Boolchand, M. Zhang, and B. Goodman, *Phys. Rev.* B53:11488 (1996).
14. B. Halfpap, Ph.D Thesis, Arizona State University, (1990); B. Halfpap, S. M. Lindsay and C. A. Angell (to be published).
15. See Chapter 3 of A. Feltz, "Amorphous Inorganic Materials and Glasses," VCH Verlag., Weinheim, N.Y. 1993.
16. M. Goldstein, *J. Chem. Phys.* 51:3728 (1969).
17. F. H. Stillinger and T. A. Weber, *Science* 225:983 (1984).
18. See articles in the special issue of *Science* on amorphous materials, 267: 1924-1953 (1995).
19. S. Sastry, P. G. Debenedetti, and F. H. Stillinger, *Nature* 393:554-557 (1998).
20. C. A. Angell, APS Symposium Proceedings, *J. Res. NIST* 102:171 (1997); C. A. Angell, in "Complex Behavior of Glassy Systems" (Proc. 14th Sitges Conference on Theoretical Physics, 1966) Ed. M. Rubi, Springer, 1967, p. 1.
21. (a) D. Turnbull, *Contemp. Phys.* 10:473 (1969).
    (b) C. A. Angell, *J. Am. Ceram. Soc.* 51:117-124 (1968).

22. G. Adam and J. H. Gibbs, *J. Chem. Phys.* 43:139-146 (1965).
23. Y. Privalko, *J. Phys. Chem.* 84:3307 (1980).
24. C. Alba, L. E. Busse and C. A. Angell, *J. Chem. Phys.* 92:617-624 (1990).
25. C. T. Moynihan, *J. Am. Ceram. Soc.* 76:1081 (1993).
26. K. Ito, C. T. Moynihan and C. A. Angell (to be published).
27. Jacques Lucas, Hong Li Ma, X. H. Zhang, Hema Senapati, Roland Böhmer and C. A. Angell, *J. Sol. State Chem.* 96:181 (1992).
28. S. S. Chang and A. B. Bestul, *J. Chem. Thermodyn.* 6:325 (1974); S. S. Chang (private communication).
29. C. A. Angell, E. Williams, K. J. Rao and J. C. Tucker, *J. Phys. Chem.* 81:238 (1977).
30. (a) C. A. Angell, *J. Phys. Chem.* 75:3698 (1971).
    (b) C. A. Angell and K. J. Rao, *J. Chem. Phys.* 57:470-481 (1972).
31. W. Schottky
32. P. B. Macedo, W. Capps, and T. A. Litovitz, *J. Chem. Phys.* 44:3357 (1966).
33. J. Perez, *J. Phys.* 46:C10:427 (1985).
34. J. Kieffer, J.E. Masnik, O. Nickolayev, and J.D. Bass, Phys. Rev. B 58, 694 (1998)
35. C.H. Wang and E.W. Fischer, *J. Chem. Phys.* 105, 7316 (1996): E. W. Fischer (preprint)
36. (a) M. H. Cohen and D. Turnbull, *J. Chem. Phys.* 31:1164 (1959).
    (b) H. Van Damme and J. J. Fripiat, *J. Chem. Phys.* 62:3365 (1978).
37. H. Frauenfelder *et al* (eds) *Landscape Paradigms in Physics and Biology Physica D*, special issue, 107 (1997).
38. J. Colmenero *et al*, *Phys. Rev. Lett.* 78:1928-1931 (1997).
39. W. Kob and H. C. Andersen, *Phys. Rev. E.* 51:4626 (1995).
40. (a) E. Rossler and A. P. Sokolov, *Chem. Geol.* 128:143 (1996).
    (b) V. N. Novikov, E. Rössler, V. K. Malinovsky and N. V. Surovtsev, *Europhys. Lett.* 35:289 (1996).
41. We note again [30] how the bond lattice treatment in the "independent bond" approximation, denies the existence of a finite Kauzmann temperature, though shows how an operationally defined ground state temperature (which would be the T_0 of Eq. (2)) is obtained by short extrapolation of the excitation profile. The transcendental equation, Eq. (11), fits transport and relaxation time data as well as does Eq. (2) [42] without requiring a finite ground state temperature. However, it was shown in ref. 30 that the glass transition phenomenon in fragile liquids is significantly sharper than can be accounted for by Eq. (10) and it is possible that cooperative effects, neglected in the zeroth order model, may produce a genuine singularity in principle though there is yet no evidence for it. The possibilities are illustrated in ref. 43.
42. C. A. Angell and R. D. Bressel. *J. Phys. Chem.* 76:3244 (1972).
43. C. A. Angell, *Proc. National Academy of Sciences*, 92:6675-6682

44. K. Vollmayr, W. Kob and K. Binder, *Phys. Rev. B* 54:15808 (1996).
45. P. Harrowell and C. A. Angell, (to be published).
46. E. Rössler, *Phys. Rev. Lett.* 65: 1595 (1990).
47. C. Hansen, F. Stickel, T. Berger, R. Richert, and E. W. Fischer, *J. Chem. Phys.* 107(4):22 (1997).
48. R. Richert and C. A. Angell, *J. Chem. Phys.* 108:9016 (1998).
49. P. Lucas and C. A. Angell, (to be published).
50. F. Sciortino, S. Sastry and P Tartaglia, (preprint)
51. R. J. Speedy and P. G. Debenedetti, *Mol. Phys.* 86:1375 (1995).
52. R. J. Speedy, *J. Phys. Condensed Matter,* 10:4185, (1998).
53. R. J. Speedy, *Mol. Phys.*, Barker Memorial issue,1998 (in press)
54. R. J. Speedy, (private communication).
55. S. Coffa, J. M. Poate, and D. C. Jacobson, *Phys. Rev. B.* 45:8355 (1992).
56. A. Feltz and F. J. Lippmann, *Z. Amorg. Allgem. Chem.* 398:157 (1973).

ENTROPIC RIGIDITY

Béla Joós[1], Michael Plischke[2], D.C. Vernon[2] and Z. Zhou[1]

[1]Department of Physics
University of Ottawa
150 Louis Pasteur
Ottawa, Ontario, K1N 6N5
Canada

[2]Department of Physics
Simon Fraser University
Burnaby, B.C., V5A 1S6
Canada

INTRODUCTION

There is now, as evidenced by several papers in this conference proceedings, a significant body of work devoted to the mechanical rigidity of networks. We define rigidity as the ability of a system to resist shear. The results that have been obtained apply to diluted central force networks, glasses, and tensegrity networks, and they focus on rigidity criteria and the nature of the transition from the floppy to the rigid phase.

One important result is that the onset of mechanical rigidity in diluted central force networks does not occur at either the site or bond connectivity percolation concentration but at a higher concentration. Mechanical rigidity in such systems requires multiple connectivity. Other results are mean field predictions on the minimum coordination number of bonds required to have rigidity in glasses. Discussions of these issues have so far been in terms of arguments based on geometrical and mechanical considerations. They are intrinsically zero temperature theories and they measure the part of the rigidity whose origin can be said to be energetically derived.

Temperature can qualitatively change the picture. Temperature introduces vibrations in the network. These can modify the existing elasticity in two ways. Firstly, the force constants have to be redefined according to the new average separation between interacting components of the system. Moreover, it produces a second component that has an intrinsically different origin than the one so far discussed: one that could be called a thermodynamic rigidity. It arises from the change in entropy of the system upon deformation. It is usually small in mechanically rigid systems but can be significant in soft materials and even dominant, as in the case of rubber where the energetic part of the shear modulus is insignificant in comparison with the entropic part[1].

Entropically derived rigidity differs from energetically derived rigidity in several ways. It is strongly dependent on temperature, *i.e.*, the leading term is linear in T. It seems to be scalar in nature, rather than vectorial as will be discussed further below. It is similar in nature to springs of zero length and therefore the entropic solid behaves like a material under tension. The first and third points lead to a negative contribution to the coefficient of thermal expansion.

Random networks in the neighborhood of their mechanical rigidity point are soft materials with high entropies. We show below that an entropically derived rigidity exists in these networks and that its onset occurs at the connectivity percolation point, significantly below

Rigidity Theory and Applications
Edited by Thorpe and Duxbury, Kluwer Academic/Plenum Publishers, 1999

the mechanical rigidity point. It dominates the elasticity up to the mechanical rigidity point, where a crossover occurs to energetically derived elasticity.

After a brief summary of findings on mechanical rigidity, we introduce entropic rigidity through a few examples and show that it is a measurable quantity that can be extracted from the temperature dependence of the elastic constants. Then we present and discuss previously published and some new results on entropic rigidity in model diluted central force networks. These results, and some models used to explain them, form the basis of a further discussion of entropic rigidity and its relevance to glasses and other materials.

A number of issues are of interest in the subject of entropic rigidity. One is an argument by de Gennes[2] who predicted that the elastic constants should vanish at the percolation threshold p_c as $\mu \sim (p - p_c)^f$, where f is the exponent describing the conductance of a random resistor network near percolation, and p is the concentration of bonds or sites. Although this is known not to be true for mechanical rigidity[3], it seems to apply to the case of entropic rigidity.

MECHANICAL RIGIDITY

A typical rigid body at $T = 0$ will settle into a ground state configuration which minimizes its internal energy. Deformation away from this equilibrium leads, for infinitesimal or small displacements, to linear restoring forces. This is what is understood as energetically derived elasticity. The conditions that determine whether or not a network is actually rigid form a subject matter that has been studied extensively by mathematicians and physicists, in models where temperature effects are absent. These theories focus on mechanical rigidity.

For our purposes we focus on systems with no bond bending forces. These require multiple connectivity for the existence of a mechanically rigid network. For this reason, the onset of rigidity in diluted networks occurs above the percolation threshold. The nature of this onset has been discussed by a number of authors in the context of central force networks at zero temperature[3,4,5].

Feng and Sen[4] were the first to point out that randomly diluted central force networks are incapable of withstanding shear or compression below a concentration of particles p_r — the rigidity percolation concentration — that is *considerably* higher than the concentration p_c at which an infinite connected cluster first appears. In the extreme case of the simple cubic lattice, $p_r = 1$, and even the perfect lattice has no resistance to shear. At the simplest level, this result can be understood in terms of the number of constraints imposed on the system of particles by the nearest neighbor forces. When this number becomes less than the remaining number of degrees of freedom, a soft mode appears. A straightforward mean field theory[5] produces the remarkably accurate result $p_r = 2d/z$ where d is the spatial dimensionality and z the coordination number of the network. For a cubic lattice, $d = 3$, $z = 6$, and $p_r = 1$. It is important to state immediately that these results only apply to unstressed materials. The perfect cubic lattice, for instance, under hydrostatic expansion pressure, has a non-zero shear modulus which is proportional to the applied pressure.

The fact that percolation and rigidity have separate onsets in dilute systems is particular to central force networks: If there are bond-bending forces present, rigidity sets in at the percolation point[6].

ENTROPICALLY DERIVED RIGIDITY

Entropy is a concept central to thermodynamics. In the microcanonical ensemble of statistical mechanics it is defined as $S = k_B \ln \Omega$ where Ω is the number of configurations available to a system at fixed energy. The equilibrium state is the state of maximum entropy consistent with the constraints imposed on the system and thus the state in which the maximum number of microscopic configurations are available to the system. For systems kept at fixed temperature, the relevant thermodynamic potential is the Helmholtz free energy $A = U - TS$ where U is the internal energy (kinetic and potential) of the system. The equilibrium state in this situation is the state of minimum A. In an isothermal process, *e.g.*, one in which the system is deformed, the amount of work necessary is $\Delta W = \Delta U - T\Delta S$.

In rigid bodies with energetically derived elasticity ΔU is typically by far the dominant term in this expression, at least at moderate temperatures. There are however systems whose internal energy changes little upon deformation. These usually have a large space of available configurations of nearly identical energy, and hence high entropy. Their equilibrium state, even at fixed T, is thus the state of maximum S. If deformations which decrease the entropy are imposed on these systems, there is a restoring force, which can be far from negligible. To show that entropically derived restoring forces can be significant and to illustrate their character, we will discuss three examples: the ideal gas, a polymer chain, and a crosslinked polymer network (or vulcanized rubber).

The first of these is obviously not a rigid material, but what makes it an interesting example is that its one nonzero modulus, the bulk modulus, is entirely entropic in origin. In an ideal gas, the molecules are non-interacting and the internal energy depends only on temperature, $U = \frac{3}{2}Nk_BT$. Upon compression at fixed T, there is an increase in pressure due to the increase in the frequency of collisions with the walls of the container. This increase in collision rate is the result of the decrease in entropy. The entropy $S = Nk_B \ln(V/N\lambda^3)$ where $\lambda(T)$ is the thermal wavelength[7]. The pressure or restoring force upon volume change is given by:

$$P = -\frac{\partial A}{\partial V} = -\frac{\partial}{\partial V}(U - TS) = \frac{\partial}{\partial V}(TNk_B \ln(V/N\lambda^3)) = \frac{Nk_BT}{V}, \qquad (1)$$

the well-known ideal gas law.

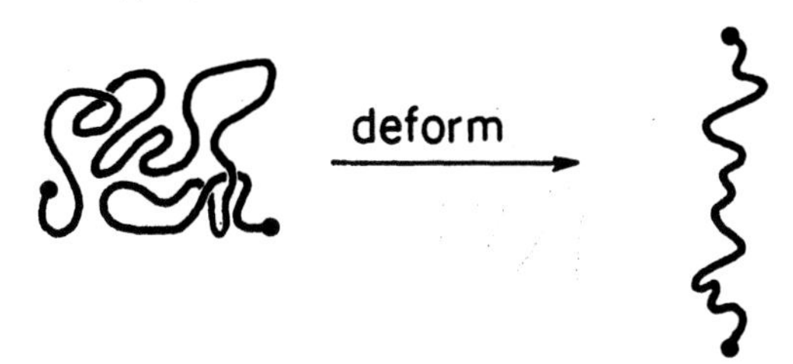

Figure 1. A polymer in two configurations: left-hand side, unstretched (high entropy), and right-hand side, stretched (low entropy). This is the archetypal example of an entropic spring. This figure, as well as Fig. 2 and 3 are inspired by similar drawings in Ref. 8.

The second is the polymer chain, which is the basic elemental component of rubber, and the simplest example of what is known as an entropic spring. A polymer may be modeled as a long flexible chain (see Fig. 1). In dilute solution it has, on a coarse-grained level, many of the properties of a self-avoiding random walk, *e.g.*, its end-to-end distance $\langle \mathbf{R}_N^2 \rangle^{1/2}$ scales as $N^\nu l_p$ where N is essentially the number of independent units, each of microscopic length l_p of the chain (l_p is the persistence length of the polymer chain, the length over which the polymer loses its orientational memory, and Nl_p, the total length of the chain)[9]. The exponent

$\nu \approx 0.6$ for a polymer in a 'good solvent' but in the relevant case of a dense melt, the same power-law behavior is seen but with $\nu = 0.5$, the Gaussian or ideal random walk exponent[9]. For such an ideal N-step random walk, one can calculate the probability $P(\mathbf{R}, N)$ that the separation between end points is $\mathbf{R}$:

$$P(\mathbf{R}, N) = \left(\frac{3}{2\pi N l_p^2}\right)^{3/2} \exp\{-3\mathbf{R}^2/2Nl_p^2\} \equiv \exp\{[S(\mathbf{R}, N) - S(N)]/k_B\}, \quad (2)$$

where we have related this probability to the entropy $S(\mathbf{R}, N)$ of this ensemble of walks. Ignoring the internal energy, we therefore obtain a Helmholtz free energy

$$A(\mathbf{R}, N) = \frac{3k_B T \mathbf{R}^2}{2Nl_p^2} + \text{const.} \quad (3)$$

This leads to an elastic restoring force equivalent to that of a spring of zero equilibrium length. The 'effective spring constant' $3k_BT/2Nl_p^2$ is a thermodynamic quantity in this case. The fact that it is proportional to T ensures that the radius of gyration R_g of the polymer is independent of T. The radius of gyration is the most probable spatial extent of the polymer. Its probability distribution is obtained by integrating over the angular degrees of freedom of the polymer. One finds that $R_g \propto \sqrt{N}$ in both two and three dimensions.

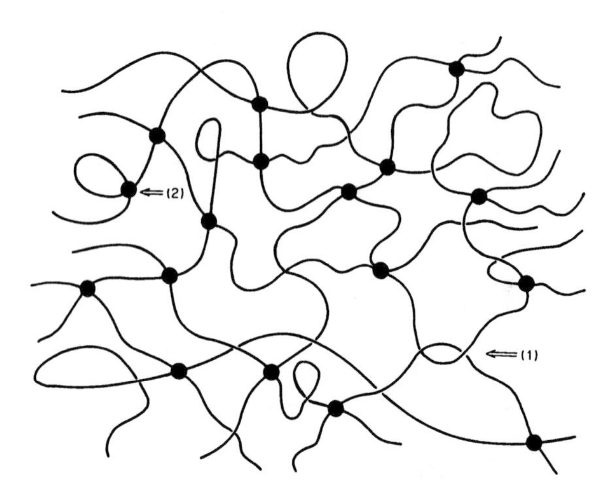

Figure 2. Sketch of a crosslinked polymer melt: The model for vulcanized rubber. (1) indicates an entanglement and (2) a permanent crosslink.

Rubber, our last example, is produced from a dense melt of polymer chains, that are substantially entangled with each other. At some instant, a number of permanent crosslinks that connect previously uncoupled monomers is imposed on the system by a chemical reaction or by radiation. These crosslinks convert the melt at least partially into a network, with chain segments of various contour lengths between the crosslinks (see Fig. 2). This is the unstressed state of the gel with the conformations of the subchains reflecting the Gaussian distribution of the melt. If this gel is distorted, *e.g.*, by stretching, the free energy of segments between two crosslinks will generically increase as the separation of the crosslinks increases and this increase is proportional to k_BT (see Fig. 3). This is the qualitative explanation for the entropic elasticity of rubber.

We will now present a simple quantitative model to put this notion on a slightly more formal basis. Consider a sample of dimension L_x, L_y, L_z with N_{el} *elastically active* chain segments, *i.e.*, segments of a given polymer between two crosslinks. Let the vector connecting the crosslinks of segment i in the undistorted state be $\mathbf{R}_i$. Suppose now that a macroscopic distortion $L_\alpha \rightarrow \lambda_\alpha L_\alpha$ is imposed on the system. In the affine distortion model that is commonly used[10], the individual elastically active segments follow this distortion so that $R_{i\alpha} \rightarrow \lambda_\alpha R_{i\alpha}$. The change in free energy of segment i is then given by

$$\Delta A_i(\lambda_x, \lambda_y, \lambda_z) = \frac{3k_BT}{2}\frac{\mathbf{R}_i^2(\{\lambda_\alpha\}) - \mathbf{R}_i^2(1)}{N_i l_p^2} \tag{4}$$

where N_i is the number of monomers in segment i. Using the Gaussian distribution (2) for the unperturbed separation $\mathbf{R}_i$ and averaging, we obtain

$$\frac{\Delta A(\lambda_x, \lambda_y, \lambda_z)}{V} = \frac{3k_BT}{2}\frac{N_{el}}{V}(\lambda_x^2 + \lambda_y^2 + \lambda_z^2 - 3) = \frac{3k_BT}{\xi^3}(\lambda_x^2 + \lambda_y^2 + \lambda_z^2 - 3) \tag{5}$$

where $\xi = (V/N_{el})^{1/3}$ characterizes the mesh size of the network. ΔA is proportional to a combination of the elastic constants of the network (which combination depends on the choice of λ_α) and we see the classical form of the theory of rubber elasticity $E \propto k_BT/\xi^3$.

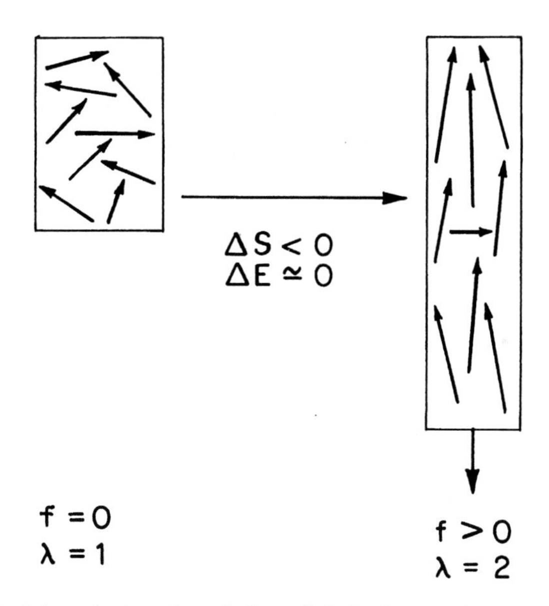

Figure 3. Schematic view of a typical crosslinked polymer melt, (a) in an isotropic state (f=0), and (b) extended (f>0). Note the increased ordering upon extension.

Experimental observation shows very little volume change under elongation. Therefore, if we assume a volume preserving($\lambda_x\lambda_y\lambda_z = 1$) stretch along a given direction, say x ($\lambda_x = \lambda, \lambda_y = \lambda_z = 1/\sqrt{\lambda}$), the restoring stress, the force per unit area, is given by

$$\sigma_{xx} = \frac{\lambda}{V}\frac{\partial \Delta A}{\partial \lambda} = \frac{6k_BT}{\xi^3}\left(\lambda^2 - \frac{1}{\lambda}\right). \tag{6}$$

The above simple free energies already contain three essential features of entropic materials. First the strength of the spring constant increases linearly with temperature. Secondly, because of the zero equilibrium length of the entropic springs, the potential is separable and the restoring forces act as *scalar* forces. And thirdly, from an elasticity point of view, the

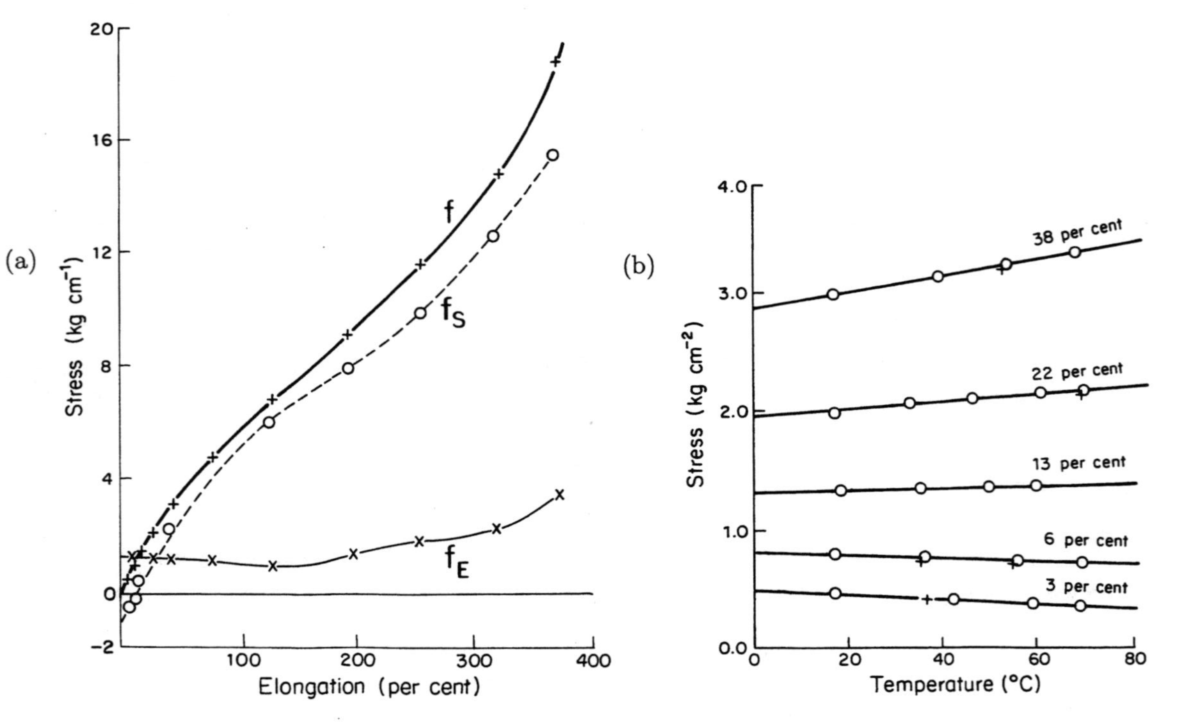

Figure 4. Typical behavior of rubber under stress: (a) Stress as function of fractional elongation separated into entropic and energetic contributions. (b) Stress as function of temperature at fixed elongation. Figure taken from Ref. 1.

materials behave as materials under tension. The third point ensures that rigidity in these networks sets in at connectivity percolation.

Lastly, vulcanized rubber shrinks upon heating for the same reason that an ideal gas will expand at constant pressure. When heat is added, the entropy is increased and it is clear from Eq. (2) that increasing the entropy implies decreasing the end-to-end distance of elastically active segments.

Experimentally, the entropic contribution can be extracted from the data if the temperature variation of the restoring force is known. Simple thermodynamic relations can be used to separate the entropic from the energetic contribution. We start from the differential form of the first law:

$$dA(L,T) = fdL - SdT \tag{7}$$

where f is the tension in the material. Therefore, at fixed T we have

$$f = f_E + f_S = \left(\frac{\partial E}{\partial L}\right)_T - T\left(\frac{\partial S}{\partial L}\right)_T, \tag{8}$$

where the first term f_E corresponds to the energetic contribution to the stress, and the second f_S to the entropic contribution. Using

$$f_S = -T\left(\frac{\partial S}{\partial L}\right)_T = T\frac{\partial}{\partial L}\left(\frac{\partial A}{\partial T}\right)_L = T\frac{\partial}{\partial T}\left(\frac{\partial A}{\partial L}\right)_T = T\left(\frac{\partial f}{\partial T}\right)_L, \tag{9}$$

we obtain

$$f_E = \left(\frac{\partial E}{\partial L}\right)_T = f - T\left(\frac{\partial f}{\partial T}\right)_L = -Tf\left\{\frac{\partial \ln(f/T)}{\partial T}\right\}_L. \tag{10}$$

These two equations are quite general. For rubber, the energetic and entropic contributions are plotted in figure 4(a) and the latter clearly dominates for all but the smallest strains. Similarly, we see from figure 4(b), for reasonably large strains, the characteristic linear increase of the stress with temperature, again an indicator of the preeminence of the entropy in the process.

ELASTICITY OF THE DILUTED CENTRAL FORCE NETWORK

As already mentioned in the introduction, diluted central force networks at $T = 0$ are soft for $p < p_r$, the rigidity percolation concentration. It is therefore of interest to investigate whether or not there is entropic rigidity at finite T in the concentration range (of bonds or sites) $p_c < p \leq p_r$. In a previous article[11] we reported the results of extensive molecular dynamics (MD) simulations for site-diluted triangular lattices. The conclusions of that study were that these systems are rigid for all nonzero temperatures in the entire range $p_c < p \leq p_r$ and that the shear modulus $\mu \sim (p - p_c)^f$ where the value of the exponent $f \approx t$ where t is the corresponding exponent for the conductivity of a diluted resistor network, consistent with the prediction of de Gennes[2]. In this article, we display some of these results and include, as well, some preliminary results for bond-diluted triangular networks and site-diluted square networks.

In this section, we first present some details regarding the models that we have used and some computational techniques. This is followed by a presentation of results and a simple theory similar to the affine theory for rubber.

The models

Our models are two-dimensional networks of particles joined by unbreakable bonds, with linear central restoring forces when deformed from an equilibrium length r_0. These bonds join nearest neighbours only and have the functional form $V_{nn}(r_{ij}) = \frac{1}{2}k(r_{ij} - r_0)^2$. Most of our results are for the site diluted triangular lattice, where we start from a perfect triangular lattice with six-fold coordination at each site and remove sites at random (for a picture see Fig. 6 in Ref. 12). This system has connectivity percolation at a probability of site occupation $p = p_c = 0.5$. Rigidity percolation (at $T = 0$) occurs at $p_r^{(ng)} \approx 0.71$ for non-generic networks. For generic networks, $p_r^{(g)} = 0.6975$ [3]. Since we remove sites randomly without checking whether the resulting network is generic or not, at $T = 0$ we may have non-generic networks. At finite temperatures thermal vibrations presumably make most configurations generic. There is therefore some uncertainty as to which p_r applies. This is however not of much concern for this paper since the onset of entropic rigidity occurs at p_c.

We will also report on some simulations for bond dilution in the triangular lattice for which $p_c = 2\sin(\pi/18) \approx 0.34729$, $p_r^{(ng)} \approx 0.641$[13] and $p_r^{(g)} \approx 0.66$[3].

Molecular dynamics simulations are done on finite size lattices with periodic boundary conditions and with the area kept constant. The starting perfect lattice is unstressed, but as sites or bonds are removed an effective tensile stress holds it at the same area. These being networks we need not worry about the generation of defects, or any topological changes. Bonds cannot break and reform.

Methods of calculation of the shear modulus

When calculating the shear modulus of soft inhomogeneous materials some care is required in the application of standard methods, in particular when, as is the case here, we are interested in the behaviour at high temperature. A direct approach is a stress-strain method (method 1) where a deformation of some kind is imposed on the computational box and the macroscopic restoring force is measured. For isotropic materials, one may impose a pure shear deformation, i.e., an area preserving stretch/compression on the two sides of the computational box of dimensions $L_x \times L_y$. Within linear elasticity theory, the shear modulus μ is then given by $\mu = (p_{yy} - p_{xx})/4\epsilon$ for a distortion in which $L_x \to (1+\epsilon)L_x$, $L_y \to (1-\epsilon)L_y$. Here p_{xx}, p_{yy} are the diagonal elements of the pressure tensor. If the system is not isotropic, as in the case of square networks, one may impose a *simple* shear, for example by shifting the boundaries of the computational box to $x_{\text{min}}(y) = \epsilon y$, $x_{\text{max}}(y) = \epsilon y + L$ where the undeformed box is an $L \times L$ square. In this situation, the shear modulus is given by $\mu = 2p_{xy}/\epsilon$.

A second method would be to determine the ground state configuration of the unstressed system, calculate the dynamical matrix, and then determine the elastic constants using standard harmonic theory from an appropriate sum in terms of phonon frequencies. This method is inappropriate for soft entropically rigid materials. In such systems, there are floppy regions that undoubtedly give rise to zero-frequency soft modes. The elastic constants determined in this way correspond to the zero temperature elastic constants and would vanish below p_r. Even if there are no floppy modes, only soft modes, at temperatures of interest the amplitudes of oscillations of the modes will be most likely out of the harmonic regime, and essentially act as floppy modes.

The third method, known as equilibrium fluctuation method also starts from the undeformed system. The elastic constants are obtained directly from the microscopic fluctuations, using a formal expression of the second derivative of the free energy[14].

The first and third methods are suitable for molecular dynamics simulations. The first method calculates the changes in the pressure tensor upon deformation. Since these changes are only a small percentage of the total, this method is not the most accurate for energetically

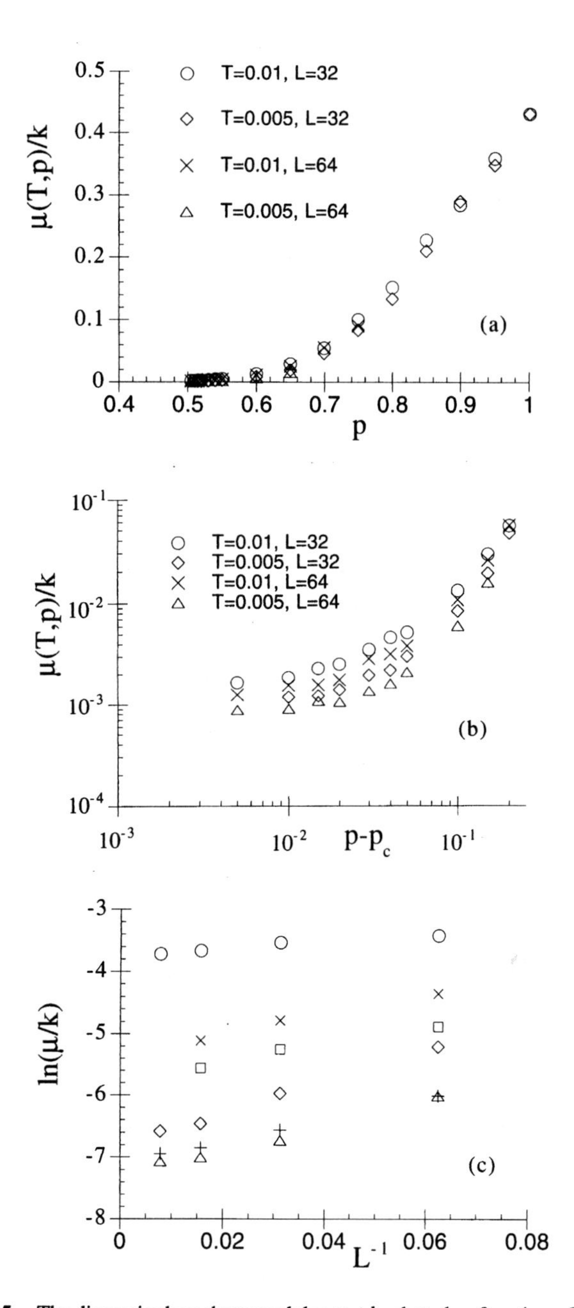

Figure 5. The dimensionless shear modulus μ / k plotted as function of concentration p of particles for site-diluted triangular lattices of size $L = 32$ and $L = 64$ for reduced temperatures $T = .005kr_0^2/k_B$ and $T = .01kr_0^2/k_B$. Part (a) contains data over the entire concentration range $p_c < p \leq 1$ whereas part (b) focuses on the regime $p_c < p < p_r$. Part (c): The logarithm of the shear modulus of the site-diluted triangular network plotted as function of L^{-1} for $T = .01kr_0^2/k_B$ for $p = 0.65$ (circles), $p = 0.55$ (squares), $p = 0.515$ (diamonds) and $T = .005kr_0^2/k_B$ for $p = 0.6$ ($\times$), $p = 0.52$ (crosses) and $p = 0.51$ (triangles). The values of L are 16, 32, 64, and 128. Note that all concentrations are below p_r. (From Ref. 11)

rigid materials. In addition deformations for identical regions may depend on the size of the computational cell within which they are imbedded, leading to noticeable finite size effects. But it is a very robust method since it measures a macroscopic average quantity which is a direct sum of measurables, force and velocity terms. The third method which is based on a fluctuation - dissipation theorem is the most acurate for rigid materials, because no deformations are made to the system and averaging is done at the microscopic level. This has the added advantage of reducing finite size effects. There are, however, intrinsic difficulties in applying it to soft inhomogeneous materials; if the ground state of the system is not unique, fluctuations are large and convergence is very slow[15]. In this paper we report the results from the first method, the stress-strain method.

The shear modulus of finite size-samples

As mentioned above, most of the calculations were done on the site diluted triangular network. With MD simulations at constant volume the changes in the stress tensor were calculated for samples of varying sizes, ranging from 16×16 to 128×128. For the smaller systems the entire concentration range $0.5 < p \leq 1.0$ of interest was studied and for the larger systems primarily in the range of concentrations close to geometric percolation. For a given p, the largest cluster was identified and all smaller clusters discarded. The equations of motion were then integrated for either 10^6 or 2×10^6 time steps. A time step of $\delta t = 0.0016\sqrt{k/m}$ was chosen and simulations have been carried out for temperatures from $k_B T = 0.00125 k r_0^2$ to $k_B T = 0.01 k r_0^2$. Although our potential conserves the connectivity of the particles, it is worth noting that for a piecewise linear force function of the same strength but of range about 15% larger than r_0, the latter temperature is very close to the melting point [16].

Results for the shear modulus as a function of concentration are displayed in Fig. 5 for diluted lattices of size 32×32 and 64×64 for temperatures $T = 0.01 k r_0^2/k_B$ and $T = 0.005 k r_0^2/k_B$. Fig 5a shows the full range of p. Clearly at $p = p_r$ the shear modulus has not reached zero. Above p_r there is little dependence on temperature or lattice size. Fig. 5b focuses on the region close to p_c where both variables have a significant impact on μ, sufficiently so that it is still clearly visible in a logarithm plot. Since for any sample, percolating or not, the shear modulus is positive definite, μ should approach a finite limit as $p \to p_c$. Indeed, precisely at p_c the percolation probability for any finite size L is 0.5 and therefore half the samples presumably make a positive contribution to the estimate of μ whereas the non-percolating ones serve only to reduce the mean value.

The shear modulus in the thermodynamic limit

The logarithm of the shear modulus plotted as a function of L^{-1} shows a non-zero limit for $L = \infty$ for all concentrations studied, the closest to p_c being $p = 0.51$ (see Fig. 5c). This fact alone strongly suggests that at finite temperature the rigidity onset coincides with the geometric or connectivity percolation. Physical arguments can also be made to support that assertion. The network near percolation can be viewed as a system of blobs, links and nodes with the overall symmetry of the original lattice[17,18]. Its elastic response is that of a network of entropic springs, which respond as discussed earlier as a lattice under tension. Therefore as a simple calculation using a stretched string shows, simple connectivity suffices to have a non-zero response to a simple shear, which is proportional to the tension in the string. It is therefore very natural to take p_c as the onset of entropic rigidity.

With this fact established, the behaviour of μ in the neighborhood of p_c can be more conveniently studied. One expects some power law dependence $\mu(T,p) \propto (p - p_c)^f$. This behaviour is masked in our data by finite size effects. With the availability of data for several sizes, a scaling argument can be used to obtain μ in the thermodynamic limit. The relevant length scales are reasonably assumed to be L, and the correlation length ξ, which approaches infinity at percolation as $(p-p_c)^{-\nu}$. These choices lead to the ansatz $\mu(L,p) = L^{-\alpha}\Phi(L/\xi(p))$ where for large x, the scaling function $\Phi(x) \sim x^{\beta}$, with α and β constants to be determined. Requiring that at large L, $\mu(T,p) \propto (p-p_c)^f$, leads to $\mu(L,p) = L^{-f/\nu}\Phi(L/\xi(p))$ and the asymptotic behaviour for $\Phi(x) \sim x^{f/\nu}$.

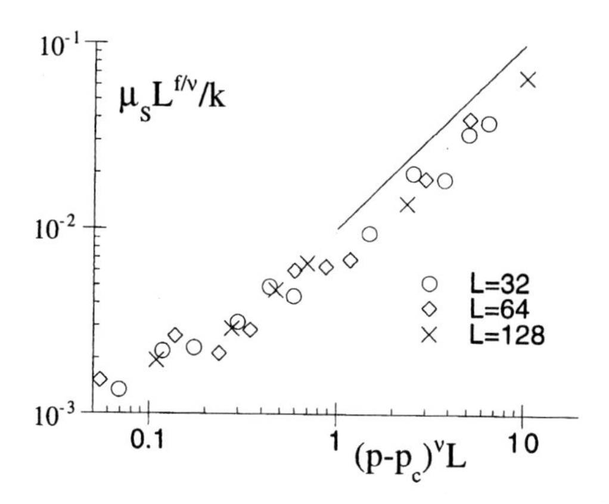

Figure 6. Finite-size scaling analysis of the entropic piece of the shear modulus for lattices of size L = 32, 64, and 128. The scaling ansatz is $\mu_S(L,p) = L^{-f/\nu}\Phi\left((p-p_c)L^{1/\nu}\right)$. The choice $f = \nu = 4/3$ produces a very reasonable collapse of the data. As well, the straight line is the expected form of the scaling function $\Phi(x) \sim x^{f/\nu}$ as $x \rightarrow \infty$ with $f = \nu$. (From Ref. 11)

Since $p_c = 0.5$ and $\nu = 4/3$ exactly, the finite size scaling analysis requires only the exponent f to be varied. The results of such an analysis are shown in Fig. 6 where $L^{f/\nu}\mu_S(L,p)/k$ is plotted as function of $L(p-p_c)^{\nu}$ for $f = \nu = 4/3$. The quantity μ_S is the *entropic* contribution to the shear modulus which is given by [1]

$$\mu_S(L,p) = T\left(\frac{\partial \mu}{\partial T}\right)_{p,L} . \qquad (11)$$

The piece of the shear modulus due to the internal energy $\mu_E = \mu - \mu_S$ is much smaller than μ_S for $p < p_r$ and should be at least second order in the temperature T. In any case, subtracting off this piece of the shear modulus has a smoothing effect on the data and improves the scaling analysis. Although the data are rather noisy, the collapse of the data in Fig. 6 becomes noticeably worse if the exponent f is increased or decreased by more than 0.1. We therefore conclude that $f = 1.33 \pm 0.10$. The expected power law for the asymptotic form of $\Phi(x) \sim x^{f/\nu}$ is also shown in the form of the solid line again for $f = \nu$. It is clear that the data are at least consistent with this behavior. We note, however, that the currently accepted value of the exponent t for the conductivity of two-dimensional random resistor networks is $t \approx 1.3$ and our data clearly cannot distinguish between the alternatives $f = \nu$ or the de Gennes prediction $f = t$.

As further evidence for the conclusion that entropic rigidity persists to $p = p_c$, we display the shear modulus of triangular bond-diluted networks. These data come from work in progress[19] and are not yet extensive or well converged enough to permit the finite size scaling analysis discussed above. However, it is clear from Fig. 7 that there is entropic elasticity for $p > p_c$.

Theoretical models

One can construct a simple theory for the entropic elasticity of diluted networks in close analogy with the classical theory of rubber elasticity. One ingredient is the blobs, links and nodes picture [17,18] of a diluted network near the percolation concentration. The nodes are connected by strands made of blobs separated by links of single bonds. The network of nodes has on average the symmetry of the original lattice. A second ingredient is that the entropic springs behave as springs of zero length, and hence the forces are separable. This turns it into a one-dimensional problem, or a scalar problem. These two aspects link the entropic elasticity problem to the random resistor problem. Here we will discuss the elasticity of the diluted network within the affine approximation of the classical theory of rubber elasticity.

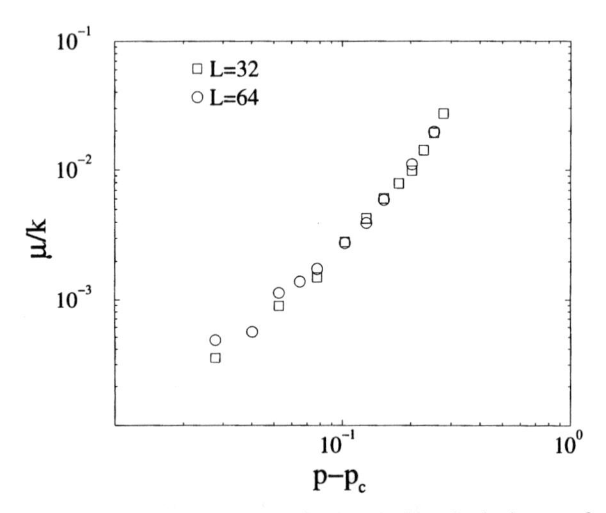

Figure 7. Shear modulus for the bond diluted triangular lattice as a function of bond concentration p for two lattice sizes, 32×32 and 64×64 at the temperature $T = .005 k r_0^2 / k_B$.

The nodes at which different links or filaments are joined are the analog of permanent crosslinks in a system of randomly crosslinked macromolecules. The links themselves consist in part of single strands and in part of more rigid blobs. If such a tenuous system is deformed, one expects that the principal effect will be to lengthen (shorten) the distance between nodes and thus to modify the entropy of the filaments. Consider two nodes i, j with relative position X_{ij}, Y_{ij}. If a distortion characterized by stretching factors (λ_x, λ_y) is applied to the system and if the nodes follow this distortion in an affine manner, we expect that the entropy change of the filament connecting these nodes due to this distortion will be

$$\Delta S_{ij} = -\frac{k_B}{a^2} \frac{R_{ij}^2(\lambda_x, \lambda_y) - R_{ij}^2(1,1)}{N} \tag{12}$$

where a is the nearest neighbor distance, N the number of links (single bonds) between the nodes and $R^2 = X_{ij}^2 + Y_{ij}^2$. This expression is obtained by treating the filament as a Gaussian random walk as is usually done in the classical theories of rubber elasticity. Averaging over all pairs of nodes for $\lambda_x = 1 + \epsilon$, $\lambda_y = 1 - \epsilon$, we obtain the estimate

$$\mu = \frac{\Delta F}{2\epsilon^2 A} = \frac{k_B T \Delta S}{2\epsilon^2 A} = \frac{k_B T}{2a^2} \frac{N_{st}}{A} \left\langle \frac{R^2(1,1)}{N} \right\rangle \tag{13}$$

where N_{st} is the number of links, A the area of the system and ΔF the change in Helmholtz free energy obtained by ignoring any changes in internal energy.

From studies of percolation clusters [18,20], we have $N_{st}/A \sim \xi^{-2} \sim (p-p_c)^{2\nu}$ and $< R^2/N > \sim \xi^2/N \sim (p-p_c)^{1-2\nu}$ [20]. Therefore, on the basis of this very simple picture, we obtain $\mu \sim (p-p_c)$, *i.e.* $f=1$ which is *not* consistent with the results of our simulations. Of course, there are at least two aspects of the above argument that are suspect. First, the assumption that the deformation of the system is affine clearly ignores fluctuations in density and therefore of local rigidity. Second and probably more important is the fact that we have ignored the self-similarity of percolation clusters [21] and modeled the known fractal structure by a network of nodes that effectively has fractal dimension $D = 2$. It is perhaps worth noting that an analogous calculation [2,6] of the conductivity of random resistor networks near the percolation point also predicts $\sigma \sim (p-p_c)$ which is similarly inconsistent with the corresponding simulation data.

DISCUSSION AND CONCLUSION

In this paper we introduced a new dimension to the problem of the rigidity of random networks, that brought about by the effects of temperature. These will be important in the region between the connectivity percolation and the rigidity percolation, where the network is soft, and its entropy is large.

The onset of entropic rigidity has to occur at the geometric percolation point. It is a thermodynamic rigidity tied to the change in entropy upon deformation. The closest that one can come to think in terms of mechanical rigidity is to note that entropic springs are equivalent to stretched springs, and therefore a single spring under shear will have a lateral linear restoring force. This is indeed quite contrary to the situation of an unstretched spring — the whole issue of mechanical rigidity changes if the system is stressed [22].

The displacement of the critical point from p_r to p_c at finite temperature, raises a number of interesting issue about the critical behaviour of this system. At finite temperature, there will be a crossover from entropic to mechanical rigidity in the neighborhood of p_r.

The models that we have studied to date are two-dimensional. Exploring three dimensional equivalents would obviously help to distinguish between the possibilities $f=\nu$ and $f=t$: for the 3D $f.c.c$ lattice, the conductivity exponent $t \approx 2$ and $\nu \approx 0.88$.

Glasses are random networks, usually with bond-bending forces. Near percolation with strands longer than the persistence length of connecting bonds, we will have a situation similar to the one we studied, albeit three dimensional. The onset of entropic rigidity is therefore expected to occur at the percolation threshold. If data was available on the temperature variation of the elastic constants, the entropic contribution to the elastic constants could be extracted, as discussed in the section titled Entropic Rigidity. The entropic contribution should rise linearly with temperature, and will have an effect on the coefficient of thermal expansivity which is expected to decrease with temperature. Temperature increases the force constant and therefore for equal tension the elongation will be smaller: entropic solids like rubber shrink with increasing temperature.

ACKNOWLEDGEMENTS

The support of the Natural Sciences and Engineering Research Council (Canada) is gratefully acknowledged.

REFERENCES

1. L.R.G. Treloar, *The Physics of Rubber Elasticity*, Clarendon, Oxford (1975).
2. P.G. de Gennes, *J. Phys. (Paris) Lett.* 37:L1 (1976).
3. C. Moukarzel and P.M. Duxbury, *Phys. Rev. Lett.* 75:4055 (1995); D.J. Jacobs and M.F. Thorpe, *Phys. Rev. E* 53:3682 (1996).
4. S. Feng and P.N. Sen, *Phys. Rev. Lett.* 52:1891 (1984).
5. S. Feng, M.F. Thorpe and E. Garboczi *Phys. Rev. B* 31:276 (1985).
6. Y. Kantor and I. Webman, *Phys. Rev. Lett.* 52:1891 (1984).
7. F. Reif, *Fundamentals of Stratistical and Thermal Physics*, McGraw-Hill, New York, N.Y. (1965).
8. J.E. Mark and B. Erman, *Rubberlike Elasticity: a Molecular Primer*, John Wiley and Sons, New York (1988).
9. P.G. de Gennes, *Scaling Concepts in Polymer Physics*, Cornell University Press, Ithaca, N.Y. (1979).
10. P.J. Flory, *Principles of Polymer Chemistry*, Cornell University Press, Ithaca, N.Y. (1953).
11. M. Plischke and B. Joós, *Phys. Rev. Lett.* 80:4907 (1998).
12. C. Moukarzel and P.M. Duxbury, *Comparison of Connectivity and Rigidity Percolation*, an article in this volume.
13. S. Arbabi and M. Sahini, *Phys. Rev. B* 47:695 (1993).
14. Z. Zhou and B. Joós, *Phys. Rev. B* 54:3841 (1996).
15. Z. Zhou and B. Joós, unpublished.
16. J.A. Combs, *Phys. Rev. Lett.* 61:714 (1988); Phys. Rev. B 38:6751 (1988).
17. H.E. Stanley, *J. Phys. A* 10:L211 (1977).
18. A. Coniglio, *Phys. Rev. Lett.* 46:250 (1981).
19. D.C. Vernon, M. Plischke and B. Joós, to be published.
20. A. Coniglio, *J. Phys. A* 15:3829 (1982).
21. H.J. Herrmann and H.E. Stanley, *Phys. Rev. Lett.* 53:1121 (1984).
22. S. Alexander, *Phys. Rep.* 296:65 (1998).

MOLECULAR DYNAMICS AND NORMAL MODE ANALYSIS OF BIOMOLECULAR RIGIDITY

David A. Case

Department of Molecular Biology
The Scripps Research Institute
La Jolla, CA 92037

INTRODUCTION

The past decade has seen an impressive advance in the application of molecular simulation methods to problems in chemistry and biochemistry. As computer hardware has become faster and software environments more sophisticated, the amount of detailed information available and its expected level of accuracy has grown steadily.[1-3] There is strong evidence that molecular dynamics or Monte Carlo simulations of biomolecules in the liquid state provide a powerful and usefully accurate model of the real behavior of proteins and nucleic acids, and permit a detailed analysis of events at a microscopic level. Often, the "easy" part of a simulation project involves setting up and carrying out the calculations, and the hard part generally lies in extracting useful data from among the very many things that can be calculated from a trajectory or Monte Carlo simulation. Normal mode analysis provides an approximate but analytical description of the dynamics, and has long been recognized as an important limiting case for molecular dynamics in condensed phases. A principal limitation arises from that fact that normal modes are defined by an expansion about a particular point on the potential energy surface, and hence have difficulty describing transitions from one local minimum to another. The quasiharmonic and "instantaneous" mode theories discussed below attempt to ameliorate some of this neglect of the "rugged" nature of biomolecular energy landscapes. Yet there remains a "paradoxical aspect"[4] of biomolecular dynamics that is still the subject of considerable study: even though the energy surface contains many local minima, proteins behave in some ways as though the energy surface were harmonic, and normal mode analyses are often more correct that one might expect. In this article I give an overview of various analysis tools that are based loosely on the normal mode approach, along with a few examples of their applications to proteins and nucleic acids.

FUNDAMENTALS OF NORMAL MODE ANALYSIS

Normal mode analysis basically involves a linearization of the problem of classical dynamics, leading to a set of "modes" (directions in configuration space) that can be useful in characterizing fluctuations from a stable equilibrium structure. The basic idea is to expand the potential function $V(\mathbf{x})$ in a Taylor series expansion about some point $\mathbf{x}_0$:

$$V(\mathbf{x}) = V(\mathbf{x}_0) + \mathbf{g}^T(\mathbf{x}-\mathbf{x}_0) + \tfrac{1}{2}(\mathbf{x}-\mathbf{x}_0)^T\mathbf{F}(\mathbf{x}-\mathbf{x}_0) + \cdots \quad (1)$$

If the gradient $\mathbf{g}$ of the potential vanishes at this point and one ignores third and higher-order derivatives, it is straightforward[5,6] to show that the dynamics of the system can be described in terms of the normal mode directions and frequencies $\mathbf{Q}_i$, ω_i, which satisfy

$$\mathbf{M}^{-1/2}\mathbf{F}\mathbf{M}^{-1/2}\mathbf{Q}_i = \omega_i^2\mathbf{Q}_i$$

$$\mathbf{Q}_i \cdot \mathbf{Q}_j = \delta_{ij} \quad (2)$$

In Cartesian coordinates, the matrix $\mathbf{M}$ contains atomic masses on its diagonal, and the Hessian matrix $\mathbf{F}$ contains the second derivatives of the potential energy evaluated at $\mathbf{x}_0$. The time evolution of the system is then:

$$x_i(t) = x_i(0) + (2)^{\frac{1}{2}} \sum_k Q_{ik} m_i^{-\frac{1}{2}} \sigma_k cos(\omega_k t + \delta_k) \quad (3)$$

where σ_k is an amplitude, ω_k the angular frequency and δ_k the phase of the kth normal mode of motion. The phases and amplitudes depend upon the positions and velocities at time t=0. It is conventional in molecular problems to divide the frequencies ω_i by the speed of light to report results in cm^{-1} units.

A straightforward computation of normal modes in Cartesian coordinates thus involves a numerical diagonalization of a matrix of size 3N x 3N, for a molecule with N atoms. With present-day computers, it is not difficult to study proteins up to about 200 amino acids with an all-atom model, or for a nucleic acid with 100 base pairs. For example, computation of the normal modes for the 159 amino acid protein dihydrofolate reductase, used as an example below, required about 4 hours on a Silicon Graphics R10000 workstation.

A common approximation for larger systems assumes that bond lengths and angles are fixed. This can reduce the size of the matrix involved by about an order of magnitude. Calculations can be carried out by direct construction of the potential and kinetic energy matrices in (curvilinear) internal coordinates,[7,8] or through matrix partitioning techniques that start from Cartesian derivatives.[9-11] In general, reductions of the dimensionality of the expansion space have noticeable but not overwhelming effects on the resulting normal mode description of the dynamics. The directions of the lower-frequency modes are largely preserved, but frequencies in general are higher in the lower-dimensional space,[11,12] suggesting that small fluctuations in bond lengths and bond angles have the effect of allowing the dihedral angles to become more flexible. Many practical aspects of computing modes for large molecules are available elsewhere.[8,10,11]

The thermal averages of the second moments σ_i^2 of the amplitude distributions can be calculated for both classical and quantum statistics:[13]

$$\sigma_{i,class}^2 = kT/\omega_i^2; \qquad \sigma_{i,qm}^2 = \frac{h}{4\pi\omega_i}\coth\frac{h\omega_i}{4\pi kT} \quad (4)$$

where h and k are the Planck and Boltzmann constants. The two statistics coincide in the limits of low frequency or high temperature. For biomolecules, the most important difference is generally that higher frequency modes have little amplitude in classical statistics but have non-negligible zero-point motion in quantum statistics. Harmonic models thus provide one of the few practical ways for including quantum effects in biomolecular simulations.

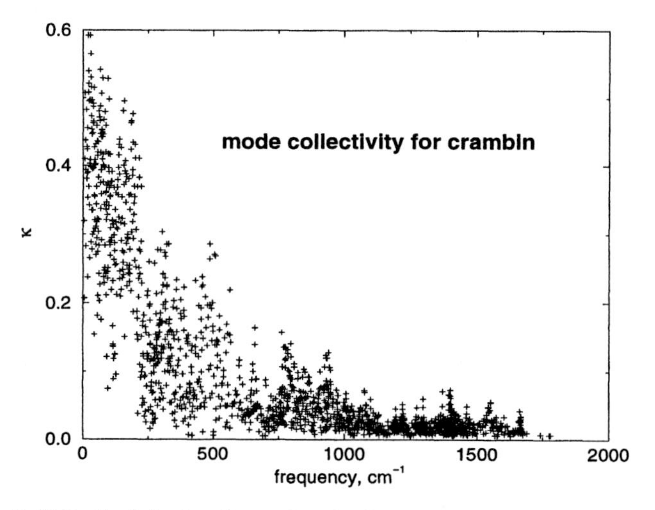

Fig. 1. Collective behavior of normal modes for crambin. Adapted from Ref. 14

There is a close connection between mode frequency and the collective character of the atomic motions. Figure 1 plots a "collectivity index" κ_i against frequency for crambin.[14] This index is proportional to the exponential of the "information entropy" of the eigenvector:

$$\kappa_i = N^{-1} \exp\left[-\sum_n u_{i,n}^2 \log u_{i,n}^2\right] \tag{5}$$

where the $u_{i,n}$ measures the extent of motion of atom n in mode i, normalized such that the sum of its squares is unity.[14] The κ_i can can vary from $1/N$ to 1, where a value of 1 indicates maximal collectivity where all eigenvector amplitudes are the same (as for the modes describing global translations). There is a clear drop-off in collectivity as the frequency increases, with large collectivity only found in modes below 200 cm^{-1}.

ELABORATIONS OF THE NORMAL MODE APPROACH

In the next few sections, I outline some that build on normal mode ideas, extending them in ways that are important for the analysis of motions in proteins and nucleic acids; these mainly involve ways to include the effects of solvent, and to take at least some account of conformational disorder, where there are "jumps" between local minima on the potential energy surface.

Langevin modes

It is also possible to solve for normal mode dynamics in the presence of viscous damping by a continuum "solvent".[15] In this approach, Newton's equations are replaced by Langevin equations that include terms describing viscous damping and random (white) noise.

$$m\ddot{\mathbf{x}} = -V'(\mathbf{x}) - \zeta\mathbf{v} + \mathbf{r}(t) \tag{6}$$

Here $\mathbf{v}$ is the velocity vector, ζ is the friction matrix and $\mathbf{r}(t)$ is a vector of random numbers. The random numbers follow Gaussian distribution with the following properties:

$$< r_i(t) > = 0$$

$$< r_i(t) r_j(t') > = 2\zeta_{ij}\delta(t-t')/k_B T \tag{7}$$

Eq. (7) is a fluctuation-dissipation relation that ensures that the long-time scale behavior of the system converges to an equilibrium one characterized by the temperature T. Expanding the potential to quadratic terms as in Eq. (1), and defining for convenience $\alpha \equiv \mathbf{M}^{\frac{1}{2}}(\mathbf{x} - \bar{\mathbf{x}})$ and $\nu \equiv \dot{\alpha}$, yields a matrix version of the Langevin dynamics:

$$\begin{pmatrix} \dot{\alpha} \\ \dot{\nu} \end{pmatrix} = \begin{pmatrix} 0 & 1 \\ -\mathbf{M}^{-\frac{1}{2}}\mathbf{F}\mathbf{M}^{-\frac{1}{2}} & -\mathbf{M}^{-\frac{1}{2}}\zeta\mathbf{M}^{-\frac{1}{2}} \end{pmatrix} \begin{pmatrix} \alpha \\ \nu \end{pmatrix} + \begin{pmatrix} 0 \\ R(t) \end{pmatrix}$$

$$\equiv \mathbf{A} \begin{pmatrix} \alpha \\ \nu \end{pmatrix} + \begin{pmatrix} 0 \\ R(t) \end{pmatrix} \tag{8}$$

The random numbers $R_i(t)$ now satisfy

$$< R_i(t) > = 0$$

$$< R_i(t) R_j(t') > = 2 m_i^{-\frac{1}{2}} \zeta_{ij} m_j^{-\frac{1}{2}} \delta(t - t') / k_B T \tag{9}$$

If the macromolecule has N atoms, then $\mathbf{A}$ is a $6N \times 6N$ matrix and is non-symmetric. It is useful to construct the distributions arising from these stochastic differential equations from solutions to the homogeneous (ordinary) equations that are obtained when R(t) vanishes.[15] These solutions involve the propagator exp($\mathbf{A}$t), which can be expressed in terms of an eigen-analysis of the matrix $\mathbf{A}$.

The coupling between solvent and solute is often represented by a "bead" model in which each atom is a source of friction, with some corrections to represent the effects of burial or of hydrodynamic interactions between atoms;[15] alternatively, effective friction couplings can be extracted from molecular dynamics simulations.[17,18] Computational details can be found in some of the early publications.[15,19] Calculations on small proteins and nucleic acids using a bead model for fractional coupling to solvent indicated that most modes with vacuum frequencies below about 75 cm^{-1} would become overdamped, and that frequency shifts could be significant.[19] These frictional models, however, may overestimate solvent damping, especially for lower frequency motions.[20] A typical result for short time-scale motion is given in Fig. 2, which compares solvated molecular dynamics, normal mode and Langevin mode predictions for a time correlation function related to NMR relaxation in a zinc-finger peptide.[16] Here the short-time oscillations predicted by the gas-phase normal modes reproduce the behavior of the solvated MD simulation for about half a picosecond, but beyond that the dephasing behavior arising from collisions with solvent molecules and from anharmonic interactions within the protein itself lead to divergent behavior. The simple Langevin treatment shown here (using a "bead" model for frictional interactions) forces the oscillations in the correlation function to decay (as they will never do in the pure normal mode treatment), but this damping takes place on much to short a time scale.

An analysis of molecular dynamics simulations of BPTI in water suggested a model in which the frictional coupling was nearly the same in all modes, with a value near 47 cm^{-1}, so that modes with effective frequencies below about 23 cm^{-1} would become overdamped.[18] Analysis of inelastic neutron scattering data[21] led a a model in which the effective friction is a Gaussian function of the frequency, so that the lower-frequency collective modes experience a greater frictional damping than do more localized, higher-frequency modes (*cf.* Fig. 1). In this model, modes below 15 cm^{-1} are overdamped, and some frictional damping effects are noticeable up to about 75 cm^{-1}.

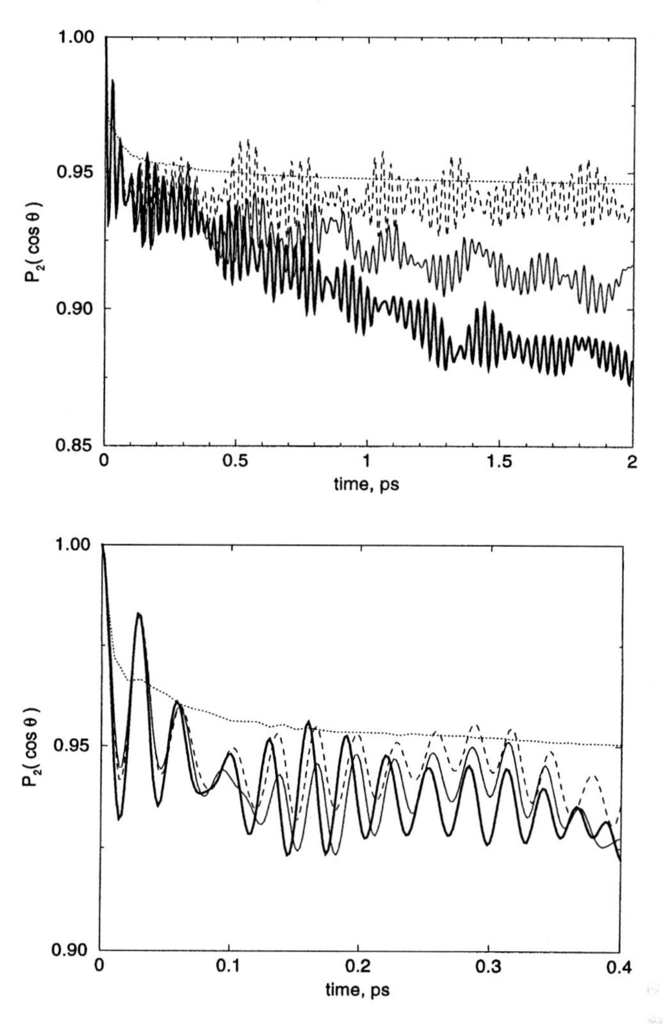

Fig. 2. Correlation function of the Cα–Hα of Ala 15 in a zinc finger peptide.[16] Here θ is the angle between the Cα–Hα vector at time 0 and at some later time. Heavy solid curve: solvated MD simulation; light solid curve: vacuum MD simulations; dashed line: vacuum normal modes; dotted line: Langevin modes. Bottom figure is an expansion of the upper one.

A classic study of the effects of solvent damping on vibrational motions involves the "hinge-bending" motion between two domains of lysozyme, which was originally analyzed in terms of the energy profile along an assumed bending coordinate, and found to be overdamped.[22] More recent normal mode investigations of this system provide a detailed description of the nature of the hinge-bending coordinate; projections of Monte Carlo or molecular dynamics trajectories onto the normal mode coordinates support the basic features of the normal mode analysis and allow the dynamics to be analyzed in terms of harmonic and anharmonic contributions.[4,23,24]

Instantaneous normal modes

It has been recognized for some time that one of the salient features of protein dynamics is existence of many local conformational minima, and that transitions from (the vicinity of) one local minimum to another represent an important feature macromolecular dynamics.[25,26] The normal mode model, which expands the potential about a single local minimum, does not directly include contributions from such transitions on a rough or corrugated surface. Recent studies on peptide and protein systems show that between 30 and 70% of the total atomic fluctuation arises from transitions between minima, with normal mode theories working better for the more tightly constrained systems.[27,28] An interesting approach to understanding the dynamics of fluid systems that do not oscillate about a single (or small number) or conformational minima involves calculations of modes about the instantaneous configurations sampled in a simulation. Since these are in general not local minima, the frequency spectrum contains both real and imaginary components, and the nature and distribution of these "unstable" modes can be related to dynamical quantities.[29-32] Straub and Thirumalai have applied such ideas to the ribonuclease S-peptide, computing the instantaneous normal mode spectrum between 40 and 500K.[33,34] The number and character of the unstable modes can be used to characterize the distribution of barriers between "conformational substates." At room temperature, about 4% of the modes are unstable, and this value is predicted to increase to about 10% in the high temperature limit. A distribution of barrier heights between conformers that fits the frequency data has the following form:

$$g(E_B) = a\theta(E_{low} - E_B) + bE_B e^{-E_B/E_0} \tag{10}$$

Here there is a constant density of low-energy barriers for $E_B < E_{low}$ ($\theta(E)$ is the Heaviside function), and a Poisson distribution of higher energy barriers with a maximum at E_0. For the S-peptide, a is 0.325 (kcal/mol)$^{-1}$, E_{low} = 0.2 kcal/mol, b = 0.13 (kcal/mol)$^{-2}$, and E_0 = 1 kcal/mol. This fit yields a broad distribution of barrier heights that includes many very small barriers. It should be of considerable interest to see what distributions are obtained in other biomolecular systems.

Recently, an elaboration of this scheme has been proposed, in which the imaginary frequency modes are further subdivided into those that belong to a double-well profile and those that correspond to a "shoulder" with a nearly local minium on one side but not the other.[35] This can have important implications for diffusion in simple molecular liquids; it is not yet clear how these ideas will track into biomolecular problems.

Quasiharmonic analysis

Another important extension of normal modes is to "quasiharmonic" behavior, in which effective modes are computed such that the second moments of the amplitude distribution match those found in a Monte Carlo or molecular dynamics simulation using the complete, anharmonic force field.[36,37] The basic idea is to compute the fluctuation matrix from a dynamics or Monte Carlo simulation:

$$\sigma_{ij} = \langle (x_i - \bar{x}_i)(x_j - \bar{x}_j) \rangle \tag{11}$$

and to assume that the complete conformational probability distribution is approximately a multivariate Gaussian:

$$P(\mathbf{x}) = (2\pi)^{-n/2} |\det\sigma|^{-1/2} \exp\left[-\tfrac{1}{2}(\mathbf{x}-\bar{\mathbf{x}})^T \sigma^{-1} (\mathbf{x}-\bar{\mathbf{x}})\right] \tag{12}$$

The probability distribution can also be related to the potential energy:

$$P(\mathbf{x}) \approx \exp[-V(\mathbf{x})/kT] \tag{13}$$

In the quasiharmonic model, V is a quadratic function of position:

$$V(\mathbf{x}) = \tfrac{1}{2}(\mathbf{x}-\bar{\mathbf{x}})^T \mathbf{F}^{quasi}(\mathbf{x}-\bar{\mathbf{x}}) \tag{14}$$

so that the effective force constant matrix becomes the inverse of the fluctuation matrix found in the simulation:

$$\mathbf{F}^{quasi} = kT\,[\sigma]^{-1} \tag{15}$$

Since $\mathbf{F}^{quasi}$ and σ have common eigenvectors, the quasiharmonic modes can be determined from the mass-weighted fluctuation matrix, and it is not necessary to explicitly construct $\mathbf{F}^{quasi}$.

It is important to recognize that this sort of quasiharmonic analysis is based on a static analysis of the fluctuation matrix and not on any time series analysis of actual motions. Many features of the correlation matrix, including aspects of the "low-frequency" behavior, are present even in fairly short molecular dynamics simulations.[28,37] The convergence characteristics of these modes (and, indeed, of molecular simulations in general) are still the subject of study, and I will return to this subject below.

The quasiharmonic assumption that the distribution of configurations is a multivariate Gaussian provides an analytical form that permits the calculation of quantities such as the vibrational entropy, which are otherwise hard to estimate.[36,38-40] As with true normal mode analysis, a contribution to thermodynamic quantities can be associated with each mode; the overall entropy can be expressed in terms of the logarithm of the determinant of σ, which is less expensive to determine than the full eigenvalue analysis. This approach has been used recently, for example, to estimate energetic consequences of cross-links to protein stability,[41] or entropic effects associated with protein oligomerization.[42]

The quasiharmonic approach includes some effects of anharmonic terms in the potential, at least to the extent that they influence the mean-square displacements, but it still assumes distributions that are unimodal in character. Tests of this assumption using MD simulations give mixed results. The atomic displacements in an α helix appear to be approximately Gaussian over a wide temperature range,[43] and MD simulations on lysozyme suggest that most atoms have fluctuations that are highly anisotropic but only slightly anharmonic.[44] Other studies, including those on myoglobin,[45] crambin,[46] mellitin,[47] and lysozyme,[48] have found distribution functions with more than one maximum, especially along low-frequency directions. It is possible to account approximately for such results in estimates of the entropy,[39] but any significant deviations from unimodal behavior are very difficult to accommodate into a quasi-harmonic description.

Figure 3 shows the frequency spectrum for a quasiharmonic calculation of dihydrofolate reductase, an enzyme containing 159 amino acids. The underlying molecular dynamics simulation was performed in a periodic box, containing one protein molecule, 11 Na^+ ions, and 6810 water molecules. The Amber all-atom potential energy function was used,[49] along with the particle-mesh Ewald method to handle long-range electrostatic interactions. (Details will be presented elsewhere.) Fluctuations were collected from a 0.5 ns simulation following 0.2 ns of equilibration, and the matrix of second moments of the fluctuations was diagonalized to construct the quasiharmonic frequencies. The modes below 20 cm^{-1} or so contribute in an important way to overall collective motion. The figure also shows the spectrum computed by a conventional (gas-phase) normal mode calculation on the same system. Here there are fewer low-frequency modes, and correspondingly more at higher frequencies. Many of the low-frequency modes in the quasiharmonic spectrum reflect extended motions over a rough

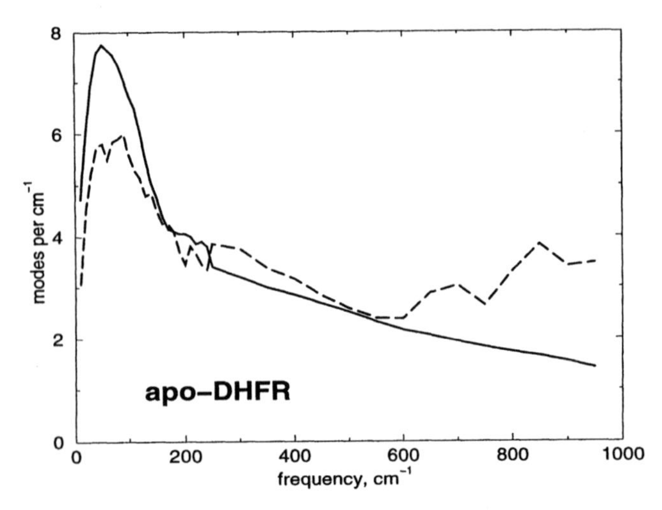

Fig. 3. Frequency distributions for dihydrofolate reductase (in the "apo" form with no substrates bound). Solid line: quasiharmonic frequency distribution; dashed line, gas-phase normal mode distribution.

energy landscape, motions that may not be captured by true normal modes, where even small barriers might prevent motion in the harmonic approximation to the potential.

As I indicated above, considerable attention has been given to using normal mode or quasiharmonic analysis to probe low-frequency, longer-time scale motions in biopolymers. Since the amplitude of mode fluctuations is inversely proportional to frequency (*cf.* Eq. 4), normal modes have the attractive feature of describing motions that contribute most to atomic fluctuations in terms of a relatively small number of mode directions and frequencies: a rough rule of thumb is that 1% of the modes contribute up to 90% of the overall root-mean-square atomic fluctuations. This has inspired an interest in characterization of low-frequency modes along with hopes that interesting domain movements[50] might appear as identifiable modes, or as a combination of a small number of modes.[51] A further hope is that the "essential dynamics"[48] of proteins might involve motions confined to a subspace of low-frequency normal modes.

A fairly large literature now exists in which large-scale collective motions of proteins have been studied with normal mode calculations.[52] In addition to the analyses of the lysozyme hinge bending modes mentioned above, inter-domain motions of G-actin,[53] an epidermal growth factor,[54] ras P21,[55] and hemoglobin[56] have recently been characterized through normal mode analyses.

An obvious problem in evaluating harmonic models for long-time scale motions is a lack of a secure standard for comparison. It is generally not possible to run molecular dynamics simulations for long enough periods of time to obtain statistically meaningful information about nanosecond scale motions. For short peptides of four to six amino acids, simulations from 10 to 100 ns sometimes appear to be approaching an equilibration among various conformational states,[57] and studies using simplified solvent models (such as Brownian dynamics or mean field simulations) can study much longer time scales.[58] But it is clear that most picosecond-to-nanosecond simulations of proteins and nucleic acids in water are not well-equilibrated, at least for some aspects of interatomic correlations,[59] and that most all simulations on this time scale contain "rare events" that complicate straightforward analyses based on the assumption of an equilibrated sample.[60-62] The nature of the protein energy landscape is such that there are likely to be conformational transitions on nearly all timescales[26] so that any individual time segment of a simulation will probably not be in equilibrium with respect to some types of motion. Slow motions that involve relatively large or correlated conformational changes may dominate the lowest frequencies in a quasiharmonic analysis. While individual low frequency quasiharmonic modes are sensitive to the details of the trajectory that produced them, it is less clear how well the subspace spanned collectively by the large-amplitude eigenvectors is determined. An approximate analysis can generally be made based on a sub-nanosecond trajectory[63] (as in the DHFR work used here for illustration), but true convergence in these properties may be difficult or nearly impossible to achieve.[61] The quasiharmonic analysis tends to mix ordinary normal modes together, so that the density of states is a smoother function of frequency than for a true normal mode distribution, and much of the distinction between local and global character (*cf.* Fig. 1) is lost.[28] Further study will be required to ascertain the extent to which quasiharmonic directions are useful in describing or rationalizing the global behavior of biomolecular motions.

Simple network models

I mention finally a recent very simple and intriguing idea. Many features of the *collective* motions of proteins and nucleic acids do not depend greatly on any details of atomic potential energies, but more on the size, shape and general connectivity of the system. It turns out that many aspects of large-scale molecular motions can be reproduced by a (parameterized) network of simple springs, with just one connection per amino acid residue or nucleic acid base.[64-67] Observed aspects of molecular motion, such as crystallographic thermal parameters (discussed below) can be obtained through this model, although it is not yet clear how transferable the parameters of this sort of model might be. This idea does lead, however, to simple models that can be readily applied to a wide range of systems, and which may provide a language for important correlations of biochemical function to molecular structure and motions.

MAKING CONNECTIONS TO EXPERIMENTS

There are many ways that models for collective molecular motions can be tested against experimental data. These range from detailed fast-time scale spectroscopic measurements through to models for functional or kinetic behavior that can arguably probe fluctuations. There is no space here for even a cursory examination of these probes, but I will give illustrative connections for three types of measurements. Important experimental techniques not covered here include inelastic neutron scattering,[21] hydrogen exchange kinetics,[68] and fluorescence energy transfer measurements.[69]

Crystallographic thermal parameters

Perhaps the easiest and most straightforward way to compare computed motions and fluctuations with experiment is through comparison to thermal parameters extracted from crystallographic analysis. Because the scattering power of an atom changes as its mobility increases, information about the root-mean-square deviations of atoms can be extracted from observed scattering factors, and is typically reported as "B factors". For an ideal crystal, these are related to mean-square atomic fluctuations:

$$\langle (\Delta \mathbf{r}_i)^2 \rangle = 3B_i/8\pi^2 \tag{16}$$

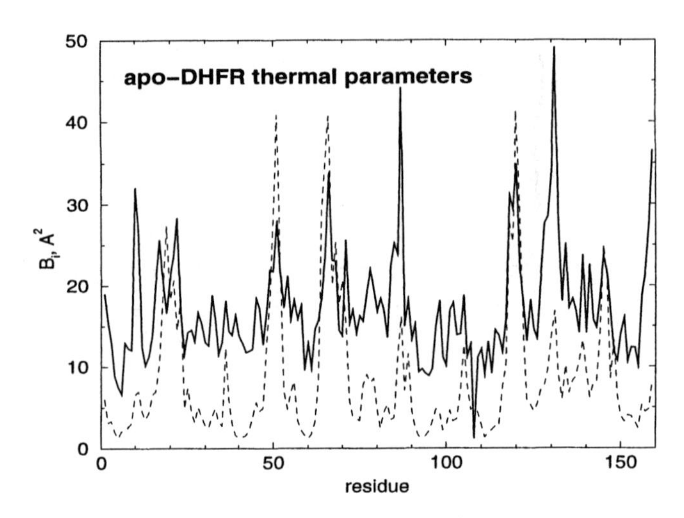

Fig. 4. Crystallographic thermal parameters for the Cα atoms in the backbone of dihydrofolate reductase. Solid line: experimental results, from pdb code 1RA2; dashed line: computed results, as discussed in the text.

This can provide important information about which sections of a protein are mobile, and which are more rigid; at high resolution, correlations and anisotropies of atomic motions can also be discerned.

Some typical results for dihydrofolate reductase are shown in Fig. 4. The solid line shows crystallographic results, and the dashed line is from the quasiharmonic simulation discussed above. Only the 50 lowest modes were used to compute the simulation results, but this will give results close to that obtained if all modes were used.[37] It is clear that in almost every case, local maxima in the crystallographic results correspond well to peaks in the computed results. There is an overall offset, with the observed thermal parameters being larger than those calculated. This may in part arise from the limited (0.5 ns) sampling used, but may also arise contributions not considered here, as discussed below. The relative magnitudes of the computed peaks are not converged at this sampling time, and will continue to change somewhat as the simulation time increases. Nevertheless, current simulation techniques generally to do give quantitative agreement with observed thermal parameters, although there is little question that overall the simulations are giving a correct picture of the relative amounts of rigidity in various portions of the protein backbone.

Although a qualitative interpretation of thermal parameters is often straightforward, it can be difficult to push much beyond this. In addition to true internal motions, thermal parameters depend upon overall librational motion of the biomolecule, and on crystallographic disorder. It is generally quite difficult to untangle these in any straightforward manner, making comparisons to computed behavior (which generally consider only "true" internal motions) difficult.[70] High-resolution crystallographic data certainly helps, but this often involves going to low temperatures, where the fluctuations are less interesting from a biochemical point of view. Some very interesting aspects of protein motion can be discerned from analyses of diffuse scattering (away from the Bragg locations),[71,72] but there is not space to discuss this topic here.

NMR spin relaxation

At most currently available magnetic fields, NMR spin relaxation is mediated by the dipolar coupling between two spins depends upon the motions in the laboratory reference

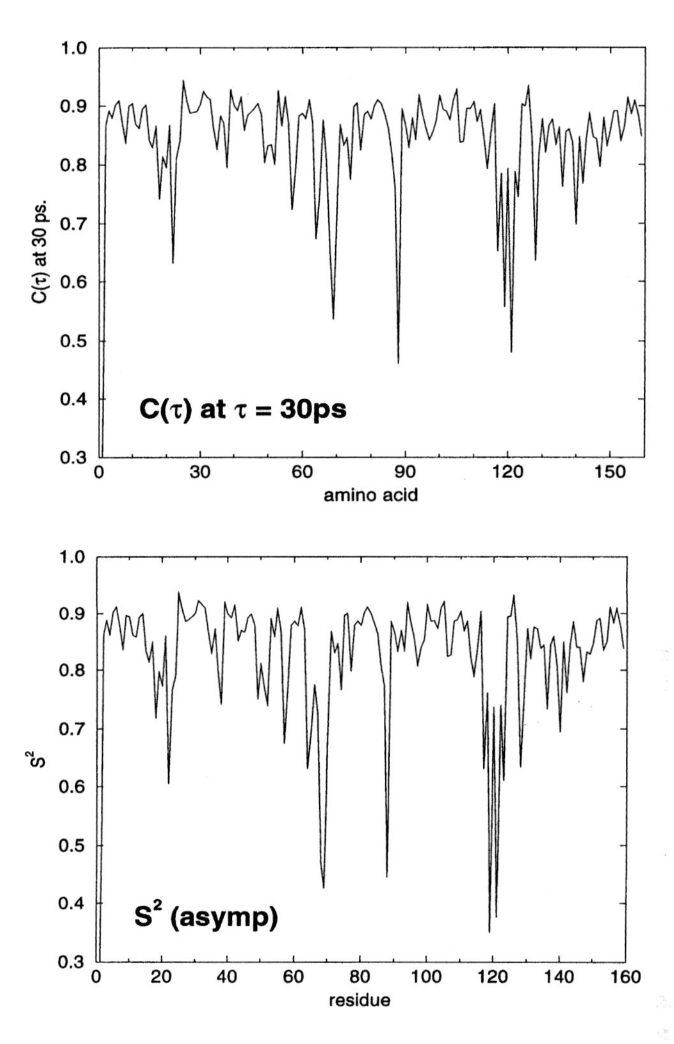

Fig. 5. Top: Order parameters (Eq. 20) for the N–H vectors of the peptide groups in dihydrofolate reductase, evaluated from the time correlation function at 30 ps. Bottom: Order parameters from the "asymptotic formula", Eq. 23.

frame of a vector connecting the two spins. If the internal and overall motions are assumed to be statistically independent (or time-scale separated), correlation functions can be expressed as the product of a correlation function for overall rotational diffusion and a one for internal motions in a fixed molecular frame of reference. For two spins at a fixed distance, the internal angular correlation function that governs NMR spin relaxation is given by[73]

$$C(\tau) = \langle P_2[\hat{\mu}(0) \cdot \hat{\mu}(\tau)] \rangle \tag{17}$$

in which $\hat{\mu}(t)$ is a unit vector in the direction of the vector **r**(t) connecting the spins at time t in the molecular reference frame, and $\langle\,\rangle$ indicates an ensemble average. These correlation functions can be calculated directly from molecular dynamics simulations, or from normal and Langevin modes using a Taylor series expansion[15,74] that expresses the correlation functions in terms of the mean-square deviations of the Cartesian coordinates from their equilibrium positions. For a classical system connected to a heat bath of temperature T, these fluctuations are given by:[13]

$$< q_i(0)\, q_j(\tau) > \; = \; \delta_{ij} \frac{kT}{\omega^2} \cos(\omega\tau) \tag{18}$$

in which q_i is the mass-weighted deviation of the i-th coordinate from its equilibrium position in a mode with frequency ω_i. For Langevin modes, the frequency may be complex, which leads to damped oscillatory decay. For quantum statistics, on the other hand,[13]

$$< q_i(0)\, q_j(\tau) > \; = \; \delta_{ij}\, (\hbar/2\omega) \coth(\hbar\omega/2kT) \cos(\omega\tau) \tag{19}$$

In the present case, the most important consequence of this change is that fluctuations arising from modes of high frequencies never drop below their zero-point values. Zero-point motion effects can be significant for some types of relaxation behavior.

Experimental results are usual reported in terms of an order parameter, S^2, defined as

$$S^2 = \lim_{\tau \to \tau_c} C(\tau) \tag{20}$$

in which $C(\tau)$ is given by Eq. 17 and τ_c is an intermediate time, much less than the overall rotation time, by which $C(\tau)$ becomes constant. The existence of such a plateau assumes that the internal correlation function decays to its limiting value on a time scale short compared to overall rotation. This follows the "model-free" *ansatz* of Lipari and Szabo in postulating a two-term parameterization of correlation functions:[75]

$$C_{MF}(\tau) = < r^{-6} > \left\{ S_{MF}^2 + (1 - S_{MF}^2) exp(-\tau/\tau_e) \right\} \tag{21}$$

in which the square of the order parameter, S_{MF}^2, and the effective internal correlation time, τ_e, are obtained by analysis of experimental data. For convenience, S^2 and S_{MF}^2 will be referred to simply as order parameters. Order parameters can also be estimated by using the addition theorem for spherical harmonics to expand Eq. 17 together with the property that

$$\lim_{\tau \to \infty} < Y_m^2[\Omega(0)]\, Y_m^2[\Omega(\tau)] > = < Y_m^2[\Omega(t)] >^2 \tag{22}$$

in which $\Omega(t)$ are the polar angles defining the orientation of $\mathbf{r}(t)$. Using the above properties in Eq. 5 yields an "asymptotic" estimate of the order parameter:

$$S_a^2 \; = \; \sum_{m=-2}^{2} < Y_m^2[\Omega(t)] >^2 \tag{23}$$

Figure 5 shows typical results for a 159-residue protein, dihydrofolate reductase. Here, the time-correlation function for each N–H vector along the protein backbone is computed from a molecular dynamics simulation. Results for the "plateau" and "asymptotic" order parameters are in good qualitative agreement with each other. Here, the more floppy parts of the molecule have order parameters further removed from unity. In most, but not all, cases, there is a good correlation of motion determined from this measure and that determined from crystallographic thermal parameters. For example, features near residues 20, 64, 82 and 120 are clearly seen both here and in Fig. 4. There is by now an extensive literature on the extent of agreement between NMR measurements and simulations, and on prospects for further detailed characterization of motions by NMR.[76-78]

Fluorescence depolarization

A third method that can provide detailed experimental checks on models for molecular motion uses the decay of fluorescence anisotropy for a fluorescent probe attached at some point in the biomolecule.[79,80] A molecule raised to an excited state by an incoming photon will re-emit light (as it returns to the ground electronic state) whose polarization reflects that of the excitation light. As molecular motions take place, and the molecule tumbles in a liquid, it loses its memory of the initial orientation, the polarization properties of the fluorescence decay to zero. Fast laser techniques enable these measurements to be made on a very short time scale. This is a direct method of probing a time-correlation function, as opposed to the NMR approach, which is sensitive to the Fourier transforms of these correlation functions. The vector whose decay is probed is essentially determined by the electronic structure of the chromophore, and it follows its molecular motion.

A "classic" example involves following fluorescence of the drug ethidium, which will intercalate between base pairs in DNA.[81,82] Fig. 6 illustrates the effects of solvent viscosity on this process for a very short DNA, modeled using Langevin modes.[19] Without solvent damping, the internal vibrations never decay, and one obtains an oscillatory correlation function. (The decay of the correlation function due to overall molecular tumbling will occur, but is on a much longer time scale than that of the figure.) As solvent viscosity is added to the theory, the decay becomes smooth, with a time constant that depends significantly on the nature of the coupling model used and the solvent viscosity. For this example, there is relatively little internal motion, but there will be much more for longer (and more interesting) segments of DNA; the basic physical ideas illustrated here should continue to be valid, however.

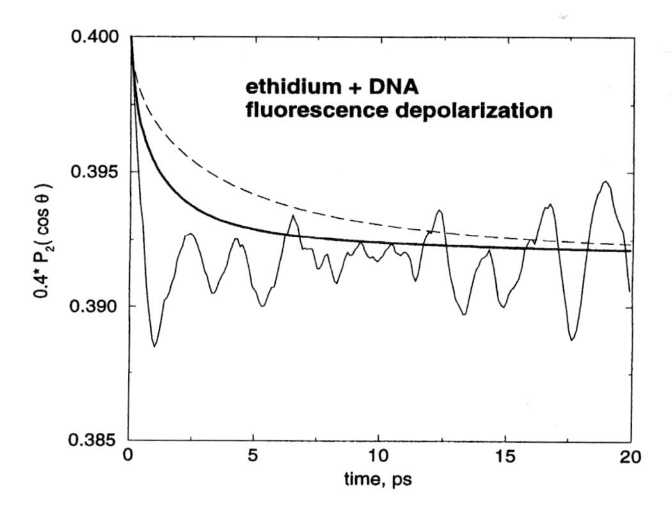

Fig. 6. Fluorescence depolarization functions for ethidium complexed to $d(CGCGCG)_2$, as a function of viscosity η.[19] The vector being followed is the C3–C6, or "long" axis of the intercalating drug. Light solid line: $\eta = 0$; heavy solid line, $\eta = 0.3$ cp; dashed line, $\eta = 0.9$ cp (the approximate viscosity of water at room temperature).

CONCLUSIONS

Normal modes analyses continue to occupy an important niche in dynamical analyses of biomolecules by providing a compact and analytical representation of an important limiting case. The directions of the low-frequency modes often provide useful quantities for description of correlated motion even in the presence of significant anharmonicity. Extensions to disordered or significantly anharmonic systems provide interesting insights into protein dynamics, and suggest new approaches to the analysis of experiments and molecular dynamics simulations.

ACKNOWLEDGMENTS

This work was supported by NIH grant GM 56879. I thank Charles Brooks and Jennifer Radkiewicz for helpful discussions.

REFERENCES

1. M.P. Allen and D.J. Tildesley. *Computer Simulation of Liquids*, Clarendon Press, Oxford (1987).
2. W.F. van Gunsteren, P.K. Weiner, and A.J. Wilkinson, eds.. *Computer Simulations of Biomolecular Systems, Vol. 2.*, ESCOM Science Publishers, Leiden (1993).
3. W.F. van Gunsteren, P.K. Weiner, and A.J. Wilkinson, eds.. *Computer Simulations of Biomolecular Systems, Vol. 3.*, ESCOM Science Publishers, Leiden (1997).
4. T. Horiuchi and N. Gō, *Proteins: Str. Func. Gen.* 10:106-116 (1991).
5. H. Goldstein. *Classical Mechanics*, Addison-Wesley, Reading, MA (1980).
6. M. Levitt, C. Sander, and P. S. Stern, *J. Mol. Biol.* 181:423-447 (1985).
7. S. Sunada and N. Gō, *J. Computat. Chem.* 16:328-336 (1995).
8. H. Wako, S. Endo, K. Nagayama, and N. Gō, *Comp. Phys. Commun.* 91:233-251 (1995).
9. M.-H. Hao and S.C. Harvey, *Biopolymers* 32:1393-1405 (1992).
10. B.R. Brooks, D. Janezic, and M. Karplus, *J. Computat. Chem.* 16:1522-1542 (1995).
11. S. Hayward and N. Go, *Annu. Rev. Phys. Chem.* 46:223-250 (1995).
12. D. Janezic and B.R. Brooks, *J. Computat. Chem.* 16:1543-1553 (1995).
13. D. A. McQuarrie. *Statistical Mechanics*, Harper and Row, New York (1976).
14. R. Brüschweiler, *J. Chem. Phys.* 102:3396-3403 (1995).
15. G. Lamm and A. Szabo, *J. Chem. Phys.* 85:7334-7348 (1986).
16. A.G. Palmer and D.A. Case, *J. Am. Chem. Soc.* 114:9059-9067 (1992).
17. A. Kitao, F. Hirata, and N. Gō, *Chem. Phys.* 158:447-472 (1991).
18. S. Hayward, A. Kitao, F. Hirata, and N. Gō, *J. Mol. Biol.* 234:1207-1217 (1993).
19. J. Kottalam and D.A. Case, *Biopolymers* 29:1409-1421 (1990).
20. F. Liu, J. Horton, C.L. Mayne, T. Xiang, and D.M. Grant, *J. Am. Chem. Soc.* 114:5281-5294 (1992).
21. J.C. Smith, *Quart. Rev. Biophys.* 24:227-291 (1991).

22. J.A. McCammon, B. Gelin, M. Karplus, and P.G. Wolynes, *Nature* 262:325-326 (1976).
23. J.-F. Gibrat and N. Gō, *Proteins: Str. Func. Gen.* 8:258-279 (1990).
24. S. Hayward, A. Kitao, and H.J.C. Berendsen, *Proteins* 27:425-437 (1997).
25. F.H. Stillinger and T.A. Weber, *Science* 225:983-989 (1984).
26. H. Frauenfelder, S.G. Sligar, and P.G. Wolynes, *Science* 254:1598-1603 (1991).
27. A. Thomas, B. Roux, and J.C. Smith, *Biopolymers* 33:1249-1270 (1993).
28. D. Janezic, R.M. Venable, and B.R. Brooks, *J. Computat. Chem.* 16:1554-1566 (1995).
29. R.A. LaViolette and F.H. Stillinger, *J. Chem. Phys.* 83:4079-4085 (1985).
30. G. Seeley and T. Keyes, *J. Chem. Phys.* 91:5581-5586 (1989).
31. R.M. Stratt and M. Maroncelli, *J. Phys. Chem.* 100:12981-12996 (1996).
32. T. Keyes, *J. Phys. Chem. A* 101:2921-2930 (1997).
33. J.E. Straub and D. Thirumalai, *Proc. Natl. Acad. Sci. USA* 90:809-813 (1993).
34. J.E. Straub, A.B. Rashkin, and D. Thirumalai, *J. Am. Chem. Soc.* 116:2049-2063 (1994).
35. W. Li, T. Keyes, and F. Sciortino, *J. Chem. Phys.* 108:252-260 (1998).
36. R.M. Levy, M. Karplus, J. Kushick, and D. Perahia, *Macromolecules* 17:1370-1374 (1984).
37. M.M. Teeter and D.A. Case, *J. Phys. Chem.* 94:8091-8097 (1990).
38. M. Karplus and J.N. Kushick, *Macromolecules* 14:325-332 (1981).
39. A. DiNola, H.J.C. Berendsen, and O. Edholm, *Macromolecules* 17:2044-2050 (1984).
40. A. Amadei, M.E.F. Apol, and H.J.C. Berendsen, *J. Chem. Phys.* 106:1893-1912 (1997).
41. B. Tidor and M. Karplus, *Prot. Str. Func. Gen.* 15:71-79 (1993).
42. B. Tidor and M. Karplus, *J. Mol. Biol.* 238:405-414 (1994).
43. D. Perahia, R.M. Levy, and M. Karplus, *Biopolymers* 29:645-677 (1990).
44. T. Ichiye and M. Karplus, *Proteins: Str. Func. Gen.* 2:236-259 (1987).
45. J. Kuriyan, G.A. Petsko, R.M. Levy, and M. Karplus, *J. Mol. Biol.* 190:227-254 (1986).
46. A.E. Garcĭa, *Phys. Rev. Lett.* 68:2696-2699 (1992).
47. A. Kitao, F. Hirata, and N. Gō, *J. Phys. Chem.* 97:10231-10235 (1993).
48. A. Amadei, A.B.M. Linssen, and H.J.C. Berendsen, *Proteins* 17:412-425 (1993).
49. W.D. Cornell, P. Cieplak, C.I. Bayly, I.R. Gould, K.M. Merz, Jr., D.M. Ferguson, D.C. Spellmeyer, T. Fox, J.W. Caldwell, and P.A. Kollman, *J. Am. Chem. Soc.* 117:5179-5197 (1995).
50. M. Gerstein, A.M. Lesk, and C. Chothia, *Biochemistry* 33:6739-6749 (1994).
51. B. de Groot, S. Hayward, and H. Berendsen, *Proteins* 31:116-127 (1998).
52. D.A. Case, *Curr. Opin. Struct. Biol.* 4:285-290 (1994).
53. M.M. Tirion and D. ben-Avraham, *J. Mol. Biol.* 230:186-195 (1993).
54. T. Ikura and N. Gō, *Proteins* 16:423-436 (1993).

55. J. Ma and M. Karplus, *J. Mol. Biol.* 274:114-131 (1997).

56. L. Mouawad and D. Perahia, *J. Mol. Biol.* 258:393-410 (1996).

57. C.L. Brooks, III and D.A. Case, *Chem. Rev.* 93:2487-2502 (1993).

58. H. De Loof, S.C. Harvey, J.P. Segrest, and R.W. Pastor, *Biochemistry* 30:2099-2113 (1991).

59. J.B. Clarage, T. Romo, B.K. Andrews, B.M. Pettitt, and G.N. Phillips, Jr., *Proc. Natl. Acad. Sci. USA* 92:3288-3292 (1995).

60. I. Chandrasekhar, G.M. Clore, A. Szabo, A.M. Gronenborn, and B.R. Brooks, *J. Mol. Biol.* 226:239-250 (1992).

61. M.A. Balsera, W. Wriggers, Y. Oono, and K. Schulten, *J. Phys. Chem.* 100:2567-2572 (1996).

62. P.H. Hünenberger, A.W. Mark, and W.F. van Gunsteren, *J. Mol. Biol.* 252:492-503 (1995).

63. B.L. de Groot, M.F. van Aalten, A. Amadei, and H.J.C. Berendsen, *Biophys. J.* 71:1707-1713 (1996).

64. D. ben-Avraham, *Phys. Rev. B* 47:14559-14560 (1993).

65. M.M. Tirion, *Phys. Rev. Lett.* 77:1905-1908 (1996).

66. T. Haliloglu, I. Bahar, and B. Erman, *Phys. Rev. Lett.* 79:3090-3093 (1997).

67. I. Bahar, A. Wallqvist, and R.L. Jernigan, *Biochemistry* 37:1067-1075 (1998).

68. S.W. Englander. in *Encyclopedia of Nuclear Magnetic Resonance*, D.M. Grant and R.K. Harris, Ed. John Wiley, London (1996). pp. 2415-2420.

69. D.P. Millar, *Curr. Opin. Struct. Biol.* 6:322-326 (1996).

70. J. Kuriyan and W.I. Weiss, *Proc. Natl. Acad. Sci. USA* 88:2773-2777 (1991).

71. F. Paure, A. Micu, D. Pérahia, J. Doucet, J.C. Smith, and J.P. Benoit, *Nature Struct. Biol.* 1:124-128 (1994).

72. J.-P. Benoit and J. Doucet, *Quart. Rev. Biophys.* 28:131-169 (1995).

73. R. Brüschweiler and D.A. Case, *Prog. NMR Spectr.* 26:27-58 (1994).

74. E.R. Henry and A. Szabo, *J. Chem. Phys.* 82:4753-4761 (1985).

75. G. Lipari and A. Szabo, *J. Am. Chem. Soc.* 104:4546-4559 (1982).

76. A.G. Palmer, III, J. Williams, and A. McDermott, *J. Phys. Chem.* 100:13293-13310 (1996).

77. A.G. Palmer, III, *Curr. Opin. Struct. Biol.* 7:732-737 (1997).

78. M.W.F. Fischer, L. Zeng, Y. Pang, W. Hu, A. Majumdar, and E.R.P. Zuiderweg, *J. Am. Chem. Soc.* 119:12629-12642 (1997).

79. A. Szabo, *J. Chem. Phys.* 81:150-167 (1984).

80. D.P. Millar, *Curr. Opin. Struct. Biol.* 6:637-642 (1996).

81. M.D. Barkley and B.H. Zimm, *J. Chem. Phys.* 70:2991-3007 (1979).

82. J. Duhamel, J. Kanyo, G. Dinter-Gottlieb, and P. Lu, *Biochemistry* 35:16687-16697 (1996).

EFFICIENT STOCHASTIC GLOBAL OPTIMIZATION FOR PROTEIN STRUCTURE PREDICTION

Yingyao Zhou and Ruben Abagyan*

Skirball Institute of Bimolecular Medicine
Biochemistry Department
New York University Medical Center
540 1st Avenue, New York, NY 10016

INTRODUCTION: Why not MD or MC?

Biological macromolecules, large chain molecules with hundreds of torsion angles, adopt compact, uniquely folded and rigid conformations that correspond to their global free energy minimum. Predicting this unique conformation from a vast number of alternatives, for the whole protein or its parts, is the biggest challenge of computational biology. One of the difficulties is conceptual. To evaluate the free energy correctly we need to account for the dynamic nature of the entire system, including mobile water molecules, flexible side-chains and soft vibrational modes of a solute. Molecular Dynamics (MD, reviewed in Ref. 1-4) or Monte Carlo simulations (MC, reviewed in Ref. 4-8) in water can be applied to sample the conformational space and evaluate the free energy. However, these methods are still too slow to reach the biologically relevant folding times for proteins or even large peptides[2,9].

Fortunately, the free energy of surrounding water molecules can be implicitly evaluated through the electrostatic and surface effects[10], the side chain mobility contribution to the free energy can be roughly estimated through its solvent exposure, and the vibrational contribution can be considered comparable in different folded conformations. Therefore, the computationally expensive MC and MD methods, aimed at the generation of a Boltzmann ensemble, can be replaced by much more efficient stochastic global optimization methods aimed at identification of a unique global minimum in the smallest number of iterations. Global optimization methods can be classified into zero-order and first-order algorithms depending on whether a local minimization step is performed after each iteration[11]. Two reasons account for the clear superiority of the first-order methods for peptides and proteins[12]. The first reason is the energy improvement due to local minimization, which is often comparable to the variation of the energy values between different local minima. Second, an adequate standard local optimization method,

* Corresponding author

using analytical energy derivatives, is the most efficient way to identify the nearest local minimum, and such algorithms as MC, MD, or random sampling will be far inferior in performing the same task.

Here we describe the principle of optimal sampling bias as an algorithm for generation of the random moves in a stochastic global optimization method and demonstrate a drastic improvement of the efficiency due to the optimal bias. The principle of optimal bias was first introduced in 1994[13] as a linear sampling bias; in this essay, we consider another optimization model and arrive at the square-root sampling bias rule. This algorithm is general and applicable to stochastic global optimization of any function, both continuous and discrete.

GLOBAL OPTIMIZATION: How to find a global minimum of a function?

In the Introduction we argued that free energy can be assigned to a single polypeptide chain conformation, and, therefore a unique native folded conformation can be predicted by global energy optimization. The global optimization algorithm is not bound by the trajectory continuity or Boltzmann ensemble generation requirements, and, therefore, has a larger potential to do what an optimization algorithm does best, i.e. find the minimum in the minimal number of function evaluations.

Global optimization is used in many fields[14,15], but in protein structure prediction it is additionally complicated by high dimensionality (the smallest protein has about 100 essential degrees of freedom), and small separation between the native energy minimum and the abundant false energy minima. The high dimensionality of the problem makes any systematic search impossible, a problem known as the Levinthal paradox[16]. To make matters worse, the optimization problem can not be considered at the discrete rotamer level since small 'continuous' angular adjustments are essential for favorable packing. Finally, the small energy difference between the correct and incorrect minima and the exponential growth of the density of the non-native states with energy impose strict requirements on the accuracy of energy evaluation (less than about 1 kcal/mol)[5].

Numerous approaches have been used to attack the global optimization problem in protein structure prediction, with some success[1-8] (Table 1). These methods are initially classified according to whether they are deterministic or not; stochastic methods are further subdivided according to the degree of similarity between conformations generated in consecutive iterations of the search algorithm.

Table 1. Classification of global optimization methods based on the degree of history dependence.

Global optimization Methods			
Deterministic search	Stochastic search		
	History independent	Intermediate history dependent	Maximum history dependent
Systematic search[17,18]*, build-up method[19-21]*, diffusion equation method (DEM)[24], packet annealing method (PA)[23].	Randomize all variables in a step: Local minimization from multiple random starting conformations.	Large change of one/a group of variables in a step: Genetic algorithm (GA)[36], lattice model MC[37], MC-Minimization (MCM)[11], standard Metropolis MC[26,27], electrostatically driven MC (EDMC)[38], extended scaled collective variable (ESCV) MC[33], restrictive MC[39], biased MC (BMC)[40,41], optimal-bias MC-Minimization (OBMCM)[13].	Changes all variables continuously in a step: Molecular dynamics (MD)[1-4,28-30], Local energy minimization[11,31], scaled collective variable (SCV) method[32], extended SCV MC[33]. High directional MC (HDMC)[34], some side chain MC[35].

* Only search the combinations of pre-calculated local minima.

In principle, deterministic methods arc guaranteed to find the global minimum. In practice, however, such methods require the adoption of certain simplifying assumptions that compromise their accuracy. Systematic search[17,18] and the build-up methods[19-21] assume that the global minimum of a complete structure is a combination of a relatively small number of local minima of structural fragments. Both assumptions turn out to be wrong; many intramolecular interactions are nonlocal (about 50% by the contact area estimate[22]), the globally optimal conformation may contain strained fragments far from their local minima, and the number of local conformations to be retained is exceedingly large. The packet annealing (PA)[23] and the diffusion equation (DEM)[24] methods introduce an elegant concept of smoothing the probability distribution and the energy surface, respectively, and reduce the global optimization problem to a series of local minimizations. However, the deterministic character of these methods is something of an illusion. DEM procedure encounters numerous bifurcation points during the annealing process and a slight inaccuracy in the free energy function can lock the search into the wrong path[25].

We will distinguish between the MC or MD methods, which are designed to generate a Boltzmann ensemble, and global optimization algorithms (such as simulated annealing[26,27]) which attempt to identify a single conformation corresponding to the global minimum of a free energy function (in the pseudo-potential energy form).

Most of the MC-like stochastic global optimization strategies employ a three-step iteration: (i) modify the current conformation by means of a random move; (ii) evaluate its energy; (iii) accept or reject the new conformation according to an acceptance criterion. The random moves can be ranked by magnitude of change with respect to the current conformation (Table 1). The first group contains algorithms in which the generated conformations do not depend on the previous ones. The second group keeps maximum memory by changing all variables quasi-continuously according to certain rules or by some small amplitude random deviations. This category contains molecular dynamics (MD)[28-30], local energy minimization methods[31], scaled-collective-variable (SCV) method[32], extended SCV Monte Carlo (ESVC)[33], high directional MC (HDMC)[34], and some side chain MC methods[35]. The third group takes an intermediate approach by changing one variable or a group of variables (generally correlated variables) at a time. This group contains most of the global optimization methods including genetic algorithm (GA)[36]-based methods, lattice model MC[37], and most other MC methods[4-8].

HISTORY-DEPENDENCE OF CONFORMATIONAL SEARCHES

Different history-dependent protocols inherit current structural information to varying degrees. Genetic-algorithm (GA) methods[36] make a single random change with each 'mutation', and conformational recombination extends the random change to a wider range. Various lattice MC methods[37] make local elemental jumps, which may involve modifying three to five bonds, and translation/rotation of a portion of the chain as well. In a global step of the MCM method[11], a random change of one angle is accompanied by a local minimization with respect to all torsion angles. Some methods[26] make sequential change to one variable at a time in standard Metropolis MC (MMC) implementation, the amplitude of randomization being tuned to ensure a sufficiently high acceptance ratio. Some other MMC methods[27] randomly change one angle with an amplitude of 90°. Electrostatically driven MC (EDMC) method[38] switches between a random prediction, where one dihedral angle is randomized with an amplitude of 180°, and an electrostatically driven move, where two coupled dihedral angles are changed with an amplitude estimated from the local electric field. Restricted MC methods[39] replace continuous side chain orientations by discrete rotamer values. Biased MC (BMC)[40,41] makes three- or four-residue backbone move at once, the statistical distributions of backbone dihedral angles and rotamer libraries for side chain angles are taken into account in the conformation generation. Optimal-bias-MC-with-minimization (OBMCM, also referred to as Biased Probability MC[13]) modifies groups of correlated backbone or side chain variables according to optimal statistical distributions. MD, local energy minimization methods, SCV/ESCV, HDMC make small amplitude changes to all variables determined by dynamic equations or local energy landscape. Some side chain MC methods[35] change all side chain torsion angles simultaneously by 0° or ±10°.

How similar should the next conformation be to the previous one? Virtually identical as in a MD method, or totally unrelated as in a random search? In the following section we investigate this question.

COMPARISON OF GLOBAL OPTIMIZERS OF ZERO ORDER (WITHOUT MINIMIZATION)

The performance of the global optimization methods can be tested on small peptides. Met-enkephalin, the Tyr-Gly-Gly-Phe-Met pentapeptide, has been extensively studied and frequently used as a test peptide before[11,12,19,32,34,42], but it is too small and conformationally unusual for a good protein-like benchmark. Two other test peptides were used instead: an α-helix and a β-hairpin. The selected helix is a 12-residue synthetic peptide Acetyl-Glu-Leu-Leu-Lys-Lys-Leu-Leu-Glu-Glu-Leu-Lys-Gly-COOH crystallized and solved by Hill *et al*[43]. The second peptide is a 13-residue ubiquitin fragment (residue number 3-15) suggested to be an independent β-hairpin fold by circular dichroism and NMR studies[44,45].

We performed a series of Metropolis Monte Carlo (MMC) simulations without minimization from random starting conformations for four different move generation algorithms. (1) Change one randomly selected variable at each step, with amplitude of 30°, 90° and 180°. (2) Change two coupled variables such as backbone φ-ψ angles or χ_1-χ_2 angles in a randomly selected residue with 180° amplitude. (3) Change all variables of a randomly selected residue, (φ, ψ and χ), with 180° amplitude. (4) Randomize all variables with 2° amplitude after each step.

A simulation temperature of 600K was used for all simulations to ensure the same 'energetic accuracy' of 1.2 kcal/mol. Each type of simulation was repeated ten times and the conformational energies were recorded. Average angular RMSDs of conformations generated in adjacent steps represent the scale of a random move. The average best energies after a certain number of energy evaluations (1×10^5 for the α-helix and 5×10^4 for the β-hairpin), as well as their standard deviations and acceptance ratios ρ, are shown in table 2.

The result shows that neither smallest nor largest random moves result in good performance. In general, a good move is the one generating the largest change at a given temperature and acceptance ratio. That is exactly what a good biased move of several angles at a time allows to be accomplished. For the above two benchmarks, the optimal-bias MC algorithm (without minimization) reached E_{min} of −132 ± 5 kcal/mol, acceptance

Tabel 2. MMC simulations of the 12-residue α-helix and the 13-residue β-hairpin.

	12-residue α-helix			12-residue β-hairpin		
Test type	RMSD (°)	E_{min} (kcal/mol)	ρ	RMSD (°)	E_{min} (kcal/mol)	ρ
1	0.16	~10^4	0.09	0.15	~10^4	0.37
1	0.50	-93± 7	0.28	0.56	-41 ± 11	0.28
1	0.78	-122± 7	0.22	0.83	-76 ± 7	0.22
2	1.4	-111 ± 7	0.16	1.2	-71 ± 7	0.15
3	1.5	-89 ± 9	0.15	1.5	-65 ± 8	0.12
4	2.0	~10^5	0.00	2.0	~10^6	0.00

ratio of 0.28 for the α-helix, and E_{min} of –88 ± 8 kcal/mol, acceptance ratio of 0.28 for the β-hairpin while changing two variables at a time.

GLOBAL OPTIMIZERS WITH LOCAL MINIMIZATION ARE SUPERIOR

In 1997 Li and Scheraga introduced a new global optimization method in which each random step is followed by local energy minimization[11]. Even though they called it Monte Carlo-Minimization (MCM), the procedure did not obey the local balance condition and can only be considered as a stochastic global optimization algorithm. But how important is local energy minimization after each large random move? On the one hand, spending valuable energy evaluations on local energy optimization in basically the same conformational vicinity instead of more extensive sampling may sound wasteful. On the other hand, minimization algorithms using function derivatives are much more efficient than random sampling in finding the local energy minimum, and the unminimized values are not really representative because of the ruggedness of the energy landscape. The number of energy evaluations spent on local adjustments is typically hundreds of times larger than the number of random moves! Maybe we should use only a partial minimization thus saving the function evaluations for more random steps, given the fact that the energy drops much faster in the beginning of the minimization?

The above questions were systematically analyzed[12] and the conclusion was that allowing a full local optimization following each random step resulted in the best performance, with both partial and no minimization being clearly inferior under the constraint of the total number of energy evaluations. In other words, making 100,000 high quality moves is preferable over making 10,000,000 low quality moves.

In the MCM algorithm a randomly chosen angle was changed by a random, evenly distributed value. Introduction of the optimal bias into the random step resulted in another drastic increase of the global optimization performance[13].

OPTIMAL BIAS FOR STOCHASTIC GLOBAL OPTIMIZATION (OBMCM)

We know that the groups of torsion angles in peptides and proteins have certain preferences, i.e. some values are found more frequently than others. The preferences of the backbone angles (ϕ-ψ angles) as well as the side chain rotamer libraries have been described [46-50], and the correlations between the backbone and side chain angles have been studied as well[51]. How can we take advantage of these statistical preferences? We know that almost every protein or peptide contains some rare, unusual torsion angles; therefore, should one still use a flat probability distribution (as in the MCM method) to ensure that these rare values are sampled frequently enough? Or should we just use the discrete peaks of the distributions (the rotamers)[39] and hope that the rest will be taken care of by local minimization? The answers to these questions are important; as we will see later, the optimization efficiency is actually more sensitive to the answer to this question than to whether one uses simulated annealing or constant temperature, or whether one uses multiple independent runs or exchanges information between simulations.

There are basically two major alternatives: uniformly distributed random moves, and moves biased according to some statistical information. The statistical information may be sequence-independent *a priori* information[13] derived from the structures in the Protein Data Bank, or the statistical information accumulated during the simulation[39]. Configurational-bias Monte Carlo (CBMC) simulations have been introduced very early on[52] (a good review of CBMC methods can be found in Chapter 13 of Frenkel and Smit's book[53]), but the ability to generate a Boltzmann ensemble, an appropriate concern for a Monte Carlo algorithm, was the primary focus. However, the primary objective of a stochastic global optimization algorithm is identification of the global minimum in a minimal number of function evaluations, a different goal that is not necessarily compatible with the local balance principle. For example, the local minimization after each move violates the local balance but is necessary for efficient global optimization. Therefore, derivation of the bias which is optimal from the global optimization point of view, a problem addressed by OBMCM/BPMC algorithm[13], became an important objective.

The idea is to use the geometrical preferences of local groups of coupled torsion angles, preferences that can be pre-calculated, to guess their final values defined by all the interactions in a larger molecule, under the assumption that the global interactions are random with respect to the local preferences. Let us denote a group of coupled variables by vector $\mathbf{x}$ and its value corresponding to the global minimum state as $\mathbf{x}^0$. Therefore an arbitrary protein conformation can be represented by its n variable groups as $(\mathbf{x}_1, \mathbf{x}_2, \ldots, \mathbf{x}_i, \ldots, \mathbf{x}_n)$ and its lowest-energy conformation as $(\mathbf{x}_1^0, \mathbf{x}_2^0, \ldots, \mathbf{x}_i^0, \ldots, \mathbf{x}_n^0)$. We further assume that for all possible protein targets, $\mathbf{x}^0$ satisfies statistical ***distribution function*** $S(\mathbf{x}^0)$. Actually a separate distribution function for each type of amino acid can be generated, and the distribution function for the jth type of amino acid will be $S_j(\mathbf{x}^0)$. In MC-like algorithms, one randomly selects a vector, for instant $\mathbf{x}_i$, to change during a global move. $\mathbf{x}_i$ is to be assigned a new value $\mathbf{x}_i'$ according to a probability function $f_j(\mathbf{x}_i')$ (assuming $\mathbf{x}_i$ belongs to the jth type of amino acid), which is to be called the ***sampling function*** later. The question is what are the sampling functions resulting in identification of the correct answer in the minimal number of energy evaluations.

Unfortunately, the question does not have a clear answer unless an analytical target function is specified to measure the performance of a global sampling. We propose here two target functions. (i) $f_j(\mathbf{x})$ maximize the probability of finding the lowest-energy conformation of a randomly given protein within a global sampling; (ii) $f_j(\mathbf{x})$ minimize the average number of global sampling steps required to successfully predict a randomly given protein. In order to simplify our analyses, it is assumed that all the n variable groups are randomly re-sampled according to their corresponding sampling function at a specific global step and there is no local minimization afterwards. The proofs for the most general case involving n continuously distributed variable groups are presented in Ref. 13 for the first target function and in the Appendix for the second target function.

We will try to guess the true value $\mathbf{x}^0$ of a vector $\mathbf{x}$, with the knowledge that $\mathbf{x}^0$ takes the value $\mathbf{x}_1$ with probability S_1, takes the value $\mathbf{x}_2$ with probability S_2, ..., takes the value $\mathbf{x}_n$ with probability S_n. In a biased guess, the sampling function f allows one to sample the value $\mathbf{x}_1$ with probability f_1, sample the value $\mathbf{x}_2$ with probability f_2, ..., sample the value $\mathbf{x}_n$ with probability f_n. The game and some possible strategies are illustrated in Figure 1, where the current target value of $\mathbf{x}$ is marked by a star and the distribution function S is resembled by the shaded bars. Random vectors are generated according to the sampling function until the true value $\mathbf{x}^0$ of $\mathbf{x}$ is hit. An additional condition is independence of each guess on the previous guess. This is counterintuitive in a simple guessing game, e.g., if there are only two states, and you gave the wrong answer, the next one will be right. However, in a real simulation with an MC-like calculation the global context of the same group of variables is constantly changing and the independence assumption can be justified.

Game 1: Find the optimal f_i so that the probability of correctly guessing the true value $\mathbf{x}^0$ in each guess is maximized.

If the actual value is $\mathbf{x}_i$, one will guess it correctly with the probability f_i in a step. Since such an event happens with a probability of S_i, the overall probability to be maximized is $P = \sum_i S_i f_i$, under the normalization condition $\sum_i f_i = 1$.

This is equivalent to maximize $P + \lambda\left(\sum_i f_i - 1\right)$, where λ is the Lagrange multiplier and f_i can be treated as n independent variables. It is then straightforward to derive the optimal sampling function by setting the derivatives of this target function with respect to f_i equal to zero. The conclusion is $\boxed{S_i/f_i=\lambda}$, i.e., the optimal sampling function equals the original distribution function.

Game 2: Find the optimal f_i so that the average number of unsuccessful guesses is minimized.

Let us note that if the true value is $\mathbf{x}_i$, and the probability of sampling of this particular state is f_i in each trial, therefore, it will take $1/f_i$ trials on average to find the true value. Since such an event occurs with expected probability S_i, the 'ensemble average' of the average numbers of required guesses is $\overline{N}=\sum_i S_i f_i^{-1}$, under the normalization condition $\sum_i f_i=1$.

Optimizing $\overline{N}+\lambda\left(\sum_i f_i-1\right)$, we arrive at $\boxed{S_i/f_i^2=\lambda}$, i.e., square-root sampling functions minimize the cost of global minimization.

As mentioned before, the same conclusions can be generalized for any arbitrary number of vectors with continuous distributions $S(\mathbf{x})$. The linear bias $f_j(\mathbf{x}) = S_j(\mathbf{x})$ maximizes the correct guessing probability[13], and square-root bias $f_j(\mathbf{x}) \propto \sqrt{S_j(\mathbf{x})}$ minimizes the average number of guesses required.

SUPERIOR PERFORMANCE OF THE OPTIMAL-BIAS-MCM

Comparison between the zero-order MMC and OBMC (with both the linear and the square-root bias) show that both biased sampling algorithms out-performed the uniform random sampling scheme. Both linear and square-root bias result in comparable performance on both previous benchmarks. However, because the square-root bias allows sampling of the rarely populated zones of the torsion space much more frequently than the linear biasing functions, we expect that less standard benchmarks would reveal a better performance of the square-root bias.

We also compared the first-order method such as unbiased MCM and linear-bias MCM algorithms using the 12-residue α-helix[13] and a more realistic ββα peptide (results are not shown) as a benchmark. The performance increase due to the optimal bias varies

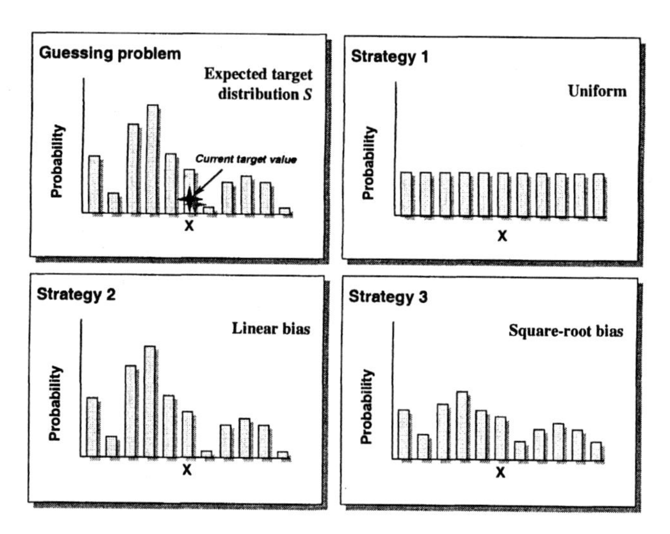

Figure 1. Schematic diagram of various sampling strategies.

but on average is about ten fold for a mixed α/β topology. However, these calculations take several days even for the OBMCM algorithm and we were not able to reach the solution with the MCM algorithm in a reasonable time.

Waiting until each algorithm reaches its global minimum may take a lot of time, and this time varies strongly between simulations. Previously, we used a more stable performance criterion, which was a fraction of the set of many low energy minima visited after the fixed number of function evaluations[12]. Here we returned to the old measure of the number of iterations until the global minimum was reached, but we softened the minimum identification criterion and averaged this number with up to 10 independent simulations. $R(n)$ is the fraction of systems that have reached the global minimum after n energy evaluations. By reaching the global minimum, we mean that a simulation hits a conformation of correct secondary structure and also has energy within 3 kcal/mol above the lowest energy found by pre-simulations. Success rate, also called cumulative distribution function (CDF), has been used before to study the folding time of the simulated annealing algorithm[54].

$R(n)$ can be approximately described by a Poisson distribution[54]. Taking the simulation cost for the early stage of forming compact globular conformations into account, we use the following expression to describe the success rate:

$$R(n) = 1 - e^{-q(n-n_0)}, \text{ with } n > n_0,$$

where q is a constant. $1\text{-}e^{-q}$ ($\approx q$, for $q << 1$) can be interpreted as the probability of hitting a global minimum conformation per energy evaluation. Since n_0 is the average number of energy evaluations required to lower the system energy to a plateau and $1/q$ is the mean value of the Poisson distribution, $n_0 + 1/q$ is the measurement of overall simulation cost including both early and latter stages in a simulation.

The benchmarks used here are the 12-residue α-helix[43] and a 12-residue β-hairpin[55]. Their global minimum energies were –185.0 kcal/mol and -198.6 kcal/mol, respectively. Three algorithms were analyzed: (i) Lee *et al.*[40], biased MC (BMC) with linear sampling function but without minimization. We used the distributions derived in Ref. 13 for

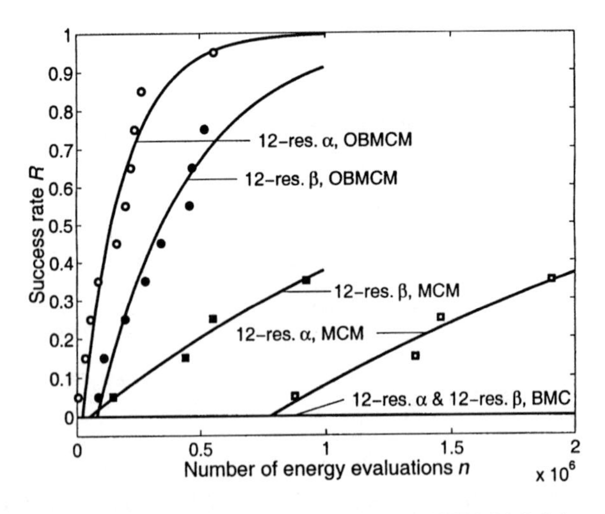

Figure 2. The success rate of BMC (Lee et al., 1993), MCM (Li & Scheraga, 1987) and OBMCM (Abagyan & Totrov, 1994) simulations. Results for the BMC simulations are denoted by the horizontal line, since no successful simulation was found within 4×10^6 energy evaluations under this scheme.

Table 3. The performance measurements of MMC and OBMCM methods

	12-residue α-helix		12-residue β-hairpin	
	MCM	OBMCM	MCM	OBMCM
n_0	7.80×10^5	2.22×10^4	5.06×10^4	8.20×10^4
$1-e^{-q} \approx q$, for $q << 1$	3.77×10^{-7}	5.81×10^{-6}	5.02×10^{-7}	2.65×10^{-6}
$n_0 + 1/q$	3.43×10^6	1.94×10^5	2.04×10^6	4.60×10^5

backbone sampling, but rotamer libraries for the side chain sampling; (ii) Li & Scheraga[11], MCM with uniform sampling and minimization; (iii) Abagyan & Totrov[13], OBMCM with the linear-bias and minimization.

Ten simulations for each case were initiated under constant simulation temperature 600K. q and n_0 values were then derived from the data. The results are shown in Figure 2 and Table 3. We found that OBMCM is 18 times faster compared to the unbiased MCM in the alpha-helix simulation, and 4.4 times faster in the beta-hairpin simulation. No successful simulations were found for the BMC case, the lowest energies reached by this protocol within 4×10^6 functional calls were –151.1 kcal/mol and –171.2 kcal/mol for the α-helix and the β-hairpin, respectively (therefore $n_0 > 4\times10^6$).

SUMMARY

The native structure of a protein may be described with reasonable accuracy as the global minimum of the free energy (in the pseudo-potential energy form), only as a function of free torsion angles. Therefore, global optimization methods might be preferable over methods designed to create dynamic ensembles, such as MD or MC that are bound by the trajectory continuity requirement or the local balance requirement.

The Monte Carlo Minimization (MCM) method outperforms zero order MC-like stochastic global optimization protocols.

The Optimal-Bias-MCM method further improves the sampling efficiency by an order of magnitude by incorporating the optimal-bias into MC conformation generation. The square-root bias derived in this work and the linear bias[13] are two possible strategies.

The OBMCM algorithm can predict a 23-residue ββα peptide[56], with 70 essential torsion angles and 385 atoms, starting from completely random conformations. (Figure 3).

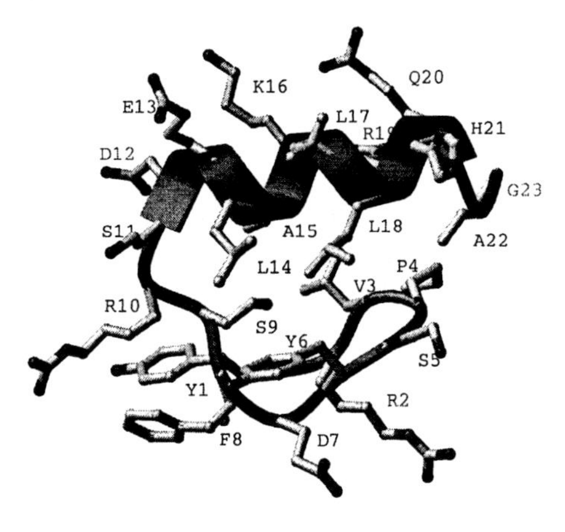

Figure 3. The predicted structure of a 23-residue ββα peptide (Ref. 56) with OBMCM method.

APPENDIX: OPTIMAL CONTINUOUS SAMPLING FUNCTIONS IN GLOBAL SAMPLING

If a group of coupled torsion angles of a residue (φ-ψ, and/or χ_1-χ_2) is denoted by vector $\mathbf{x}$, the lowest-energy conformation of protein consisting of n such groups of variables can be denoted as $(\mathbf{x}_1^0, \mathbf{x}_2^0, \ldots, \mathbf{x}_i^0, \ldots, \mathbf{x}_n^0)$, where $\mathbf{x}_i^0$ is the values of the ith group of torsion angles in the lowest-energy protein conformation. $\mathbf{x}_i^0$ has an *a priori* continuous probability distribution $S_j(\mathbf{x}_i^0)$ in the subspace formed by the vector, where j denotes the type of amino acid $\mathbf{x}_i$ belongs to. We sample each variable group $\mathbf{x}_i$ according to a sampling function $f_j(\mathbf{x})$. Randomly given a representative protein, we consider the problem of finding the optimal sampling functions that minimize the average number of energy evaluations required for successful structure prediction.

Following the same assumption made in Ref.13, i.e., the probability of finding the true value $\mathbf{x}_i^0$ of variable group $\mathbf{x}_i$ is proportional to $f_j(\mathbf{x}_i^0)$, when a global sampling is made for $\mathbf{x}_i$ according to the sampling function f_j. Since the probability of finding the true conformation at this specific conformation generation step reads

$$P = c\prod_{i=1}^{n} f_j(\mathbf{x}_i^0),$$

where c is a constant, it takes $1/P$ steps to find the true conformation on average. The S-ensemble average is the mean number of iterations:

$$\overline{N} = \frac{1}{c}\int_{\mathbf{x}_1^0}\int_{\mathbf{x}_2^0}\cdots\int_{\mathbf{x}_n^0}\prod_{i=1}^{n} S_j(\mathbf{x}_i^0)\prod_{i=1}^{n} f_j^{-1}(\mathbf{x}_i^0)d\mathbf{x}_i^0,$$

$$\overline{N} = \frac{1}{c}\prod_{i=1}^{n}\int S_j(\mathbf{x}_i^0) f_j^{-1}(\mathbf{x}_i^0)d\mathbf{x}_i^0.$$

Since $\overline{N}$ is always positive, maximizing $\overline{N}$ is equivalent to maximize

$$\ln\overline{N} = \sum_{i=1}^{n}\ln\int S_j(\mathbf{x}_i^0) f_j^{-1}(\mathbf{x}_i^0)d\mathbf{x}_i^0 - \ln c$$

to find the optimal sampling functions, we set $\delta\ln\overline{N}$ to zero:

$$\delta\ln\overline{N} = -\sum_{i=1}^{n}\int S_j(\mathbf{x}_i^0) f_j^{-2}(\mathbf{x}_i^0)\delta f_j(\mathbf{x}_i^0)d\mathbf{x}_i^0\cdot\left(\int S_j(\mathbf{x}_i^0) f_j^{-1}(\mathbf{x}_i^0)d\mathbf{x}_i^0\right)^{-1} = 0.$$

Given the normalization conditions for the sampling functions, $\int\delta f_j(\mathbf{x}_i^0)d\mathbf{x}_i^0 = 0$, in order for the above equation to hold for any arbitrary function δf_j, we have

$$f_j(\mathbf{x}) = \frac{1}{c'}\sqrt{S_j(\mathbf{x})},$$

where c' is the normalization constant equal to $\int\sqrt{S_j(\mathbf{x})}d\mathbf{x}$.

ACKNOWLEDGEMENTS

We thank NIH (Grant R01 GM55418-01) and DOE (Grant DE-FG02-96ER62268) for financial support (this does not constitute an endorsement by both agencies of the views expressed in the article). We also thank Alex Morrill and Sheila Silvestein for careful reading of the manuscript.

REFERENCES

1. R. Elber, *Curr. Opin. Struct. Biol.*, 6:232 (1996).
2. C.L. Brook III, *Curr. Opin. Struct. Biol.*, 5:211 (1995).
3. T. Schlick, E. Barth, and M. Mandzink, *Ann. Rev. Biophys. Biolmol. Struct.*, 16:179 (1997).
4. B.J. Berne and J.E. Straub, *Curr. Opin. Struct. Biol.*, 7:181 (1997).
5. R.A. Abagyan, Eds. W.F. van Gunsteren, P.K. Weiner, and A.J. Wilkinson, *Computer Simulation of Biomolecuar Systems, Theoretical and Experimental Applications*, Kluwer Academic Publisher, vol. 3 pp.363-394 (1997).
6. H.A. Scheraga, *Biophys. Chem.*, 59:329 (1996).
7. M. Karplus and A. Šali, *Curr. Opin. Struct. Biol.*, 5:58 (1995).
8. G. Nemethy and H.A. Scheraga, *FASEB J.*, 4:3189 (1990).
9. E. Demchuk, D. Bashford, and D.A. Case, *Fold. Des.*, 2:35 (1997).
10. B. Honig and A. Nicholls. *Science*, 268:1144 (1995).
11. Z. Li and H.A. Scheraga, *Proc. Natl. Acad. Sci. U.S.A.*, 84:6611 (1987).
12. R. Abagyan and P. Argos, *J. Mol. Biol.*, 225:519 (1992).
13. R. Abagyan and T. Totrov, *J. Mol. Biol.*, 235:983 (1994).
14. B. Berg, *Nature*, 361:708 (1993).
15. S. Kirkpatrick, C.D. Gellatt, and M.P. Vecchi, *Science*, 220:671 (1983).
16. C. Levinthal, Eds. P. Debruner, J.C.M. Tsibris, and E. Munck, *Mossbauer Spectroscopy in Biological Systems*, Univ. Illinois Press, pp. 22-24 (1969).
17. R.E. Bruccoleri and M. Karplus, *Biopolymers*, 26:137 (1987).
18. T. Schaumann, W. Braun, and K. Wüthrick, *Biopolymers*, 29:679 (1990).
19. M. Vásquez and H.A. Scheraga, *Biopolymers*, 24:1437 (1985).
20. S. Vajda and C. DeLisi, *Biopolymers*, 29:1755 (1990).
21. I. Simon, L. Glasser, and H.A. Scheraga, *Proc. Nat. Acad. Sci., U.S.A.*, 88:3661 (1991).
22. R.A. Abagyan and M.M. Totrov, *J. Mol. Biol.*, 268:678 (1997).
23. D. Shalloway, C.A. In Floudas, and P.M. Pardalos, Eds. *Recent Advances in Global Optimization*, Princeton, Vol. 1, pp. 433-648 (1991).
24. J. Kostrowicki and H.A. Scheraga, *J. Phys. Chem.*, 96:7442 (1992).
25. R.A. Abagyan, *FEBS Letters*, 325:17 (1993).
26. H. Kawai, T. Kikuchi, and Y. Okamoto, *Protein Eng.*, 3:85 (1989).
27. S.R. Wilson and W. Cui, *Biopolymers*, 29:225 (1990).
28. J.A. McCammon, B.R. Gelin, and M. Karplus, *Nature*, 267:585 (1977).
29. M. Levitt and R. Sharon, *Proc. Nat. Acad. Sci., U.S.A.*, 85:7557 (1988).
30. R.E. Bruccoleri and M.A. Karplus, *Biopolymers*, 29:1847 (1990).
31. M.J.D. Powell, *Math. Programming*, 12:241 (1997).
32. T. Noguti and N. Go, *Biopolymers*, 24:527 (1985).
33. A. Kidera, *Proc. Natl. Acad. Sci. U.S.A.*, 92:9886 (1995).
34. J.K. Shin and M.S. Jhon, *Biopolymers*, 31:177 (1991).
35. C. Lee and S. Subbiah, *J. Mol. Biol.*, 217:373 (1991).
36. R. Unger and J. Moult, *J. Mol. Biol.*, 231:75 (1993).
37. J. Skolnick, A. Kolinski, and R. Yaris, *Biopolymers*, 28:1059 (1989).
38. D.R.Pipoll and H.A. Scheraga, *Biopolymers*, 27:1283 (1988).
39. L. Holm and C. Sander, *J. Mol. Biol.*, 218:183 (1991).
40. H.S. Kang, N.A. Kurochkina, and B. Lee, *J. Mol. Biol.*, 229:448 (1993).
41. B. Lee, N. Kurochkina, and H.S. Kang, *FASEB J.*, 10:119 (1996).
42. Purisima and Scheraga, *J. Mol. Biol.*, 196:697 (1987).
43. C.P. Hill, D.H. Anderson, L., Wesson, W.F. DeGrado, and D. Eisenberg, *Science*, 249:543 (1990).
44. M.S. Briggs and H. Roder, *Proc. Natl. Acad. Sci. U.S.A.*, 89:2017 (1992).
45. J.P. Cox, P.A. Evans, L.C. Packman, D.H. Williams, aznd D.N. Woolfson, *J. Mol. Biol.*, 234:483 (1993).
46. B. Robson and R.H. Pain, *Biochem. J.*, 141:869 (1975).

47. M.J. Rooman, J.P.A. Kocher, and S.J. Wodak, *J. Mol. Biol.*, 221:961 (1991).
48. R. Abagyan and T. Totrov, *J. Mol. Biol.*, 235:983 (1994).
49. J.W. Ponder and F.M. Richards, *J. Mol. Biol.*, 193:775 (1987).
50. J. Janin, S. Wodak, M. Levitt, and B. Maigret, *J. Mol. Biol.*, 125:357 (1978).
51. R.L. Roland and M. Karplus, *J. Mol. Biol.*, 230:543 (1993).
52. M.N. Rosenbluth and A.W. Rosenbluth, *J. Chem. Phys.*, 23:356 (1955).
53. D. Frenkel and B. Smit, *Understanding Molecular Simulation from Algorithms to Applications*, Academic Press, (1996).
54. R.C. Brower, G. Vasmatzis, M. Silverman, and C. DeLisi, *Biopolymers*, 33:329 (1993).
55. M. Ramirez-Alvarado, F.J. Blanco, H. Niemann, and L. Serrano, *J. Mol. Biol.*, 273:898 (1997).
56. M.D. Struthers, J.J. Ottesen, B. Imperiali, *Fold. & Des.*, 3:95 (1998).

FLEXIBLE AND RIGID REGIONS IN PROTEINS

Donald J. Jacobs[1], Leslie A. Kuhn[2] and Michael F. Thorpe[1]

[1]Department of Physics and Astronomy
[2]Department of Biochemistry
Michigan State University
East Lansing, MI 48824, U.S.A.

INTRODUCTION

We represent the microstructure of a protein as a generic bar-joint truss framework, where the hard covalent forces and strong hydrogen bonds are modeled as distance constraints. The mechanical stability of the corresponding bar-joint network is then analyzed using graph theoretical techniques. The computer program for analyzing the rigidity of substructures within macromolecules[1] is referred to as FIRST (**F**loppy **I**nclusion and **R**igid **S**ubstructure **T**opography). This program provides a real-time tool for evaluating the intrinsic flexibility within a protein by applying a new combinatorial constraint counting algorithm. Unlike many methods for parsing protein folds, this new approach gives exact mechanical properties of a protein structure (or other macromolecules) under a given set of distance constraints. These properties include; counting the number of independent degrees of freedom, locating overconstrained regions where internal strain arises, partitioning the protein structure into rigid clusters that are separated by flexible joints and identifying underconstrained regions where continuous deformations can take place.

This article proceeds as follows. We begin by discussing some basic notions about intrinsic flexibility in proteins. We then highlight the underlying physical assumptions of our approach, introducing concepts from graph rigidity and sketching the computational algorithm. Some example applications are given, including the FIRST analysis of a lysine-binding protein.

Intrinsic Flexibility of Proteins

Over seven thousand protein structures have been determined to date using X-ray crystallography[2]. Such crystallographic structures are deduced from sophisticated

Rigidity Theory and Applications
Edited by Thorpe and Duxbury, Kluwer Academic/Plenum Publishers, 1999

refinement analysis in conjunction with stereochemical modeling[3], and they represent the best average structure over many realizations of individual protein molecules making up the solid crystal. Frequently in crystallographic structural studies, independent crystal forms trap the protein in different conformational states, suggesting an intrinsic flexibility within the protein[4]. The study of protein motion in solution using NMR spectroscopy shows that many different conformational states can be explored by a protein in its natural environment[5]. The intrinsic flexibility of a protein is often manifested in conformational changes[6] that are responsible for the large-scale rearrangements of domains or the relative motion of smaller fragments within individual domains.

A collection of X-ray crystal structures for a protein in different conformations allows a limited number of structural comparisons, from which some information about the intrinsic flexibility can be deduced. Prediction of the flexibility of a protein structure that is known only in one conformation can sometimes be inferred from other, similar protein structures. Molecular dynamics (MD) simulations can also be used to observe the dynamical evolution of a given protein as it explores various conformational states having energies near that of its native state. In practice, simulations cannot be performed for long enough times to explore the full range of available conformations.

Some computational procedures have been developed[7–12] to characterize the intrinsic flexibility and rigidity within a protein. These procedures fall into two classes. One makes use of comparing different conformational states[7–9] while the other class deals with identifying protein domains or characterizing protein structure[10–12] for any given conformation. These methods are generally fast (ranging from hours to minutes of CPU time), and give considerable insight into the mechanical stability of a protein. Nevertheless, each method has its limitations. In a broad sense, the methods in the first class are limited to the diversity (and accuracy) of the conformational states that are available from experiment for comparison, whereas the methods in the second class are limited to the correlation between the selected empirical criteria such as packing density or structural protrusions. FIRST falls into the second class, but it has the advantage that it runs in a fraction of a second, and it is based on a systematic distance constraint approach where rigidity can be directly calculated.

Characterizing Mechanical Stability

What would the *ideal* scenario be for an automated computational procedure to determine flexibility and rigidity, or more specifically the *mechanical stability* of a protein? Besides being fast and efficient, the method should only require knowing the protein structure. The output should consist of a universal scheme for characterizing mechanical stability that gives a robust quantified description of the degree of stability throughout the protein.

Perhaps the ideal universal way to characterize mechanical stability of a protein is via a hierarchical scheme of substructuring, reflecting in a self-consistent way the interaction strengths between the identified substructures. Within this characterization, gross features of mechanical stability between different protein conformations should be the same with some local differences. Of course, when comparing experimentally observed protein structures in different conformations, differences in mechanical stability due to binding different substrates or to different packing within the crystal lattice can occur.

Developing such an ideal automated procedure is our goal. A distance constraint approach is used to characterize mechanical stability of a protein, where only knowing the underlying *connectivity* of the structure is required. Overconstrained regions, rigid substructures and underconstrained regions can be determined within a hierarchical scheme. Distance constraints are used to model the covalent bond forces (bond-stretching and bond-bending) as well as any selected set of hydrogen bonds. Selection of the imposed distance constraints for the hydrogen bonding is determined by introducing a cut-off criterion on the interaction strength. A hierarchical characterization of the mechanical stability can be constructed by changing the chosen cut-off.

Applications to Drug Design and Protein Engineering

Rigidity and structural stability in proteins is commonly associated with close packing of amino acids into modules or domains[6,10,11,13]. Multifunctional proteins are often built from several closely packed domains linked by flexible regions which can allow large scale, hinge-type conformational changes[6,14]. There can also be shear-type conformational changes within densely packed regions[6], which are more subtle to identify. Flexible regions in proteins are often associated with loop structures, which are thought to be functionally important. For example, as a means of regulating interactions between proteins, one region in the CksHs1 protein switches between loop and beta-strand conformations, resulting in its dimerization with a protein kinase and regulation of the cell cycle in humans[15].

Conformational changes within a protein are often observed during and after ligand binding. In addition to the flexible regions, rigid motifs are important in several areas of protein structure and function. For many proteins, active or ligand-binding sites are described as rigid templates, which do not undergo significant conformational change upon ligand binding.

MICROSCOPIC FORCES AS DISTANCE CONSTRAINTS

Imposing distance constraints to replace important bonding forces between atoms reduces the total number of degrees of freedom available to the protein. If we imagine atoms interacting pairwise within a protein, the effect of a distance constraint is to fix the interaction between a pair of atoms. Applying pairwise distance constraints is the *input*, and the *output* is the set of rigid clusters, separated by flexible joints.

Obviously, in reality no set of atoms within a protein will ever define a *perfectly* rigid cluster. By imposing distance constraints on the strongest interactions, to quench mainly higher frequency motions, it is expected that defining a rigid cluster is meaningful only on sufficiently long time scales. Floppy regions consist of atoms that continue to have relative motion on these long time scales. The inherent assumption that we make in our distance constraint approach is to neglect vibrational modes within a rigid cluster and the coupling between inter cluster vibrational modes. The lowest frequency from the intra cluster vibrational modes will set the relevant time scale for which the rigid and floppy regions become meaningful. The cleanest situation occurs when the separation between strong and weak interactions has a significant gap.

Separation Between Strong and Weak Forces

The hard covalent bonding within the protein, consisting of the bond-stretching and bond-bending forces, defines a natural set of distance constraints. The energies associated with central bond-bending and torsional forces are of the order 25 (Kcal/mol)/Å^2,

4 (Kcal/mol)/rad^2 and 0.06 Kcal/mol respectively[16]. Note that although the energy units are different, a direct comparison in magnitudes can roughly be made upon realizing that the covalent bond lengths are about 1.5Å. It is common practice to fix the covalent bond lengths and bond angles while allowing the dihedral angles to be free to rotate. Using the dihedral angles as a set of internal coordinates, the number of degrees of freedom to describe the motion of a protein is typically reduced by a factor of about seven[17].

In addition to the central and bond-bending forces, some torsional forces associated with the peptide or resonant bonds lock rotational motion about the bond axis. In this case, the dihedral angle can be fixed by using a third neighbor distance constraint. Modeling only the strongest set of forces associated with covalent bonding as distance constraints will not define any large rigid region. Instead, the protein flexibility will be described by a floppy set of connected small rigid clusters that can be identified by inspection. This happens because in all proteins, the covalent bonds form a tree-like structure (i.e no loops) with a few exceptions consisting of rings found in some residues (proline, histidine, phenylalanine, tyrosine and tryptophan) or perhaps more interestingly, crosslinking from a few disulfide bonds.

Groups of atoms tend to cluster into well defined substructures such as alpha helices and beta sheets (common secondary structures[18]). Furthermore, many proteins are made up of a collection of stable fragments[4], that range in size from a small part of a domain to an entire domain. Protein domains can be defined in various ways[11,19], either by function or some particular structural characteristic, such as packing density. In general a protein domain will consist of many atoms (typically 1000 atoms or more) and will contain common secondary structures. From empirical evidence, it is believed that conserved substructures exist within domains during the course of conformational changes. These are the rigid clusters we seek to identify a priori.

We must apply distance constraints associated with other (weaker) forces in order to determine larger rigid regions. We could of course (erroneously) produce one entire rigid region by placing *all* third neighbor distance constraints associated with torsional forces, thereby fixing all dihedral angles. The same trivial result would occur by modeling van der Waals interactions by distance constraints. Keep in mind, however, that a reasonably clean separation between strong and weak forces is desired to make distinctions between rigid and floppy regions. Therefore we need to apply distance constraints to bonding forces that are stronger than torsional forces (not associated with resonant bonds), but these forces will still be weaker than the covalent bond-bending forces.

Hydrogen Bonding: A Hierarchical Approach

It is natural to consider hydrogen bonds as the next set of (weaker) forces, after the covalent forces, and to model them also as distance constraints. The strength of a hydrogen bond varies from nearly as strong as the covalent bonds to as weak as the van der Waals interactions[20,21]. This broad range of variation in strength indicates that the hydrogen bond, unlike covalent bonding, is very sensitive to its local environment. We know hydrogen bonding is very important in proteins because secondary structures, such as alpha helices, beta sheets and hairpin turns, can be readily identified in terms of hydrogen bond patterns that form along the mainchain of the protein[18]. These substructures frequently occur in proteins (evident from the Brookhaven Protein Data Bank[2], which is an on line computer-based archive for bio-macromolecular structures), and their relative arrangements further define interesting structural motifs.

The hydrogen bond patterns within the mainchain, associated with the nitrogen (donor) and oxygen (acceptor) atoms provide crosslinking between residues. The crosslinking of hydrogen bonds is responsible for the high degree of mechanical stability found within secondary structures. Similarly, we expect that the additional crosslinking provided by hydrogen bonds outside of the mainchain will support larger mechanically stable substructures, which may become as large as an entire domain. Modeling these hydrogen bonds as distance constraints gives a simple way to characterize the degree of stability within the protein. Since placing a distance constraint between two atoms does not make a distinction between a strong or weak interaction, a weak hydrogen bond becomes just as important as a strong hydrogen bond or a covalent bond in this regard. Therefore, at our discretion, we must choose whether or not to model certain bonding forces with distance constraints.

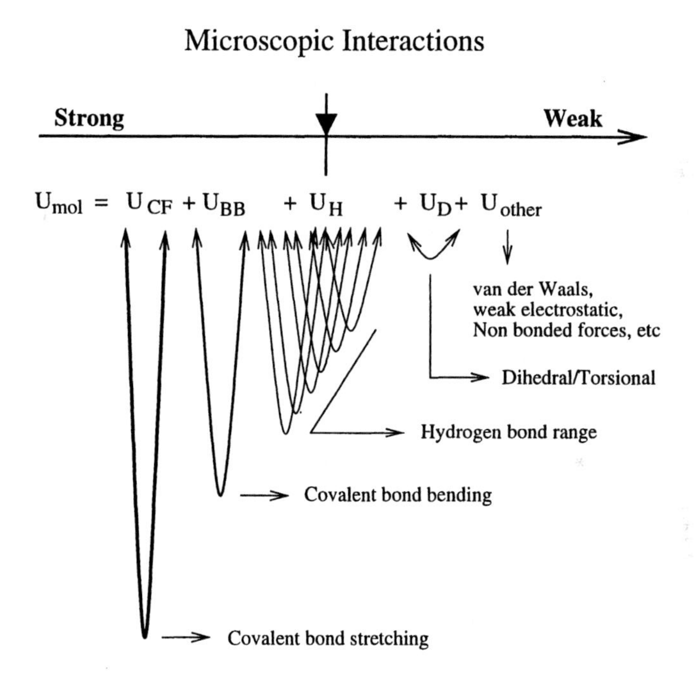

Figure 1. A schematic representation of ordering the microscopic forces from strongest to weakest. Distance constraints are used to model strong bonding forces to the left of a sliding pointer. This approach defines a system of interacting rigid clusters.

Consider modeling only those hydrogen bonds having an interaction strength *greater* than some predefined cut-off, as distance constraints. This will divide the hydrogen bonds into two groups, but a gap between strong and weak forces will *not* be found in this case. Unlike covalent network glasses[22], where there are no hydrogen bonds, obtaining a clean separation in bond strength is not possible in proteins, because the hydrogen bonds cover a broad range of interaction strengths. As a result, no justification can be given for selecting a single set of hydrogen bonds to be modeled as distance constraints.

The schematic diagram shown in Figure 1 represents a hierarchical approach that we have adopted for selecting which interactions to model as distance constraints. Naturally, the covalent bond-stretching and angular or bond-bending forces will be modeled as distance constraints. The nearest neighbor bond-stretching force defines a nearest neighbor distance, and the angular or bond-bending force defines a second nearest neighbor constraint. In addition, we introduce a *pointer* on an interaction scale such that any hydrogen bonds interacting more strongly are modeled as distance constraints, while those that have a weaker interaction strength are ignored. The pointer is allowed to slide down the scale starting from an interaction strength just below that of covalent bonding. As the pointer slides down, more and more hydrogen bonds are included. This approach allows us to uncover various rigid substructures, that will merge together as more crosslinks are added, albeit via weaker hydrogen bonds.

Intuitively, one expects that a large rigid cluster, consisting of many weak hydrogen bonds, will not be as *rigid* as some internal parts that form rigid substructures involving strong hydrogen bonds. Of course, the strongest rigid substructures within any large rigid region will be the set of small rigid clusters defined by the covalent bonding. The hierarchical approach of gradually selecting weaker and weaker hydrogen bonds allows us to access the relative degree of stability (as a continuum measure) between different regions in the protein. However, before we accomplish this task, we first construct a quantitative measure for the degree of stability applied locally, rather than globally.

Van der Waals and Hydrophobic Forces

In proteins there are both short- and long-ranged non-bonding forces. Individually these forces are generally weak, but collectively they are important in governing the dynamics of a protein. For example, each van der Waals interaction is too weak to model as a distance constraint, yet collectively the van der Waals interactions play an important role in determining steric conformational constraints. The hydrophobic force is a thermodynamic force that is generally regarded as a dominant contribution that drives a protein to fold[23], and it plays an important role in stablizing a protein structure in the native state. It is not possible to apply distance constraints to model the hydrophobic force, because it is an entropic force, depending on the ensemble of configurations available to all atoms of the protein and solvent molecules.

During conformational changes, certain regions of the protein are essentially preserved as a rigid body. Our viewpoint is that the rigid clusters, defined by the covalent bonding and the crosslinking hydrogen bonds, will interact with one another via the weaker non-bonding forces as a system of coupled rigid bodies. Therefore, rather than modeling the protein stucture as a system of interacting particles, we focus on the long time dynamics where conformational changes are controlled by weaker, but collectively important, forces acting between coupled rigid bodies.

The van der Waals and hydrophobic forces can be regarded as the most important for driving and stablizing protein folds, because they will essentially determine the most probable conformations having the lowest thermodynamic free energy. As a result, it should not be surprising to experimentally observe thermodynamically stable regions within the protein that are identified as floppy regions using a simple distance constraint approach. These floppy regions locate low energy pathways that enable a protein to explore different conformations. These pathways are important to facilitate domain rearrangement or allow some local flexibility for the protein to successfully bind with a ligand. Thus, we are focusing on the mechanical stability of a network of rigid clusters, rather than the thermal stability of the protein as a whole.

GENERIC RIGIDITY

A brute force method to *identify all rigid clusters* in a system with N atoms would involve diagonalizing a dynamical matrix N^2 times to check for redundant distance constraints between all pairs of atoms[24]. This leads to a polynomial time algorithm that scales as $O(N^5)$. Other methods to check for redundant distance constraints, involving the rigidity matrix for example, can be used instead of diagonalizing the dynamical matrix, but the final result will still lead to a polynomial time algorithm that also scales as $O(N^5)$. Finding stressed regions requires an additional numerical calculation involving a stress-strain relaxation of the network, using for example a conjugate gradient method, that scales as $O(N^2)$.

The distance constraint approach will only be useful in as much as the calculation for identifying floppy and rigid regions is fast and scales linearly or nearly linearly with protein size. By considering the protein as a generic structure, one can deduce the rigidity properties using concepts from graph rigidity[25]. This is the contribution of this paper.

A generic structure is one that has no special symmetries, such as parallel bonds or bond angles of 180 degrees, that could create geometrical singularities[26]. Under these conditions, the study of rigidity becomes much easier to deal with because the properties of network rigidity depend only on the *connectivity* of the network, as determined by an underlying graph. Although removing concerns about the *particular* coordinates of atoms gives a great simplification conceptually, the calculational problem at hand is still far from trivial.

There is a theorem by Laman[27] that states how generic rigidity of any *two* dimensional bar-joint network can be completely characterized by applying constraint counting to *all* subgraphs. Applying Laman's theorem directly leads to a combinatorial calculation that scales as the exponential in the number of atoms within the network. However, by applying Laman's theorem recursively, a very fast and efficient algorithm (the so called *pebble game*) has been constructed[28,31,32] for identifying rigid clusters and stressed regions. The *pebble game* uses computer memory that scales linearly with the size of the network, and has a performance that scales in the worst case as $O(N^2)$ for pathological networks, but in practice performs nearly linearly with N.

In three dimensional generic bar-joint networks, constraint counting over all subgraphs is known to *fail* in general. However, for the special class of truss frameworks, the theorem of Laman can be generalized[29] to three dimensions. Again constraint counting over all subgraphs is enough to completely characterize generic rigidity. We will model the microstructure of a protein using distance constraints that define a truss framework, which is also called a bond-bending network.

Three Dimensional Bond-Bending Networks

Covalent bonding naturally defines bond distances and angles. As shown in Figure 2, the covalent bonding defines a graph, where each vertex represents an atom, nearest neighbor distances represent central forces and next nearest neighbor distances represent bond-bending forces. This type of graph is called a *squared graph*[30], a truss framework or a bond-bending network. The network connectivity is completely described by the nearest neighbor central-force constraints, which are viewed as inducing the bond-bending next nearest neighbor distance constraints. The only elementary floppy element that exists within a bond-bending network is a *hinge joint.* Rotations

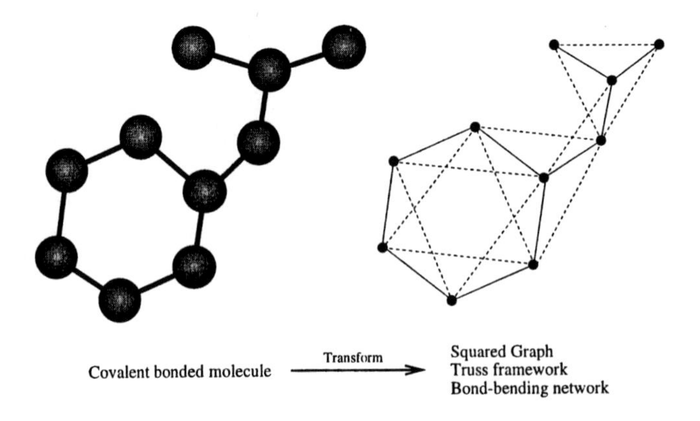

Figure 2. Once a covalent bonded molecule or network is regarded as generic, the rigidity of the network is completely determined by the connectivity of the underlying structure. A transformation is made from a physical molecule to a mathematical graph, where vertices represent atoms, solid lines represent distance constraints between nearest neighbor atoms associated with central forces and dashed lines represent distance constraints between next nearest neighbor atoms associated with bond-bending forces. The graph is of a special kind, called a squared graph. This structure is also referred to as a truss framework or bond-bending network.

through a dihedral angle about the axis of a central-force constraint is a priori possible, but may be locked because of the network properties.

As shown in Figure 3 the hydrogen bond is modeled in a similar way to a covalent bond with generic geometry, in which the donor, hydrogen and acceptor atoms are *not co-linear.* Typically, each hydrogen bond will introduce three distance constraints, corresponding to one central force between the hydrogen and acceptor atoms and two bond-bending forces associated with the hydrogen and acceptor atoms. If an acceptor atom is involved in n_h hydrogen bonds, there will be $(n_h - 1)$ additional bond-bending force constraints between hydrogen-acceptor-hydrogen atoms. This particular model for a hydrogen bond is mathematically convenient because it allows the protein structure to be described as a bond-bending network. Physically, the model is also reasonable because hydrogen bonds are almost never linear and the three dihedral angle degrees of

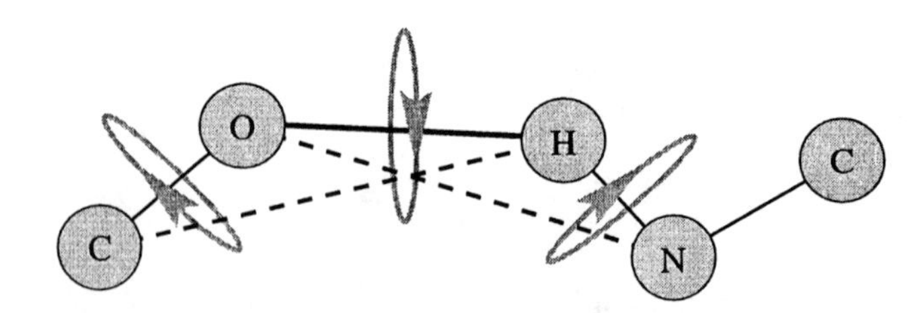

Figure 3. A diagram showing a hydrogen bond involving a donor and acceptor atom taken here as nitrogen and oxygen respectively. It is modeled as three generic distance constraints, consisting of a nearest neighbor central-force constraint shown as a thick solid line, and two next nearest neighbor bond-bending force constraints shown as dashed lines. Each constrained hydrogen bond is also associated with three a priori rotatable dihedral angles indicated by the arrows.

freedom associated with the hydrogen bond allows it to be relatively flexible. Although modeling the hydrogen bond to be more or less constrained is easy to do, we consider the model in Figure 3 to strike a good balance between neither over- nor under-representing the effectiveness of a hydrogen bond.

The 3D Pebble Game

At the heart of the FIRST computer program is the *3D pebble game* algorithm that is constructed in a very similar way as the two dimensional *pebble game*[28,29,31,32]. Here, three pebbles, representing three translational degrees of freedom, are assigned to each vertex in the graph. For each independent distance constraint between two vertices, a pebble from either one of the incident vertices must be used to *cover* the constraint. Pebbles associated with vertices are called *free* pebbles, and they represent the independent degrees of freedom remaining within the network. Each independent distance constraint uses up one independent degree of freedom. Then the following covering rule is applied: once an independent constraint is covered by a pebble, it must always remain covered by any of the pebbles associated with either of its incident sites. Rearrangement of pebbles throughout the network is possible provided this covering rule is not violated.

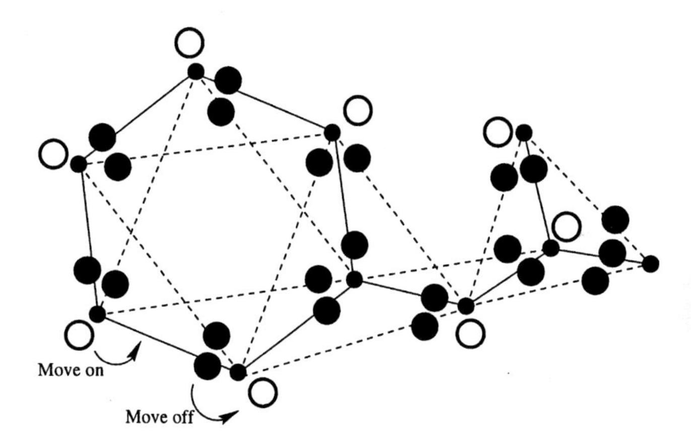

Figure 4. A diagram showing the final pebble covering of a simple network. The open big circles represent free pebbles that are placed directly on vertices and denote free degrees of freedom available to the network. The filled big circles represent pebbles that are covering a distance constraint, and are placed directly on the edges of a graph. The pebble covering is not unique, because pebbles can be rearranged according to a few simple rules as explained in the text. Here, we show an example of how an elementary pebble exchange works, as indicated by the two arrows. A free pebble can be moved on to a covered edge (off from a vertex) provided a corresponding pebble, which is presently covering the edge and is associated with the neighboring vertex, is moved off (on to a vertex).

In Figure 4 an example of a pebble covering is given for the graph shown in Figure 2 (rotated by 45 degrees). The pebble covering is a convenient way to represent a dynamically changing directed graph[28] facilitated by the rearrangement of pebbles through a series of elementary pebble exchanges. Note that if another constraint is added between a pair of vertices within the six fold ring, it would be found to be redundant, because there are not enough free pebbles in this region to cover the constraint. This physically corresponds to the fact that the six fold ring is generically rigid.

The *3D pebble game* is a recursive algorithm like its 2D counterpart. The network is built up by placing one distance constraint at a time. An essential feature of the *3D pebble game* is that each central-force distance constraint associated with vertices v_1 and v_2 must have associated with it angular (i.e. second nearest neighbor) constraints around both vertices v_1 and v_2. For each new independent distance constraint that is introduced, pebbles are rearranged in a way to test if the new distance constraint is independent or not. If the new distance constraint is found to be independent, it is then covered, otherwise it is not covered. This process continues until all distance constraints within the network have been completely placed in the network. The algorithm can be sketched in the following way:

1. Place a central-force distance constraint between vertices v_1 and v_2 as appropriate and as described in the previous sections.
2. Rearrange the pebble covering to collect three pebbles on vertex v_1.
3. Rearrange the pebble covering to collect the maximum number of pebbles on vertex v_2 while holding the three pebbles at vertex v_1.
4. If the number of pebbles on vertex v_2 is two, the distance constraint is redundant. Otherwise, three pebbles reside at vertices (v_1 and v_2).
   Continue to rearrange the pebble covering:
   (a) Hold the three pebbles on both vertices v_1 and v_2.
   (b) For each neighbor of vertex v_2: Attempt to collect a pebble.
   (c) If for any neighbor of vertex v_2 a pebble cannot be obtained, then that distance constraint is redundant.
5. If the distance constraint is not redundant, cover it with a pebble from vertex v_2.

Unlike the 2D *pebble game*, the distance constraints cannot be placed in *any* random order. There is an additional rule about the placement of distance constraints. The first distance constraint that is introduced must correspond to a central-force constraint. After each central-force distance constraint is placed, all of its associated induced angular or bond-bending constraints (next nearest neighbor distance constraints) must be placed before another central-force constraint can be placed. Within this restriction, the order of placing either central force or the induced bond-bending constraints is completely arbitrary. The restriction on the order of placing distance constraints *nearly* maintains the form of the network to be that of a truss framework throughout the building process. The entire network is a truss framework, except in the local region where the additional central-force bond is initially placed. The network is restored to a truss framework *everywhere*, after all the induced bond-bending constraints are added. This local deviation from a truss framework does not create any pivot points or implied hinge joints. Therefore this restriction on recursively placing constraints is sufficient[29] for combinatorial constraint counting to remain valid in characterizing generic network rigidity.

Finally, torsional constraints for the peptide and resonant bonds are fixed by *third* nearest neighbor distance constraints. These are most conveniently placed after the central and bond-bending distance constraints have been placed. No induced bond-bending constraints are associated with the torsional constraints, because these are auxiliary distance constraints for locking in certain dihedral angles.

After all these distance constraints have been placed, the number of free pebbles remaining on the vertices gives the total number of degrees of freedom required to describe the motion of the network. This includes the six trivial rigid body translational and rotational degrees of freedom of the whole network. The free pebbles can be rearranged, but are restricted to certain regions because of the rule for pebble covering. For example, no more than six free pebbles can be found within a rigid cluster. Based on the location and number of free pebbles throughout the network, one can identify overconstrained regions, rigid clusters and underconstrained regions.

Identifying Overconstrained Regions

A redundant constraint is identified when a failed pebble search occurs. A failed pebble search consists of a set of vertices that have no extra free pebbles to give up. This physically corresponds to the region of vertices and distance constraints that predefines the length between a pair of vertices. As we are considering generic networks, placing a distance constraint between this pair of vertices will cause a length mismatch. Physically, this means that the bond lengths and angles within this region of the failed pebble search will become distorted as this region will be internally stressed. Thus, by recording the failed pebble searches, which we call Laman subgraphs as in two dimensions, we automatically identify the overconstrained regions.

Overconstrained regions will always consist of closed loops. As distance constraints are added to the network, more overconstrained regions will be found, and generally these regions will overlap. Overlapping overconstrained regions merge together into a single overconstrained region. As these networks are generic, stress will propagate throughout such a merged set of overconstrained regions.

A subtle point that does not occur in two dimensions is that stress can propagate from one floppy region to another in three dimensions. In bond-bending networks, the effect is a trivial propagation of stress between neighboring atoms that are both four (or more) coordinated. This occurs because a four coordinated atom has six angular constraints, but only five are independent. As a result, every four (or more) coordinated atom is part of at least a locally stressed region consisting of itself and its neighbors. The loops are formed by central-force and bond-bending constraints.

As an example, consider a long floppy chain constructed such that along the mainchain, there are carbon atoms, each connected to two other carbon atoms along the mainchain, and two hydrogen atoms forming dangling ends, so that each carbon atom has four neighbors. Although the chain is floppy, having a rotatable dihedral angle between each pair of carbon atoms in the mainchain, it will carry stress from one end to the other! Again the reason for this effect is in modeling the local chemistry of a four coordinated covalently bonded atom with six angular constraints. By explicitly monitoring this type of overconstrained region, we can include or exclude this effect. This effect can be excluded by removing *any* one of the six angular constraints at each four fold coordinated atom in the model. For the remainder of this article we will not be concerned with this type of locally induced stress. Instead, we are interested in network induced overconstrained regions caused by loops formed by central-force constraints.

Rigid Cluster Decomposition

The vertices within each rigid cluster are divided into two types. A vertex is classified as a *bulk vertex* when all of its neighboring vertices also belong to the same rigid cluster, otherwise it is classified as a *surface vertex.* Thus a surface vertex has at least one neighbor belonging to a different rigid cluster than itself. Isolated vertices and the vertices within dimers are regarded as bulk vertices. This is a very useful classification scheme because in general a vertex can belong to more than one rigid cluster simultaneously, yet a bulk vertex can be assigned to one and only one rigid cluster. A unique rigid cluster labeling scheme can be constructed by labeling only the bulk vertices. This labeling scheme for rigid clusters is appropriate to three dimensional bond-bending networks[1,28], where it is possible to take advantage of some special properties. In general this labeling scheme will not work. For example, this labeling scheme cannot be used at all in two dimensional networks[28,31,32].

Each rigid cluster consists of a set of bulk vertices, all having the same *cluster label,* and a set of surface vertices all having a different cluster label than the bulk vertices. Any non-bulk vertex that is nearest neighbor to a bulk vertex for a particular rigid cluster is actually a surface vertex for that rigid cluster. There are two additional Properties about the rigid clusters within a bond-bending network[29] that will be needed here.

P1 For each vertex, v_1, all its neighbors are automatically mutually rigid with respect to one another as well as with vertex v_1 itself because of the bond-bending constraints.

P2 All vertices within a rigid cluster are connected via a path of central-force constraints.

Using the above properties, an algorithm to obtain the rigid cluster decomposition within bond-bending networks is given as:

1. Initialize rigid cluster counter. Label all vertices null.
2. For each vertex, v_1, not already associated with a rigid cluster:
   (a) If an isolated vertex: Increment counter and label vertex.
   (b) If the vertex belongs to a dimer: Increment counter, label both vertices.
   (c) If vertex is one fold coordinated; return to (2).
   (d) Rearrange pebbles: Collect 3 pebbles on vertex v_1, 2 pebbles on its 1st neighbor, v_2, and 1 pebble on its 2nd neighbor, v_3.
   (e) By property **P1**, use vertices $\{v_1, v_2, v_3\}$ as a rigid base, and place them in a stack defining the rigid cluster.
   (f) By property **P2**, use a breadth first search[33] via nearest neighbors, $\{v_{new}\}$, to grow the rigid cluster stack to completion.
      i. For each vertex v_{new}: Rearrange pebbles attempting to collect a pebble while holding the six pebbles on the rigid base.
      ii. If attempt fails: Include new vertex in the rigid cluster stack.
   (g) Increment cluster label: Label all bulk vertices within the rigid cluster.

After the above algorithm is finished, every bulk vertex will have a cluster label assigned ranging from 1 to the number of rigid clusters in the network. One fold coordinated vertices are regarded as bulk vertices.

Identifying Underconstrained Regions

After the rigid cluster decomposition is made, locating hinge joints is an easy task. A hinge joint can never occur about a bond-bending distance constraint[29]. To find all the hinge joints within the network, one needs to check the two incident vertices associated with each central-force distance constraint. If the two vertices have different cluster labels, then a dihedral rotation is possible and the central-force constraint is a hinge joint, otherwise the dihedral angle motion is locked, as it is part of a rigid cluster.

The number of hinge joints will generally be more than the number of residual internal degrees of freedom in the network. This means that not all the rotatable dihedral angles associated with the hinge joints are *independent.* We call a hinge joint independent if its dihedral angle can be changed without affecting other dihedral angles within the network. Collective motions will take place in underconstrained regions within the protein. It is convenient to partition hinge joints into distinct underconstrained regions, which may include only one independent hinge joint.

Let the set of dihedral angles associated with the hinge joints define a reference set of internal coordinates, not all of which are independent. We will partition the independent degrees of freedom of the network into subsets, which correspond to underconstrained regions. In a similar way that there are redundant constraints within overconstrained regions, there are floppy modes within underconstrained regions. Physically, collective motions can occur within a particular underconstrained region without affecting internal coordinates outside of the region. The underconstrained regions are identified by attempting to specify a value for each dihedral angle. Specifying a dihedral angle is equivalent to placing an external torsional constraint to lock in this choice of angle. Independent externally imposed torsional constraints represent independent degrees of freedom available to the system, while redundant externally imposed constraints indicate the angle is predetermined as part of a collective motion. Therefore, the algorithm for identifying the underconstrained regions is given by:

1. Initialization: Record null Laman subgraphs by regarding all previously identified redundant constraints within the network as not present.

2. Continue to play the *pebble game* by placing a third nearest neighbor distance constraint to lock the dihedral angle for each hinge joint in the network.

3. After completion: The set of newly formed Laman subgraphs identifies the underconstrained regions.

After all torsional constraints are placed on each hinge joint, a completely connected network will become totally rigid having six degrees of freedom that are represented by six remaining free pebbles. The Laman subgraphs within this bond network now define the underconstrained regions, which consist of loops involving central-force bonds. The number of independent dihedral angles within an underconstrained region is given by the number of torsional constraints covered by a pebble. Independent hinge joints are associated with a covered torsional constraint not belonging to a Laman subgraph.

FLOPPY INCLUSION AND RIGID SUBSTRUCTURE TOPOGRAPHY

A large amount of detailed information about the mechanical stability of a protein under a fixed set of distance constraints becomes available from a FIRST analysis. All overconstrained regions, rigid clusters and underconstrained regions are determined. We discuss in detail the output of such a FIRST analysis for a hand made 32 atom example shown in Figure 5, which ties together the discussion of the previous section. In addition to partitioning a structure into different types of regions, a continuous measure for the degree of stability is introduced and is worked out for the network shown in Figure 5. The FIRST analysis is then applied to protein structure, where the floppy inclusion and rigid substructure topography is described by a one dimensional representation in terms of what we define as a *stability index.*

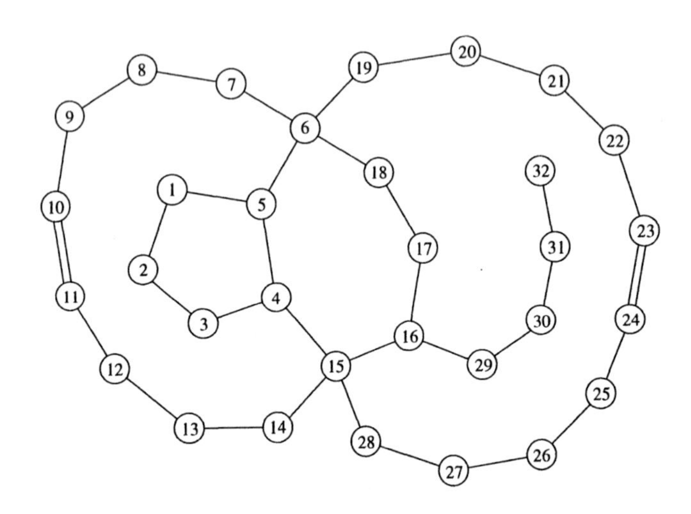

Figure 5. A simple hand made 32 atom network. Each atom is labeled from 1 to 32. Single lines between labeled atoms represent central-force distance constraints that have a priori rotatable dihedral angles, while the double lines represent double bonds where the dihedral angle is locked.

3D Graphical Display

Constructing a three dimensional graphical display for the different types of mechanical regions amounts to devising a systematic labeling scheme. We take our basic set of reference labels to be the atom numbers. For the network shown in Figure 5, it is possible to obtain the complete solution for the floppy inclusion and rigid substructure topography by inspection. Each part of the solution will be discussed in turn.

Recall that overconstrained regions are identified within the *3D Pebble Game* as a result of failed pebble searches. Each time a failed pebble search is encountered a Laman subgraph is identified, which contains more distance constraints than needed to make that region rigid. A convenient labeling system for identifying Laman subgraphs is to initially assign each atom a Laman subgraph label. The atom numbers are used as the *initial* set of Laman subgraph labels.

When a Laman subgraph is uncovered from a failed pebble search, all atoms found within are re-assigned the lowest Laman subgraph label encountered. This allows Laman subgraphs to naturally merge together. Since Laman subgraphs (representing overconstrained regions) consist of loops, stress will reside in the central-force bonds connecting pairs of atoms having the same Laman subgraph label. Loops of atoms with

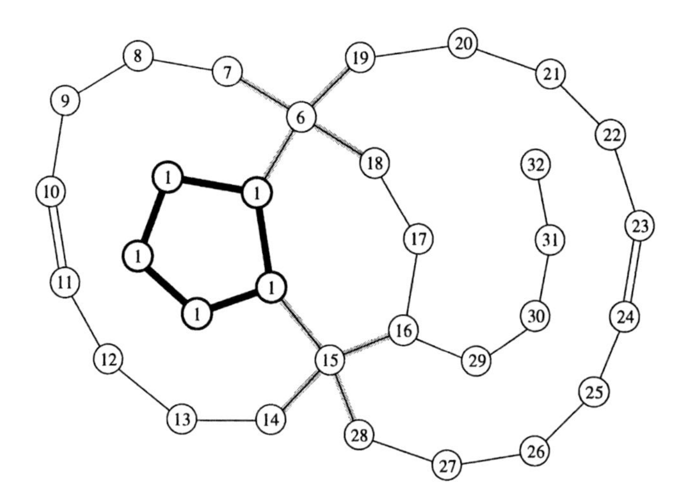

Figure 6. The FIRST analysis for internally stressed regions. Each atom is assigned a Laman subgraph label. The thick solid black lines show a network induced overconstrained region where there will be internal stress. In addition, there are two localized regions that are internally stressed (highlighted using thick grey lines) each associated with the presence of a four-coordinated atom.

the same Laman subgraph label define a particular overconstrained region. As shown in Figure 6 only one network induced overconstrained region is present, consisting of a five fold ring with one redundant constraint. Since the incident atoms about all other central-force bonds are assigned different Laman subgraph labels, no network induced stress is present anywhere else.

Induced stressed regions, localized around four (or higher) coordinated atoms, can also be monitored. Four coordinated atoms, such as atoms 6 and 15, produce a local Laman subgraph with one redundant constraint within a set of edge sharing loops consisting of the central and bond-bending constraints associated with the four coordinated atom and its nearest neighbor atoms. The Laman subgraph labeling scheme works for both kinds of induced stressed regions, where the former is identified by checking the atom coordination number. We eliminate this effect by removing one of the six angular constraints, chosen arbitrarily, at each four fold coordinated site.

The coloring of rigid clusters is associated with the bulk atoms, each assigned a particular color according to its cluster label. The assignment of colors is restricted by the requirement that two nearest neighbor bulk atoms having different cluster labels must have different colors. As a general property of bond-bending networks, each central-force bond belongs to either two rigid clusters when it is a hinge joint[29], or one rigid cluster otherwise. Therefore, the first half of each central-force bond stemming from a bulk atom is half colored. A central-force bond is completely colored only when both incident atoms have the same cluster label. As a result, only hinge joints appear as half colored bonds.

Empirically we find that a complete color assignment is possible using as little as four colors, although we generally use more colors by having a dark and bright set of colors to denote rigid and floppy regions respectively. The relationship between uniquely coloring the rigid cluster decomposition and the chromatic coloring of a graph, such as the famous four color problem on planar graphs[30,34] has not been investigated.

In Figure 7, we show the rigid cluster decomposition of our example 32 atom

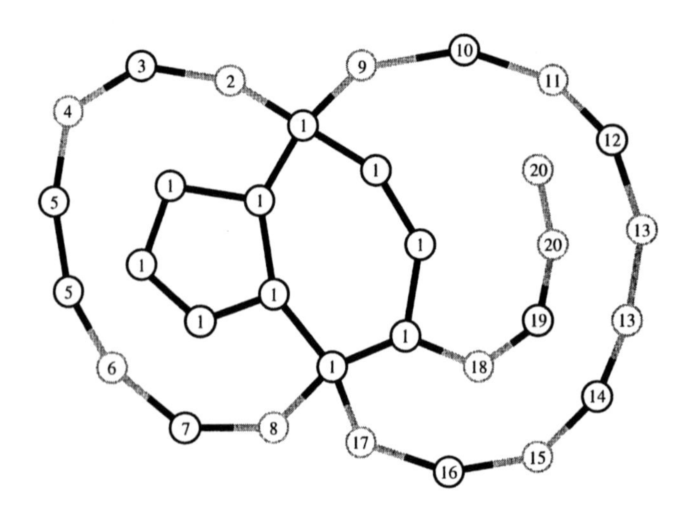

Figure 7. The FIRST analysis for the rigid cluster decomposition. Each atom is assigned a cluster label. Each rigid cluster is assigned a color, with the constraint that neighboring clusters must be assigned a different color. Half colored bonds represent hinge joints.

network. Here we are able to completely color the rigid cluster decomposition using only two colors (ie. black and grey). Comparing Figure 5 with Figure 7 indicates that rigid cluster #1 consist of 14 atoms (bulk atoms $\{1, 2, 3, 4, 5, 6, 15, 16, 17, 18\}$ and surface atoms $\{7, 14, 19, 28\}$), rigid cluster #2 consist of 3 atoms (bulk atom 7 and surface atoms $\{6, 8\}$), rigid cluster #5 consist of 4 atoms (bulk atoms $\{10, 11\}$ and surface atoms $\{9, 12\}$), etc. Within rigid cluster #1, only the five fold ring is overconstrained as shown in Figure 6 while the seven fold ring is isostatically rigid. Each half colored bond identifies a hinge joint where its dihedral angle can be rotated, but this does not indicate the inter dependence of rotating a set of dihedral angles in the floppy regions.

Each underconstrained region is displayed with a unique color to indicate an interdependent set of hinge joints. The labeling scheme is applied directly to the hinge joints, where each hinge joint is *initially* assigned a distinct label representing the underconstrained regions. We denote the hinge joint between atoms i and j as $H_{i,j}$. As external torsional constraints are placed on each hinge joint, all H_{ij} found within a failed pebble search are re-assigned the lowest common label encountered. Therefore, underconstrained regions merge together whenever failed pebble searches have overlapping H_{ij}. Only central-force bonds that were originally hinge joints (labeled by the set $\{H_{ij}\}$ and actually are hinge joints within the protein) are colored. Therefore, underconstrained regions are generally not contiguous.

In Figure 8 it can be seen that the 32 atom example network has two large underconstrained regions and three other underconstrained regions, each consisting of one independent hinge joint. The left underconstrained region consist of 8 hinge joints, each colored black, while it has only two independent dihedral angles indicated by the circular arrows. Comparing with Figure 7 it is seen how intermediate rigid clusters #1 and #5 cause the underconstrained region to be non-contiguous. Similarly, the underconstrained region on the right is also non-contiguous. It consist of 10 hinge joints, shown in grey, while it has four independent dihedral angle rotations. The three independent hinge joints are indicated only by circular arrows.

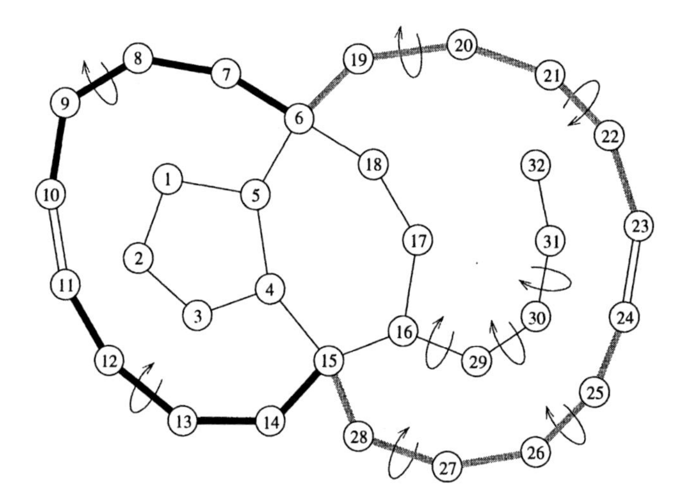

Figure 8. The FIRST analysis for the underconstrained regions. Each atom is numbered from 1 to 32. Two underconstrained regions are found. The black thick lines on the left represent one underconstrained region consisting of two floppy modes extending over eight hinge joints. The grey thick lines on the right represent another underconstrained region consisting of four floppy modes extending over ten hinge joints. The nine arrows represent internal independent dihedral angles. All central-force bonds that are not part of the two identified underconstrained regions or have a circular arrow have locked dihedral angles. Three independent hinge joints are also shown in the section from 16 to 32.

The independent dihedral angles within an underconstrained region correspond to free pebbles in the the *3D pebble game.* The choice of which hinges are taken as independent is arbitrary, and is analogous to choosing which bonds are redundant within an overconstrained region.

In general, rigid clusters will subdivide a floppy region into multiple underconstrained regions, where independent collective motions can take place. Figure 8 shows an example of how rigid cluster #1, shown in Figure 7, partitions free degrees of freedom into two localized regions. It is seen that once the two independent dihedral angles are specified in the underconstrained region on the left, the relative position of atoms $\{8, 9, 10, 11, 12, 13\}$ are all fixed relative to rigid cluster #1. The specification of these two angles, however, has no affect on the relative position of atoms $\{20, 21, 22, 23, 24, 25, 26, 27\}$ within the underconstrained region on the right. Notice that if atoms $\{1, 2, 3\}$ were removed from the network, rigid cluster #1 would fall apart. Moreover, the two underconstrained regions would merge together with the seven fold ring to form a single underconstrained region with 25 hinge joints and 11 independent dihedral angles. This illustrates how a local structural change can have important consequences to the overall mechanical stability of the network.

Quantifying the Degree of Flexibility

With the above system of coloring, we are able to make a 3D multi-color map for overconstrained regions, the rigid cluster decomposition and the underconstrained regions. Although it is insightful to view the three dimensional multi-color rendering of these regions, one must exercise caution not to be mislead about the stability within a protein.

Modeling only a few additional hydrogen bonds as distance constraints will often make some floppy regions (or parts within) rigid. Likewise, the removal of distance constraints associated with just a few hydrogen bonds will sometimes make some rigid regions break apart like a house of cards. Modeling just a *few* more or less hydrogen bonds as distance constraints can sometimes dramatically change the mechanical stability of a protein. However these observed changes in the 3D graphical displays do not necessarily reflect the actual change in mechanical stability that is occurring physically.

A floppy region consisting of many interconnected rigid clusters may define a collective motion having only a few independent degrees of freedom. This floppy region, although underconstrained, would be quite stable mechanically as it is nearly rigid. An isostaticly rigid region is not expected to be as stable as an overconstrained region. For this reason a continuous index of stability is useful. As the hydrogen bond selection criteria is relaxed using a hierarchical protocol, the observed changes in the mechanical stability of the protein can be tracked in a continuous fashion.

The total number of floppy modes in a protein, denoted by F, corresponds to the number of *internal* independent degrees of freedom. To obtain F, the six trivial rigid body degrees of freedom must be subtracted out from the total number of independent degrees of freedom. For purposes of simplifying the discussion, it has been assumed that all atoms in the protein are connected via covalent bonding. The global count of the number of floppy modes gives a good sense of intrinsic flexibility. However, a better measure for the degree of floppyness can be obtained by tracking how the total number of floppy modes are spatially *distributed* throughout the protein. In particular we are interested in locating underconstrained regions and the number of floppy modes contained within each of these regions.

Regions containing more constraints are regarded as being more stable than regions with less constraints. Overconstrained regions have more constraints than necessary to be rigid, and therefore are considered to be more stable. A global count for the number of redundant constraints, denoted by R, gives a sense of the overall stability of a protein. However, a better measure for the degree of stability can be obtained by tracking how the total number of redundant constraints are distributed throughout the protein. In a similar way as done with the floppy modes, we will be interested in where the overconstrained regions are located and count the number of redundant constraints present in each region.

We will define a quantity s_i, as a *stability index* characterizing the i-th central-force bond in the protein. Let H_k and F_k respectively denote the number of hinge joints and the number of floppy modes (internal independent degrees of freedom) within the k-th underconstrained region. Let C_j and R_j respectively denote the number of central-force bonds and the number of redundant constraints within the j-th overconstrained region. Combining a quantitative measure for both the degree of floppyness and stability associated with the distribution of floppy modes and redundant constraints respectively, the definition for the stability index is given by:

$$s_i \equiv \begin{cases} \frac{-F_k}{H_k} & \text{in an underconstrained region} \\ 0 & \text{in an isostatically rigid region} \\ \frac{R_j}{C_j} & \text{in an overconstrained region.} \end{cases} \tag{1}$$

When the i-th central-force bond is a hinge joint, the stability index is defined to be a negative quantity with magnitude given by the the number of floppy modes divided by the total number of hinges within the underconstrained region. The number of floppy modes correspond to the number of independent dihedral angle rotations that can be made within the underconstrained region. Since the number of independent dihedral angle rotations must be less than or equal to the number of hinge joints, the stability index can never be less than -1. For an independent hinge joint the stability index works out to be -1. Furthermore, the stability index for a hinge joint must always be less than zero because there will always be at least one floppy mode within an underconstrained region.

When the i-th central-force bond is not a hinge joint, it must be part of a rigid cluster, although it may or may not be part of an overconstrained region. If the central-force bond is within an overconstrained region, the stability index is assigned a positive value given by the number of redundant constraints divided by the total number of central-force bonds within the region. Although there is no useful upper bound for the stability index it rarely goes above unity and generally stays below 1/2.

A stability index of zero is assigned to central-force bonds within isostatic rigid regions, because this is the limiting value for an underconstrained region that is nearly rigid and for an overconstrained region that is nearly isostatic. The intrinsic flexibility of a protein can be studied using the stability index within a one-dimensional representation. Here, the stability index for central-force bonds along the mainchain of a protein, where a plot is made against residue number along the protein sequence, gives a good sense for the mechanical stability.

The number of independent degrees of freedom and redundant constraints within the protein vary gradually as the number of hydrogen bond constraints is varied. Likewise, the stability index also varies gradually when hydrogen bond constraints are added or removed. It is worth mentioning that two independent sum rules result directly from the definition of the stability index. It follows that adding the stability index over all hinge joints sums to the negative of the total number of floppy modes, while summing over all non-hinge joints results in the total number of redundant constraints.

As a simple example, consider a single n-fold ring of atoms that are connected by covalent bonds. From simple constraint counting, the number of degrees of freedom less the number of constraints is given by $F = n - 6$. The number of hinge joints (and central-force bonds) is simply given by n. Therefore the stability index for a n-fold ring is given by

$$s_i = \frac{6-n}{n} \quad \text{for each central-force bond in a } n\text{-fold ring.} \tag{2}$$

Notice that as the ring becomes very large, the stability index goes to the limit of -1, as each dihedral angle is nearly independent and is almost as flexible as a linear chain. For a six fold ring, the stability index is zero, indicating that the generic ring structure is isostatically rigid. For a three fold ring, the stability index works out to be $+1$ and is highly overconstrained. Note that for the case of a n-fold ring, the stability index is bounded between -1 and 1.

As another example, consider once again the hand made 32 atom network. From Figure 8 the assigned value for the stability index to each hinge joint within the black and grey underconstrained regions are $-1/4$ and $-2/5$ respectively, indicating that the underconstrained region on the right (grey) is less stable, or more floppy. Each independent hinge joint has a stability index of -1. From Figure 6 the stability index for each central-force constraint within the network induced overconstrained region is 1/5. All other central-force bonds, including the resonant bonds, have a stability index of zero.

The Alpha Helix

The alpha helix is a secondary structure[18] that is frequently found in proteins, and they are generally regarded as rigid substructures. The structure of an alpha helix is that of a helix with mainchain hydrogen bonds forming between every fourth consecutive residue as shown in Figure 9. Each turn, through 2π along the alpha helix consists on average of 3.6 residues.

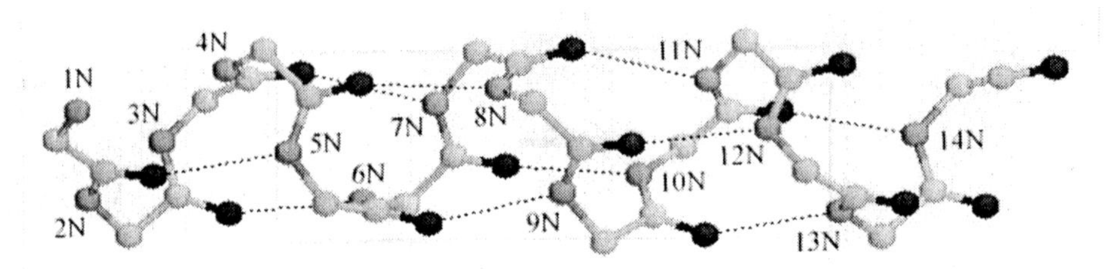

Figure 9. An example of an alpha helix made from 14 glycine residues. The hydrogen bonds are shown as thin black dashed lines.

We will analyize the intrinsic mechanical stability of an alpha helix using our distance constraint approach. Underconstrained, isostaticly rigid and overconstrained regions can easily be calculated because of the linear repeating pattern of the mainchain hydrogen bonds. The stability index will be worked out for an alpha helix made up of entirely glycine residues because we are interested in understanding the effects of the crosslinking hydrogen bonds along the mainchain. The same effects will be present for any alpha helix substructure within a protein regardless of its constituent residue types, although other rigiditifying affects may result from (additional) sidechain or end cap hydrogen bonding.

The internal independent degrees of freedom (floppy modes) and redundant constraints will be counted explicitly by hand. The distance constraints associated with the covalent bonding between atoms along the mainchain allow two free dihedral angles within each residue as shown in Figure 10. Recall that the dihedral angles associated with peptide bonds are locked by third nearest neighbor distance constraints. The two free dihedral angles are conventionally referred to as the $\{\phi_i, \psi_i\}$ angles within the i-th residue. The number of floppy modes for the alpha helix is given by

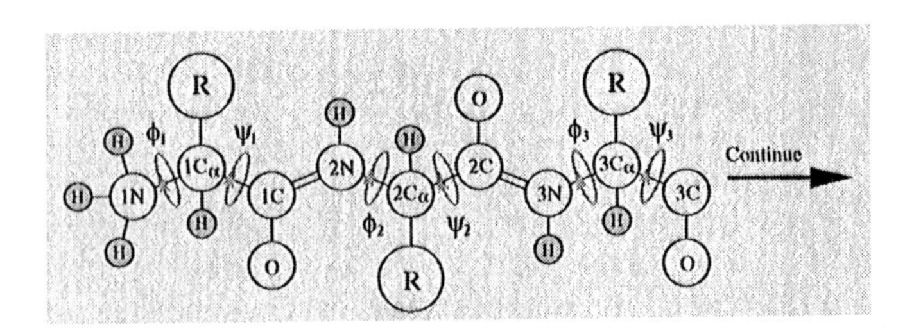

Figure 10. A simple illustration for the mainchain of a protein. Double lines represent the peptide bond for which we do not allow any dihedral rotation. There are two a priori free dihedral angles denoted by ϕ and ψ respectively per residue. In our simple alpha helix, all residues (denoted by R) are glysine, which is equivalent to replacing R by a hydrogen atom.

$$F(n) = (6 + 2n) - 3N_h - 6 \tag{3}$$

where $(6 + 2n)$ is the number of degrees of freedom for n residues, $3N_h$ is the number of crosslinking distance constraints (three per hydrogen bond), and the 6 trivial rigid body motion degrees of freedom are subtracted out.

The regular repeating pattern of the alpha helix gives $N_h = n - 4$ hydrogen bonds for $n \geq 4$ residues. Since we want the alpha helix to have at least one turn, we will only consider an alpha helix consisting of four or more residues. The hydrogen bond pattern does not crosslink the very ends of the alpha helix. As a result, two independent dihedral angle motions are allowed at both ends of the alpha helix regardless of its length. The region that is of interest to us is along the main body of the helix. Depending on the number of residues in the alpha helix, the region of interest will form one underconstrained, isostatically rigid, or overconstrained region.

The two independent hinge joints at each end of the helix do not belong to the region of interest and must be accounted for separately. To prove that the main body of the alpha helix does not break into multiple regions (assuming all hydrogen bonds are present) requires one to perform an all subgraph constraint counting exercise either manually or with the aid of the FIRST program. Since the structure is quite regular, one can prove this by combining inspection with mathematical induction as an alpha helix is built up one residue at a time.

A global count in the number of floppy modes within a n-residue alpha helix is given by $F(n) = 12 - n$ where Eq. 3 has been used. However, there are five distinct regions within the alpha helix, consisting of the two hinge joints at each end with independent dihedral angles and the single region of interest within the main body. Therefore, the number of floppy modes within the main body of a n-residue alpha helix, $F_\alpha(n)$, is given by:

$$F_\alpha(n) = 8 - n \quad \begin{cases} > 0 \text{ for } n < 8 & \text{an underconstrained region} \\ = 0 \text{ for } n = 8 & \text{an isostaticly rigid region} \\ < 0 \text{ for } n > 8 & \text{an overconstrained region} \end{cases} \tag{4}$$

The rigid cluster decomposition is shown in Figure 11 for a seven residue alpha helix. From Eq. 4 the main body of the alpha helix is floppy with one floppy mode, describing a twisting motion along the axis. In this floppy region there are 21 interconnected rigid clusters that form an underconstrained region consisting of nineteen hinge joints, which is also shown in Figure 11. For a n-residue alpha helix, the number of hinge joints in the region of interest can be counted as

$$H_\alpha = 2n - 4 + 3N_h \quad = \quad 5n - 16 \tag{5}$$

where there are $2n$ hinge joints consisting of the ϕ and ψ rotatable dihedral angles in total. Four hinge joints correspond to the two independent hinge joints at each end of the helix, and recall there are 3 dihedral angles associated with each hydrogen bond as shown in Figure 3.

The stability index for each of the rotatable dihedral angles within the main body of the alpha helix (ie. not those associated with the peptide bond) is given by $s_\alpha = -F_\alpha/H_\alpha = -1/19$. The stability index associated with each of the central-force bonds in the peptide bond is zero because the third neighbor distance constraint

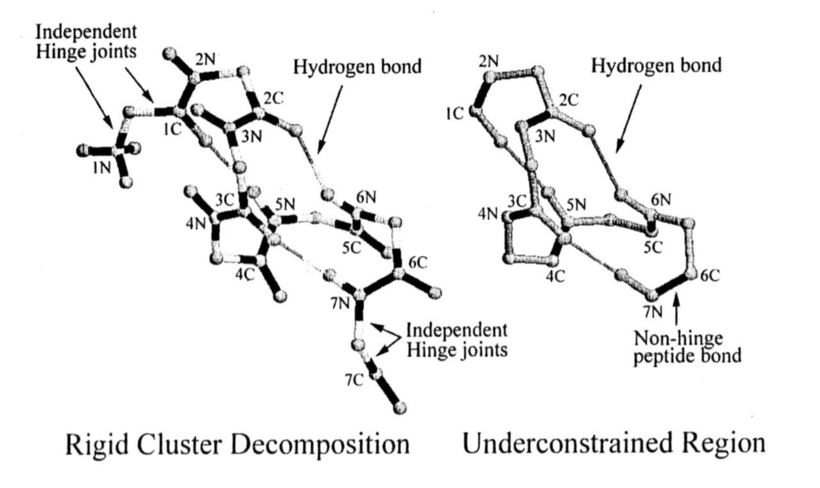

Figure 11. The rigid cluster decomposition (right panel) and the underconstrained region (left panel) along the main body of an alpha helix of 7 residues is shown.

used to lock the dihedral angle makes a small isostatic rigid cluster. A one dimensional representation of the degree of flexibility along the mainchain of the seven residue alpha helix is shown in Figure 12. A topological representation of the corresponding rigid cluster decomposition and the main body underconstrained region is also shown. Unlike the 32 atom example network worked out earlier, a protein structure can be unfolded into a linear covalent polymer chain and sequenced according to residue in a one dimensional fashion, otherwise known as the primary structure. Therefore, in Figure 12 we plot the degree of flexibility for the central-force bonds along the mainchain of the protein structure against residue number.

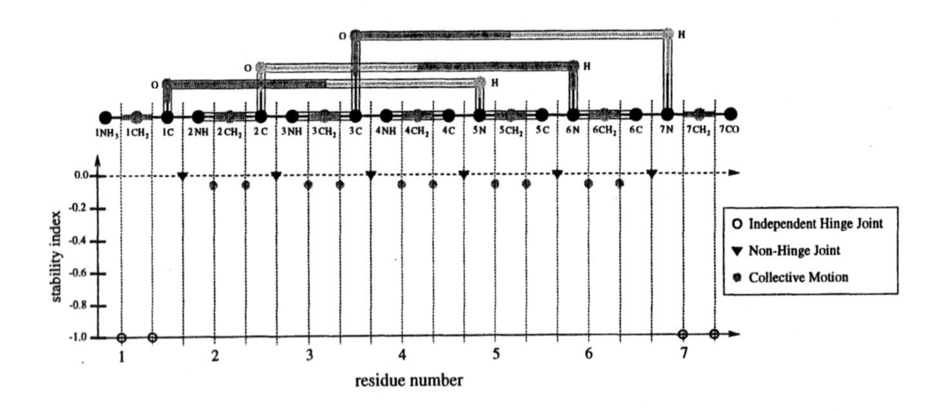

Figure 12. The stability index is shown along the mainchain of the 7 residue alpha helix. To show the correspondence clearly, a schematic diagram for the rigid cluster decomposition (black, grey and light grey regions) and the single underconstrained region (outlined by solid lines) are also shown.

From Eq. 4 it is seen that the main body of the alpha helix will be rigid with

$(n - 8)$ redundant constraints for an alpha helix having eight or more residues. In this region, the total number of central-force constraints is given by

$$C_\alpha = H_\alpha + (n-1) \;\; = \;\; 6n - 17 \tag{6}$$

where H_α (from Eq. 5) is used because it represents the number of central-force constraints associated with a priori rotatable dihedral angles, and the addition of $(n - 1)$ corresponds to the number of (non-rotatable) peptide bonds between residues. Unlike within underconstrained regions where hinge joints are counted, we count all central-force bonds within a rigid region. Using Eq. 1, the stability index for each central-force bond associated with the $\{\phi, \psi\}$ angles along the mainchain and within the main body of a n-residue alpha helix works out to be

$$s_\alpha(n) = \begin{cases} \frac{n-8}{5n-16} & \text{for } 4 \leq n \leq 8 \\ \frac{n-8}{6n-17} & \text{for } n > 8. \end{cases} \tag{7}$$

It is worth noting that for long alpha helices (ie. in the limit $n \rightarrow \infty$), the stability index approaches 1/6 corresponding to one redundant constraint per residue.

Lysine/Arginine/Ornithine-Binding Protein

The lysine/arginine/ornithine binding protein is a bacterial periplasmic protein found in *Salmonella typhimurium*, and is responsible for the transport of various substrates (or ligands) such as lysine from the periplasm to the cytoplasm[35]. The X-ray crystal structures of the lysine binding protein in the closed conformation with lysine bound and in its open conformation without a ligand are known[35] to within 1.9 Å. This protein has two globular domains connected by a double flexible linkage, which allows a large-scale hinge-type ("clam shell") movement of its two domains[6,14]. Each globular domain essentially define the two halfs of the "clam shell" separated by loop hinges as defined by the crystallographers[35]. Domain 1 contains residues between 1-87 and 195-237, consisting of parts of the N- and C-termini. Domain 2 contains all residues between 94-181.

By comparing the ϕ and ψ mainchain dihedral angles in the protein structure in its closed and open conformations, given in the *1lst* and the *2lao* PDB files, (from the Brookhaven Protein Data Bank[2] of atomic resolution macromolecular structures) it is ascertained that the large-scale domain motion is localized to a few dihedral angles within a flexible linkage consisting of two loops[35]. From kinetic studies[36], the time scale for the opening and closing of the structure is found to be 10 nanoseconds without the substrate and 10 miliseconds with the substrate – a difference of six orders of magnitude. We selected this protein as an initial test case in applying FIRST to predict intrinsic flexibility.

Hydrogen atoms are generally not resolved in a protein structure derived from crystallography, because their low electronic densities do not contribute much to the scattering of X-rays. As we consider generic networks, the exact position of the hydrogen atoms is not important. However, the placement of hydrogen atoms is useful for identifying hydrogen bonds with good geometry[37] within the protein structure. Therefore, we use the WhatIf software[38] for geometrically correct optimal placement of hydrogen atoms for hydrogen bonding.

Once the hydrogen atom positions are explicitly provided, we model hydrogen bonds as distance constraints within the bond network analyzed by FIRST, provided they satisfy the following criteria:

1. The hydrogen bond is intra-molecular.

2. The donor-hydrogen-acceptor angle must be greater than 120 degrees.

3. Either the donor-acceptor distance is less than 3.5 Å or the hydrogen-acceptor distance is less than 2.5 Å.

It is possible to consider additional hydrogen bonds that form between the protein and solvent (water) molecules, and between the protein and other ligands. However, here we will focus on the effect of intra-protein hydrogen bonds on rigidity.

In Figures 13 and 14, we show the one dimensional representation for the intrinsic flexibility of the lysine-binding protein in the closed conformation, with lysine bound, and in the open conformation without lysine. In both figures, the stability index associated with the $\{\phi_i, \psi_i\}$ angles in each residue along the mainchain is plotted against residue number. The one dimensional representation alone gives a clear overview of the rigid and flexible regions, and can be compared directly with the three-dimensional molecular graphics of those regions.

It can be seen that the gross features in mechanical stability of each conformation are the same. Most of the rigid substructures and underconstrained regions identified in the two conformations correspond to one another.

Domain 1, consisting of both the N- and C-terminus of the lysine binding protein, is seen to be mechanically more stable than domain 2 consisting of residues between 94 to 181 (between the inner two vertical dashed lines). Both conformations show that domain 1 is almost completely rigid and made up almost entirely by rigid cluster R1. Domain 2 contains a large underconstrained region (labeled as U2) with several rigid substructures (labeled as R2, R3, ... R8) that are mainly alpha helices.

For the same hydrogen bond criteria, it is seen that the intrinsic flexibility is greater for domain 2 than that for domain 1. The large underconstrained region (labeled as U2) consists of the known double flexible linkage, residues 88-93 and 182-194, in addition to a beta sheet in domain 2. Thus, domain 2 apparently does not have enough sidechain hydrogen bonding to rigidify the beta-sheet together with other rigid substructures to form a large rigid domain. Instead, the various rigid substructures move collectively. The underconstrained region U2 decreases in size as the hydrogen bond selection criteria are relaxed and more hydrogen bonds are modeled as constraints. Eventually, the beta-sheet region within domain 2 rigiditifies in both the closed and open conformations, making the floppy motion in the double flexible linkage more pronounced. These results taken together suggest that during conformational changes of the lysine binding protein, some motion will take place within domain 2.

Differences found in the stability index between the closed and open conformations are caused predominately by differences in hydrogen bonding. This must be the case since the covalent bonding is identical in each conformation with the sole exception of the presence of the lysine substrate in the closed conformation, which itself is non-covalently bound. It is important to discriminate between real physical differences as a result of the presence of the lysine substrate, in contrast to the sensitivity of our approach in selecting a particular set of hydrogen bonds that are modeled as distance constraints. We consider differences as physical when the results are robust against small perturbations in the hydrogen bond selection criteria.

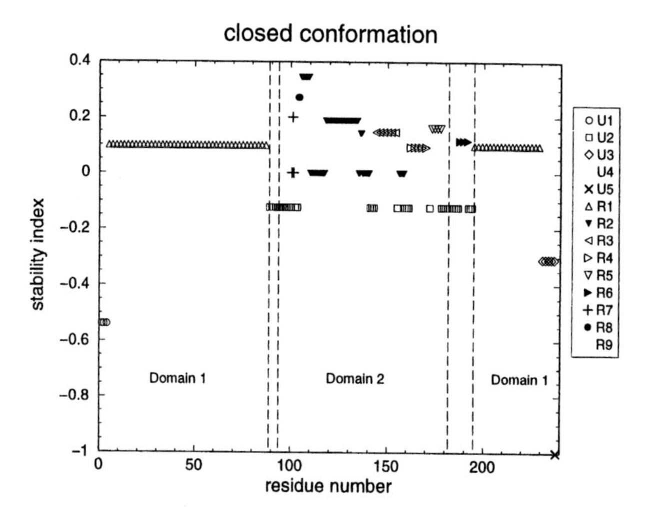

Figure 13. The stability index associated with the ϕ and ψ dihedral angles (hinge joints) along the mainchain is shown for the lysine binding protein in its closed conformation (PDB code 1lst). In the legend on the right, U1-U5 represent underconstrained regions and R1-R9 represent rigid substructures that FIRST identified. The labeling scheme and choice of symbols used for the underconstrained regions and rigid substructures are consistent with the open conformation shown below to aid in direct comparisons. Note that a single rigid region (eg. R1) can be discontinuous in sequence number along the mainchain. The two sets of dashed lines locate the double flexible linkage between residues 88-93 and 182-194

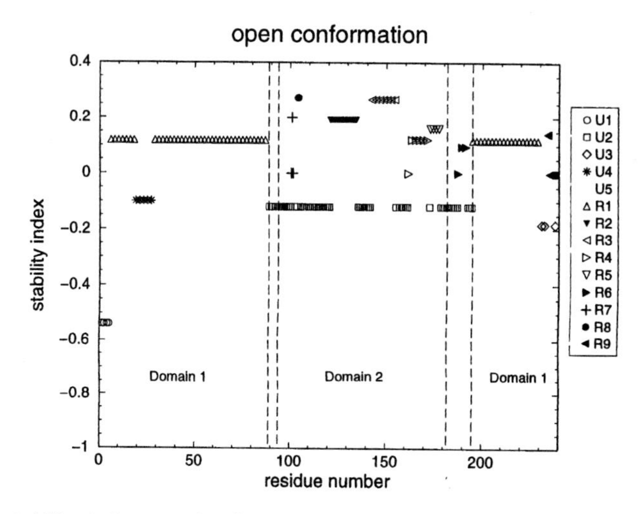

Figure 14. The stability index associated with the ϕ and ψ dihedral angles (hinge joints) along the mainchain is shown for the lysine binding protein in its open conformation (PDB code 2lao). The legend on the right and the dashed lines are the same as used above for the closed conformation.

There is one main difference in the intrinsic flexibility between the closed and open conformations that is worth mentioning here. This difference is associated with the underconstrained region involving the double flexible linkage (labeled U2) that extends into domain 2. There is also a corresponding difference in a rigid substructure (labeled as R2). In both conformations rigid substructure R2 consists of a relatively stable region that is overconstrained. In the closed conformation it has some additional regions that are (mostly) isostaticly rigid to the stable base, while in the open conformation these regions remain floppy and are part of underconstrained region U2. Thus in the open conformation, underconstrained region U2 is larger in size, consisting of 300 rotatable dihedral angles from which only 36 are independent. In the closed conformation underconstrained region U2 has 219 rotatable dihedral angles from which only 27 are independent. Note that the count in dihedral angles include the sidechain covalent bonding as well as hydrogen bonding.

As the hydrogen bond selection criteria is relaxed or tightened slightly, it is observed that underconstrained region U2 in the closed conformation is always smaller than that in the open conformation. We therefore suspect that this result is a robust signature for the presence of the lysine substrate, and that domain 2 is mechanically more stable when a substrate is present. This rigiditifying effect from the substrate could be a contributing factor in explaining why the time scale is six orders of magnitude slower for the hinge motion when the substrate is present. This is an exciting prospect as this suggests the possibility of identifying allosteric effects directly from characterizing the mechanical properties of a structure.

In our preliminary investigation on the lysine-binding protein, the FIRST analysis already has provided new insights as well as correlated with known motions within the protein. The next step is to implement the hierarchical protocol in modeling hydrogen bonds as distance constraints from strongest to weakest using a hydrogen bonding energy term.

SUMMARY

In this article, we have introduced a novel distance constraint approach for characterizing the intrinsic flexibility of a protein. The conceptual framework is to develop a hierarchical protocol for adding more distance constraints that model successively weaker bonding forces. The underlying physical and mathematical assumptions behind this idea were layed out. We then focused on the first step toward our goal of developing a robust automated program, FIRST, that can characterize the mechanical stability of a protein. Namely, we have shown the application of concepts from graph rigidity to identify a series of mechanical properties of a protein structure under a given set of bonding forces modeled as distance constraints.

The initial implementation of FIRST determines the floppy inclusion and rigid substructure topography of a given protein structure based on a single set of distance constraints as determined by covalent and hydrogen bonding. The *stability index* is introduced as a *continuous* measure for quantifying the local mechanical stability of each dihedral angle. When the measure falls between $(-1, 0)$, this physically corresponds to a dihedral angle within an underconstrained region that is anywhere from being completely independent to nearly isostatically rigid. A zero value indicates that the dihedral angle is within an isostaticly rigid region. A stability index that is greater than zero physically corresponds to a dihedral angle within an overconstrained region, where stability is enhanced as the number of redundant constraints within the region increases.

There are several advantages of FIRST relative to previous methods for analyzing protein flexibility. The FIRST analysis requires calculations that can be done in real-time (fraction of a second). For a given set of distance constraints the rigidity is calculated exactly, which includes identifying overconstrained and underconstrained regions in addition to obtaining a rigid cluster decomposition. The ability to determine underconstrained regions gives the FIRST analysis a distinctive advantage over other methods, because these regions localize important collective motions, whereby changing one dihedral angle will generally change other dihedral angles within the same underconstrained region. The stability index is introduced as a continuous measure to give meaningful one-dimensional representations for the mechanical stability of a protein structure, and this has already provided new insights on the lysine-binding protein.

Acknowledgements

The work by DJ and MT was supported in part by NSF grant DMR-96 32182, the work by LK and MT was supported in part by The Center for Protein Structure, Function, & Design at Michigan State University. DJ and MT would also like to thank John Grinstead for help in working out the stability index for regular secondary structures.

REFERENCES

1. D.J. Jacobs and M.F. Thorpe. US Patent pending: *Computer-implemented System for Analyzing Rigidity of Substructures Within a Macromolecule* (1998)
2. www.pdb.bnl.gov is the electronic data base. Also see, E.E. Abola, F.C. Bernstein, S.H. Bryant, T.F. Koetzle and J. Weng, *Protein Data Bank* 107-132, in *Crystallographic Databases - Information Content, Software Systems, Scientific Applications*, Editors: F.H. Allen, G. Bergerhoff and R. Sievers, (Data Commission of the International Union of Crystallography) (1987)
3. D. Ringe and G.A. Petsko, *A Consumer's Guide to Protein Crystallography* 210-229, in *Protein Engineering and Design*, Editor: P.R. Carey. (Academic Press) San Diego (1996)
4. W.S. Bennett and R. Huber, *Crit. Rev. Biochem.* **15**, 291 (1984)
5. K. Wuthrich and G. Wagner, *Trends Biochem. Sci.* **3**, 227 (1978)
6. M. Gerstein, A.M. Lesk and C. Chothia, *Biochem.* **33**, 6739 (1994)
7. W.L. Nichols, G.D. Rose, L.F. Ten Eyck and B.H. Zimm, *Proteins* **23**, 38 (1995)
8. A.S. Siddiqui and G.J. Barton, *Protein Sci.* **4**, 872 (1995)
9. N.S. Boutonnet, M.J. Rooman and S.J. Wodak, *J. Mol. Biol.* **253**, 633 (1995)
10. L. Holm and C. Sander, *Proteins* **19**, 256 (1994)
11. M.H Zehfus and R.D. Rose, *Biochem.* **25**, 5759 (1986)
12. P.A. and G.E. Schulz, *Naturwissenschaften* **72**, 212 (1985)
13. B.W. Matthews, *Ann. Rev. Biochem.* **62**, 139 (1993)
14. V. Maiorov and R. Abagyan, *Proteins* **27** 410 (1997)
15. Y. Bourne, M.H. Watson, M.J. Hickey, W. Holmes, W. Rocque, S.I. Reed and J.A. Tainer, *Cell* **84**, 863 (1996)
16. A. Askar, B. Space and H. Rabitz, *J. Phys. Chem.* **99**, 7330 (1995)
17. R. Abagyan, M. Totrov and D. Kuznetsov, *J. Comp. Chem.* **15**, 488 (1994)
18. C. Branden and J. Tooze. *Introduction to Protein Structure* Garland Publishing, New York and London (1991)
19. J. Janin and S.J. Wodak, *Prog. Biophys. Molec. Biol.* **42**, 21 (1983)
20. G.A. Jeffrey and W. Saenger. *Hydrogen Bonding in Biological Structures* Springer-Verlag, Germany (1991)

21. A.R. Fersht, *Trends Biochem. Sci.* **87**, 301 (1987)
22. M.F. Thorpe, B.R. Djordjevic and D.J. Jacobs, *Amorphous Insulators and Semiconductors* M.F. Thorpe and M.I. Mitkova, ed. NATO ASI Series 3: High Technology **23**, Kluwer Academic (1997): M.F. Thorpe, D.J. Jacobs and B.R. Djordjevic, *Insulating and Semiconducting Glasses* P. Boolchand, ed, World Scientific (1998): Also see article within this book. M.F. Thorpe, *et al.*
23. K.A. Dill, *Biochem.* **29**, 7133 (1990)
24. D. Franzblau, Private communication (1996)
25. J. Graver, B. Servatius and H. Servatius *Combinatorial Rigidity (Graduate Studies in Mathematics)* American Mathematical Society, Providence RI (1993)
26. E. Guyon, S. Roux, A. Hansen, D. Bibeau, J.P. Troadec and H. Crapo, *Rep. Prog. Phys.* **53** 373 (1990)
27. G. Laman, *J. Eng. Math.* **4** 331 (1970)
28. D.J. Jacobs and B. Hendrickson, *J. Comp. Phys.* **137** 346 (1997)
29. D.J. Jacobs, *J. Phys. A: Math. Gen.* **31** 6653 (1998)
30. F. Harary. *Graph Theory* Addison-Wesley, Reading MA (1969)
31. D.J. Jacobs and M.F. Thorpe, *Phys. Rev. Lett.* **75** 4051 (1995)
32. D.J. Jacobs and M.F. Thorpe, *Phys. Rev. E* **53** 3683 (1996)
33. D. C. Kozen. *The Design and Analysis of Algorithms* Chapter 4, Springer-Verlag, New York (1992)
34. J. Saks, American Mathematical Society, #441, 91 (1991)
35. B. Oh, J. Pandit, C. Kang, K. Nikaido, S. Gokcen, G.F. Ames and S. Kim, *J. Biol. Chem.* **268** 11348 (1993)
36. D.M. Miller III, J.S. Olson, J.W. Pflugrath and F.A. Quiocho, *J. Biol. Chem.* **258** 13665 (1983)
37. S.M. Habermann and K.P. Murphy, *Protein Science* **5** 1229 (1996)
38. http://swift.embl-heidelberg.de/whatif/

FLEXIBLY SCREENING FOR MOLECULES INTERACTING WITH PROTEINS

Volker Schnecke and Leslie A. Kuhn

Protein Structural Analysis and Design Laboratory
Department of Biochemistry
Michigan State University
East Lansing, MI 48824-1319, U.S.A.
http://www.bch.msu.edu/labs/kuhn

INTRODUCTION

Flexibility in Proteins and Their Interactions

The flexibility of proteins and their *ligands* (molecules specifically bound by proteins) has a major influence on the ways they interact. Generally, protein molecules are thought of as primarily rigid structures, with chemically specific and somewhat flexible side chains attached to a main chain of fixed structure. The tendency to think of proteins as rigid is reinforced by the fact that X-ray crystallography, the most widely-used technique for analyzing protein structures at atomic resolution, traps the copies of a protein molecule in the crystalline lattice into a single state. However, as seen in Figure 1, rotatable single bonds in the main as well as side chains of a protein provide significant potential for flexibility (*conformational change*). For a number of proteins, such as the HIV protease, lysine-arginine-ornithine binding protein, and adenylate kinase, the protein is known to undergo significant conformational change upon binding its natural ligand or drugs designed to inhibit its activity. Flexibility is thus a biologically essential feature of proteins.

Despite the importance and widespread interest in characterizing protein flexibility, this remains a challenge both experimentally and computationally (see papers by David Case, Ruben Abagyan, and Mark Gerstein in this volume). Our laboratory's goal has been to develop computational methods that incorporate realistic modeling of protein flexibility into the design of new ligands for proteins. In collaboration with Jacobs and Thorpe (see accompanying paper), we have shown that graph-theoretic analysis of the covalent and hydrogen-bond networks in proteins using the FIRST algorithm provides an extremely fast way of assessing large-scale flexibility in proteins, e.g., when large, independently folded regions of the protein (*domains*) are attached by hinge joints, resulting in clamshell-like motion. In the present paper, we review the state of the art in template-based algorithms for analyzing protein–

Rigidity Theory and Applications
Edited by Thorpe and Duxbury, Kluwer Academic/Plenum Publishers, 1999

ligand interactions, which reduce the orientational search for the optimal placement of the ligand relative to the protein, and discuss how ligand flexibility is modeled in some of these methods. Finally, we address how side-chain motion in the protein, coupled with flexibility in the ligand, can be modeled in an algorithm that allows fast and effective computational screening of hundreds of thousands of compounds for ligands. This method, SPECITOPE, is the first to model protein, as well as ligand, flexibility, in the context of screening. The identification of favorable ligand candidates is a first and crucial step in modern drug design, with major potential to develop new therapeutics for AIDS, cancer, bacterial infection, arthritis, and other diseases.

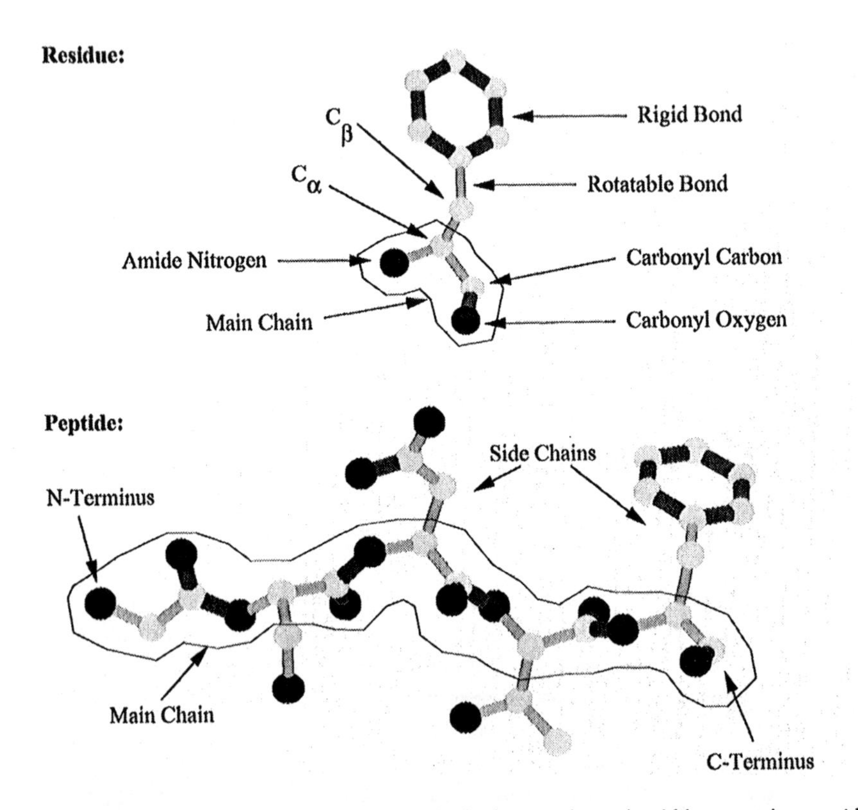

Figure 1. *Protein structure and intrinsic flexibility.* The basic repeating unit within a protein, a *residue*, is shown at top. The side chain of the residue in this case (phenylalanine) consists of a phenyl ring (at top) connected to the protein main chain (polymer *backbone*) at the alpha-carbon. Rigid, unrotatable bonds with partial or full double-bond character are shown by black tubes, and rotatable single bonds are shown as thin grey tubes; carbon atoms appear in light grey and polar (oxygen or nitrogen) atoms in black. In the bottom panel, a *peptide* (fragment of protein) is shown, consisting of several linked residues with various side chains. The main chain of the polymer, formed by the repeating (amide nitrogen)–(alpha-carbon)–(carbonyl carbon and oxygen) motif, is outlined. The N-terminus is the start of the peptide chain, and the C-terminus is its end. In reality, not all of the flexibility implied by the single bonds is accessible, since van der Waals, hydrogen-bond, and electrostatic interactions typically lock the protein chain into a unique *conformation* (molecular shape) or set of conformations dependent on its sequence of amino acid residues.

Ligand Docking and Database Screening

With the increase in computational power and availability of structural information for proteins and small molecules, computer-based drug design has become a competitive methodology to identify new *inhibitors* (ligands that block the function of proteins)[1–4]. Although computational methods do not replace the *in vitro* and *in vivo* tests during drug development, they can be very efficient in identifying and optimizing the structures of ligand candidates as *lead* compounds for further development, and thus accelerate the early design stages. Generally, there are two tasks involved in identifying leads in computational drug design, *screening* the structures of compounds for potential ligands, and *docking*, or optimally fitting, these potential inhibitors into the binding site of the target (typically a protein). Screening is important for reducing the vast number of potential ligands to an experimentally testable number. Furthermore, developing methods to predict protein–ligand interactions – in particular, predicting ligands and their *binding modes*, or specific orientation and conformation upon interaction with the protein – provides significant insights into the way proteins work, which is always by interacting with other molecules. Docking combined with modeling structural modifications of the protein or ligand also provides a valuable design tool for developing new ligands with greater specificity for a given protein and *vice versa.*

Computational screening is often closely associated with docking, as the final evaluation of ligand candidates requires a detailed evaluation of how well the ligand and protein fit together. Docking approaches can be classified based on how they characterize the ligand-binding site of the protein. *Grid-search techniques* fill the space around the binding site with a three-dimensional grid, precompute the potentials (van der Waals, electrostatic, etc.) at each grid point without the ligand, then sample different ligand conformations and orientations on the grid and compute the resulting binding free energy. An example for this approach is AutoDock, which used simulated annealing in its previous releases[5,6], but now applies a hybrid genetic algorithm to sample over the feasible binding modes of the ligand relative to the protein[7,8]. The advantages of grid-based docking are that a template of favorable interactions in the ligand-binding site does not need to be defined, reducing bias in modeling the protein–ligand interactions, and evaluation of binding modes is made more efficient by precomputing protein potentials on the grid. However, the accuracy and timing of this approach depends on the grid fineness, making this approach too computationally intensive for database screening, in which thousands of molecules (as well as ligand orientations and shapes, or *conformers*) need to be tested. Furthermore, precomputation of the protein grid potentials limits this approach to rigid binding sites.

When the ligand-binding site in the protein is known, this can be utilized by constructing a *template* for ligand binding based upon favorable interaction points in the binding site. During the search for a favorable ligand-binding mode, different conformations of the ligand can be generated and subsets of its atoms matched to complementary template points, as a basis for docking the ligand into the binding site. Advantages to this template-based approach are that it can incorporate known features of ligand binding (for example, conserved interactions observed experimentally for known ligands), and it reduces the docking search space to matching N ligand atoms onto N template points, rather than the 6-dimensional orientational search space (3 degrees of rotational freedom and 3 degrees of translational freedom) required in other approaches for sampling and evaluating ligand binding.

In the well-known docking tool DOCK[9], the template typically consists of up to 100 spheres that generate a negative image of the binding site. During the search, subsets of ligand atoms are matched to spheres, based on the distances between ligand atoms. DOCK has been extended to consider chemistry[10] and include hydrogen-bonding interaction centers[11] in addition to the shape template. Other current template approaches specify a set of interaction points defining favorable positions for placing polar ligand atoms or *hydrophobic* (nonpolar) centers, e.g., aromatic rings. Such a template can be generated automatically, e.g., by placing probe points on the solvent accessible surface of the binding site[12], or interactively by superimposing known protein–ligand complexes to identify favorable interaction points based on observed binding modes for known ligands. FlexX[13,14] uses a template of 400 to 800 points when docking drug-size molecules (up to 40 atoms, not including hydrogens) to define positions for favorable interactions of hydrogen-bond donors and acceptors, metal ions, aromatic rings, and methyl groups. The ligand is fragmented and incrementally constructed in the binding site and matched to template points based on geometric *hashing* (indexing) techniques, bond torsional flexibility is modeled discretely, and a tree-search algorithm is used to keep the most promising partially constructed ligand conformations during the search. Hammerhead[15] uses up to 300 hydrogen-bond donor and acceptor and *steric* (van der Waals interaction) points to define the template, and the ligand is incrementally constructed, as in FlexX. A fragment is docked based on matching ligand atoms and template points with compatible internal distances, similar to the matching algorithm used in DOCK. If a new fragment is positioned closely enough to the partially constructed ligand, the two parts are merged, and the most promising placements kept. Other successful docking approaches, such as GOLD[16,17] and the method of Oshiro *et al.*[18], use genetic algorithms to sample over possible matchings of conformationally flexible ligands to the template. However, a drawback of genetic algorithm approaches, including AutoDock, is the high computation time, especially in comparison to fragment-based docking approaches.

When screening databases of more than 10^5 compounds to identify potential ligands, the computational efficiency of the search process becomes a significant concern. Docking a small flexible molecule with high accuracy takes at least several minutes on a desktop workstation for the fastest of the recent algorithms[11,13,14,19–21]. Spending only one minute to dock each molecule when screening a dataset of 100,000 compounds results in a computation time on the order of two months, which is unacceptably slow, particularly when improving and validating the method. Recent screening tools can identify potential ligands from up to 150,000 compounds within a few days when ligand flexibility is modeled[15,22–26]; however, none of these methods models protein conformational change upon ligand binding (also called *induced complementarity*).

For a number of structurally characterized protein–ligand complexes, induced complementarity of protein side chains is known to be important for ligand binding[27]. Thus, for the development of a new screening procedure, SPECITOPE, our goal has been to model protein side-chain flexibility as well as ligand flexibility when evaluating their interaction. SPECITOPE narrows down the vast number of ligand candidates to several dozen molecules with good shape and chemical complementarity to a protein ligand-binding site within 3 hours on a typical desktop workstation[28]. It is difficult to make direct comparisons between the timing of SPECITOPE and other screening algorithms, because the other methods assume the protein is rigid, differ in their modeling of ligand flexibility, and some require manual scoring (molecular graphics assessment by a structural biologist) outside the algorithm. However, methods screening rigid ligands typically take several hours, whereas methods modeling full ligand flexibility typically take several days, plus time spent for external scoring.

SPECITOPE's relative speed results from adapting distance geometry techniques[29] (see also the paper by Havel in this volume) to perform quick feasibility checks on each ligand based on comparing its interatomic distances and number of hydrogen-bond donors and acceptors with those in a template representing the binding site. We have shown that protein and ligand side-chain flexibility can be modeled while screening a large database of peptide structures for inhibitors to three diverse proteins, an aspartyl proteinase, a serine protease, and a DNA repair enzyme[28]. This approach was successful in identifying the known peptidyl inhibitors within the top five of 140,000 ligand candidates screened, and for two of the three proteins, the known ligand received the top score, based on shape complementarity and favorable hydrophobic and hydrogen-bond interactions with the protein. In each case, pro-

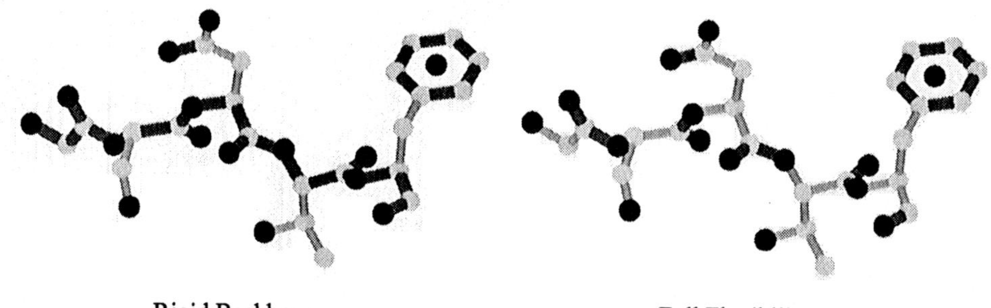

Figure 2. *Side-chain versus full flexibility in a peptide.* For the same peptide shown in Figure 1, two representations of peptide flexibility used by screening algorithms are shown: a model in which the backbone (main chain) is a rigid unit and the side chains are free to rotate, and a fully flexible model, where the backbone and side-chain dihedral angles are free to rotate. Rigid bonds are shown as black tubes and flexible bonds as grey tubes; in the full-flexibility model, rigid bonds arise from the partial-double and double bonds inherent in the structure. Black spheres indicate polar (nitrogen, oxygen) atoms to be compared with a hydrogen-bonding template representing the ligand-binding site, whereas the black sphere in the ring center (top right) indicates a hydrophobic ring center to be matched with hydrophobic centers in the template. Grey spheres indicate carbon atoms.

tein side-chain motion was known to be important for ligand binding and was appropriately modeled by SPECITOPE during docking.

Here we present and validate improvements of SPECITOPE to probe the ligand-binding site more thoroughly; this approach will also enable screening of fully flexible ligands in the near future. Geometric hashing techniques are employed so that the exhaustive checking of different matchings of the ligand to the protein's ligand-binding site is reduced to checking only those matchings that are feasible, based on distance and chemistry indices stored for the template in a look-up table. This provides time savings and linear scaleability, allowing sampling over more template points within the binding site and modeling of full flexibility for peptidyl and small organic ligand candidates. Hydrophobic interaction centers are now considered, in addition to hydrogen-bonding centers, in the ligand and in the binding-site template. Figure 2 compares the difference between peptidyl ligand flexibility, as previously modeled by SPECITOPE, where side chains were flexible but main-chain dihedral angles were held fixed, with the degree of flexibility that will be enabled by the hashing approach, where both side-chain and main-chain dihedrals are rotatable (within the limitations imposed by van der Waals contacts). Many organic compounds are not polymeric and thus do not have a clear main chain/side chain distinction, though they often have rigid frameworks provided by ring systems. To illustrate how the same concepts applied here to peptides can be applied to more general organic compounds, a goal for our future ligand design work, Figure 3 shows the interaction centers and flexible bonds for two compounds arbitrarily chosen from crystallographic structures in the Cambridge Structural Database (http://www.ccdc.cam.ac.uk).

METHODS

SPECITOPE shares with other current docking and screening approaches the use of a binding site template to limit the orientational search for each prospective ligand, and differs in the use of distance geometry techniques to avoid the computationally intensive fitting of infeasible ligands into the binding site. The speed gained by distance geometry allows the sec-

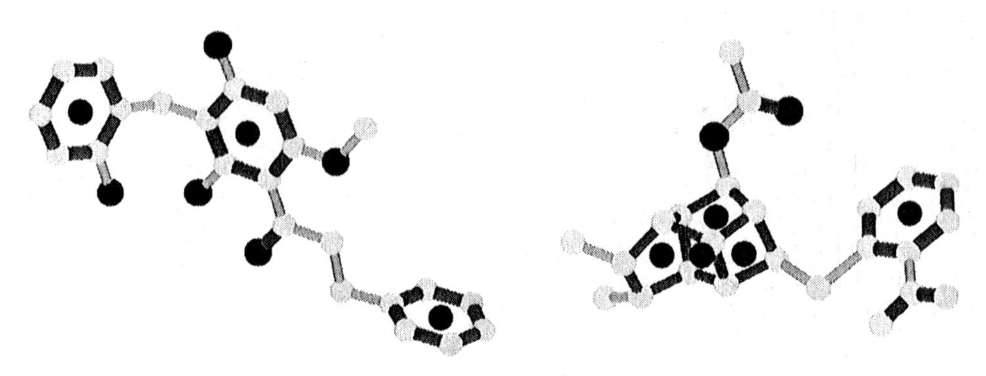

Figure 3. *Flexibility in organic compounds.* Two compounds from the Cambridge Structural Database are shown, indicating how the representation of rigid and flexible bonds and polar and hydrophobic (nonpolar) atom centers carries from peptides to more general compounds (bonds and atoms colored as in previous figure). Notice the similar degree of flexibility of the compound at left to the fully flexible peptide (previous figure), whereas the compound at right has significantly more rigid and hydrophobic clusters than are found in peptides.

ond advantage over other screening methods, modeling protein side-chain flexibility during docking. Here we overview the algorithm and present the use of hashing techniques adapted from fragment-based docking to allow sampling over more interaction sites as well as future modeling of full ligand flexibility. We now include hydrophobic centers for template matching, in addition to the hydrogen-bonding centers previously used. While hydrogen bonds are important for providing specificity to most protein–ligand interactions, hydrophobic interactions are especially significant for some organic ligands, and contribute favorably to the binding free energy[17,30,31]. The main steps of SPECITOPE, as described below, are: template design; distance geometry and hashing steps for screening out geometrically infeasible ligand candidates and efficiently matching ligands to the template; rigid-body translations of the ligand coupled with ligand and protein side-chain flexibility to resolve steric overlaps in the complex; and ligand scoring based on interactions with the protein.

Template Design

For SPECITOPE, the template consists of key interaction points (hydrogen-bond donors, hydrogen-bond acceptors, and hydrophobic centers) where ligand atoms with matching character can make favorable interactions with the protein. Design of the template can be based on observed interactions in the structure of a known ligand in complex with the protein; for two of our recent applications, uracil-DNA glycosylase and cyclodextrin glycosyltransferase, templates were based on the positions of polar atoms in the ligand that formed hydrogen bonds to protein atoms[28] (see Figure 4). When structures are available for the protein in complex with several different ligands, as for aspartyl proteinase, interactions shared by the ligands can be used to develop a consensus template. When ligand-bound protein structures are unavailable, templates can be based on the positions of crystallographically observed water molecules bound in the ligand-binding site of the protein, representing favorable sites for making hydrogen bonds to the protein[32]; this approach was taken for screening subtilisin inhibitors[28]. Template-design methods from other docking and screening approaches (discussed in the Introduction) may also be employed.

In the prior version of SPECITOPE, the template was limited to about five interaction points, due to the combinatorics of sampling every possible matching of five ligand atoms

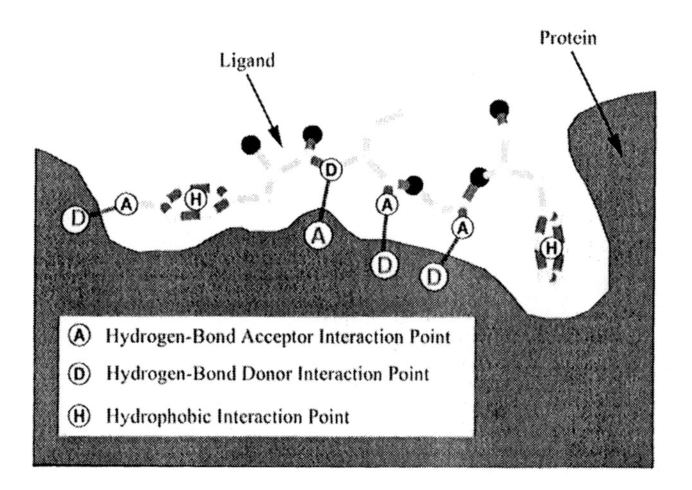

Figure 4. *Example of a binding site template.* The template is comprised of sites above the protein circled in black and labeled as hydrophobic or hydrogen-bond interaction points. This template represents sites where similarly-labeled ligand atoms can make favorable interactions with the protein (whose hydrogen-bond acceptor and donor sites are indicated below by grey letters).

(out of a larger number, typically ~15 for five-residue peptides) onto five template points; the computational complexity of this step is factorial, due to enumerating all permutations (matchings) of the ligand atoms onto template points (see equation below). In the present version, we avoid this complexity by considering 3-point subsets of larger (10 or more point) templates and using *triangle hashing* to evaluate only the geometrically feasible subsets of these triangles. As seen in Table 1, the number of potential matchings of 3-point ligand atom subsets onto a template with T points scales linearly as T increases, whereas the previously-used *complete enumeration*, in which all N-point subsets of the ligand are permuted onto the T-point template, scales poorly in both N and T. The scaling is linear in practice because the number of feasible triangles in each bin in the hash table (see Distance Geometry, Hashing, and Docking section below) has been found experimentally to scale linearly in T, effectively reducing the first term in the product below from the binomial coefficient, T-choose-N, to T.

$$\left(\begin{array}{c}\#\text{ of potential}\\ \text{matchings}\end{array}\right)=\left(\begin{array}{c}\#\text{ of ways of}\\ \text{choosing}\\ N\text{ template pts}\\ \text{from }T\text{ pts}\end{array}\right)\cdot\left(\begin{array}{c}\#\text{ of ways of}\\ \text{choosing}\\ N\text{ ligand pts}\\ \text{from }L\text{ pts}\end{array}\right)\cdot\left(\begin{array}{c}\#\text{ of ways of}\\ \text{matching}\\ N\text{ ligand pts to}\\ N\text{ template pts}\end{array}\right)$$

$$=\left(\frac{T!}{N!(T-N)!}\right)\cdot\left(\frac{L!}{N!(L-N)!}\right)\cdot(N!)$$

Matching triangles of ligand interaction centers onto triangular subsets of the template makes sense for two other reasons. The previous matching of four or five interaction centers in a ligand to the same number of template points in the binding site effectively rigidified the ligand, since most bonds between those atoms in the ligand could not be rotated without

Table 1. *Combinatorics of complete enumeration versus hashing approaches for matching N ligand points to T template points.*

	T	N	Number of Potential Matchings
Complete Enumeration	5	5	360,360
	10	5	90,810,720
	10	3	327,600
	20	3	3,112,200
Triangle Hashing	5	3	13,650
	10	3	27,300
	20	3	54,600
	50	3	136,500

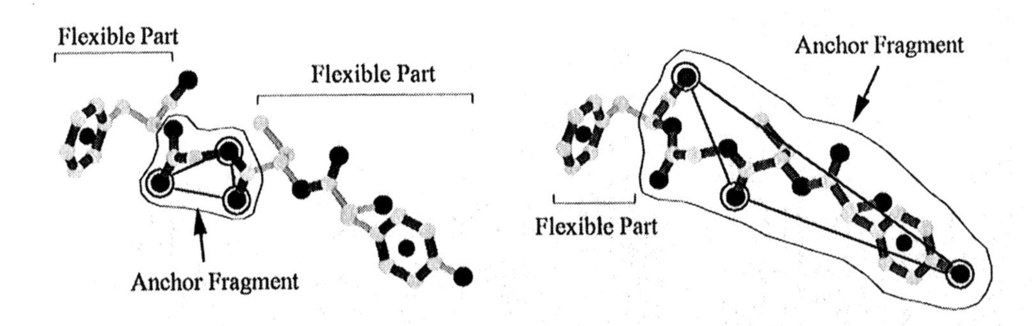

Figure 5. *The rigid and flexible regions in a ligand, using triangle-based docking.* During the template matching described below, every possible triangle of hydrogen-bonding and hydrophobic interaction points in the ligand is matched to every possible triangle of template points; however, the hashing procedure focuses directly on those matchings with feasible geometry and chemistry. The resulting triangle-based docking essentially rigidifies the triangular *anchor fragments*, which may be small (as in the left panel) or large (right panel), while maintaining the flexibility of the other parts of the ligand. Docking is based on matching the ligand triangle to the template and adjusting the flexible parts of the ligand and/or protein to remove intra- or intermolecular van der Waals collisions (overlaps between atoms). Then, all non-colliding ligand dockings are scored according to their hydrogen-bond and hydrophobic complementarity with the protein.

disrupting the template match. When, alternatively, all chemically and geometrically feasible ligand triangles are tested for docking to the protein via hashing, much smaller rigid fragments of the ligand are also tested (Figure 5). Furthermore, organic compounds, a major future application, tend to have fewer hydrogen-bonding atoms than peptides and are more chemically diverse. Thus, it is useful to be able to screen and dock these compounds based on fewer interaction points (3, which still uniquely define an orientation with respect to the protein), include hydrophobic as well as hydrogen-bond interaction centers, and consider shape complementarity as another major contributor to specificity. For compounds with more than 3 interaction points, each triangle can be docked independently, by optimizing the rotations in the linkages between them. An interface to read organic compounds and identify their flexible bonds, hydrophobic centers (interpreted as carbon-ring centers), and hydrogen-bond donors and acceptors has been developed in our lab. This interface is based on the generic mol2 molecular data file (of the Tripos Sybyl software) frequently used with other databases and modeling tools, making SPECITOPE portable and compatible with other systems.

Distance Geometry, Hashing, and Docking

SPECITOPE first uses simple distance geometry[29] techniques to screen out ligands with incompatible geometry relative to a template specifying the positions of hydrophobic and hydrogen-bond interaction centers. Many of the ligand–template matchings can be ruled out by distance geometry alone, based on the incompatibility between ligand interatomic distances and inter-template-point distances. Given a set of N (in this case, 3) ligand interaction centers, a sorted list of the $N \cdot (N-1)/2$ distances, l_i, between ligand interaction centers is compared to the sorted list, t_i, of distances between the N template points. With d_i equal to the difference between distances l_i and t_i, the root-mean-square deviation between distances in the two lists, is defined as:

$$RMSD_{list} = \sqrt{\frac{2}{N \cdot (N-1)} \sum_{i=1}^{N \cdot (N-1)/2} (d_i)^2}$$

The $RMSD_{list}$ gives a measure for the compatibility of distances between the ligand atoms and between points in the template. A more exact measure for their compatibility is the distance matrix error:

$$DME = \sqrt{\frac{2}{N \cdot (N-1)} \sum_{i=1}^{N-1} \sum_{j=i+1}^{N} (D_{ij})^2}$$

where $D_{ij} = L_{ij} - T_{ij}$ is defined as the matrix of differences between matched distances in the L and T matrices containing the distances between ligand interaction centers and distances between template points, respectively. The $RMSD_{list}$ can be proven to give a lower bound for the DME for any matching of the two sets. Hence, if the $RMSD_{list}$ is above a given threshold for the current set of interaction centers and template points, this set can be ruled out, since the DME for any one-to-one matching of these centers to the template points can only exceed this value. An advantage of using the $RMSD_{list}$ as a screening criterion before the DME check is that the factorial complexity of specifying one-to-one correspondences between ligand and template points can be avoided for the majority of cases.

Aside from the major time savings from comparing intra-template and intra-ligand distances via sorted distance lists rather than by docking the molecules together, SPECITOPE screens out infeasible ligands by a series of quick, initial distance and chemistry checks: Does the longest distance between interaction centers in the ligand significantly exceed the longest distance in the template? If so, they cannot match. Are there enough hydrogen-bond and hydrophobic interaction points in the ligand to match the template? If not, they do not match. Overall, these distance geometry and chemistry checks, through the DME step, typically rule out 70% of the infeasible ligand candidates before the time-intensive docking steps[28].

Figure 6 compares the current, hashing-based strategy used by SPECITOPE with the steps used in the previous complete-enumeration screening[28]. The distance geometry and docking (least-squares fit) steps remain the same, with the major difference being that triangles of ligand interaction centers are matched by hashing to the triangles of template points. The use of triangles, rather than more complex geometric objects, provides a convenient basis for screening ligand–template matches based on simple chemical and geometric characteristics of the triangles. This is done in three stages of hashing, or indexing, and is extremely efficient because a ligand triangle is only compared to template triangles with similar characteristics. As shown in Figure 7, the hash table allows direct access to all template triangles having

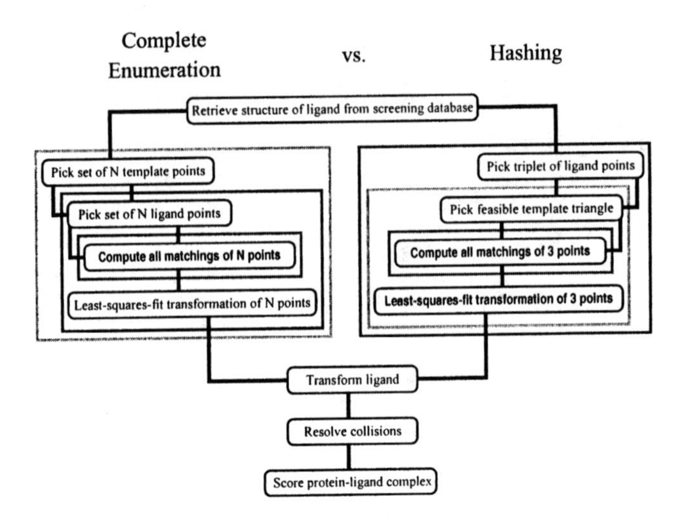

Figure 6. *Comparison of the steps in hashing versus complete enumeration screening of ligands by* SPECITOPE.

the same number and type of interaction points as the ligand triangle (e.g., one hydrogen-bond acceptor and two donors, or ADD), a similar perimeter, and a similar length for the longest triangle side. The distance geometry and hashing steps ensure that the relatively time-intensive DME and docking (least-squares fit) calculations are only done for feasible template triangles. The $RMSD_{list}$ and DME are then calculated for each matching of triangles, using a cutoff of 0.2 Å, reflecting a much closer match than could be required for matching a larger number of points. This close match to the template also preserves the possibility of hydrogen bonding and was found to retain known ligands during screening. Docking involves taking the minimal-DME matching between the current ligand interaction centers and the template, then transforming the ligand into the protein's ligand-binding site based on a least-squares fit of these matched points. If the root-mean-square deviation of this transformation is below a fixed threshold (0.2 Å), the entire ligand is transformed into the ligand-binding site.

Van der Waals Collisions and Flexibility

This ligand transformation into the protein's ligand-binding site results in a close fit of the ligand interaction points with the template, but also may result in van der Waals collisions between ligand and protein atoms. Because of the flexibility of the protein and ligand, many such collisions are ultimately resolvable. In its current incarnation, halfway to full ligand flexibility, SPECITOPE employs triangle hashing for matching the ligand and template, while modeling the peptidyl ligand backbone as rigid and the ligand and protein side chains as flexible. The following discussion also covers the more general case in which the ligand backbone is flexible, which triangle hashing will make possible. If the collisions are between the rigid part of the ligand (backbone or anchor fragment) and the rigid backbone of the protein, the minimal translation vector for resolving these collisions is calculated, and the ligand is translated accordingly. If new rigid-body collisions result, this procedure is iterated up to 100 times, effectively shaking the ligand inside the ligand-binding site. If these collisions cannot be resolved, this ligand matching to the template is discarded. When the collisions can be resolved, then van der Waals overlaps involving flexible parts of the ligand and the

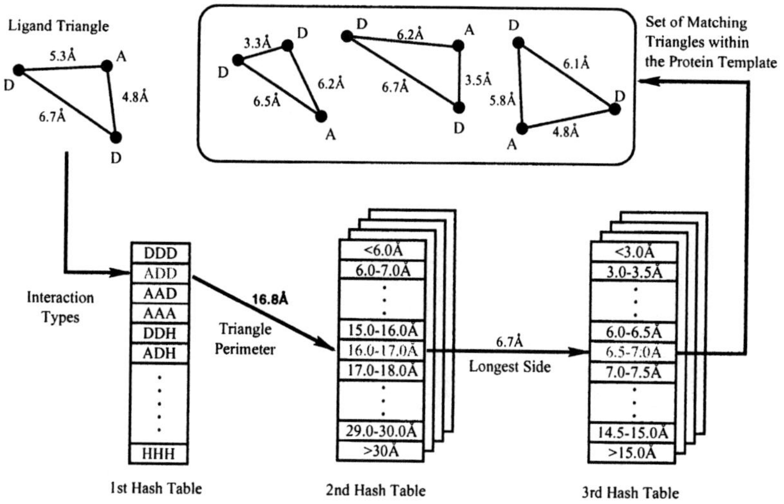

Figure 7. *Three-stage hashing applied to ligand–template matching.* Hashing enables direct access to those sets of template triangles (shown at top center, with length and interaction-type data precomputed and exhaustively listed in a table) that match a given set of three ligand points (upper left). The index into the first table is based on the interaction types of the ligand atoms (in this case, two hydrogen-bond donors and one acceptor), which points to the subset of template triangles with the same label (ADD). The second index, the perimeter of the ligand triangle, locates those template triangles that are labeled ADD and have a similar perimeter. The third index, the length of the longest side of the triangle, points to those ADD template triangles with similar perimeter and longest side. This results in more efficient, but still exhaustive, checking of ligand–template matches.

protein are addressed.

Each flexible part in the ligand is checked for overlaps with protein atoms, which are cleared by rotating this part of the ligand through the minimal angle that resolves the overlaps. The single bond closest to the colliding atoms in the ligand is used first to resolve the overlap. If a collision-free conformation cannot be generated with this rotation, the next rotatable bond closer to the rigid part of the ligand is rotated. When it is not possible to resolve an overlap by rotations within the ligand, the same approach is applied to the protein side chain involved in the collision. If an intermolecular collision remains, despite testing all protein side-chain and ligand single-bond rotations, this ligand matching to the template is deemed too close and rejected. For ligand matchings in which all intermolecular overlaps have been resolved, both molecules are checked for intramolecular collisions. If a rotation has caused an internal clash, then the flexible group is rotated back to its original conformation, and the next single bond closer to the backbone or anchor fragment is rotated. This procedure is followed by rechecking for inter- and intramolecular collisions, until either a collision-free conformation is found, or all possibilities have been exhausted and this ligand matching is excluded. The aim of this step in SPECITOPE is not to predict the optimal ligand conformation, but to ensure that a collision-free conformation of the molecules exists for this matching.

Scoring

A scoring function is then used to rank the relative complementarity of the ~100 ligand candidates (passing the previous screening steps) to the protein's ligand-binding site. Because the conformation and orientation of each ligand candidate could likely be optimized by fine docking, scoring is mainly intended to recognize molecules that lack chemical complementarity to the protein and to emphasize molecules that fit well in the given binding mode. The scoring function weighs the dominant factors, the number of hydrogen bonds between the protein and ligand and their hydrophobic complementarity. For hydrogen-bond donors and acceptors separated by 2.8 to 3.5 Å, SPECITOPE computes the optimal position of the shared hydrogen atom (because most X-ray structures do not provide hydrogen atom positions) to identify intermolecular hydrogen bonds with good geometry. The hydrophobicity measure is based on a statistical survey of atomic hydration in 56 protein structures[33] and compares the hydrophobicity value of each ligand atom to the average hydrophobicity value of nearby protein atoms. The overall complementarity of the protein-ligand complex, SCORE (protein,ligand), is given by a weighted sum of the number of hydrogen bonds and the hydrophobic complementarity:

$$\mathrm{SCORE}(\mathrm{protein}, \mathrm{ligand}) = \mathrm{A} \cdot \mathrm{HBONDS}(\mathrm{protein}, \mathrm{ligand}) + \mathrm{B} \cdot \mathrm{HPHOB}(\mathrm{protein}, \mathrm{ligand})$$

Based on the functions of Böhm[34] and Jain[35], a ratio of 1:1.2 is assumed for the relative contributions of the hydrogen bond and hydrophobic interaction terms to the overall stability of the protein-ligand complex, with the weights A and B tuned accordingly, based on the values of HBOND and HPHOB from 30 structures of protein complexes with small peptidyl ligands[28] from the Protein Data Bank (PDB)[36]. These structures have 6.3 hydrogen bonds between protein and ligand, on average.

RESULTS

Given the new implementation of triangle hashing, our goal here is to assess whether hashing identifies ligands with similar or better protein complementarity relative to those identified by complete enumeration.

Screening for Aspartic Proteinase Ligands

To compare ligands from the hashing and complete enumeration approaches, SPECITOPE was used to screen for peptidyl ligands of rhizopuspepsin, an aspartic proteinase. Rhizopuspepsin is a relative of medically important inhibitor design targets including renin, which regulates blood pressure, and HIV protease, which is essential to the virus life cycle and a major target for AIDS drug design[37]. The template for rhizopuspepsin was designed by superimposing three complexes of this protein with pepstatin (PDB structure 6apr) or pepstatin-like renin inhibitors (4apr, 5apr) onto the ligand-free structure of rhizopuspepsin (2apr). Figure 8 shows the structure of rhizopuspepsin bound to a known peptidyl ligand.

For complete enumeration screening, the average positions of four hydrogen-bond donors and one acceptor in the three inhibitors were selected as the template points. For hashing, 9 such template points were chosen, and the characteristics of each 3-point subset of these template points were listed in the hashing tables (see Figure 7). In both cases, all 5-residue peptides were screened from 140,000 peptides occurring in known protein structures with low similarity (<25% sequence identity)[38].

Complete enumeration identified 117 possible peptidyl ligands for rhizopuspepsin in 96 minutes on a desktop workstation (Sun SPARC Ultra 140), whereas hashing identified 357 possible ligands in 75 minutes. For this case, the 10-fold or more computational savings possible from hashing were not realized because more peptides passed the distance geometry steps; however, for another protein tested, human uracil-DNA glycosylase, screening was 3 times as fast. More importantly, the peptidyl ligands identified by hashing tended to have higher complementarity to the protein than the ligands identified by complete enumeration (Table 2). The feasible ligands from triangle hashing had an average complementarity score

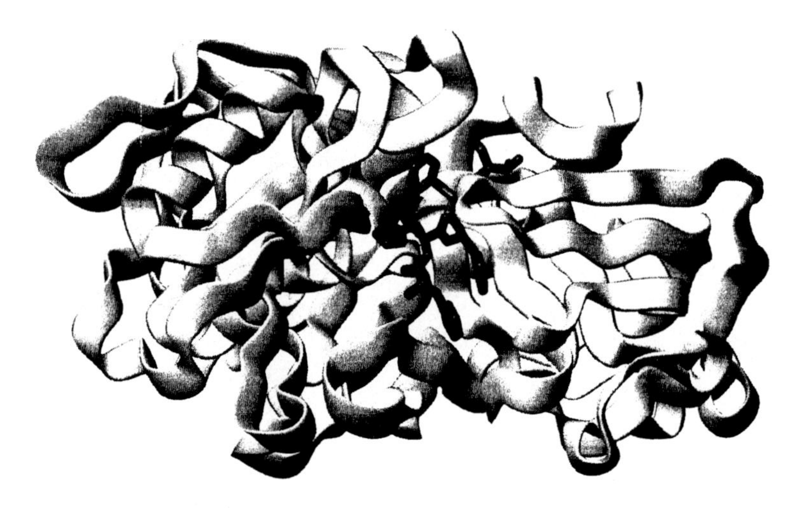

Figure 8. *Rhizopuspepsin in complex with a peptidyl ligand.* The rhizopuspepsin backbone is shown as a grey ribbon, with its ligand-binding site forming the vertical cleft at center. Bound in this site is the known peptidyl ligand, FHFFV, shown in black tubes. The crystallographic structure of the protein-ligand complex is from PDB entry 3apr.

of 1369.8 and an average number of 3.5 hydrogen bonds between protein and ligand, whereas complete enumeration resulted in ligands with an average score of 816.8 and an average of 2.2 protein-ligand hydrogen bonds. The top-five ligands found by hashing had scores 200-300 points higher and typically had one more hydrogen bond than those found by complete enumeration. In both cases, a known rhizopuspepsin ligand, the peptide with amino-acid sequence FHFFV (PheHisPhePheVal), was identified as the top-scoring ligand, with similar scores for complete enumeration and hashing; this reflects similar dockings despite the different number of ligand atoms matched to template points, 5 for complete enumeration and 3 for hashing. The docking based on hashing also resulted in a very similar ligand-binding mode to that observed in the crystallographic structure of the rhizopuspepsin-peptide complex (Figure 9). The backbones of the ligand from the crystal structure and as docked by SPECITOPE (backbones running horizontally in Figure 9) are essentially superimposed, with some side-chain reorientation (in the rings at top).

CONCLUSIONS AND FUTURE DIRECTIONS

Triangle-based hashing has been implemented in our ligand screening algorithm, SPECITOPE, and provided time savings in ruling out infeasible ligand candidates, as well as more thoroughly sampling the ligand binding site. Screening for rhizopuspepsin ligands showed a 28% speed increase using hashing as compared with the previous complete enumeration approach, while finding ligands with higher complementarity to the protein. Our goal, nearing completion, is to screen fully flexible peptidyl and small organic ligands against proteins with side-chain flexibility. Hashing enables this by sampling over all possible rigid fragments of ligands during screening and docking. Recent construction of a structural database

Table 2. *Summary of the top-five rhizopuspepsin ligands identified by* SPECITOPE *using complete enumeration and triangle hashing.*

Complete Enumeration (5 template points)				Triangle Hashing (9 template points)			
Rank	Sequence	H-Bonds	Score	Rank	Sequence	H-Bonds	Score
1	FHFFV	5	3178.8	1	FHFFV	5	3324.1
2	KTVTD	2	3150.6	2	YYTAL	4	3280.3
3	ETTSF	2	3067.2	3	NLKFG	3	2982.4
4	LWCNG	3	2573.0	4	LYIDS	3	2797.8
5	YGLSV	3	2385.0	5	GYYTA	4	2770.4
1—117	Average	2.2	816.8	1—357	Average	3.5	1369.8
	Std.Dev.	1.5	769.2		Std.Dev.	1.2	509.7

Figure 9. *Close-up view of the known ligand-binding mode for rhizopuspepsin and that predicted by* SPECITOPE *based on triangle hashing.* The protein backbone is shown in grey ribbons, with the peptidyl ligand FHFFV, from its crystal structure in complex with the protein (PDB 3apr), shown in black tubes. The similar binding mode for this peptide predicted by SPECITOPE is shown in grey tubes, and the template points (to which triangles were matched) are shown as grey spheres.

interface to SPECITOPE, including flexible and rigid bond information as well as hydrogen-bond donors and acceptors and hydrophobic interaction sites for each ligand candidate, will facilitate screening of organic compounds as well as peptides. Ultimately, we anticipate a merging of the capabilities of SPECITOPE and FIRST, which can predict regions of protein backbone flexibility (see companion paper by Jacobs, Kuhn, and Thorpe), to model the interactions between fully-flexible ligands *and* fully-flexible proteins. This problem of modeling the induced shape complementarity between proteins and ligands upon binding remains one of the most difficult and important problems in structural biology.

ACKNOWLEDGMENTS

We thank Mike Thorpe and Phil Duxbury for organizing and involving us in this exciting workshop, and the Deutsche Forschungsgemeinschaft (postdoctoral fellowship SCHN 576/1-1 to V.S.) and the MSU Vice Provost for Libraries, Computing, and Technology for supporting this research.

REFERENCES

1. I. D. Kuntz. *Science* 257:1078–1081 (1992).
2. C. L. Verlinde and W. G. Hol. *Structure* 2(7):577–587 (1994).
3. I. D. Kuntz, E. C. Meng, and B. K. Shoichet. *Acc. Chem. Res.* 27(5):117–123 (1994).
4. D. A. Gschwend, A. C. Good, and I. D. Kuntz. *J. Mol. Recog.* 9:175–186 (1996).
5. D. S. Goodsell and A. J. Olson. *Proteins* 8:195–202 (1990).
6. G. M. Morris, D. S. Goodsell, R. Huey, and A. J. Olson. *J. Comput. Aided Mol. Des.* 10:293–304 (1996).
7. C. D. Rosin, R. S. Halliday, W. E. Hart, and R. K. Belew. In *Proc. 7th Int. Conf. on Genetic Algorithms (ICGA),* T. Bäck, editor, 221–228 (Morgan Kaufmann Publishers, Inc., San Francisco, CA, 1997).
8. G. M. Morris, D. S. Goodsell, R. S. Halliday, R. S. Huey, W. E. Hart, R. K. Belew, and A. J. Olson. *J. Comput. Chem., in press* (1998).
9. I. D. Kuntz, J. M. Blaney, S. J. Oatley, R. Langridge, and T. E. Ferrin. *J. Mol. Biol.* 161:269–288 (1982).
10. B. K. Shoichet and I. D. Kuntz. *Protein Eng.* 6(7):723–732 (1993).
11. A. R. Leach and I. D. Kuntz. *J. Comput. Chem.* 13(6):730–748 (1992).
12. J. Ruppert, W. Welch, and A. N. Jain. *Protein Science* 6:524–533 (1997).
13. M. Rarey, S. Wefing, and T. Lengauer. *J. Comput. Aided Mol. Des.* 10:41–54 (1996).
14. M. Rarey, B. Kramer, T. Lengauer, and G. Klebe. *J. Mol. Biol.* 261:470–489 (1996).
15. W. Welch, J. Ruppert, and A. N. Jain. *Chem. Biol.* 3:449–462 June (1996).
16. G. Jones, P. Willett, and R. C. Glen. *J. Mol. Biol.* 245:43–53 (1995).
17. G. Jones, P. Willett, R. C. Glen, A. R. Leach, and R. Taylor. *J. Mol. Biol.* 267:727–748 (1997).
18. C. M. Oshiro, I. D. Kuntz, and J. Scott Dixon. *J. Comput. Aided Mol. Des.* 9:113–130 (1995).
19. D. K. Gelhaar, G. M. Verkhivker, P. A. Reijto, C. J. Sherman, D. B. Fogel, L. J. Fogel, and S. T. Freer. *Chem. Biol.* 2:317–324 (1995).
20. R. M. A. Knegtel, I. D. Kuntz, and C. M. Oshiro. *J. Mol. Biol.* 266:424–440 (1997).
21. N. C. J. Strynadka, M. Eisenstein, E. Katchalski-Katzir, B. K. Shoichet, I. D. Kuntz, R. Abagyan, M. Totrov, J. Janin, J. Cherfils, F. Zimmerman, A. Olson, B. Duncan, M. Rao, R. Jackson, M. Sternberg, and M. N. G. James. *Nature Struct. Biol.* 3(3):233–239 (1996).
22. M. C. Lawrence and P. C. Davis. *Proteins* 12:31–41 (1992).
23. H.-J. Böhm. *J. Comput. Aided Mol. Des.* 8:623–632 (1994).
24. B. K. Shoichet, R. M. Stroud, D. V. Santi, I. D. Kuntz, and K. M. Perry. *Science* 259:1445–1450 (1993).
25. D. A. Gschwend, W. Sirawaraporn, D. V. Santi, and I. D. Kuntz. *Proteins* 29:59–67 (1997).
26. D. M. Lorber and B. K. Shoichet. *Protein Science* 7(4):938–950 (1998).
27. A. R. Friedman, V. A. Roberts, and J. A. Tainer. *Proteins* 20:15–24 (1994).
28. V. Schnecke, C. A. Swanson, E. D. Getzoff, J. A. Tainer, and L. A. Kuhn. *Proteins* 33(1):74–87 (1998).
29. G. M. Crippen and T. F. Havel. *Distance Geometry and Molecular Conformation.* John Wiley & Sons, New York, (1988).
30. A. R. Fersht. *Trends Biochem. Sci.* 12:301–304 (1987).
31. J. Janin and C. Chothia. *J. Biol. Chem.* 265(27):16027–16030 (1990).

32. M. L. Raymer, P. C. Sanschagrin, W. F. Punch, S. Venkataraman, E. D. Goodman, and L. A. Kuhn. *J. Mol. Biol.* 265:445–464 (1997).
33. L. A. Kuhn, C. A. Swanson, M. E. Pique, J. A. Tainer, and E. D. Getzoff. *Proteins* 23:536–547 (1995).
34. H.-J. Böhm. *J. Comput. Aided Mol. Des.* 8:243–256 (1994).
35. A. N. Jain. *J. Comput. Aided Mol. Des.* 10:427–440 (1996).
36. E. E. Abola, F. C. Bernstein, S. H. Bryant, T. F. Koetzle, and J. Weng. In *Crystallographic Databases – Information Content, Software Systems, Scientific Applications*, F. H. Allen, G. Bergerhoff, and R. Sievers, editors, 107–132. Data Commission of the International Union of Crystallography, Bonn/Cambridge/Chester (1987).
37. E. A. Lunney. *Network Science* 1:http://www.awod.com/netsci/Issues/Sept95/feature1.html (1995).
38. U. Hobohm and C. Sander. *Protein Science* 3(3):522–524 (1994).

STUDYING MACROMOLECULAR MOTIONS IN A DATABASE FRAMEWORK: FROM STRUCTURE TO SEQUENCE

Mark Gerstein,[1] Ronald Jansen,[1]
Ted Johnson,[1] Jerry Tsai,[2] and Werner Krebs[1]

[1]Department of Molecular Biophysics and Biochemistry
Yale University, 266 Whitney Avenue
New Haven, CT 06511

[2]Department of Structural Biology, Stanford University
Stanford, CA 94305

ABSTRACT

We describe database approaches taken in our lab to the study of protein and nucleic acid motions. We have developed a database of macromolecular motions, which is accessible on the World Wide Web with an entry point at http://bioinfo.mbb.yale.edu/MolMovDB. This attempts to systematize all instances of macromolecular movement for which there is at least some structural information. At present it contains detailed descriptions of more than 100 motions, most of which are of proteins. Protein motions are further classified hierarchically into a limited number of categories, first on the basis of size (distinguishing between fragment, domain, and subunit motions) and then on the basis of packing. Our packing classification divides motions into various categories (shear, hinge, other) depending on whether or not they involve sliding over a continuously maintained and tightly packed interface. We quantitatively systematize the description of packing through the use of Voronoi polyhedra and Delaunay triangulation. In addition to the packing classification, the database provides some indication about the evidence behind each motion (i.e. the type of experimental information or whether the motion is inferred based on structural similarity) and attempts to describe many aspects of a motion in terms of a standardized nomenclature (e.g. the maximum rotation, the residue selection of a fixed core, etc). Currently, we use a standard relational design to implement the database. However, the complexity and heterogeneity of the information kept in the database makes it an ideal application for an object-relational approach, and we are moving it in this direction. The database, moreover, incorporates innovative Internet cooperatively features that allow authorized remote experts to serve as database editors. The database also contains plausible representations for motion pathways, derived from restrained 3D interpolation between known endpoint conformations. These pathways can be viewed in a variety of movie formats, and the database is associated with a server that can automatically generate these movies from submitted coordinates. Based on the structures in the database we have developed sequence patterns for linkers and flexible hinges and are currently using these for the annotation of genome sequence data.

INTRODUCTION

Motion is frequently the way macromolecules (proteins and nucleic acid) carry out particular functions; thus motion often serves as an essential link between structure and function. In particular, protein motions are involved in numerous basic functions such as catalysis, regulation of activity, transport of metabolites, formation of large assemblies and cellular locomotion. In fact, highly mobile proteins have been implicated in a number of diseases—e.g., the motion of gp41 in AIDS and that of the prion protein in scrapie[1-5]. Another reason for the study of macromolecular motions results from their fundamental relationship to the principles of protein and nucleic acid structure and stability.

Macromolecular motions are amongst the most complicated biological phenomena that can be studied in great quantitative detail, involving concerted changes in thousands of precisely specified atomic coordinates. Fortunately, it is now possible to study these motions in a database framework, by analyzing and systematizing many of the instances of protein structures solved in multiple conformations. We summarize here some recent work in our laboratory relating to the construction of a database of protein motions[6]) and the use of Voronoi polyhedra to study packing[7]. We also present some preliminary results relating to creating sequence patterns for hinges and flexible linkers.

Table 1. Statistics for the Mechanism of the Motions. This table cross-tabulates the two main classifying attributes of motions: their size (row heads) and their packing characteristics (column heads). We define a known motion to be a motion with two or more solved conformations, and a suspected motion is defined to have only one or fewer solved conformations. (Adapted from Gerstein and Krebs (1998).[6])

Size Mechanism	Domain		Fragment		Subunit		Total	
Hinge	38	51%	16	59%			54	45%
Shear	14	19%	3	11%			17	14%
Partial Refolding	5	7%					5	4%
Allosteric					8	57%	8	7%
Other/Non-Allosteric	2	3%	1	4%	6	43%	9	7%
Unclassifiable	15	20%	7	26%			22	18%
Notably Motionless							1	1%
Complex							2	2%
Nucleic Acid							3	2%
Known / % category	53	72%	25	93%	11	79%	94	78%
Suspected / % category	21	28%	2	7%	3	21%	27	22%
Totals / % DB	74	62%	27	23%	14	12%	121	100%

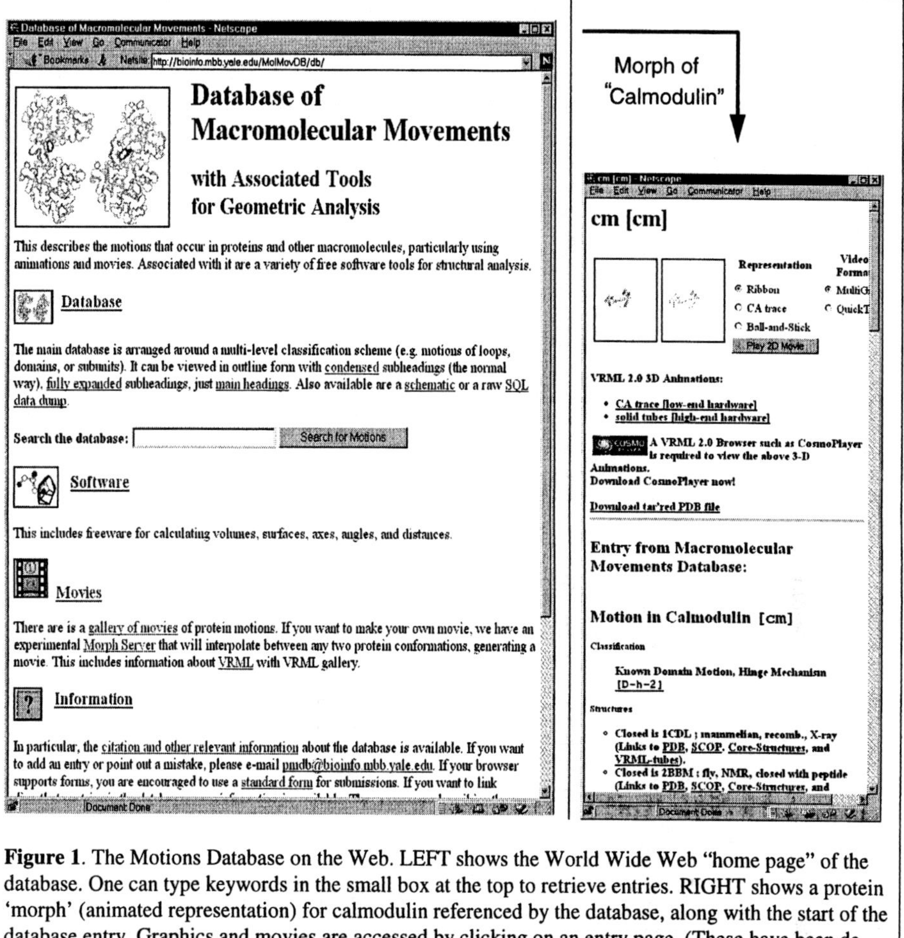

Figure 1. The Motions Database on the Web. LEFT shows the World Wide Web "home page" of the database. One can type keywords in the small box at the top to retrieve entries. RIGHT shows a protein 'morph' (animated representation) for calmodulin referenced by the database, along with the start of the database entry. Graphics and movies are accessed by clicking on an entry page. (These have been deliberately segregated from the textual parts of the database since the interface was designed to make it easy to use on a low-bandwidth, text-only browser, e.g. lynx or the original www_3.0.) The main URL for the database is http://bioinfo.mbb.yale.edu/MolMovDB. Beneath this are pages listing all the current movies, graphics illustrating the use of VRML to represent endpoints, and an automated submission form to add entries to the database. The database has direct links to the PDB for current entries (http://www.pdb.bnl.gov); the obsolete database (http://pdbobs.sdsc.gov) for obsolete entries; scop (http://scop.mrc-lmb.cam.ac.uk); Entrez/PubMed (http://www.ncbi.nlm.nih.gov/PubMed/medline.html); and LPFC (http://smi-web.stanford.edu/projects/helix/LPFC). Through these links one can easily connect to other common protein databases such Swiss-Prot, Pro-Site, CATH, RiboWeb, and FSSP [8-15].

THE DATABASE

The primary public interface to the database consists of coupled hypertext documents available over the World Wide Web at http://bioinfo.mbb.yale.edu/MolMovDB. As shown in Figure 1, use of the web interface is straightforward and simple. The database may be browsed either by typing various search keywords into the main page or by navigating through an outline. Either way brings one to the entries. Thus far, the database has ~120 entries, which reference over 240 structures in the Protein Databank (PDB) (Table 1).

Table 2. Standard Statistics for the Magnitude of the Motions. The motions in the database range greatly in size, with maximum mainchain displacements between 1.5 and 60 Å. All the statistics are for version 1.7 of the database, based on the relatively small set of values culled from the literature. The averages are only approximate given the sparse nature of the data. We are developing software tools to extract these values automatically from structural data. (Adapted from Gerstein and Krebs (1998).[86])

Value	Num. Entries	min	max	average
Maximum Cα displacement	11	1.5	60	12
Maximum Atomic Displacement	3	8.8	10	9.3
Maximum Rotation	12	5	148	24
Maximum Translation	2	0.7	2.7	1.7

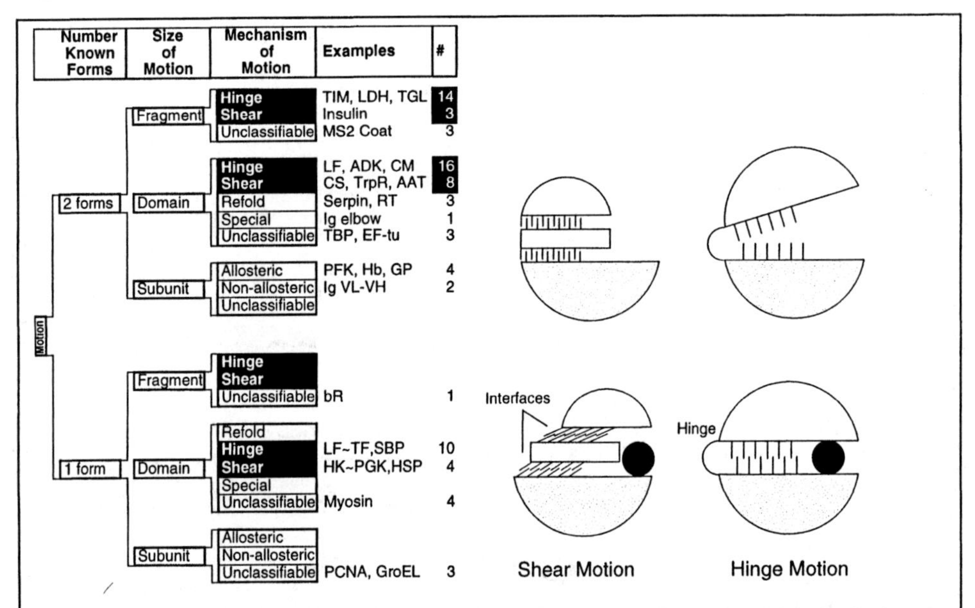

Figure 2. Schematic Showing the Overall Classification Scheme for Motions. TOP-LEFT, the database is organized around a hierarchical classification scheme, based on size (fragment, domain, subunit) and then packing (hinge or shear). Currently, the hierarchy also contains a third level for whether or not the motion is inferred. TOP-RIGHT is a schematic showing the difference between shear (sliding) and hinge motions. Figure adapted from[20,45]. It is important to realize that the hinge-shear classification in the database is only "predominate" so that a motion classified as shear can contain a newly formed interface and one classified as hinge can have a preserved interface across which there is motion. The essential characteristics of the various motions are summarized below. (Adapted from Gerstein and Krebs (1998).[86])

Unique Motion Identifier

Each entry is indexed by a *unique motion identifier*, rather than around individual proteins and nucleic acids. This is necessary because a single macromolecule can not only have a number of motions, but the essential motion can be shared amongst a number of different macromolecules.

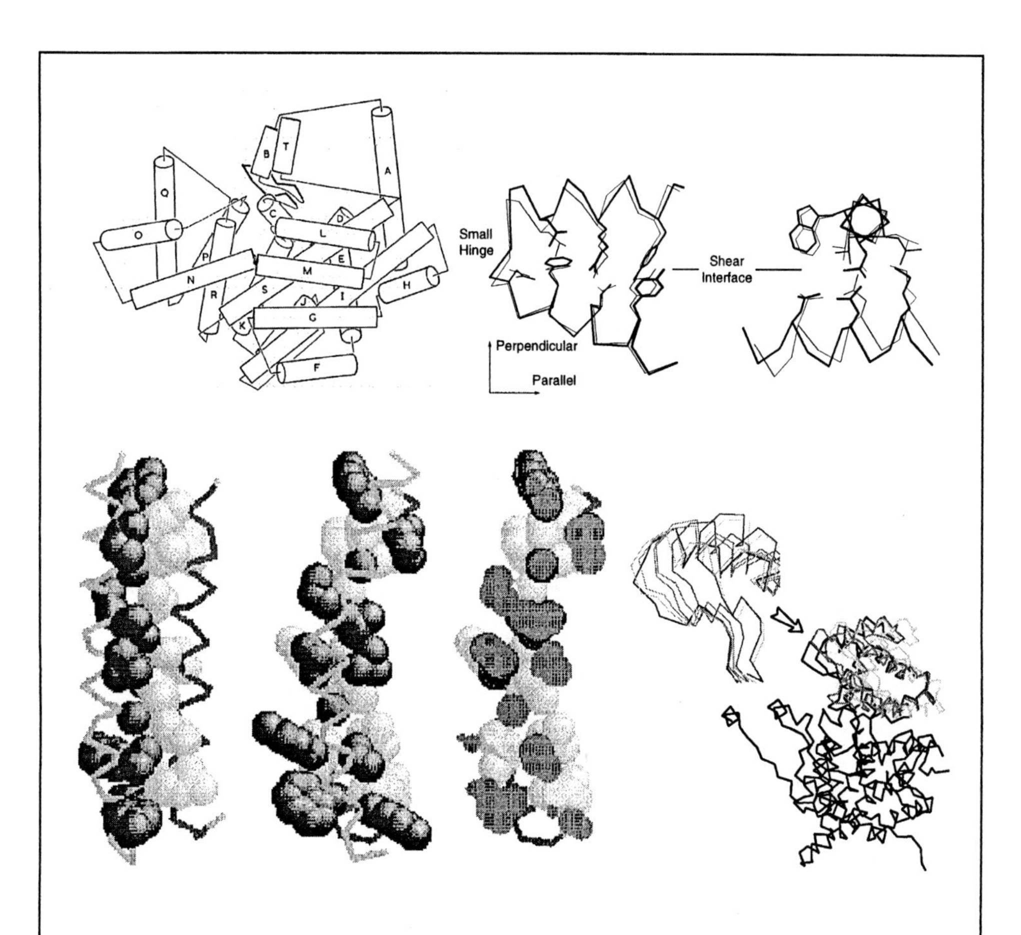

Figure 3. Closeup on the Shear Mechanism. The figure gives a close up illustrating shear motion in one protein, citrate synthase[20,93]. TOP-LEFT, Cartoon of one subunit of citrate synthase (1CTS) gives an overall view of the protein showing that it is composed of many helices. The adjacent one is related by two-fold axis shown. The small two-stranded sheet is omitted to improve clarity. a-helices are represented by cylinders. The small domain contains helices N, O, P, Q, and R. TOP-MIDDLE and TOP-RIGHT show representative shear motions between close-packed helices. Note how the mainchain only shifts by a small amount and the sidechains stay in the same rotamer configuration. BOTTOM-LEFT highlights the "knobs into holes" interdigitation of two close-packed helices. BOTTOM-RIGHT shows how these small motions can be added together to produce a large overall motion. Specifically, many small motions add up to shift helix O by 10.1 Å and rotate it by 28°. The incremental motion in shear domain closure is shown by Ca traces of the whole protein and of a closeup of the OP loop. BLACK is the apo form; WHITE, holo form; GRAY, cumulative effect of motion over the K, P, and then Q helix-helix interfaces. (The apo form was fit to the holo form, first on the core, and then on the K, P, and Q helices.) (Parts adapted from Gerstein and Krebs (1998).[86])

* At the time of writing, the PDB contained in excess of 6600 protein structures, but less than 600 nucleic acids structures.

† There is, of course, also the motion (i.e. rotation) of individual sidechains, often on the protein surface. However, this is on a much smaller scale than the motion of fragments or domains. It also occurs in all proteins. Consequently, sidechain motions are not considered to constitute individual motions in the database, being considered here a kind of background, intrinsic flexibility, common to all proteins.

insulin[31-33]. Often domain and fragment motions involve portions of the protein closing around a binding site, with a bound substrate stabilizing a closed conformation. They, consequently, provide a specific mechanism for induced-fit in protein recognition[34,35]. In enzymes this closure around a binding site has been analyzed in particular detail[36-40]. It serves to position important chemical groups around the substrate, shielding it from water and preventing the escape of reaction intermediates.

Subunit motion is distinctly different from fragment or domain motion. It affects two large sections of polypeptide that are *not* covalently connected. It is frequently part of an allosteric transition and tied to regulation[41,42]. The relative motions of the subunits in the transport protein hemoglobin and the enzyme glycogen phosphorylase change the affinity with which these proteins bind to their primary substrates[43,44] and are good examples.

Packing Classification: Hinge and Shear

For protein motions of domains and smaller units, we have systematized the motions on the basis of packing, using a scheme developed previously[6,28]. This is because the tight packing of atoms inside of proteins provides a most fundamental constraint on protein structure[45-50]. Unless there is a cavity or packing defect, it is usually impossible for an atom inside a protein to move much without colliding with a neighboring atom[51,52].

Internal interfaces between different parts of a protein are packed very tightly[7,28,53]. Furthermore, they are not smooth, but are formed from interdigitating sidechains. Common sense consideration of these aspects of interfaces places strong constraints on how a protein can move and still maintain its close packing. Specifically, maintaining packing throughout a motion implies that the sidechains at the interface must maintain their same relative orientation and pattern of inter-sidechain contacts in both conformations (e.g. open and closed).

These straightforward constraints on the types of motions that are possible at interfaces allow an individual movement within a protein to be described in terms of two basic mechanisms, shear and hinge, depending on whether or not it involves sliding over a continuously maintained interface[28] (Figure 2). A complete protein motion (which can contain many of these smaller "movements") can be built up from these basic mechanisms. For the database, a motion is classified as *shear* if it predominately contains shear movements and as *hinge* if it is predominately composed of hinge movements. More detail on the characteristics of the two types of motion follows.

Shear. As shown in Figure 3, the shear mechanism basically describes the special kind of sliding motion a protein must undergo if it wants to maintain a well-packed interface. Because of the constraints on interface structure described above, individual shear motions have to be very small. Sidechain torsion angles maintain the same rotamer configuration[54] (with $<15°$ rotation of sidechain torsions); there is no appreciable mainchain deformation; and the whole motion is parallel to the plane of the interface, limited to total translations of ~2 Å and rotations of 15°. Since an individual shear motion is so small, a single one is not sufficient to produce a large overall motion, and a number of shear motions have to be concatenated to give a large effect — in a similar fashion to each plate in a stack of plates sliding slightly to make the whole stack lean considerably. Examples include the Trp repressor and aspartate amino transferase[55,56].

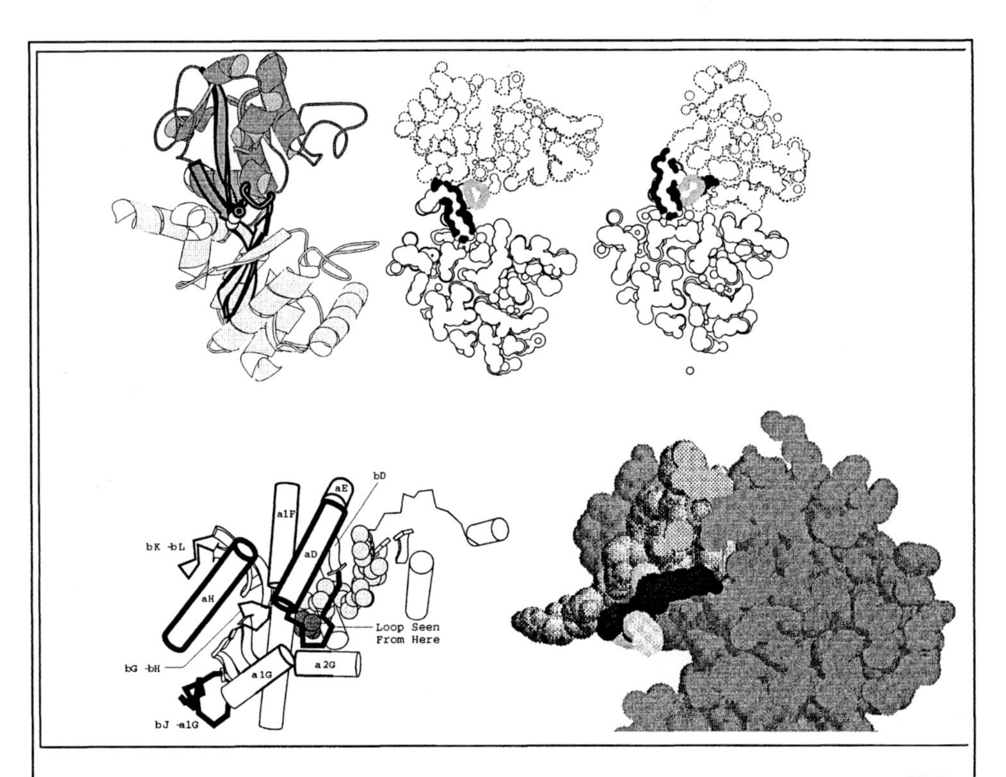

Figure 4. Close-up on the Hinge Mechanism. The figure shows the hinge motion in lactoferrin[20,45]. FAR-LEFT shows a ribbon drawing of the protein in the open conformation. The view is down the screw-axis, which is indicated in the figure by the circle with the dot in it. The screw-axis passes very close to the hinge region, which occurs in the middle of two beta strands (highlighted in bold). MIDDLE-LEFT and MIDDLE-RIGHT show the open and closed conformations in terms of space filling slices. The hinge region is highlighted by a thick black line. Note how few packing constraints there are on the hinge in contrast to the other atoms in the protein. (Figure adapted from Gerstein (1993).[45]) BOTTOM-LEFT shows the placement of a mobile loop in another protein, lactate dehydrogenase.
BOTTOM-RIGHT shows a close-up of this loop that highlights the absence of close-packing at the base of the hinge. Hinge mainchain is shown in black (first hinge) and almost white (second hinge). Rest of protein is shown in shades of gray.

Hinge. As shown in Figure 4, hinge motions occur when there is *no* continuously maintained interface constraining the motion. These motions usually occur in proteins that have two domains (or fragments) connected by linkers (i.e. hinges) that are relatively unconstrained by packing. A few large torsion angle changes in the hinges are sufficient to produce almost the whole motion. The rest of the protein rotates essentially as a rigid body, with the axis of the overall rotation passing through the hinges. The overall motion is always perpendicular to the plane of the interface (so the interface exists in one conformation but not in the other, as in the closing and opening of a book) and is identical to the local motion at the hinge. Examples include lactoferrin and tomato bushy stunt virus (TBSV)[57,58].

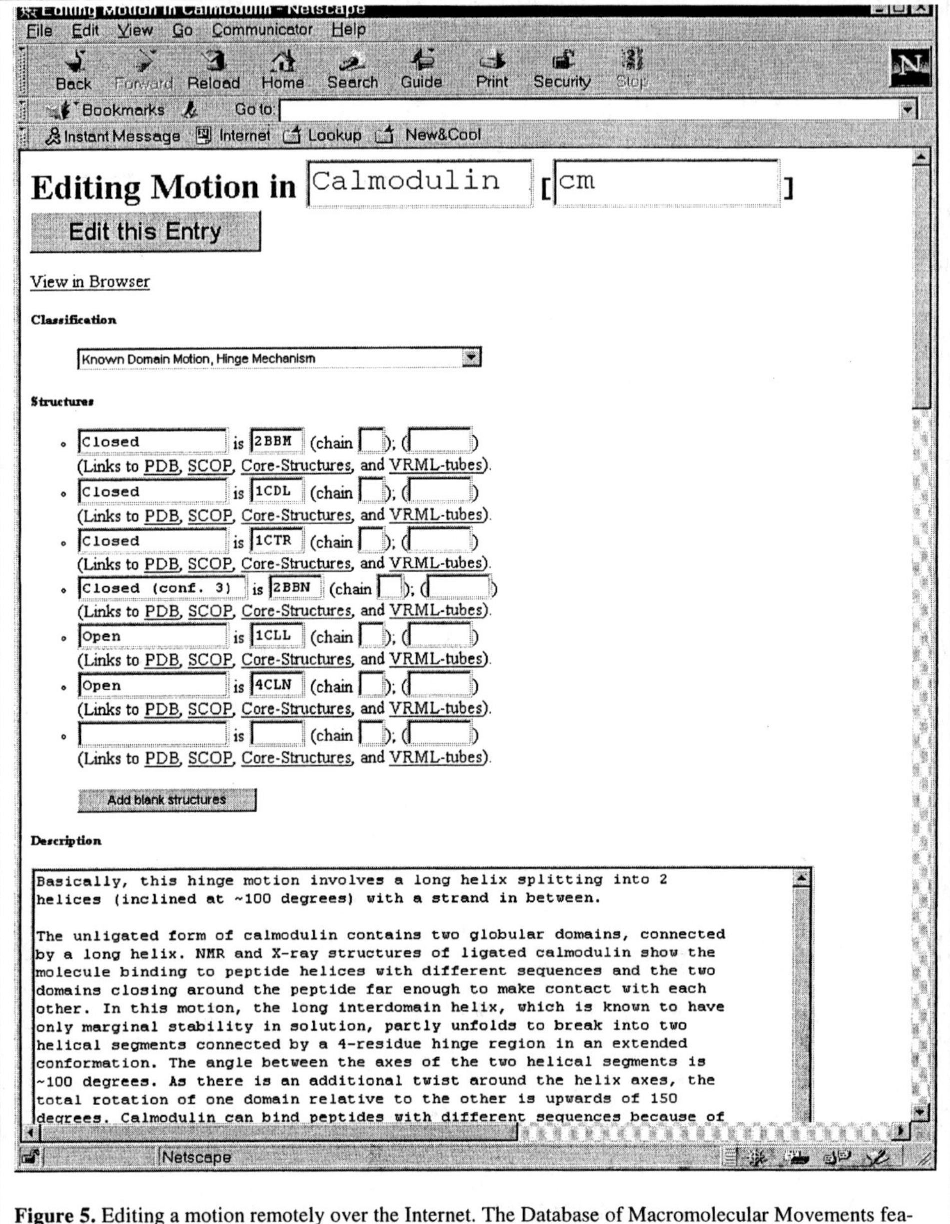

Figure 5. Editing a motion remotely over the Internet. The Database of Macromolecular Movements features an innovative Web form (shown here) that allows authorized remote users to collaborate and edit motions from remote sites around the world. Saved changes to motions may be previewed to see how they would appear to an end user and then applied to the database. If desired, saved changes can be made to appear immediately in the public Web interface to the database.

Gerstein et al.[53,59] analyzed the hinged domain and loop motion in specific proteins (lactate dehydrogenase, adenylate kinase, lactoferrin). These studies emphasized how critical the packing at the base of a protein hinge is (in the same sense that the "packing" at the base of an everyday door hinge determines whether or not the door can close). Protein hinges are special regions of the mainchain in the sense that they are exposed and have few packing constraints on them and are thus free to sharply kink (Figure 4). Most mainchain atoms, in contrast, are usually buried beneath layers of other atoms (usually sidechain atoms), precluding large torsion angle changes and hinge motions.

It is important to note that because most shear motions do, in fact, contain hinges, (joining the various sliding parts) the existence of a hinge is not the salient difference between the two basic mechanisms. Instead, it is the existence of a continuously maintained interface.

Other Classification

Most of the fragment and domain motions in the database fall within the hinge-shear classification. However, we have created additional categories to deal with the small number of exceptions.

Data Entry

One innovative feature of the database is that it allows authorized remote researchers to enter motions in their area of expertise directly into the database via a Web form. Authorization to edit a given motion entry, if necessary, works in conjunction with the standard password feature built into modern Web browser systems. The layout of the Web form is analogous to that of a normal HTML page describing a motion in the database, except that the various fields have been replaced by textboxes and pull-down selectors to make the Web page editable. The user retrieves either a blank form or a form corresponding to a pre-existing motion entry, makes appropriate changes remotely over the Internet via his or her Web browser, and then simply clicks the 'Submit' button to save changes into the database. Depending on whether or not the user has editing privileges over a particular motion entry, the changes may be published immediately or upon further approval by the database maintainers. The remote user may immediately preview the edited motion entry to see what it will look like once it becomes public.

The Web form system (Figure 5) takes advantage of advanced features of the Informix Dynamic Server with Universal Option to enable user previews. The Web Datablade module allows database content to be dynamically and rapidly translated into Web content with little additional overhead compared to static pages. Because updates to the database can be translated instantaneously into updated Web content, remote editors are able to preview their changes as it will appear to the end database user instantaneously before submitting or publishing them. Previously, we stored the database using the MSQL database software package, which is freely available to academic users. Unlike the commercial Informix system, the MSQL package does not support Application Program Interfaces (APIs) that allow for an efficient, rapid translation of database content into Web content. Consequently, it was necessary to store the Web interfaces as static HTML files on the server. For Web content to remain current, these pages would need to be rebuilt each time the database changed, a time-consuming process that would have prevented accurate previews. In addition, the Informix database system also features state-of-the-art transaction concurrency and logging, important features when multiple users are simultaneously updating the database.

In this way, the database takes full advantage of the cooperatively features of the Internet and modern database software, allowing experts in distant parts of the world to collaborate simultaneously on macromolecular motions. In addition to accelerating the rate at which the database may be populated, this feature improves the accuracy and timeliness of existing database entries by allowing them to be edited, revised, and updated, if necessary, by experts in the field.

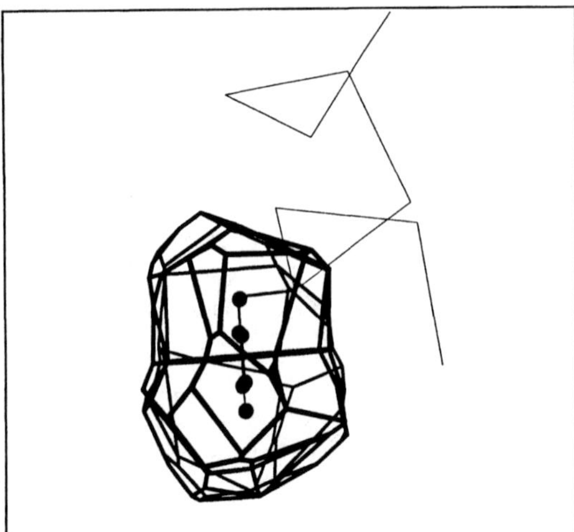

Figure 6. Voronoi Polyhedra. Two representative Voronoi polyhedra from 1CSE (subtilisin). On the left is shown the polyhedron around the sidechain hydroxyl oxygen (OG) of a serine. On right is shown the six polyhedra around the atoms in a Phe ring.

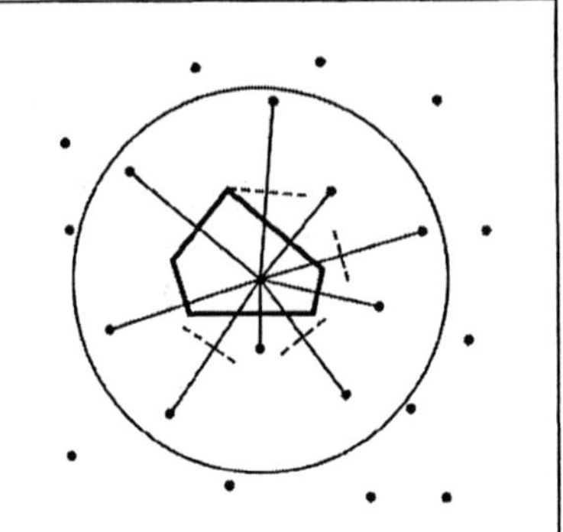

Figure 7. The Voronoi Polyhedra Construction. A schematic showing the construction of a Voronoi polyhedron in 2-dimensions. The asymmetry parameter is defined as the ratio of the distances between the central atom and the farthest and nearest vertex.

Internet Hits

The database is currently receiving over 65,000 hits from over 45,000 sites each month. Internet traffic on the database's main web server grew approximately exponentially between November, 1997, and February 1998, with database usage doubling approximately every other month during this period. In recent months, database usage has continued to grow, albeit at a somewhat reduced rate. We expect this trend to continue as the database becomes established in the structural biology community.

STANDARDIZED TOOLS FOR PROTEIN MOTIONS

Quantification of packing using Voronoi polyhedra

Packing clearly is an essential component of the motions classification. Often this concept is discussed loosely and vaguely by crystallographers analyzing a particular protein structure—for instance, "Asp23 is packed against Gly38" or "the interface between domains appears to be tightly packed." We have attempted to systematize and quantify the discussion of packing in the context of the motions database through the use of particular geometric constructions called Voronoi polyhedra and Delaunay triangulation.[53]

Voronoi polyhedra are a useful way of partitioning space amongst a collection of atoms. Each atom is surrounded by a single convex polyhedron and allocated the space within it (Figure 6). The faces of Voronoi polyhedra are formed by constructing dividing planes perpendicular to vectors connecting atoms, and the edges of the polyhedra result from the intersection of these planes.

Voronoi polyhedra were originally developed (obviously enough) by Voronoi[60] nearly a century ago. Bernal and Finney[61] used them to study the structure of liquids in the 1960s. However, despite the general utility of these polyhedra, their application to proteins was limited by a serious methodological difficulty: while the Voronoi construction is based around partitioning space amongst a collection of "equal" points, all protein atoms are not equal: some are clearly larger than others (e.g. sulfur versus oxygen). Richards[62] found a solution to this problem and first applied Voronoi polyhedra to proteins in 1974. He has, subsequently, reviewed their use in this application[48,49].

Voronoi polyhedra are particularly useful in studying the packing of the protein interior. This is because the construction of Voronoi polyhedra allocates all space amongst a collection of atoms; there are no gaps as there would be if one, say, simply drew spheres around the atoms. Thus, the volume of cavities or defects between atoms are included in their Voronoi volume, and one finds that the packing efficiency is inversely proportional to the size of the polyhedra. This indirect measurement of cavities contrasts with other types of calculations that measure the volume of cavities explicitly[63]. Moreover, since protein interiors are tightly packed, fitting together like a jig-saw puzzle, the various types of protein atoms occupy well-defined amounts of space. This fact has made the calculation of standard volumes for residues in proteins[46,64] a worthwhile proposition.

Voronoi polyhedra calculations have been applied to other aspects of packing in protein structure. In particular, they have been used to study protein-protein recognition[65], protein motions[53], and the protein surface[7,66-68]. As the Voronoi volume of an atom is a weighted average of the distances to all its neighbors (where the contact area with a neighbor is the weight), Voronoi polyhedra are very useful in assessing interatomic contacts[68-70]. Furthermore, the faces of Voronoi polyhedra have been used to characterize protein accessibility and to assess the fit of docked substrates in enzymes[71,72].

Voronoi polyhedra have many uses beyond the analysis of protein structures. For instance, they have also been used in the analysis of liquid simulations[73] and in weighting sequences to correct for over- or under-representation in an alignment[74]. In non-biological applications, they are used in "nearest-neighbor" problems (trying to find the neighbor of a query point) and in finding the largest empty circle in a collection of points[75]. The dual of a Voronoi diagram is a Delaunay triangulation. Since this triangulation has the "fattest" possible triangles, it is convenient for such procedures as finite element analysis. Furthermore, the border of Delaunay triangulation is the convex hull of an object, which is useful in graphics[75].

The simplest method for calculating volumes with Voronoi polyhedra is to put all atoms in the system on a grid. Then go to each grid-point (i.e. voxel) and add its volume to the atom center closest to it. This is prohibitively slow for a real protein structure, but it can be made somewhat faster by randomly sampling grid-points. It is, furthermore, a useful approach for high-dimensional integration[74] and for the curved dividing surface approach discussed later.

More realistic approaches to calculating Voronoi volumes have two parts: (1) for each atom find the vertices of the polyhedron around it and (2) systematically collect these vertices to draw the polyhedron and calculate its volume.

In the basic Voronoi construction (Figure 7), each atom is surrounded by a unique limiting polyhedron such that all points within an atom's polyhedron are closer to this atom than all other atoms. Points equidistant from two atoms are on a plane; those equidistant from three atoms are on a line, and those equidistant from four centers form a vertex. One can use this last fact to easily find all the vertices associated with an atom. With the coordinates of four atoms, it is straightforward to solve for possible vertex coordinates using the equation of a sphere.* One then checks whether this putative vertex is closer to these

* That is, one uses four sets of coordinates (x,y,z) to solve for the center (a,b,c) of the sphere: $(x-a)^2+(y-b)^2+(z-c)^2=r^2$. (This method can fail for certain pathological arrangements of atoms that would

four atoms than any other atom; if so, it is a vertex.

In the procedure outlined above, all the atoms are considered equal, and the dividing planes are positioned midway between atoms (Figure 6). This method of partition, called bisection, is not physically reasonable for proteins, which have atoms of obviously different size (such as oxygen and sulfur). It chemically misallocates volume, giving an excess to the smaller atom.

Two principal methods of re-positioning the dividing plane have been proposed to make the partition more physically reasonable: method B[62] and the radical-plane method[77]. Both methods depend on the radii of the atoms in contact (R_1 and R_2) and the distance between the atoms (D).

Representing Motion Pathways as "Morph Movies"

One of the most interesting of the complex data types kept in the database are "morph movies" giving a plausible representation for the pathway of the motion. These movies can immediately give the viewer an idea of whether the motion is a rigid-body displacement or involves significant internal deformations (as in tomato bushy stunt virus versus citrate synthase). Pathway movies were pioneered by Vorhein et al.[78], who used them to connect the many solved conformations of adenylate kinase.

Normal molecular-dynamics simulations (without special techniques, such as high temperature simulation or Brownian dynamics[79-81]) cannot approach the timescales of the large-scale motions in the database. Consequently a pathway movie cannot be generated directly via molecular simulation. Rather, it is constructed as an interpolation between known endpoints (usually two crystal structures). The interpolation can be done in a number of ways.

Straight Cartesian interpolation. The difference in each atomic coordinate (between the known endpoint structures) is simply divided into a number of evenly spaced steps, and intermediate structures are generated for each step. This was the method used by Vorhein et al. It is easy to do, only requiring that the beginning and ending structures be intelligently positioned by fitting on a motionless core. However, it produces intermediates with clearly distorted geometry.

Interpolation with restraints. This is the above method where each intermediate structure is restrained to have correct stereochemistry and/or valid packing. One simple approach is to minimize the energy of each intermediate (with only selected energy terms) using a molecular mechanics program, such as X-PLOR[82]. This technique will be described more fully in a forthcoming paper (Krebs & Gerstein, manuscript in preparation). The database, furthermore, is currently home to an experimental server that applies this interpolation technique to two arbitrary structures, generating a movie.

ANALYSIS OF AMINO ACID COMPOSITION OF LINKER SEQUENCES

Now that we have developed a database of protein motions, an essentially structure-orientated database, we want to use this to help interpret the mass of sequence data coming out of genome sequencing projects. In this way we are extrapolating ideas developed on the (relatively) smaller structure database to the much larger sequence database. We propose to do this through the calculation of two propensity scales for amino acids to be in linkers or flexible hinges.

---

not normally be encountered in a real protein structure; see Proacci and Scateni[76]).

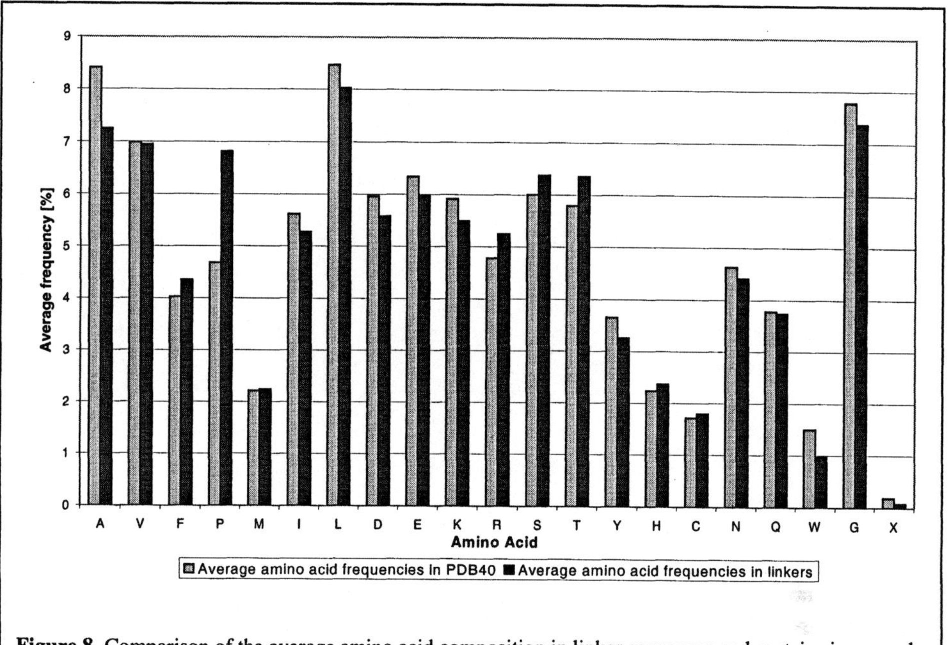

Figure 8. Comparison of the average amino acid composition in linker sequences and proteins in general (as represented by the PDB40 database).

Solved protein structures typically reveal different domains of proteins and linker regions between these domains. Linker regions are typically flexible, and, as such, form the basis for the hinge regions that allow two protein domains or fragments to move relative to each other as a part of a hinge mechanism.

Information about the amino acid composition of linker sequences can potentially be used to predict protein domains in protein sequences of unknown structure. In particular, a profile of flexible linker regions might be used to predict the location of domain hinges, for structural annotation of genome sequences.[93] Here we present some preliminary results involving two methods for statistical analysis of linker sequences.

Propensities for Linkers in General

Our first method of analysis of linker sequences includes both flexible as well as inflexible linkers. In this method we have arbitrarily defined a linker sequence as the 16 residue region centered around the peptide bond linking two domains.

The analysis of the amino acid composition of linker sequences is an example of deriving sequence information from structural information. The structural information (i.e., the location of protein domains) can be found in the Structural Classification of Proteins (SCOP)[19,20]. SCOP contains several databases of amino acid sequences of protein domains. In our study, the PDB40 database provided by SCOP has been used to create a database of linker sequences. The PDB40 database comprises a subset of proteins in the Protein Data Bank (PDB) with known structure selected so that, when aligned, no two proteins in the subset show a sequence identity of 40% or greater. Thus, the data set is not biased towards protein structures listed multiple times in the PDB. We were able to extract 234 linker sequences from the PDB40 database, although the PDB40 database itself contains about 1,500 protein sequences. This mainly reflects the fact that many proteins consist of only a single domain and therefore contain no linker region.

Figure 8 compares the average amino acid composition of the linker sequences with the average amino acid composition of the PDB40 database, while Table 3 shows in more detail the profile of the amino acid composition at each of the sixteen positions in the linker sequence. For an interpretation of these results it is important to compute two-sided P-values to determine which amino acids show statistically different frequencies in linkers than in the database as a whole. (A two-sided P-value represents the probability that, in a data set of equal size drawn at random from the PDB40 database, a given amino acid would have a frequency of occurrence as different as or more different from its occurrence in the entire PDB40 database than what was actually observed in the linker subset.) Figure 9 shows the P-values for the average amino acid composition in the linkers. We are able to conclude, with better than 98% confidence, that linker regions are proline-rich and alanine- and trypthophan-poor. In particular, the statistical evidence that linkers are proline-rich is unusually strong and is significant at better than the hundredth-of-a-percent level. Table 4 shows the P-values of the amino acids at each of the sixteen linker positions.

In Table 4 and Figure 9 the amino acids have been roughly grouped according to the attributes hydrophobic, charged, and polar (following the classification of Branden and Tooze[83]). As shown in Table 4 and Figure 9, the frequencies of the remaining amino acids in linkers are not statistically different from the database as a whole at the 5% significance level.

The statistical significance of the results of the computed amino acid averages can be assessed by comparing the composition of the linker sequences with random data sets of

Table 3. Profile of the amino acid composition in linker sequences for every single linker position in detail compared with the PDB40 averages. A linker has been arbitrarily defined as the 16 residue region centered around the peptide bond (between positions 8 and 9) linking two domains. Positions where the amino acid frequency is less than the PDB40 average have a gray background.

	1	2	3	4	5	6	7	8	9	10	11	12	13	14	15	16	PDB40 average
A	8.6	7.8	4.7	5.6	6.0	8.6	9.5	5.6	4.7	6.5	5.6	7.3	6.9	9.1	9.5	9.9	8.4
V	6.0	8.2	8.2	6.0	8.2	5.6	9.1	6.0	8.2	4.7	6.0	4.7	7.3	9.1	5.2	8.6	7.0
F	4.7	3.9	6.5	3.5	2.6	2.6	6.0	2.6	4.7	3.0	4.3	6.0	5.2	4.3	4.3	5.6	4.0
P	3.9	6.5	6.0	6.0	5.2	9.1	6.9	10.8	9.1	10.3	9.9	6.0	8.6	2.6	4.7	3.5	4.7
M	4.7	1.3	1.3	2.6	2.6	0.0	1.7	1.7	4.3	3.0	1.3	1.3	2.2	1.7	3.0	3.0	2.2
I	5.6	3.5	7.3	6.5	3.9	6.0	3.9	3.5	5.2	6.9	4.7	2.6	4.7	8.6	5.6	6.0	5.6
L	11.6	9.1	11.2	6.0	16.4	7.3	4.3	6.5	8.2	3.5	7.3	5.2	7.3	6.5	10.3	7.8	8.5
D	4.7	6.5	6.0	3.9	6.0	4.7	5.6	8.6	4.3	3.9	3.5	7.3	6.9	7.3	4.3	5.6	6.0
E	5.2	5.2	3.9	6.5	4.7	4.7	7.8	4.7	6.5	4.3	6.5	9.1	7.3	5.2	8.6	5.6	6.3
K	5.2	6.5	3.9	5.6	5.2	6.9	4.7	4.7	6.0	7.8	3.9	6.5	5.2	5.2	3.0	7.8	5.9
R	5.2	3.9	4.7	9.1	6.5	5.2	5.2	5.6	5.6	4.7	6.0	5.2	5.2	4.7	3.0	4.3	4.8
S	7.8	6.0	5.2	6.9	6.5	8.2	6.9	6.5	3.5	6.0	9.5	7.8	4.3	3.9	8.6	4.7	6.0
T	4.7	5.6	3.0	5.6	6.5	9.5	6.9	6.0	6.5	11.2	7.3	6.5	6.0	4.7	8.2	3.5	5.8
Y	2.2	3.9	6.5	3.0	3.5	2.2	2.6	3.5	2.2	3.9	2.6	2.2	3.0	3.5	3.5	4.3	3.7
H	1.7	3.5	3.0	3.5	3.5	2.6	3.5	2.2	2.2	0.9	1.7	2.2	1.7	2.6	1.3	2.2	2.2
C	1.7	2.6	0.9	1.3	1.7	2.6	0.4	2.2	0.9	1.3	4.7	1.7	1.7	3.9	0.4	0.9	1.7
N	4.7	3.9	3.5	6.5	3.0	4.3	2.6	3.0	5.6	5.2	3.5	6.5	3.9	6.0	3.0	5.6	4.6
Q	3.9	5.2	3.5	5.2	2.6	0.9	3.0	2.2	3.5	4.7	3.5	2.2	6.5	4.3	4.3	4.7	3.8
W	1.3	0.9	0.9	2.6	0.4	0.9	0.4	0.9	0.4	1.3	0.0	1.3	0.4	0.9	2.2	0.9	1.5
G	6.0	6.0	9.9	4.3	5.2	8.2	9.1	13.4	8.2	6.9	8.2	8.6	5.6	6.0	6.9	5.6	7.8
X	0.4	0.4	0.0	0.0	0.0	0.0	0.0	0.0	0.4	0.0	0.0	0.0	0.0	0.0	0.0	0.0	0.2

Table 4. P-values for the profile of the amino acid composition of linker sequences for every single position in the linkers. P-values less than 0.05 are represented by a gray background. The low P-values for proline in positions 6 to 11 are most conspicuous. The classification according to the attributes hydrophobic, charged, and polar (Branden and Tooze[76]) does not provide a satisfactory explanation for the observed levels of amino acids (see also Figure 9).

	1	2	3	4	5	6	7	8	9	10	11	12	13	14	15	16	
A	.908	.728	4e-2	.125	.196	.908	.562	.125	4e-2	.293	.125	.561	.415	.729	.562	.416	hydrophobic
V	.577	.481	.481	.577	.481	.417	.224	.577	.481	.184	.577	.184	.841	.224	.285	.338	
F	.598	.911	.059	.666	.276	.276	.126	.276	.598	.449	.836	.126	.393	.836	.836	.235	
P	.573	.207	.346	.346	.737	2e-3	.114	5e-5	2e-3	1e-4	3e-4	.346	4e-3	.134	.971	.385	
M	1e-2	.366	.366	.717	.717	2e-2	.637	.637	3e-2	.433	.366	.366	.961	.637	.433	.433	
I	.990	.155	.267	.585	.257	.793	.257	.155	.772	.408	.571	4e-2	.571	5e-2	.990	.793	
L	.084	.754	.136	.186	3e-5	.541	2e-2	.280	.882	6e-3	.541	.071	.541	.280	.312	.705	
D	.442	.750	.966	.185	.966	.442	.821	.089	.296	.185	.108	.389	.556	.389	.296	.821	charged
E	.476	.476	.127	.936	.327	.327	.384	.327	.936	.211	.936	.092	.545	.476	.158	.653	
K	.638	.730	.194	.842	.638	.538	.457	.457	.945	.243	.194	.730	.638	.638	.061	.243	
R	.793	.530	.974	2e-3	.240	.793	.793	.575	.575	.974	.389	.793	.793	.974	.215	.742	
S	.269	.990	.599	.578	.774	.166	.578	.774	.101	.990	2e-2	.269	.283	.176	.095	.425	polar
T	.498	.897	.069	.897	.673	2e-2	.485	.886	.673	5e-4	.328	.673	.886	.498	.121	.127	
Y	.234	.864	2e-2	.619	.872	.234	.402	.872	.234	.864	.402	.234	.619	.872	.872	.612	
H	.619	.237	.455	.237	.237	.740	.237	.939	.939	.166	.619	.939	.619	.740	.354	.939	
C	.997	.336	.345	.647	.997	.336	.139	.634	.345	.647	2e-2	.997	.997	2e-2	.139	.345	
N	.942	.597	.404	.193	.251	.820	.143	.251	.500	.710	.404	.193	.597	.326	.251	.500	
Q	.937	.281	.804	.281	.359	2e-2	.562	.206	.804	.460	.804	.206	3e-2	.684	.684	.460	
W	.810	.459	.459	.193	.197	.459	.197	.459	.197	.810	.055	.810	.197	.459	.452	.459	
G	.324	.324	.233	5e-2	.139	.823	.482	1e-3	.823	.621	.823	.643	.218	.324	.621	.218	
X	.717	.717	.752	.752	.752	.752	.752	.752	.717	.752	.752	.752	.752	.752	.752	.752	

sequences of the same length and the same amount taken from the PDB40 database. The number of times a single amino acid occurs in multiple random data sets follows the binomial distribution according to the familiar equation:

$$P^N(k) = \binom{n}{k} p^k (1-p)^{n-k}$$

Here, p is the probability that the amino acid occurs in the PDB40 database, and $P^n(k)$ is the probability that the amino acid occurs k times in a data set of n samples (n = 234 for the distribution of every *single* of the sixteen linker positions and n = 234 x 16 for the distribution of the linker *average*). The ratio k/n represents the fraction of the amino acid in the data set. Knowledge of the distribution functions of the amino acids then allows the calculation of P-values from the cumulative distribution function:

$$CDF^n(k) = \sum_{i=0}^{k} P^n(i)$$

The value of $CDF^n(k)$ is the probability that the number of counts of an amino acid in a random data set would be less than k. Consequently, if o and e represent the observed and expected counts, then the two-sided P-value is given by $1\text{-}CDF^n(e+|o\text{-}e|) + CDF^n(e\text{-}|o\text{-}e|)$. This is simply the probability that the number of counts observed in a random subset of PDB40 would take on a value more different from what was expected than what was observed. In order to assign a P-value to an amino acid frequency in the linkers data set, the discrete values of the cumulative distribution function have been linearly interpolated. In

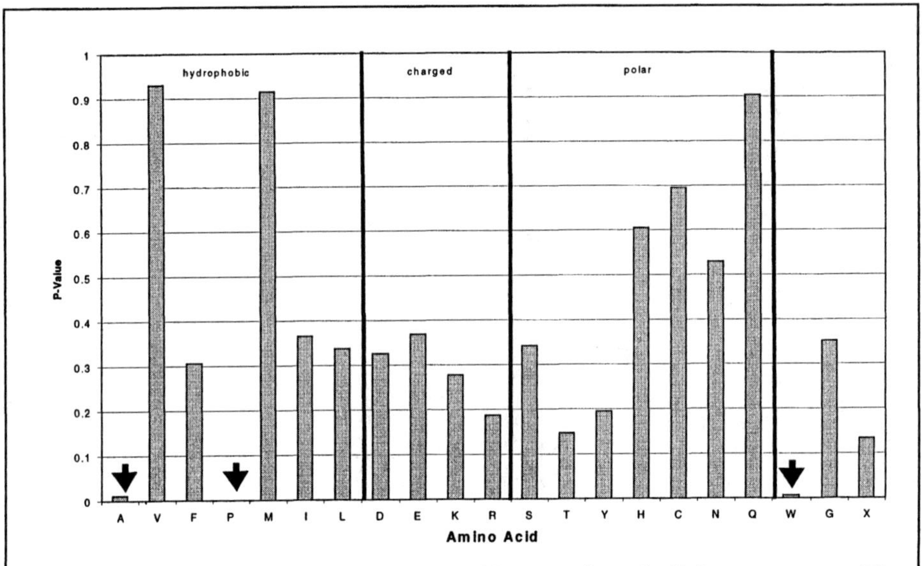

Figure 9. P-values for the average amino acid compositons in linker sequences. The P-values of alanine, proline, and tryptophan are close to zero. The difference between the content of these amino acids in linkers and protein sequences in general (as represented by the PDB40 database) is statistically significant at better than 98% confidence.

most cases, it is also possible to obtain a satisfactory approximation to the P-values by applying the two-sided significance test to the Normal approximation of the Binomial distribution.

Towards Propensities for Flexible Linkers

A variant on this procedure involves focusing just on linkers that are known to be flexible. Our Database of Macromolecular Motions contains residue selections for known protein hinge regions (i.e., flexible linkers) that have been culled from the scientific literature. These sequences have been verified manually to be true flexible linker regions, and thus this database constitutes a potential "gold standard" free from algorithmic biases that can be used as a starting point in the development of propensity scales and other research leading towards algorithmic techniques. By expanding these residue selections slightly with a predetermined protocol and extracting the corresponding sequences from the PDB, a series of sequences of known flexible linkers may be obtained. A FASTA search with a suitable cutoff (e.g., e-value 0.001) may then be performed on known linker sequence to obtain a series of near homologues (Table 6).These homologues can then be arranged into a multiple alignment (via the CLUSTALW) program[84,85] and the multiple alignment can be fused into a variety of consensus pattern representations, such as Hidden Markov Models or simply consensus sequences[86-90]. A sample multiple alignment for the hinge in calmodulin is shown in Table 6 and a number of consensus sequences are shown in Table 5. The amino acid composition may be averaged over all the different hinges and different positions within a hinge to give a single composition vector for flexible hinges. Finally, this can be compared to the overall amino acid composition or that of linkers to obtain a preliminary scale of amino acid propensity in mobile linkers, as shown in Table 7. This can be compared with the scale of amino acid propensities in linkers as obtained by the procedure previously described and shown in Table 3.

Table 5. Example of protein flexible linker consensus sequences extracted from the Macromolecular Movements Database. The database contains residue selections for known hinge regions (flexible linkers) culled from the scientific literature. Sixteen of these residue selections were then "grown" slightly in both directions according to a fixed protocol. Each selection was assigned a linker ID, which is based either on a PDB ID or on the macromolecular movements database motion ID plus possible an optional additional numeric suffix to identify the specific residue selection used. A FASTA search with a cutoff of 0.01 was then performed on each sequence to obtain near homologues. The consensus sequence corresponding to each linker ID is given here.

Linker ID	Linker Consensus Sequence
4cln	MARKMKDTDSE
6ldh	AGARQQEGESRLNLVQRNVNIFKF
adenkin1	VPFEVI
adenkin2	LRLTA
adenkin3	GEPLIQRDDDKE
adenkin4	AYHAQTE
anxbreat	MKGAGT
anxtrp1	YEAGELKWG
anxtrp2	EETIDRET
dt	LFQVVHNS
enolase	GASTGIY
enolase2	SDKS
lfh_hinge1	QTHY
lfh_hinge2	RVPS
ras	AGQEEYSAMRDQYMR
tbsv	PQPTNTL

CONCLUSION AND FUTURE DIRECTIONS

We have developed a number of database-based techniques for the study of macromolecular motions. We have constructed a database of macromolecular motions, which currently documents ~120 motions, and have developed a classification scheme for the database based on size then packing (whether or not there is motion across a well-packed interface). The database incorporates innovative cooperatively features, allowing authorized remote experts to act as database editors via the Internet. We also developed a standardized nomenclature, such as maximum atomic displacement or degrees of rotation. We are developing automated tools to analyze protein and nucleic acid structures and sequences with possible motions, to extract standardized statistics on macromolecular motions from structural data, and allow the database to be more readily populated.

We expect that the number of macromolecular motions will greatly increase in the future, making a database of motions somewhat increasingly valuable. Our reasoning behind this conjecture is as follows: The number of new structures continues to go up at a rapid rate (nearly exponential). However, the increase in the number of folds is much slower and is expected to level off much more in the future as the we find more and more of the limited number of folds in nature, estimated to be as low as 1000[91,92]. Each new structure solved that has the same fold as one in the database represents a potential new motion -- i.e. it is often a structure in a different liganded state or a structurally perturbed homologue. Thus, as we find more and more of the finite number of folds, crystallography and NMR will increasingly provide information about the variability and mobility of a given fold, rather than identifying new folding patterns.

Databases potentially represent a new paradigm for scientific computing. In an (over-simplified!) cartoon view, scientific computing traditionally involved big calculations on fast computers. The aim in these often was prediction based on first principles -- e.g. prediction of protein folding based on molecular dynamics. These calculations naturally emphasized the processor speed of the computer. In contrast, the new "database paradigm" focuses on small, inter-connected information sources on many different computers. The aim is communication of scientific information and the discovering of unexpected relationships in the data – e.g. the finding that heat shock protein looks like hexokinase. In contrast to their more traditional counterparts, these calculations are more dependent on disk-storage and networking rather than raw CPU power.

Table 6. Example of FASTA results. This table gives an example of sequences that might be obtained from a FASTA run on a known flexible linker sequence. In this case, the output of one FASTA run on the OWL database using the flexible linker region from Calmodulin (4cln) with a cutoff (e-value) of 0.001.

OWL ID	Sequence
CALN_CHICK	MARKMKDTDSE
MUSCAMC	MARKMKDTDSE
CALM_PATSP	MARKMKDTDSE
CALM_PYUSP	MARKMKDTDSE
CALM_METSE	MARKMKDTDSE
CALM_STIJA	MARKMKDTDSE
CALM_HUMAN	MARKMKDTDSE
CALM_DROME	MARKMKDTDSE
HSCAM3X1	MARKMKDTDSE
CALM_EMENI	MARKMKDTDSE
CALM_NEUCR	MARKMKDTDSE
CALM_ELEEL	MAKKMKDTDSE
NEUCLMDLN	MARKMKDTDSE
SSO4B01	MARKMKDTDSE
CALL_ARBPU	MARKMKETDSE
CALM_PLECO	MARKMRDTDSE
CALL_HUMAN	MARKMKDTDNE
CALS_CHICK	MARKMRDSDSE
CALM_PHYIN	MARKMKDTDSE
CALM_PNECA	MARKMKDVDSE
CALM_TRYBB	MARKMQDSDSE
CALM_TRYCR	MARKMQDSDSE
S53019	MARKMKDTDSE
TRBCMRSG	MARKMQDSDSE
CALM_HORVU	MARKMKDTDSE
JC1033	MARKMKDTDSE
CAL1_PETHY	MARKMKDTDSE
CAL6_ARATH	MARKMKDTDSE

Table 7. Preliminary Flexible Linker Propensity Scale. A FASTA search with a cutoff of 0.01 was performed on sixteen flexible linker sequences, as described in the text. Amino acid frequency in the flexible linker sequences and their near homologues obtained in the FASTA search were tabulated and divided by the amino acid sequence frequency in the PDB to obtain the preliminary propensities given in this table. (The high propensity shown for Methionine may be an artifact arising from Methionine's presence as the first residue in many proteins.)

Residue	Propensity
A	1.3268
C	0.1097
D	1.1684
E	1.4702
F	0.5624
G	1.2972
H	0.4806
I	0.4462
K	1.0519
L	0.5303
M	2.6603
N	0.7729
P	0.4051
Q	1.8076
R	1.8013
S	0.8269
T	0.9002
V	0.6865
W	0.308
Y	1.3375

ACKNOWLEDGEMENTS

The authors gratefully acknowledge the financial support of the National Science Foundation (Grant DBI-9723182) and the numerous people who have either contributed entries or information to the database or have given us feedback on what the user community wants. The authors also wish to thank Informix Software, Inc. for providing a grant of its database software.

All correspondence to Mark.Gerstein@yale.edu.

REFERENCES

1. N. Wade, *Scientists Find A Key Weapon Used by H.I.V.*, in *New York Times*. 1997: New York. p. A1.
2. D.G. Donne, *et al.*, *Proc Natl Acad Sci USA*. **94**:13452–13457 (1997).
3. D.C. Chan, *et al.*, *Cell.* **89**:(2):263-73 (1997).
4. D. Peretz, *et al.*, *J Mol Biol.* **273**:(3):614-22 (1997).
5. P.M. Harrison, *et al.*, *Curr Opin Struct Biol.* **7**:(1):53-9 (1997).
6. M. Gerstein and W. Krebs, *Nucl Acids Res* (In press) (1998).
7. M. Gerstein and C. Chothia, *Proc Natl Acad Sci USA*. **93**:10167-10172 (1996).
8. A. Bairoch and B. Boeckmann, *Nucl Acids Res.* **20**:2019-2022 (1992).
9. L. Holm and C. Sander, *Nuc Acid Res.* **22**:3600-3609 (1994).
10. G.D. Schuler, *et al.*, *Meth Enz.* **266**:141-162 (1996).
11. E. Abola, *et al.*, *Meth Enz.* **277**:556-571 (1997).
12. C.A. Orengo, D.T. Jones, and J.M. Thornton, *Nature.* **372**:631-634 (1994).
13. R.B. Altman, N.F. Abernethy, and R.O. Chen, *Ismb.* **5**:15-24 (1997).
14. R.O. Chen, R. Felciano, and R.B. Altman, *Ismb.* **5**:84-7 (1997).
15. A. Bairoch, P. Bucher, and K. Hofmann, *Nucleic Acids Research.* **24**:(1):189-196 (1996).
16. H.M. Berman, *et al.*, *Biophys J.* **63**:(3):751-759 (1992).
17. J.A. Epstein, J.A. Kans, and G.D. Schuler, *2nd Ann Int WWW Conf.* :(in press) (1994).
18. C.W. Hogue, H. Ohkawa, and S.H. Bryant, *Trends Biochem Sci.* **21**:(6):226-9 (1996).
19. A. Murzin, *et al.*, *J Mol Biol.* **247**:536-540 (1995).
20. T.J.P. Hubbard, *et al.*, *Nucleic Acids Res.* **25**:(1):236-9 (1997).
21. W.G. Scott, J.T. Finch, and A. Klug, *Cell.* **81**:(7):991-1002 (1995).
22. H.W. Pley, K.M. Flaherty, and D.B. McKay, *Nature.* **372**:(6501):68-74 (1994).
23. J.H. Cate, *et al.*, *Science.* **273**:(5282):1678-85 (1996).
24. B. Rees, J. Cavarelli, and D. Moras, *Biochimie.* **78**:(7):624-31 (1996).
25. M. Ruff, *et al.*, *Science.* **252**:(5013):1682-9 (1991).
26. S. Remington, G. Wiegand, and R. Huber, *J Mol Biol.* **158**:111-152 (1982).
27. W.S. Bennett, Jr and T.A. Steitz, *Proc Natl Acad Sci USA.* **75**:4848-4852 (1978).
28. M. Gerstein, A.M. Lesk, and C. Chothia, *Biochemistry.* **33**:6739-6749 (1994).
29. W.S. Bennett and R. Huber, *Crit Rev Biochem.* **15**:291-384 (1984).
30. J. Janin and S. Wodak, *Prog Biophys Mol Biol.* **42**:21-78 (1983).
31. C. Abad-Zapatero, *et al.*, *J Mol Biol.* **198**:445-67 (1987).
32. R.K. Wierenga, *et al.*, *Proteins.* **10**:93 (1991).
33. C. Chothia, *et al.*, *Nature.* **302**:500-505 (1983).
34. D.E. Koshland, *Sci Am.* **229**:52-64 (1973).
35. D.E. Koshland, Jr, *Proc Natl Acad Sci USA.* **44**:98-104 (1958).
36. C.M. Anderson, F.H. Zucker, and T. Steitz, *Science.* **204**:375-380 (1979).
37. J.R. Knowles, *Nature.* **350**:121-4 (1991).
38. L. Stryer. *Biochemistry.* 4th ed, W H Freeman and Company, New York (1995).
39. N.S. Sampson and J.R. Knowles, *Biochemistry.* **31**:8482-8487 (1992a).
40. J.R. Knowles, *Phil Trans R Soc Lond B.* **332**:115-121 (1991).
41. M. Perutz, *Quart Rev Biophys.* **22**:139-236 (1989).
42. P.R. Evans, *Curr Opin Struc Biol.* **1**:773-779 (1991).
43. G. Fermi and M.F. Perutz. *Haemoglobin and Myoglobin*, Claredon Press, Oxford (1981).
44. L.N. Johnson and D. Barford, *J Biol Chem.* **265**:2409-2412 (1990).
45. F.M. Richards and W.A. Lim, *Quart Rev Biophys.* **26**:423-498 (1994).
46. Y. Harpaz, M. Gerstein, and C. Chothia, *Structure.* **2**:641-649 (1994).

47. M. Levitt, *et al.*, *Ann Rev Biochem.* **66**:549-579 (1997).
48. F.M. Richards, *Methods in Enzymology.* **115**:440-464 (1985).
49. F.M. Richards, *Ann Rev Biophys Bioeng.* **6**:151-76 (1977).
50. L.M. Gregoret and F.E. Cohen, *J Mol Biol.* **211**:(4):959-974 (1990).
51. S.J. Hubbard and P. Argos, *Protein Science.* **3**:(12):2194-2206 (1994).
52. S.J. Hubbard and P. Argos, *J Mol Biol.* **261**:289-300 (1996).
53. M. Gerstein, *et al.*, *J Mol Biol.* **234**:357-372 (1993).
54. J.W. Ponder and F.M. Richards, *J Mol Biol.* **193**:775-791 (1987).
55. C.L. Lawson, *et al.*, *Proteins.* **3**:18-31 (1988).
56. C.A. McPhalen, *et al.*, *J Mol Biol.* **227**:197-213 (1992).
57. A.J. Olson, G. Bricogne, and S.C. Harrison, *J Mol Biol.* **171**:61 (1983).
58. B.F. Anderson, *et al.*, *Nature.* **344**:784-787 (1990).
59. M. Gerstein and C.H. Chothia, *J Mol Biol.* **220**:133-149 (1991).
60. G.F. Voronoi, *J Reine Angew Math.* **134**:198-287 (1908).
61. J.D. Bernal and J.L. Finney, *Disc Faraday Soc.* **43**:62-69 (1967).
62. F.M. Richards, *J Mol Biol.* **82**:1-14 (1974).
63. G.J. Kleywegt and T.A. Jones, *Acta Cryst.* **D50**:178-185 (1994).
64. C. Chothia, *Nature.* **254**:304-308 (1975).
65. J. Janin and C. Chothia, *J Biol Chem.* **265**:16027-16030 (1990).
66. J.L. Finney, *J Mol Biol.* **96**:721-732 (1975).
67. J.L. Finney, *et al.*, *Biophys J.* **32**:(1):17-33 (1980).
68. M. Gerstein, J. Tsai, and M. Levitt, *J Mol Biol.* **249**:955-966 (1995).
69. J. Tsai, M. Gerstein, and M. Levitt, *J Chem Phys.* **104**:9417-9430 (1996).
70. J. Tsai, M. Gerstein, and M. Levitt, *Protein Science.* :(in press) (1997).
71. J.L. Finney, *J Mol Biol.* **119**:415-441 (1978).
72. C.W. David, *Biopolymers.* **27**:339-344 (1988).
73. J.P. Shih, S.Y. Sheu, and C.Y. Mou, *J Chem Phys.* **100**:(3):2202-2212 (1994).
74. P.R. Sibbald and P. Argos, *J Mol Biol.* **216**:813-818 (1990).
75. J. O'Rourke. *Computational Geometry in C*, Cambridge UP, Cambridge (1994).
76. P. Procacci and R. Scateni, *Int J Quant Chem.* **42**:151-1528 (1992).
77. B.J. Gellatly and J.L. Finney, *J Mol Biol.* **161**:305-322 (1982).
78. C. Vonrhein, G.J. Schlauderer, and G.E. Schulz, *Structure.* **3**:483-490 (1995).
79. D. Joseph, G.A. Petsko, and M. Karplus, *Science.* **249**:1425-1428 (1990).
80. R.C. Wade, *et al.*, *Biophys J.* **64**:9-15 (1993).
81. J.A. McCammon and S.C. Harvey. *Dynamics of Proteins and Nucleic Acids*, Cambridge UP, (1987).
82. A.T. Brünger. *X-PLOR 3.1, A System for X-ray Crystallography and NMR*, Yale University Press, New Haven (1993).
83. C. Branden and J. Tooze. *Introduction to Protein Structure*, Garland Publishing Incorporated, New York (1991).
84. J.D. Thompson, D.G. Higgins, and T.J. Gibson, *Nuc Acid Res.* **22**:4673-4680 (1994).
85. D.G. Higgins, J.D. Thompson, and T.J. Gibson, *Methods Enzymol.* **266**:383-402 (1996).
86. E.L. Sonnhammer, *et al.*, *Nucleic Acids Res.* **26**:(1):320-2 (1998).
87. A. Krogh, *et al.*, *J Mol Biol.* **235**:1501-1531 (1994).
88. S.R. Eddy, *Curr Opin Struc Biol.* **6**:361-365 (1996).
89. S.R. Eddy, G. Mitchison, and R. Durbin, *J Comp Bio.* **9**:9-23 (1994).
90. P. Baldi, Y. Chauvin, and T. Hunkapiller, *Proc Natl Acad Sci.* **91**:(1059-1063) (1994).
91. S.E. Brenner, C. Chothia, and T.J. Hubbard, *Curr Opin Struct Biol.* **7**:(3):369-76 (1997).
92. C. Chothia, *Nature.* **357**:543-544 (1992).
93. M. Gerstein, *J Mol Biol.* **274**: 562-576 (1997).

PARTICIPANTS

Ruben Abagyan
The Skirball Institute of Biomolecular Medicine
New York University Medical Center
540 First Avenue
New York, NY 10016
abagyan@earth.med.nyu.edu

C. Austen Angell
Arizona State University
Department of Chemistry
Box 871604
Tempe, AZ 85287-1604
angell@asuchm.la.asu.edu

Sorin Bastea
Michigan State University
Physics and Astronomy Department
East Lansing, MI 48824
bastea@pa.msu.edu

Punit Boolchand
University of Cincinnati
Dept. of Electrical and Computer Engineering
Mail Location 30
Cincinnati, OH 45221
pboolcha@ececs.uc.edu

David A. Case
The Scripps Research Institute
Dept. of Molecular Biology, TPC15
10550 N. Torrey Pines Road
La Jolla, CA 92037
case@scripps.edu

Premala Chandra
NEC Research Institute
Phys. Sci. Research
4 Independence Way
Princeton, NJ 08540
premi@research.nj.nec.com

N.V. Chubynsky
Michigan State University
Physics and Astronomy Department
East Lansing, MI 48824
chubynsky@pa.msu.edu

Robert Connelly
Cornell University
Department of Mathematics
124 White Hall
Ithaca, NY 14853
rc46@cornell.edu

A. Roy Day
Marquette University
Department of Physics
P.O. Box 1881
Milwaukee, WI 53201-1881
dayroy@vms.csd.mu.edu

Martin Dove
University of Cambridge
Department of Earth Sciences
Downing Street
Cambridge CB2 3EQ, ENGLAND
martin@minp.esc.cam.ac.uk

Phillip Duxbury
Michigan State University
Physics and Astronomy Department
East Lansing, MI 48824-1116
duxbury@pa.msu.edu

Hans-Peter Eckle
University of Tours
Dept. of Physics, Faculty of Science
Parc de Grandmont
F-37200 Tours, FRANCE
eckle@celfi.phys.univ-tours.fr

Deborah S. Franzblau
College of Staten Island, CUNY
Department of Mathematics, 1S-215
2800 Victory Blvd.
Staten Island, NY 10314
franzblau@postbox.csi.cuny.edu

Mark Gerstein
Yale University
Molecular Biophysics & Biochemistry
266 Whitney Avenue
New Haven, CT 06520-8114
Mark.Gerstein@yale.edu

Paul Goldbart
Univ. of Illinois, Urbana-Champaign
Department of Physics
1110 W. Green St.
Urbana, IL 61801-3080
goldbart@uiuc.edu

Prabhat Gupta
Ohio State University
Mat. Sci. & Engr. (477 Watts Hall)
2041 College Road
Columbus, OH 43210
gupta.3@osu.edu

Gregory Hassold
Kettering University
Applied Physics Division, Science and Mathematics Dept.
1700 W. Third Ave.
Flint, MI 48504-4898
ghassold@kettering.edu

Timothy Havel
Harvard Medical School
Dept. of Biological Chemistry and Molecular Pharmacology
240 Longwood Ave., C1-103a
Boston, MA 02115
havel@menelaus.med.harvard.edu

Bruce Hendrickson
Sandia National Laboratories
MS 1110
Albuquerque, NM 87185-1110
bahendr@sandia.gov

Brandon Hespenheide
Michigan State University
Biochemistry Department
502 Biochemistry
East Lansing, MI 48824
hespenhe@pilot.msu.edu

Linn W. Hobbs
Massachusetts Institute of Technology
Dept. of Materials Science and Engr. (Rm. 13-4062)
77 Massachusetts Ave.
Cambridge, MA 02139
hobbs@mit.edu

Donald Jacobs
Michigan State University
Physics and Astronomy Department
East Lansing, MI 48824-1116
jacobs@pa.msu.edu

Esther Jesurum
Massachusetts Institute of Technology
Laboratory for Computer Science
545 Technology Square
Cambridge, MA 02139
esther@theory.lcs.mit.edu

Béla Joós
University of Ottawa
Physics Department
Ottawa, ON K1N 6N5
CANADA
joos@physics.uottawa.ca

Leslie Kuhn
Michigan State University
Dept. of Biochemistry
East Lansing, MI 48824
kuhn@agua.bch.msu.edu

Paul L. Leath
Rutgers University
Department of Physics and Astronomy
136 Frelinghuysen Road
Piscataway, NJ 08854-8019
leath@physics.rutgers.edu

Maria Mitkova
University of Cincinnati
Department of Electrical and Computer Engineering
Mail Location 30
Cincinnati, OH 45221
mmitkova@ececs.uc.edu

Camille Mittermeier
York University
Mathematics and Statistics
4700 Keele St., North York, S616Ross
North York, ON M3J 1P3
CANADA
camille@yorku.ca

Cristian Moukarzel
IF-UFF
Av. Litoranea s/n.
24210-340 Niteroi RJ
BRAZIL
cristian@if.uff.br

Sergei Obukhov
University of Florida
Department of Physics
215 Williamson Hall
Gainesville, FL 32611-2085
sergei@phys.ufl.edu

Bruce Patton
Ohio State University
Department of Physics
174 West Eighteenth Avenue
Columbus, OH 43210
patton@mps.ohio-state.edu

Jim C. Phillips
Lucent Technologies
Bell Labs Innovations (1E-245)
600 Mountain Avenue
Murray Hill, NJ 07974-0636
jcp@physics.bell-labs.com

Andrew J. Rader
Michigan State University
Physics and Astronomy Department
East Lansing, MI 48824-1116
rader@pa.msu.edu

Brigitte Servatius
Syracuse University
Department of Mathematics
Syracuse, NY 13244-1150
bservat@wpi.edu

Mark Stevens
Sandia National Laboratories
P.O. Box 5800
MS 1111
Albuquerque, NM 87185-1111
mjsteve@cs.sandia.gov

M.F. Thorpe
Michigan State University
Physics and Astronomy Department
East Lansing, MI 48824-1116
thorpe@pa.msu.edu

Daniel Vernon
Simon Fraser University
Department of Physics
Burnaby, BC V5A 1S6
CANADA
dvernon@sfu.ca

Anke Walz
Cornell University
Department of Mathematics
124 White Hall
Ithaca, NY 14853
abw1@cornell.edu

Walter Whiteley
York University
Mathematics and Statistics
4700 Keele St., North York, S616Ross
North York, ON M3J 1P3
CANADA
whiteley@mathstat.yorku.ca

Chen Zeng
Rutgers University
Department of Physics and Astronomy
136 Frelinghuysen Road
Piscataway, NJ 08854-8019
chenz@physics.rutgers.edu

PARTICIPANT PHOTO

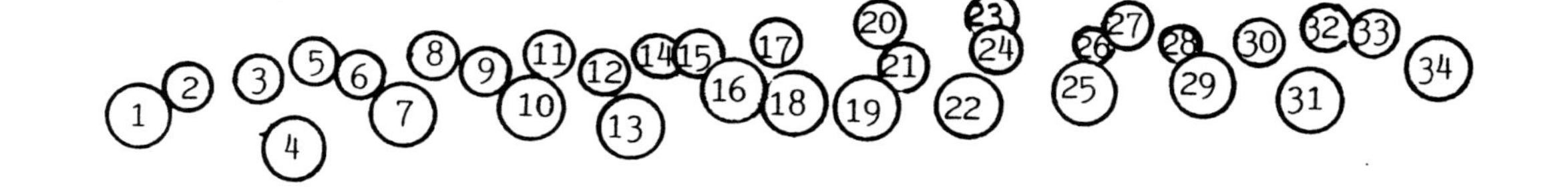

[1] C. Moukarzel, [2] L. Kuhn, [3] M. Mitkova, [4] C. Mittermeier, [5] B. Patton, [6] N. Chubynsky, [7] S. Obukhov, [8] S. Bastea, [9] A.J. Rader, [10] P.M. Duxbury, [11] D.A. Case, [12] H.P. Eckle, [13] L. Neuman, [14] C.A. Angell, [15] P.L Leath, [16] M.F. Thorpe, [17] G. Hassold, [18] B. Servatius, [19] R. Abagyan, [20] R. Connelly, [21] P. Gupta, [22] B. Joós, [23] J.C. Phillips (half-hidden), [24] T. Havel, [25] W. Whiteley, [26] M. Dove, [27] L. Hobbs, [28] A.R. Day, [29] B. Hespenheide, [30] P. Boolchand, [31] D. Franzblau, [32] M. Stevens, [33] D. Jacobs, [34] D. Vernon.

Not in photograph: P. Chandra, M. Gerstein, P. Goldbart, B. Hendrickson, Esther Jesurum, A. Walz, and Chen Zeng.

INDEX